MATHEMATICS IN INDUSTRY **13**

THE EUROPEAN CONSORTIUM
FOR MATHEMATICS IN INDUSTRY

ECMI

SUBSERIES

Wilhelmus H.A. Schilders
Henk A. van der Vorst
Joost Rommes
Editors

Model Order Reduction: Theory, Research Aspects and Applications

With 133 Figures, 53 in color and 9 Tables

Springer

Authors and Editors

Wilhelmus H.A. Schilders
NXP Semiconductors
Corporate I&T/DTF/Design Methods
Physical Design Methods – Mathematics
High Tech Campus 37
5656 AE Eindhoven
The Netherlands
wil.schilders@nxp.com

Joost Rommes
NXP Semiconductors
Corporate I&T/DTF/Design Methods
Physical Design Mothods
High Tech Campus 37
5656 AE Eindhoven
The Netherlands
joost.rommes@nxp.com

Henk A. van der Vorst
Mathematical Institute
Utrecht University
Budapestlaan 6
3508 TA Utrecht^
The Netherlands
h.a.vandervorst@uu.nl

ISBN 978-3-642-42773-2 ISBN 978-3-540-78841-6 (eBook)
DOI 10.1007/978-3-540-78841-6

Mathematics Subject Classification (2000): 65, 93

Softcover re-print of the Hardcover 1st edition 2008

Cover design: deblik, Berlin

Printed on acid-free paper

9 8 7 6 5 4 3 2 1

springer.com

Preface

The idea for this book originated during the workshop "Model order reduction, coupled problems and optimization" held at the Lorentz Center in Leiden from September 19–23, 2005. During one of the discussion sessions, it became clear that a book describing the state of the art in model order reduction, starting from the very basics and containing an overview of all relevant techniques, would be of great use for students, young researchers starting in the field, and experienced researchers. The observation that most of the theory on model order reduction is scattered over many good papers, making it difficult to find a good starting point, was supported by most of the participants. Moreover, most of the speakers at the workshop were willing to contribute to the book that is now in front of you.

The goal of this book, as defined during the discussion sessions at the workshop, is three-fold: first, it should describe the basics of model order reduction. Second, both general and more specialized model order reduction techniques for linear and nonlinear systems should be covered, including the use of several related numerical techniques. Third, the use of model order reduction techniques in practical applications and current research aspects should be discussed.

We have organized the book according to these goals. In Part I, the rationale behind model order reduction is explained, and an overview of the most common methods is described. Furthermore, in the second chapter, an introduction is given to background material from numerical linear algebra needed to assess the theory and methods presented later in the book. This is very important and useful information, as advances in numerical linear algebra often lead to new results in the area of model order reduction. Thus, the first two chapters serve as an introduction to readers who are not familiar with the subject. In Part II, model order reduction techniques and related numerical problems are described from different points of view: both frameworks for structure-preserving techniques and more specialized techniques are presented, while numerical methods for (closely) related problems and approaches for nonlinear systems are considered as well. This part serves as the theoretical backbone of the book, containing an overview of techniques used and areas covered. In Part III the focus is on research aspects and applications of model order reduction. A variety of experiments with real-life examples shows that different problems require different techniques, while application of the techniques leads to new research topics that are described as well.

Despite the fact that the workshop was organized already in 2005, this book contains many recent advances in model order reduction. Moreover, it presents several open problems for which techniques are still in development, related to both linear systems, which become larger and more complex mainly due to industrial requirements, and nonlinear systems, which demand a completely new theory. The latter illustrates the final and most important goal of this book, namely to serve as a source of inspiration for its readers, who will discover that model order reduction is a very exciting and lively field.

At this point we would like to thank all authors of the chapters in this book. Without the contributions of these experts, it would not be possible to cover the wide and rapidly developing field of model order reduction in one book.

Leiden, Utrecht, Eindhoven
September 2005 – May 2008

Wil Schilders
Henk van der Vorst
Joost Rommes

Contents

Part I Basic Concepts

Introduction to Model Order Reduction
Wil Schilders 3

Linear Systems, Eigenvalues, and Projection
Henk van der Vorst 33

Part II Theory

Structure-Preserving Model Order Reduction of RCL Circuit Equations
Roland W. Freund 49

A Unified Krylov Projection Framework for Structure-Preserving Model Reduction
Zhaojun Bai, Ren-cang Li, and Yangfeng Su 75

Model Reduction via Proper Orthogonal Decomposition
René Pinnau 95

PMTBR: A Family of Approximate Principal-components-like Reduction Algorithms
Joel R. Phillips, Zhenhai Zhu, and L. Miguel Silveira 111

A Survey on Model Reduction of Coupled Systems
Timo Reis and Tatjana Stykel 133

Space Mapping and Defect Correction
David Echeverría, Domenico Lahaye, and Piet W. Hemker 157

Modal Approximation and Computation of Dominant Poles
Joost Rommes 177

Some Preconditioning Techniques for Saddle Point Problems
Michele Benzi and Andrew J. Wathen 195

Time Variant Balancing and Nonlinear Balanced Realizations
E.I. Verriest 213

Singular Value Analysis and Balanced Realizations for Nonlinear Systems
Kenji Fujimoto and Jacquelien M.A. Scherpen 251

Part III Research Aspects and Applications

Matrix Functions
Andreas Frommer and Valeria Simoncini 275

Model Reduction of Interconnected Systems
Antoine Vandendorpe and Paul Van Dooren 305

Quadratic Inverse Eigenvalue Problem and Its Applications to Model Updating — An Overview
*Moody T. Chu*** 323

Data-Driven Model Order Reduction Using Orthonormal Vector Fitting
Dirk Deschrijver and Tom Dhaene 341

Model-Order Reduction of High-Speed Interconnects Using Integrated Congruence Transform
Emad Gad, Michel Nakhla, and Ram Achar 361

Model Order Reduction for MEMS: Methodology and Computational Environment for Electro-Thermal Models
Tamara Bechtold, Evgenii B. Rudnyi, and Jan G. Korvink 403

Model Order Reduction of Large RC Circuits
Nick P. van der Meijs 421

Reduced Order Models of On-Chip Passive Components and Interconnects, Workbench and Test Structures
Daniel Ioan and Gabriela Ciuprina 447

Index 469

List of Contributors

Ram Achar
Deptartment of Electronics
Carleton University
Ottawa, ON, Canada, K1S 5B6
achar@doe.carleton.ca

Zhaojun Bai
Department of Computer Science
University of California
One Shields Avenue
Davis, CA 95616
bai@cs.ucdavis.edu

Tamara Bechtold
NXP Semiconductors
High Tech Campus 37, WY4.042
5656 AE Eindhoven, The Netherlands
tamara.bechtold@nxp.com

Michele Benzi
Department of Mathematics
and Computer Science
Emory University
Atlanta, Georgia 30322, USA
benzi@mathcs.emory.edu

Moody T. Chu
Department of Mathematics
North Carolina State University
Raleigh, NC 27695-8205, USA
chu@math.ncsu.edu

Gabriela Ciuprina
Politehnica University of Bucharest
Romania
gabriela@lmn.pub.ro

Dirk Deschrijver
Ghent University
Sint Pietersnieuwstraat 41
9000 Gent, Belgium
dirk.deschrijver@intec.
ugent.be

Tom Dhaene
Ghent University
Sint Pietersnieuwstraat 41
9000 Gent, Belgium
tom.dhaene@intec.ugent.be

David Echeverría
CWI
Kruislaan 413
1054 AL Amsterdam, The Netherlands
D.Echeverria@cwi.nl

Roland W. Freund
Department of Mathematics
University of California at Davis
One Shields Avenue
Davis, CA 95616, U.S.A
freund@math.ucdavis.edu

Andreas Frommer
Department of Mathematics and Science
University of Wuppertal
D-42097 Wuppertal, Germany
frommer@math.uni-wuppertal.de

Kenji Fujimoto
Department of Mechanical Science and Engineering
Nagoya University
fujimoto@nagoya-u.jp

Emad Gad
School of Information Technology and Engineering (SITE)
University of Ottawa
Ottawa, ON, Canada, K1N 6N5
egad@site.uottawa.ca

Piet W. Hemker
CWI
Kruislaan 413
1054 AL Amsterdam, The Netherlands
P.W.Hemker@cwi.nl

Daniel Ioan
Politehnica University of Bucharest
Romania
lmn@lmn.pub.ro

Jan G. Korvink
IMTEK, University of Freiburg
Georges-Koehler-Allee 102
D-79110, Germany
korvink@imtek.de

Domenico Lahaye
CWI
Kruislaan 413
1054 AL Amsterdam, The Netherlands
Domenico.Lahaye@cwi.nl

Ren-cang Li
Department of Mathematics
University of Texas
Arlington, TX 76019
rcli@uta.edu

Nick P. van der Meijs
Faculty of EEMCS
Delft University of Technology
Delft, The Netherlands
N.P.vanderMeijs@tudelft.nl

Michel Nakhla
Deptartment of Electronics,
Carleton University
Ottawa, ON, Canada, K1S 5B6
msn@doe.carleton.ca

Joel R. Phillips
Cadence Berkeley Laboratories
Berkeley, CA 94704, U.S.A
jrp@cadence.com

René Pinnau
Fachbereich Mathematik
Technische Universität Kaiserslautern
D-67663 Kaiserslautern, Germany
pinnau@mathematik.uni-kl.de

Timo Reis
Institut für Mathematik, MA 4-5
Technische Universität Berlin
Straße des 17. Juni 136
10623 Berlin Germany
reis@math.tu-berlin.de

Joost Rommes
NXP Semiconductors
High Tech Campus 37, WY4.042
5656 AE Eindhoven, The Netherlands
joost.rommes@nxp.com

Evgenii B. Rudnyi
CAD-FEM GmbH
Marktplatz 2
85567 Grafing, Germany
erudnyi@cadfem.de

Jacquelien M. A. Scherpen
Faculty of Mathematics and Natural Sciences
University of Groningen
j.m.a.scherpen@rug.nl

Wil Schilders
NXP Semiconductors
High Tech Campus 37, WY4.042
5656 AE Eindhoven, The Netherlands
wil.schilders@nxp.com

L. Miguel Silveira
Cadence Laboratories/INESC ID
IST/Tech. U. Lisbon
Lisbon, Portugal
lms@inesc-id.pt

Valeria Simoncini
Dipartimento di Matematica and CIRSA
Università di Bologna
I-40127 Bologna, Italy
valeria.simoncini@unibo.it

Tatjana Stykel
Institut für Mathematik, MA 4-5
Technische Universität Berlin
Straße des 17. Juni 136
10623 Berlin, Germany
stykel@math.tu-berlin.de

Yangfeng Su
School of Mathematical Science
Fudan University
Shanghai 200433, China
yfsu@fudan.edu.cn

Antoine Vandendorpe
Fortis
Equity Derivatives Quantitative Research
Belgium
antoine.vandendorpe@fortis.com

Paul Van Dooren
Department of Mathematical Engineering
Université catholique de Louvain
Belgium
vdooren@csam.ucl.ac.be

Erik I. Verriest
School of Electrical and Computer Engineering
Georgia Institute of Technology
Atlanta, GA 30332-0250, USA
erik.verriest@ee.gatech.edu

Henk van der Vorst
Mathematical Institute
Utrecht University
The Netherlands
h.a.vandervorst@uu.nl

Andrew J. Wathen
Oxford University Computing Laboratory
Wolfson Building, Parks Road
Oxford OX1 3QD, UK
andy.wathen@comlab.ox.ac.uk

Zhenhai Zhu
Cadence Berkeley Laboratories
Berkeley, CA 94704, U.S.A
zhzhu@cadence.com

Part I

Basic Concepts

Introduction to Model Order Reduction

Wil Schilders[1,2]

[1] NXP Semiconductors, Eindhoven, The Netherlands
wil.schilders@nxp.com
[2] Eindhoven University of Technology, Faculty of Mathematics and Computer Science, Eindhoven, The Netherlands
w.h.a.schilders@tue.nl

1 Introduction

In this first section we present a high level discussion on computational science, and the need for compact models of phenomena observed in nature and industry. We argue that much more complex problems can be addressed by making use of current computing technology and advanced algorithms, but that there is a need for model order reduction in order to cope with even more complex problems. We also go into somewhat more detail about the question as to what model order reduction is.

1.1 Virtual Design Environments

Simulation or, more generally, computational science has become an important part of todays technological world, and it is now generally accepted as the third discipline, besides the classical disciplines of theory and (real) experiment. Physical (and other) experiments lead to theories that can be validated by performing additional experiments. Predictions based on the theory can be made by performing virtual experiments, as illustrated by Figure 1.

Computer simulations are now performed routinely for many physical, chemical and other processes, and virtual design environments have been set up for a variety of problem classes in order to ease the work of designers and engineers. In this way, new products can be designed faster, more reliably, and without having to make costly prototypes.

The ever increasing demand for realistic simulations of complex products places a heavy burden on the shoulders of mathematicians and, more generally, researchers working in the area of computational science and engineering (CSE). Realistic simulations imply that the errors of the virtual models should be small, and that different aspects of the product must be taken into account. The former implies that care must be taken in the numerical treatment and that, for example, a relatively fine adaptively determined mesh is necessary in the simulations. The latter explains the trend in coupled simulations, for example combined mechanical and thermal behaviour, or combined mechanical and electromagnetic behaviour.

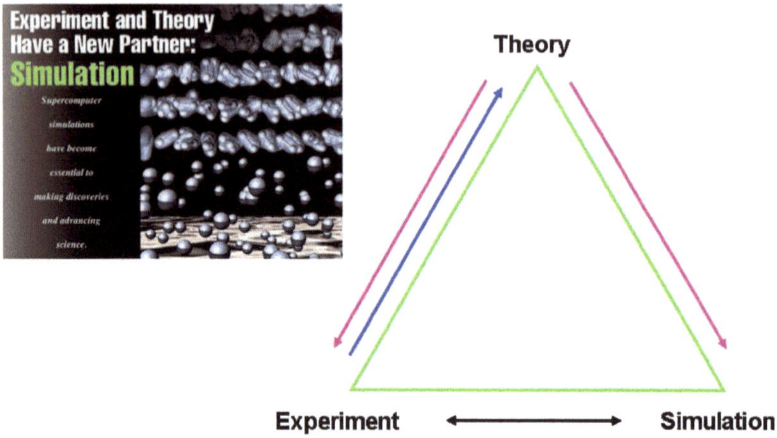

Fig. 1. Simulation is the third discipline.

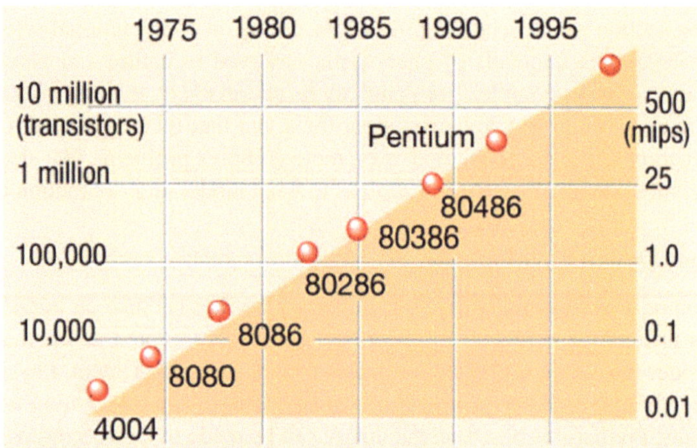

Fig. 2. Moore's law.

An important factor in enabling the complex simulations carried out today is the increase in computational power. Computers and chips are getting faster, Moores law predicting that the speed will double every 18 months (see Figure 2).

This increase in computational power appears to go hand-in-hand with developments in numerical algorithms. Iterative solution techniques for linear systems are mainly responsible for this speed-up in algorithms, as is shown in Figure 3. Important contributions in this area are the conjugate gradient method (Hestenes and Stiefel [22]), preconditioned conjugate gradient methods (ICCG [25], biCGstab [34]) and multigrid methods (Brandt [4] and [5]).

The combined speed-up achieved by computer chips and algorithms is enormous, and has enabled computational science to make big steps forward. Many problems that people did not dream of solving two decades ago are now solved routinely.

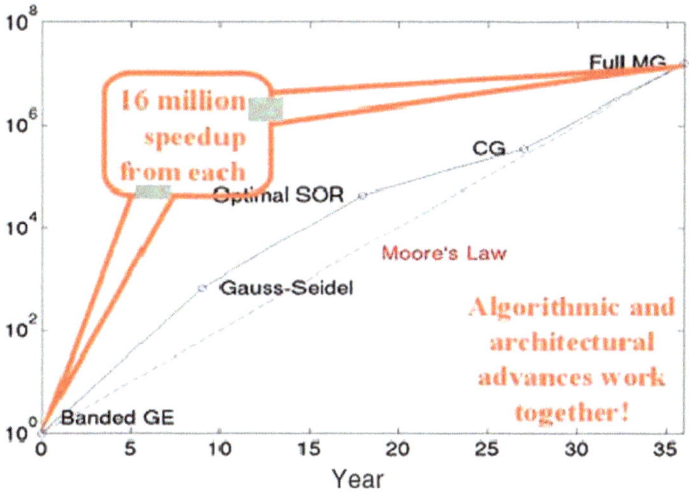

Fig. 3. Numerical version of Moore's law.

1.2 Compact Model Descriptions

The developments described in the previous section also have a counter side. The increased power of computers and algorithms reduces the need to develop smart, sophisticated solution methods that make use of properties of the underlying systems. For example, whereas in the 1960s and 1970s one often had to construct special basis functions to solve certain problems, this can be avoided nowadays by using brute force methods using grids that are refined in the right places.

The question arises whether we could use the knowledge generated by these very accurate, but time-consuming, simulations to generate the special basis functions that would have constituted the scientific approach a few decades ago. This is a promising idea, as many phenomena are described very well by a few dominant modes.

Example: electromagnetic behaviour of interconnect structures in chips

To give an example, consider the electromagnetic behaviour of interconnect structures in a computer chip, depicted in Figure 4. Such a chip consists of millions of devices, such as transistors, that need to be connected to each other for correct functioning of the chip. The individual devices are contained in the semiconductor material, their contacts being located in a two dimensional domain. Clearly, to connect these contacts in the way designers have prescribed, a three dimensional structure of wires is needed. This is the so-called interconnect structure of the chip, which nowadays consists of 7-10 layers in which metal wires are running, with so-called vias between the wires located in different metal layers.

In previous generations of chips, these interconnect structures occupied a relatively large area, and contained less wires, so that the distance between wires was large enough to justify discarding mutual influence. In recent years, however, chips have shrunk, and the number of devices has grown enormously. This means that for modern interconnect structures one needs to take into account mutual influence of wires, as this can lead to serious delay phenomena and other spurious effects. The problem is complicated even further by the use of higher frequencies.

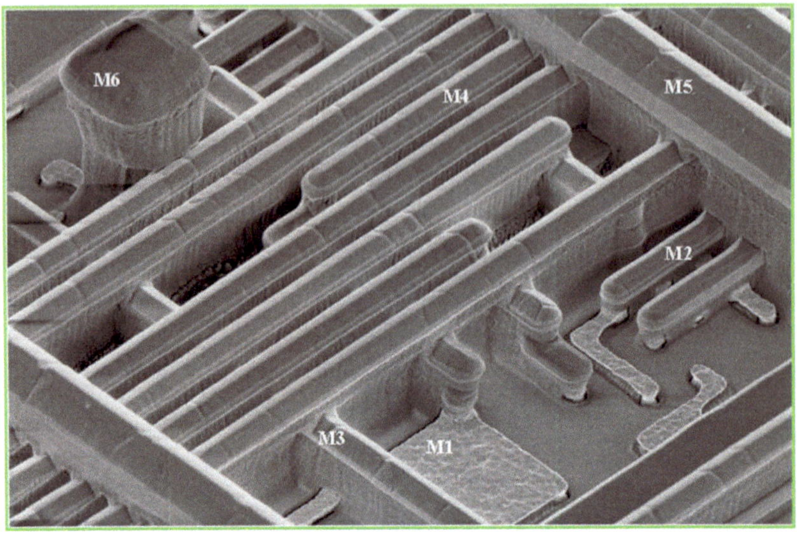

Fig. 4. Interconnect structure.

Clearly, the modelling of the mutual electromagnetic influence of interconnect wires is a gradual process. A decade ago, one did not have to take this influence into account, and could consider the wires as individual entities. Nowadays, resistive and capacitive effects are clearly noticeable, and will become more significant over the years. Because of the gradual character of this phenomenon, one can imagine that it is not necessary to include all minute detail of an electromagnetic simulation of interconnect structures. Such a simulation could easily involve millions of nodes, because of the complicated geometric structure. The simulation will probably reveal that crosstalk and signal integrity problems are quite localized, at a few places in the structure where wires are too close together.

Another point of view may be to consider the problem as an input-output model, where a time-dependent input signal is sent through the interconnect structure, and a resulting time-dependent output signal is registered. Again, to calculate the output resulting from the given input is a time-consuming exercise due to the excessive number of nodes necessary for this simulation, in the spatial and time domain. However, it is expected to be possible to delete superfluous detail, and calculate a very good approximation to the output in a much more efficient way.

The foregoing example clearly shows that it may not be necessary to calculate all details, and nevertheless obtain a good understanding of the phenomena taking place. There may be many reasons why such detail is not needed. There may be physical reasons that can be formulated beforehand, and therefore incorporated into the model before starting calculations. A very nice example is that of simulating the blood flow in the human body, as described in many publications by the group of Alfio Quarteroni (see [30], but also work of others). In his work, the blood flow in the body is split into different parts. In very small arteries, it is assumed that the flow is one dimensional. In somewhat larger arteries, two dimensional models are used, whereas in the heart, a three dimensional model is used as these effects are very important and must be modelled in full detail. This approach does enable a simulation of the blood flow in the entire human body; clearly, such simulations would not be feasible if three dimensional models would be used throughout. This approach, which is also observed in different application areas, is also termed *operational model order reduction.* It uses physical (or other) insight to reduce the complexity of models.

Another example of operational model order reduction is the simulation of electromagnetic effects in special situations. As is well known, electromagnetic effects can be fully described by a system of Maxwell equations. Despite the power of current computers and algorithms, solving the Maxwell equations in 3-dimensional space and time is still an extremely demanding problem, so that simplifications are being made whenever possible. An assumption that is made quite often is that of quasi-statics, which holds whenever the frequencies playing a role are low to moderate. In this case, simpler models can be used, and techniques for solving these models have been developed (see [32]).

In special situations, the knowledge about the problem and solutions can be so detailed, that a further reduction of model complexity can be achieved. A prominent and very successful example is the *compact modelling* [19] of semiconductor devices. Integrated circuits nowadays consist of millions of semiconductor devices, such as resistors, capacitors, inductors, diodes and transistors. For resistors, capacitors and inductors, simple linear models are available, but diodes and especially transistors are much more complicated. Their behaviour is not easily described, but can be calculated accurately using software dedicated to semiconductor device simulation. However, it is impossible to perform a full simulation of the entire electronic circuit, by using the results of the device simulation software for each of the millions of transistors. This would imply coupling of the circuit simulation software to the device simulation software. Bearing in mind that device simulations are often quite time consuming (it is an extremely nonlinear problem, described by a system of three partial differential equations), this is an impossible task.

The solution to the aforementioned problem is to use accurate compact models for each of the transistors. Such models look quite complicated, and can easily occupy a number of pages of description, but consist of a set of algebraic relations that can be evaluated very quickly. The compact models are constructed using a large amount of measurements and simulations, and, above all, using much human insight. The models often depend on as many as 40-50 parameters, so that they are widely applicable for many different types and geometries of transistors. The most prominent model nowadays is the Penn-State-Philips (PSP) model for MOS transistors (see Figure 5), being chosen as the world standard in 2007 [15]. It is very accurate, including also derivatives up to several orders. Similar developments can be observed at Berkeley [6], where the BSIM suite of models is constructed.

Using these so-called compact models, it is possible to perform simulations of integrated circuits containing millions of components, both for steady-state and time-dependent situations. Compact modelling, therefore, plays an extremely important role in enabling such demanding simulations. The big advantage of this approach is that the compact models are formulated in a way that is very appealing to designers, as they are formulated in terms of components they are very familiar with.

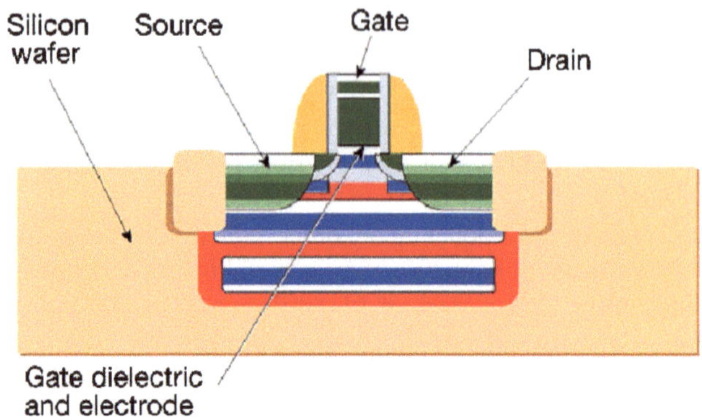

Fig. 5. MOS transistor.

Unfortunately, in many cases, it is not possible to a priori simplify the model describing the behaviour. In such cases, a procedure must be used, in which we rely on the automatic identification of potential simplifications. Designing such algorithms is, in essence, the task of the field of model order reduction. In the remainder of this chapter, we will describe it in more detail.

1.3 Model Order Reduction

There are several definitions of model order reduction, and it depends on the context which one is preferred. Originally, MOR was developed in the area of systems and control theory, which studies properties of dynamical systems in application for reducing their complexity, while preserving their input-output behavior as much as possible. The field has also been taken up by numerical mathematicians, especially after the publication of methods such as PVL [9]. Nowadays, model order reduction is a flourishing field of research, both in systems and control theory and in numerical analysis. This has a very healthy effect on MOR as a whole, bringing together different techniques and different points of view, pushing the field forward rapidly.

So what is model order reduction about? As was mentioned in the foregoing sections, we need to deal with the simplification of dynamical models that may contain many equations and/or variables ($10^5 - 10^9$). Such simplification is needed in order to perform simulations within an acceptable amount of time and limited storage capacity, but with reliable outcome. In some cases, we would even like to have on-line predictions of the behaviour with acceptable computational speed, in order to be able to perform optimizations of processes and products.

Model Order Reduction tries to quickly capture the essential features of a structure. This means that in an early stage of the process, the most basic properties of the original model must already be present in the smaller approximation. At a certain moment the process of reduction is stopped. At that point all necessary properties of the original model must be captured with sufficient precision. All of this has to be done automatically.

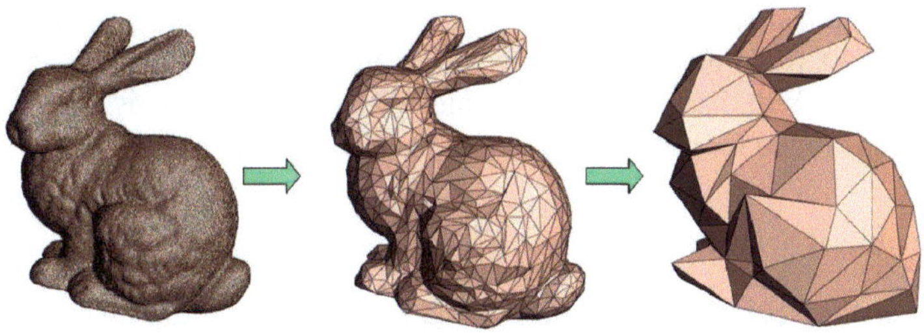

Fig. 6. Graphical illustration of model order reduction.

Figure 6 illustrates the concept in a graphical easy-to-understand way, demonstrating that sometimes very little information is needed to describe a model. This example with pictures of the Stanford Bunny shows that, even with only a few facets, the rabbit can still be recognized as such (Graphics credits: Harvard University, Microsoft Research). Although this example was constructed for an entirely different purpose, and does not contain any reference to the way model order reduction is performed mathematically, it can be used to explain (even to lay persons) what model order reduction is about.

In the history of mathematics we see the desire to approximate a complicated function with a simpler formulation already very early. In the year 1807 Fourier (1768-1830) published the idea to approximate a function with a few trigonometric terms. In linear algebra the first step in the direction of model order reduction came from Lanczos (1893-1974). He looked for a way to reduce a matrix in tridiagonal form [64, 65]. W.E. Arnoldi realized that a smaller matrix could be a good approximation of the original matrix [2]. He is less well-known, although his ideas are used by many numerical mathematicians. The ideas of Lanczos and Arnoldi were already based on the fact that a computer was available to do the computations. The question, therefore, was how the process of finding a smaller approximation could be automated.

The fundamental methods in the area of Model Order Reduction were published in the eighties and nineties of the last century. In 1981 Moore [71] published the method of Truncated Balanced Realization, in 1984 Glover published his famous paper on the Hankel-norm reduction [38]. In 1987 the Proper Orthogonal Decomposition method was proposed by Sirovich [94]. All these methods were developed in the field of systems and control theory. In 1990 the first method related to Krylov subspaces was born, in Asymptotic Waveform Evaluation [80]. However, the focus of this paper was more on finding Padé approximations rather than Krylov spaces. Then, in 1993, Freund and Feldmann proposed Padé Via Lanczos [28] and showed the relation between the Padé approximation and Krylov spaces. In 1995 another fundamental method was published. The authors of [73] introduced PRIMA, a method based on the ideas of Arnoldi, instead of those of Lanczos. This method will be considered in detail in Section 3.3, together with the Laguerre-SVD method [61].

In more recent years much research has been done in the area of the Model Order Reduction. Consequently a large variety of methods is available. Some are tailored to specific applications, others are more general. In the second and third part of this book, many of these new developments are being discussed. In the remainder of this chapter, we will discuss some basic methods and properties, as this is essential knowledge required for the remainder of the book.

1.4 Dynamical Systems

To place model reduction in a mathematical context, we need to realize that many models developed in computational science consist of a system of partial and/or ordinary differential equations, supplemented with boundary conditions. Important examples are the Navier-Stokes equations in computational fluid dynamics (CFD), and the Maxwell equations in electromagnetics (EM). When partial differential equations are used to describe the behaviour, one often encounters the situation that the independent variables are space and time. Thus, after (semi-)discretising in space, a system of ordinary differential equations is obtained in time. Therefore, we limit the discussion to ODE's and consider the following explicit finite-dimensional dynamical system (following Antoulas, see [2]):

$$\begin{aligned} \frac{d\mathbf{x}}{dt} &= \mathbf{f}(\mathbf{x}, \mathbf{u}) \\ \mathbf{y} &= \mathbf{g}(\mathbf{x}, \mathbf{u}). \end{aligned}$$

Here, $\mathbf{u}$ is the input of the system, $\mathbf{y}$ the output, and $\mathbf{x}$ the so-called *state variable*. The dynamical system can thus be viewed as an *input-output system*, as displayed in Figure 7.

The complexity of the system is characterized by the number of its state variables, i.e. the dimension n of the state space vector $\mathbf{x}$. It should be noted that similar dynamical systems can also be defined in terms of differential algebraic equations, in which case the first set of equations in (1) is replaced by $\mathbf{F}(\frac{d\mathbf{x}}{dt}, \mathbf{x}, \mathbf{u}) = 0$.

Model order reduction can now be viewed as the task of reducing the dimension of the state space vector, while preserving the character of the input-output relations. In other words, we should find a dynamical system of the form

$$\begin{aligned} \frac{d\hat{\mathbf{x}}}{dt} &= \hat{\mathbf{f}}(\hat{\mathbf{x}}, \mathbf{u}), \\ \mathbf{y} &= \hat{\mathbf{g}}(\hat{\mathbf{x}}, \mathbf{u}), \end{aligned}$$

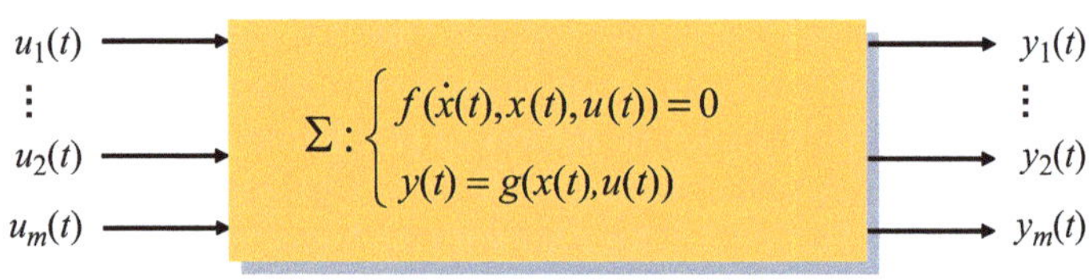

Fig. 7. Input-output system

where the dimension of $\hat{\mathbf{x}}$ is much smaller than n. In order to provide a good approximation of the original input-output system, a number of conditions should be satisfied:

- the approximation error is small,
- preservation of properties of the original system, such as stability and passivity (see Sections 2.4-2.6),
- the reduction procedure should be computationally efficient.

A special case is encountered if the functions $\mathbf{f}$ and $\mathbf{g}$ are linear, in which case the system reads

$$\frac{d\mathbf{x}}{dt} = A\mathbf{x} + B\mathbf{u},$$
$$\mathbf{y} = C^T\mathbf{x} + D\mathbf{u}.$$

Here, the matrices A, B, C, D can be time-dependent, in which case we have a *linear time-varying (LTV) system*, or time-independent, in which case we speak about a *linear time-invariant (LTI) system*. For linear dynamical systems, model order reduction is equivalent to reducing the matrix A, but retaining the number of columns of B and C.

1.5 Approximation by Projection

Although we will discuss in more detail ways of approximating input-output systems later in this chapter, there is a unifying feature of the approximation methods that is worthwhile discussing briefly: *projection*. Methods based on this concept truncate the solution of the original system in an appropriate basis. To illustrate the concept, consider a basis transformation T that maps the original n-dimensional state space vector $\mathbf{x}$ into a vector that we denote by

$$\bar{\mathbf{x}} = \begin{pmatrix} \hat{\mathbf{x}} \\ \tilde{\mathbf{x}} \end{pmatrix},$$

where $\hat{\mathbf{x}}$ is k-dimensional. The basis transformation T can then be written as

$$T = \begin{pmatrix} W^* \\ T_2^* \end{pmatrix},$$

and its inverse as

$$T^{-1} = (V \;\; T_1).$$

Since $W^*V = I_k$, we conclude that

$$\Pi = VW^*$$

is an oblique projection along the kernel of W^* onto the k-dimensional subspace that is spanned by the columns of the matrix V.

If we substitute the projection into the dynamical system (1), the first part of the set of equations obtained is

$$\frac{d\hat{\mathbf{x}}}{dt} = W^*\hat{\mathbf{f}}(V\hat{\mathbf{x}} + T_1\tilde{\mathbf{x}}, \mathbf{u}),$$
$$\mathbf{y} = \hat{\mathbf{g}}(V\hat{\mathbf{x}} + T_1\tilde{\mathbf{x}}, \mathbf{u}).$$

Note that this is an exact expression. The approximation occurs when we would delete the terms involving $\tilde{\mathbf{x}}$, in which case we obtain the reduced system

$$\frac{d\hat{\mathbf{x}}}{dt} = W^*\hat{\mathbf{f}}(V\hat{\mathbf{x}}, \mathbf{u}),$$
$$\mathbf{y} = \hat{\mathbf{g}}(V\hat{\mathbf{x}}, \mathbf{u}).$$

For this to produce a good approximation to the original system, the neglected term $T_1\tilde{\mathbf{x}}$ must be sufficiently small. This has implications for the choice of the projection Π. In the following sections, various ways of constructing this projection are discussed.

2 Transfer Function, Stability and Passivity

Before discussing methods that have been developed in the area of model order reduction, it is necessary to shed light on several concepts that are being used frequently in the field. Often, model order reduction does not address the reduction of the entire problem or solution, but merely a number of characteristic functions that are important for designers and engineers. In addition, it is important to consider a number of specific aspects of the underlying problem, and preserve these when reducing. Therefore, this section is dedicated to a discussion of these important concepts.

2.1 Transfer Function

In order to illustrate the various concepts related to model order reduction of input-output systems as described in the previous section, we consider the linear time-invariant system

$$\frac{d\mathbf{x}}{dt} = A\mathbf{x} + B\mathbf{u},$$
$$\mathbf{y} = C^T\mathbf{x}.$$

The general solution of this problem is

$$\mathbf{x}(t) = \exp\left(A(t - t_0)\right)\mathbf{x}_0 + \int_{t0}^{t} \exp\left(A(t - \tau)\right)B\mathbf{u}(\tau)d\tau. \qquad (1)$$

A common way to solve the differential equation is by transforming it from the time domain to the frequency domain, by means of a Laplace transform defined as

$$\mathcal{L}(f)(s) \equiv \int_0^\infty f(t) \exp(-st) dt.$$

If we apply this transform to the system, assuming that $\mathbf{x}(0) = 0$, the system is transformed to a purely algebraic system of equations:

$$\begin{aligned} (I_n - sA)\mathbf{X} &= B\mathbf{U}, \\ \mathbf{Y} &= C^T\mathbf{X}, \end{aligned}$$

where the capital letters indicate the Laplace transforms of the respective lower case quantities. This immediately leads to the following relation:

$$\mathbf{Y}(s) = C^T(I_n - sA)B\mathbf{X}(s). \tag{2}$$

Now define the *transfer function* $H(s)$ as

$$H(s) = C^T(I_n - sA)B. \tag{3}$$

This transfer function represents the direct relation between input and output in the frequency domain, and therefore the behavior of the system in frequency domain. For example, in the case of electronic circuits this function may describe the transfer from currents to voltages, and is then termed impedance. If the transfer is from voltages to currents, then the transfer function corresponds to the admittance.

Note that if the system has more than one input or more than one output, then B and C have more than one column. This makes $H(s)$ a matrix function. The i, j entry in $H(s)$ then denotes the transfer from input i to output j.

2.2 Moments

The transfer function is a function in s, and can therefore be expanded into a moment expansion around $s = 0$:

$$H(s) = M_0 + M_1 s + M_2 s^2 + \dots,$$

where $M_0, M_1, M_2, \dots$ are the *moments* of the transfer function. In electronics, M_0 corresponds to the DC solution.In that case the inductors are considered as short circuits, and capacitors as open circuits. The moment M_1 then corresponds to the so-called Elmore delay, which represents the time for a signal at the input port to reach the output port. The Elmore delay is defined as

$$t_{elm} \equiv \int_0^\infty t h(t) dt,$$

where $h(t)$ is the *impulse response function*, which is the response of the system to the Dirac delta input. The transfer function in the frequency domain is the Laplace transform of the impulse response function:

$$H(s) = \int_0^\infty h(t) \exp(-st) dt.$$

Expanding the exponential function in a power series, it is seen that the Elmore delay indeed corresponds to the first order moment of the transfer function.

Of course, the transfer function can also be expanded around some non-zero s_0. We then obtain a similar expansion in terms of moments. This may be advantageous in some cases, and truncation of that alternative moment expansion may lead to better approximations.

2.3 Poles and Residues

The transfer function can also be expanded as follows:

$$H(s) = \sum_{j=1}^{n} \frac{R_j}{s - p_j}, \tag{4}$$

where the p_j are the poles, and R_j are the corresponding residues. The poles are exactly the eigenvalues of the matrix $-A^{-1}$. In fact, if the matrix E of eigenvectors is non-singular, we can write

$$-A^{-1} = E\Lambda E^{-1},$$

where the diagonal matrix Λ contains the eigenvalues λ_j. Substituting this into the expression for the transfer function, we obtain:

$$H(s) = -C^T E(I + s\Lambda)^{-1} E^{-1} A^{-1} B.$$

Hence, if B and C contain only one column (which corresponds to the single input, single output or SISO case), then

$$H(s) = \sum_{j=1}^{n} \frac{l_j^T r_j}{1 + s\lambda_j},$$

where the l_j and r_j are the left and right eigenvectors, respectively.

We see that there is a one-to-one relation between the poles and the eigenvalues of the system. If the original dynamical system originates from a differential algebraic system, then a generalized eigenvalue problem needs to be solved. Since the poles appear directly in the pole-residue formulation of the transfer function, there is also a strong relation between the transfer function and the poles or, stated differently, between the behavior of the system and the poles. If one approximates the system, one should take care to approximate the most important poles. There are several methods that do this, which are discussed in later chapters of this book. In general, we can say that, since the transfer function is usually plotted for imaginary points $s = \omega i$, the poles that have a small imaginary part dictate the behavior of the transfer function for small values of the frequency ω. Consequently, the poles with a large imaginary part are needed for a good approximation at higher frequencies. Therefore, a successful reduction method aims at capturing the poles with small imaginary part rather, and leaves out poles with a small residue.

2.4 Stability

Poles and eigenvalues of a system are strongly related to the stability of the system. Stability is the property of a system that ensures that the output signal of a system is limited (in the time domain).

Consider again the system (1). The system is *stable* if and only if, for all eigenvalues λ_j, we have that $Re(\lambda_j) \leq 0$, and all eigenvalues with $Re(\lambda_j) = 0$ are simple. In that case, the corresponding matrix A is termed stable.

There are several properties associated with stability. Clearly, if A is stable, then also A^{-1} is stable. Stability of A also implies stability of A^T and stability of A^*. Finally, if the product of matrices AB is stable, then also BA can be shown to be stable. It is also clear that, due to the relation between eigenvalues of A and poles of the transfer function, stability can also be formulated in terms of the poles of $H(s)$.

The more general linear dynamical system

$$\begin{aligned} Q\frac{d\mathbf{x}}{dt} &= A\mathbf{x} + B\mathbf{u}, \\ \mathbf{y} &= C^T\mathbf{x}, \end{aligned}$$

is stable if and only if for all generalized eigenvalues we have that $Re(\lambda_j(Q, A)) \leq 0$, and all generalized eigenvalues for which $Re(\lambda_j(Q, A)) = 0$ are simple. The set of generalized eigenvalues $\sigma(Q, A)$ is defined as the collection of eigenvalues of the generalized eigenvalue problem

$$Q\mathbf{x} = \lambda A\mathbf{x}.$$

In this case, the pair of matrices (Q, A) is termed a *matrix pencil*. This pencil is said to be regular if there exists at least one eigenvalue λ for which $Q + \lambda A$ is regular. Just as for the simpler system discussed in the above, stability can also be formulated in terms of the poles of the corresponding transfer function.

2.5 Positive Real Matrices

The concept of stability explained in the previous subsection leads us to consider other properties of matrices. First we have the following theorem.

Theorem 1. *If* $Re(\mathbf{x}^* A\mathbf{x} > 0$ *for all* $x \in C^n$*, then all eigenvalues of* A *have a positive real part.*

The converse of this theorem is not true, as can be seen when we take

$$A = \begin{pmatrix} -\frac{1}{3} & -1 \\ 1 & 2 \end{pmatrix},$$

and

$$\mathbf{x} = \begin{pmatrix} 1 \\ 0 \end{pmatrix}.$$

Matrices with the property that $Re(\mathbf{x}^* A\mathbf{x} > 0$ for all $x \in C^n$ are termed *positive real*. The counter example shows that the class of positive real matrices is smaller than the class of matrices for which all eigenvalues have positive real part. In the next section, this new and restricted class will be discussed in more detail. For now, we remark that a number of properties of positive real matrices are easily derived. If A is positive real, then this also holds for A^{-1} (if it exists). Furthermore, A is positive real if and only if A^* is positive real. If two matrices A and B are both positive real, then any linear combination $\alpha A + \beta B$ is also positive real provided $Re(\alpha) > 0$ and $Re(\beta) > 0$.

There is an interesting relation between positive real and positive definite matrices. Evidently, the class of positive definite matrices is a subclass of the set of positive real matrices. But we also have:

Theorem 2. *A matrix $A \in C^{n\times n}$ is positive real if and only if the Hermitian part of A (i.e. $\frac{1}{2}(A + A^*)$) is symmetric positive definite.*

Similarly one can prove that a matrix is non-negative real if and only if its Hermitian part is symmetric positive semi-definite.

2.6 Passivity

Stability is a very natural property of physical structures. However, stability is not strong enough for electronic structures that contain no sources. A stable structure can become unstable if non-linear components are connected to it. Therefore, another property of systems should be defined that is stronger than stability. This property is called *passivity*. Being passive means being incapable of generating energy. If a system is passive and stable, we would like a reduction method to preserve these properties during reduction. In this section the principle of passivity is discussed and what is needed to preserve passivity.

To define the concept, we consider a system that has N so-called ports. The total instantaneous power absorbed by this real N-port is defined by:

$$w_{inst}(t) \equiv \sum_{j=1}^{N} v_j(t) i_j(t),$$

where $v_j(t)$ and $i_j(t)$ are the real instantaneous voltage and current at the j-th port. An N-port contains stored energy, say $E(t)$. If the system dissipates energy at rate

$w_d(t)$, and contains sources which provide energy at rate $w_s(t)$, then the energy balance during a time interval $[t_1, t_2]$ looks like:

$$\int_{t_1}^{t_2} (w_{inst} + w_s - w_d)dt = E(t_2) - E(t_1). \tag{5}$$

An N-port is termed *passive* if we have

$$\int_{t_1}^{t_2} w_{inst}dt \geq E(t_2) - E(t_1), \tag{6}$$

over any time interval $[t_1, t_2]$. This means that the increase in stored energy must be less than or equal to the energy delivered through the ports. The N-port is called lossless if (6) holds with equality over any interval. Assume that the port quantities exhibit purely exponential time-dependence, at a single complex frequency s. We may then write:

$$v(t) = \hat{v}\exp(it), i(t) = \hat{\imath}\exp((it)),$$

where $\hat{v}$ and $\hat{\imath}$ are the complex amplitudes. We define the total complex power absorbed to be the inner product of $\hat{\imath}$ and $\hat{v}$,

$$w = \hat{\imath}^*\hat{v},$$

and the average or active power as:

$$< w > = Re(w).$$

For an N-port defined by an impedance relationship, we may immediately write $< w >$ in terms of the voltage and current amplitudes:

$$< w >= \frac{1}{2}(\hat{\imath}^*\hat{v} + \hat{v}^*\hat{\imath}) = \frac{1}{2}(\hat{\imath}^* Z\hat{\imath} + \hat{\imath}^* Z^*\hat{\imath}) = \frac{1}{2}(\hat{\imath}^*(Z + Z^*)\hat{\imath}).$$

For such a real linear time invariant (LTI) N-port, passivity may be defined in the following way. If the total active power absorbed by an N-port is always greater than or equal to zero for frequencies s such that $Re(s) \geq 0$, then it is called passive. This implies that $Z + Z^* \geq 0$ for $Re(s) \geq 0$. Hence, the matrix Z must be positive real.

Given our discussion and the definition of passivity based on an energy argument, we can formulate the following theorem.

Theorem 3. *The transfer function $H(s)$ of a passive system is positive real, i.e.*

$$H^*(s) + H(s) \geq 0$$

for all s with $Re(s) > 0$.

Sometimes another definition of passivity is used, for instance in [35]. Under certain assumptions these definitions are equal.

3 A Short Account of Techniques for Model Order Reduction

Having discussed the need for model order reduction, and several essential prerequisites, this section will now give an overview of the field by discussing the most important techniques. The methods and new developments discussed in subsequent chapters of this book build on these basic algorithms. Adequate referencing is provided, so that readers can go into more detail when desired.

3.1 Asymptotic Waveform Evaluation

One of the basic and earliest methods in Model Order Reduction is Asymptotic Waveform Evaluation (AWE), proposed by Pillage and Rohrer in 1990 [7, 29]. The underlying idea of the method is that the transfer function can be well approximated by a Padé approximation. A Padé approximation is a ratio of two polynomials $P(s)$ and $Q(s)$. AWE calculates a Padé approximation of finite degree, so the degree of $P(s)$ and $Q(s)$ is finite and $deg(Q(s)) \geq deg(P(s))$. There is a close relation between the Padé approximations and Krylov subspace methods (see Chapter 2). To explain this fundamental property, consider the general system:

$$(sI_n - A)\mathbf{X}(s) = B\mathbf{U}(s).$$

Expanding $\mathbf{X}(s)$ around some expansion point $s_0 \in C$ we obtain:

$$(s_0 I_n - A + (s - s_0)I_n)(\mathbf{X}_0 + (s - s_0)\mathbf{X}_1 + \ldots) = B\mathbf{U}(s).$$

Here, the $\mathbf{X}_i(s)$ are the moments. Assuming $\mathbf{U}(s) = 1$, and equating like powers of $(s - s_0)^i$, we find:

$$(s_0 I_n - A)\mathbf{X}_0 = B,$$

for the term corresponding to $i = 0$, and for $i \geq 1$

$$(s_0 I_n - A)\mathbf{X}_i = -X_{i-1}.$$

We conclude that, in fact, a Krylov space is built up (see Chapter 2):

$$\mathcal{K}((s_0 I_n - A)^{-1}B, (s_0 I_n - A)^{-1}).$$

The process can be terminated after finding a sufficient number of moments, and the hope is that then a good approximation has been found for the transfer function. Clearly, this approximation is of the form

$$\tilde{H}(s) = \sum_{k=0}^{n} m_k (s - s_0)^k,$$

for some finite n.

Once the moments have been calculated, a Padé approximation $\hat{H}(s)$ of the transfer function $H(s)$ can be determined:

$$\hat{H}(s) = \frac{P(s - s_0)}{Q(s - s_0)}.$$

Letting

$$P(s) = \sum_{k=0}^{p} a_k (s - s_0)^k, \qquad Q(s) = \sum_{k=0}^{p+1} b_k (s - s_0)^k,$$

we find that the following relation must hold:

$$\sum_{k=0}^{p} a_k (s - s_0)^k = \left(\sum_{k=0}^{n} m_k (s - s_0)^k \right) \left(\sum_{k=0}^{p+1} b_k (s - s_0)^k \right).$$

Equating like powers of $s - s_0$ (for the higher powers), and setting $b_0 = 1$, we find the following system to be solved:

$$\begin{pmatrix} m_0 & m_1 & \dots & m_p \\ m_1 & m_2 & \dots & m_{p+1} \\ \vdots & \vdots & \ddots & \vdots \\ m_p & m_{p+1} & \dots & m_{2p} \end{pmatrix} \begin{pmatrix} b_{p+1} \\ b_p \\ \vdots \\ b_1 \end{pmatrix} = - \begin{pmatrix} m_{p+1} \\ m_{p+2} \\ \vdots \\ m_{2p+1} \end{pmatrix}, \tag{1}$$

from which we can extract the coefficients $b_i, i = 1, \dots, p+1$ of Q. A similar step can be used to subsequently find the coefficients of the polynomial P.

The problem with the above method, and with AWE in general, is that the coefficient matrix in (1) quickly becomes ill-conditioned as the number of moments used goes up. In fact, practical experience indicates that applicability of the method stops once 8 or more moments are used. The method can be made more robust by using *Complex Frequency Hopping* [29], meaning that more than one expansion point is used. However, the method remains computationally demanding and, for that reason, alternatives as described in the next subsections are much more popular nowadays.

3.2 The PVL Method

Although AWE was initially considered an attractive method for approximating the transfer function, soon the disadvantages of the method were recognized. Then, in 1993, Roland Freund and Peter Feldmann [9] published their method named *Padé-via-Lanczos* or *PVL*. In this method, the Padé approximation based on the moments is calculated by means of a two-sided Lanczos algorithm (see also Chapter 2). The algorithm requires approximately the same computational effort as AWE, but it generates more poles and is much more robust.

To explain the method, consider the transfer function of a SISO system:

$$H(s) = \mathbf{c}^T (sI_n - A)^{-1} \mathbf{b}.$$

Let $s_0 \in C$ be the expansion point for the approximation. Then the transfer function can be cast into the form

$$H(s) = \mathbf{c}^T (I_n - (s - s_0)\hat{A})^{-1} \mathbf{r},$$

where

$$\hat{A} = -(s_0 I_n - A)^{-1},$$

and

$$\mathbf{r} = (s_0 I_n - A)^{-1}\mathbf{b}.$$

This transfer function can, just as in the case of AWE, be approximated well by a rational function in the form of a Padé approximation. In PVL this approximation is found via the Lanczos algorithm. By running q steps of this algorithm (see Chapter 2 for details on the Lanczos method), an approximation of $\hat{A}$ is found in the form of a tridiagonal matrix T_q, and the approximate transfer function is of the form:

$$H_q(s) = \mathbf{c}^T\mathbf{r} \cdot \mathbf{e}_1^T(I_n - (s - s_0)T_q)^{-1}\mathbf{e}_1, \tag{2}$$

where $\mathbf{e}_1$ is the first unit vector. The moments can also be found from this expression:

$$\mathbf{c}^T\hat{A}^k\mathbf{r} = \mathbf{c}^T\mathbf{r} \cdot \mathbf{e}_1^T T_q^k \mathbf{e}_1.$$

A proof of these facts can be found in the original paper [9].

Every iteration leads to the preservation of two extra moments. This makes PVL a very powerful and efficient algorithm. Unfortunately, there are also disadvantages associated with the algorithm. For example, it is known that PVL does not always preserve stability. The reason for this is that PVL is based on a two-sided Lanczos algorithm, and uses non-orthogonal or skew projections. The problem has been studied by several authors; in [9], it is suggested to simply delete the poles which have a positive real part. However, such "remedies" are not based upon theory, and should probably be avoided. In the case of symmetric matrices, the problem can be resolved by using a one-sided Lanczos process, which would preserve stability.

Another problem associated with PVL is that the inner products $\mathbf{w}_{n+1}^T\mathbf{v}_{n+1}$ in the bi-orthogonal sequence may be zero or near to zero, in which case the Lanczos algorithm breaks down. To avoid this, a look-ahead version of the Lanczos process has been suggested in [10].

The PVL method has been very successful since 1993, and many new developments are based upon it. We mention the *matrix PVL* method, published by the inventors of PVL [10]. In [1] a more extensive version of MPVL with look-ahead and deflation is described. Another method presented by Freund and Feldmann is *SymPVL* [12–14], which is an efficient version of PVL for the case of symmetric matrices. Te method cures the stability problem observed for PVL. A similar (and earlier) development is SyPVL [11] The main idea of all these methods is to make use of the fact that the matrix is symmetric, so that it can be decomposed using a Cholesky decomposition. This then automatically leads to stability of the associated approximate models.

Another nice development worth mentioning is the *two-step Lanczos algorithm*. It splits the problem of reducing the original system into two separate phases. First a Lanczos process is performed using a Krylov subspace based upon the matrix itself, rather than its inverse. Clearly, this is much more efficient, and so it is easy to perform many steps of this procedure. This then leads to a reduction of the original problem to a system of size, say, a few thousand. In the second phase, the 'ordinary' Lanczos procedure is used to reduce the problem much further, now using the inverse of the coefficient matrix. For more details, see [35,36].

3.3 Arnoldi and PRIMA Method

The Lanczos process is most suitable for symmetric matrices (see Chapter 2), but for general matrices the Arnoldi method is a better choice. It can also be used as the basis for a model order reduction method like PVL. Similar to PVL, one can define an expansion point s_0, and work with the shift-and-invert transfer function:

$$H(s) = C^T(sI_n - A)^{-1}B = C^T(I_n + (s - s_0)\hat{A})^{-1}R,$$

with

$$\hat{A} = (s_0 I_n - A)^{-1},$$

and

$$R = (s_0 I_n - A)^{-1}B.$$

Then, in the Arnoldi process, a Krylov space associated with the matrices $\hat{A}$ and R is generated:

$$\mathcal{K}_q(R, \hat{A}) = span\{R, \hat{A}R, \ldots, \hat{A}^q R\}.$$

The main differences with PVL are that only one Krylov space is generated, namely with the (block) *Arnoldi process* [31], and that the projections are performed with orthogonal operators.

The expansion point in the above can be chosen either real or complex, leading to different approximations of the poles of the system. A real shift may be favorable over a complex shift, as the convergence of the reduced transfer function towards the original transfer function is more global.

A very important new development was published in 1998, with a method now known as *PRIMA* [26]. Up till then, the methods developed suffered from non-passivity. Odabasioglu and Celik realized that the Arnoldi method had the potential to resolve these problems with passivity. The PRIMA method, in full passive reduced-order interconnect macromodeling algorithm, builds upon the same Krylov space as in the Arnoldi method and PVL, using the Arnoldi method to generate an orthogonal basis for the Krylov space. The fundamental difference with preceding methods is, however, that the projection of the matrices is done explicitly. This is in contrast with PVL and Arnoldi, where the tridiagonal or the Hessenberg matrix is used for this purpose. In other words, the following matrix is formed:

$$A_q = V_q^T A V_q,$$

where V_q is the matrix containing an orthonormal basis for the Krylov space.

Although more expensive, the explicit projection onto the Krylov space has strong advantages. It makes PRIMA more accurate than the Arnoldi method and it ensures preservation of stability and passivity. As such, it was the first method to achieve this in a provable way. A slight disadvantage of the method, as compared to PVL, is that only one moment per iteration is preserved additionally. This is only a minor disadvantage, if one realizes that PVL requires iterating both with A and its transpose.

3.4 Laguerre Methods

The search for provable passive model order reduction techniques continued after the publication of PRIMA. A new development was the construction of approximations using the framework of Laguerre functions, as proposed in [23,24]. In these methods, the transfer function is not shifted-and-inverted, as is the case in the PVL and Arnoldi methods. Instead, it is expanded in terms of Laguerre functions that are defined as

$$\phi_k^\alpha(t) \equiv \sqrt{2\alpha}\exp\left(-\alpha t\right)\updownarrow_k(2\alpha t),$$

where α is a positive scaling parameter, and $\updownarrow_k(t)$ is the Laguerre polynomial

$$\updownarrow_k(t) \equiv \frac{\exp\left(t\right)}{k!}\frac{d^k}{dt^k}(\exp\left(-t\right)t^k).$$

The Laplace transform of $\phi_k^\alpha(t)$ is

$$\Phi_k^\alpha(s) = \frac{\sqrt{2\alpha}}{s+\alpha}\left(\frac{s-\alpha}{s+\alpha}\right)^k.$$

Furthermore, it can be shown (see [20]) that the Laguerre expansion of the transfer function is

$$H(s) = \frac{2\alpha}{s+\alpha}C^T\sum_{k=0}^{\infty}\left((\alpha I_n - A)^{-1}(-\alpha I_n - A)\right)^k(\alpha I_n - A)^{-1}B\left(\frac{s-\alpha}{s+\alpha}\right)^k.$$

Clearly, this expansion gives rise to a Krylov space again. The number of linear systems that needs to be solved is equivalent to that in PRIMA, so the method is computationally competitive.

The algorithm presented in [23] then reads:

1. select a value for α and q
2. solve $(\alpha I_n - A)R_1 = B$
3. for k=2,...,q, solve $(\alpha I_n - A)R_k = (-\alpha I_n - A)R_{k-1}$
4. define $R = [R_1, ..., R_q]$ and calculate the SVD of R: $R = V\Sigma W^T$
5. $\tilde{A} = V^TAV$
6. $\tilde{C} = V^TCV$
7. $\tilde{B} = V^TBV$

In [23] it is argued that the best choice for α is to take it equal to $2\pi f_{\max}$, where $f_{\max}$ is the maximum frequency for which the reduced order model is to be valid.

As can be seen from the algorithm, the Krylov space is built without intermediate orthogonalisation. Instead, a singular value decomposition (SVD) is performed after the process has terminated. Consequently, V is an orthonormal basis of R. SVD is known to be a very stable and accurate way to perform this orthogonalisation, on the other hand it is computationally expensive. There are good alternatives, such as the QR method or Modified Gram-Schmidt. In [21], an alternative to the Laguerre-SVD method is presented that makes use of intermediate orthogonalisation, and has been shown to have certain advantages over using an SVD.

Just as in the PRIMA method, the Laguerre-based methods make use of explicit projection of the system matrices. Consequently, these methods preserve stability and passivity. Since α is a real number, the matrices in the Laguerre algorithm remain real during projection, thereby making it suitable for circuit synthesis (see [21,24]).

3.5 Truncation Methods

As mentioned before, model order reduction has its roots in the area of systems and control theory. Within this area, methods have been developed that differ considerably from the Krylov based methods as discussed in subsections 3.1-3.5. The basic idea is to truncate the dynamical system studied at some point. To illustrate how it works, consider again the linear dynamical system (1):

$$\frac{d\mathbf{x}}{dt} = A\mathbf{x} + B\mathbf{u},$$
$$\mathbf{y} = C^T\mathbf{x} + D\mathbf{u}.$$

Applying a state space transformation

$$T\tilde{\mathbf{x}} = \mathbf{x},$$

does not affect the input-output behavior of the system. This transformation could be chosen to be based on the eigenvalue decomposition of the matrix A:

$$AT = T\Lambda.$$

When T is non-singular, $T^{-1}AT$ is a diagonal matrix consisting of the eigenvalues of A, and we could use an ordering such that the eigenvalues on the diagonal occur in order of decreasing magnitude. The system can then be truncated by restricting the matrix T to the dominant eigenvalues. This process is termed *modal truncation.*

Another truncation method is that of balanced truncation, usually known as *Truncated Balanced Realization* (TBR). This method is based upon the observation that only the largest singular values of a system are important. As there is a very good reference to this method, containing all details, we will only summarize the main concepts. The reader interested in the details of the method is referred to the book by Antoulas [2].

The controllability Gramian and the observability Gramian associated to the linear time-invariant system (A, B, C, D) are defined as follows:

$$P = \int_0^\infty e^{At}BB^*e^{A^*t}dt, \tag{3a}$$

and

$$Q = \int_0^\infty e^{At}C^*Ce^{A^*t}dt, \tag{3b}$$

respectively.

The matrices P and Q are the unique solutions of two Lyapunov equations:

$$AP + PA^* + BB^* = 0, \tag{4a}$$

$$A^*Q + QA + C^*C = 0. \tag{4b}$$

The Lyapunov equations for the Gramians arise from a stability assumption of A. Stability in the matrix A implies that the infinite integral defined in (3) is bounded. Finding the solution of the Lyapunov equation is quite expensive. There are direct ways and iterative ways to do this. One of the interesting iterative methods to find a solution is vector ADI [8, 33]. New developments are also in the work of Benner [3].

After finding the Gramians, we look for a state space transformation which balances the system. A system is called *balanced* if $P = Q = \Sigma = \text{diag}(\sigma_i)$. The transformation will be applied to the system as follows:

$$\begin{aligned} A' &= T^{-1}AT \\ B' &= T^{-1}B \\ C' &= CT \\ D' &= D. \end{aligned}$$

This transformation also yields transformed Gramians:

$$\begin{aligned} P' &= T^{-1}PT^{-*} \\ Q' &= T^*QT. \end{aligned}$$

Because P and Q are positive definite, a Cholesky factorization of P can be calculated, $P = R^{\mathrm{T}}R$, with $R \in \mathbb{R}^{n\times n}$. Then the Hankel singular values are derived as the singular values of the matrix RQR^{T}, which are equal to the square root of the eigenvalues of RQR^{T} and QP. So:

$$RQR^{\mathrm{T}} = U^{\mathrm{T}}\Sigma^2 U. \tag{5}$$

Then the transformation $T \in R^{n\times n}$ is defined as:

$$T = R^{\mathrm{T}}U^{\mathrm{T}}\Sigma^{-1/2}. \tag{6}$$

The inverse of this matrix is:

$$T^{-1} = \Sigma^{1/2}UR^{-1}. \tag{7}$$

This procedure is called balancing. It can be shown that T indeed balances the system:

$$\begin{aligned} Q' &= T^{\mathrm{T}}QT = \Sigma^{-1/2}URQR^{\mathrm{T}}U^{\mathrm{T}}\Sigma^{-1/2} = \Sigma^{-1/2}\Sigma^2\Sigma^{-1/2} = \Sigma \\ P' &= T^{-1}PT^{-\mathrm{T}} = \Sigma^{1/2}UR^{-\mathrm{T}}PR^{-1}U^{\mathrm{T}}\Sigma^{1/2} = \Sigma^{1/2}\Sigma^{1/2} = \Sigma. \end{aligned}$$

Since a transformation was defined which transforms the system according to the Hankel singular values [2], now very easily a truncation can be defined.

Σ can be partitioned:

$$\Sigma = \begin{pmatrix} \Sigma_1 & 0 \\ 0 & \Sigma_2 \end{pmatrix}, \tag{8}$$

where Σ_1 contains the largest Hankel singular values. This is the main advantage of this method, since now we can manually choose an appropriate value of the size of the reduction, instead of guessing one.

A', B' and C' can be partitioned in conformance with Σ:

$$A' = \begin{pmatrix} A_{11} & A_{12} \\ A_{21} & A_{22} \end{pmatrix} \tag{9}$$

$$B' = \begin{pmatrix} B_1 \\ B_2 \end{pmatrix} \tag{10}$$

$$C' = \begin{pmatrix} C_1 & C_2 \end{pmatrix}. \tag{11}$$

The reduced model is then based on A_{11}, B_1 and C_1:

$$\begin{aligned} \dot{\tilde{\mathbf{x}}} &= A_{11}\tilde{\mathbf{x}} + B_1\mathbf{u} \\ \mathbf{y} &= C_1\tilde{\mathbf{x}}. \end{aligned}$$

It is sometimes proposed to apply Balanced Truncation-like methods as a second reduction step, after having applied a Krylov method. This can be advantageous in some cases, and has also been done by several authors.

A remark should be made on solving the Lyapunov equation. These equations are normally solved by first calculating a Schur decomposition for the matrix A. Therefore, finding the solution of the Lyapunov is quite expensive, the number of operations is at least $O(n^3)$, where n is the size of the original model. Hence, it is only feasible for small systems. Furthermore, because we arrived at this point using the inverse of an ill-conditioned matrix we have to be careful. B can have very large entries, which will introduce tremendous errors in solving the Lyapunov equation. Dividing both equations by the square of the norm of B spreads the malice a bit, which makes finding a solution worthwhile. In recent years, however, quite a lot of progress has been made on solving larger Lyapunov equations. We refer to the work of Benner [3].

Another remark: since the matrices are projected by a similarity transform, preservation of passivity is not guaranteed in this method. In [27] a Balanced Truncation method is presented which is provably passive. Here also Poor Man's TBR [28] should be mentioned as a fruitful approach to implement TBR is a more efficient way. We refer to a later chapter in this book for more information on this topic.

3.6 Optimal Hankel Norm Reduction

Closely related to Balanced Truncation is Optimal Hankel Norm reduction [18]. In the Balanced Truncation norm it was not clear whether the truncated system of size say k was an optimal approximation of this size. It is seen that this optimality can be calculated and reached given a specific norm, the Hankel norm.

To define the Hankel norm we first have to define the Hankel operator $\mathcal{H}$:

$$\mathcal{H} : u \to y = \int_{-\infty}^{0} h(t-\tau)u(\tau), \tag{12}$$

where $h(t)$ is the impulse response in time domain: $h(t) = C\exp(At)B$ for $t > 0$. This operator considers the past input, the energy that was put into the system before $t = 0$, in order to reach this state. The amount of energy to reach a state tells something about the controllability of that state. If, after $t = 0$, no energy is put into the system and the system is stable, then the corresponding output will be bounded as well. The energy that comes out of a state, gives information about the observability of a state. The observability and controllability Gramians were defined in (3).

Therefore, the maximal gain of this Hankel operator can be calculated:

$$\|\Sigma\|_H = \sup_{u \in \mathcal{L}_2(-\infty,0]} \frac{\|y\|_2}{\|u\|_2}. \tag{13}$$

This norm is called the Hankel norm. Since it can be proved that $\|\Sigma\|_H = \lambda_{max}^{1/2}(PQ) = \sigma_1$, the Hankel norm is nothing but the largest Hankel singular value of the system.

There exists a transfer function and a corresponding linear dynamical system which minimizes this norm. In [18] an algorithm is given which explicitly generates this optimal approximation in the Hankel-norm. The algorithm is based on a balanced realization.

3.7 Selective Node Elimination

Krylov based methods build up the reduced order model by iterating, every iteration leading to a larger size of the model. Hence, the first few iterations yield extremely small models that will not be very accurate in general, and only when a sufficient number of iterations has been performed, the approximate model will be sufficiently accurate. Hence, characteristic for such methods is that the space in which approximations are being sought is gradually built up. An alternative would be to start 'at the other end', in other words, start with the original model, and reduce in it every iteration, until we obtain a model that is small and yet has sufficient accuracy. This is the basic idea behind a method termed *selective node elimination*. Although it can, in principle, be applied in many situations, it has been described in literature only for the reduction of electronic circuits. Therefore, we limit the discussion in this section to that application.

Reduction of a circuit can be done by explicitly removing components and nodes from the circuit. If a node in a circuit is removed, the behaviour of the circuit can be preserved by adjusting the components around this node. Recall that the components connected to the node that is removed, are also removed. The value of the remaining components surrounding this node must be changed to preserve the behavior of the circuit. For circuits with only resistors this elimination can be done exactly.

We explain the idea for an RC circuit. For this circuit we have the following circuit equation $Y(s)\mathbf{v} = (G + sC)\mathbf{v} = \mathbf{J}$. The vector $\mathbf{v}$ here consists of the node voltages, $\mathbf{J}$ is some input current term. Suppose the n-th node is eliminated. Then, we partition the matrix such that the (n, n) entry forms one part:

$$\begin{bmatrix} \tilde{Y} & \mathbf{y} \\ \mathbf{y}^{\mathrm{T}} & \gamma_n + s\chi_n \end{bmatrix} \begin{bmatrix} \tilde{\mathbf{v}} \\ v_n \end{bmatrix} = \begin{bmatrix} \mathbf{J}_1 \\ j_n \end{bmatrix}.$$

Then the variable v_n is eliminated, which leads to:

$$(\tilde{Y} - E)\hat{\mathbf{v}} = (\mathbf{J}_1 - \mathbf{F}), \tag{14}$$

with

$$E_{ij} = \frac{y_i y_j}{\gamma_n + s\chi_n} = \frac{(g_{in} + sc_{in})(g_{jn} + sc_{jn})}{\gamma_n + s\chi_n} \tag{15a}$$

$$F_i = \frac{y_i}{\gamma_n + s\chi_n} j_n = \frac{g_{in} + sc_{in}}{\gamma_n + s\chi_n} j_n. \tag{15b}$$

If node n is not a terminal node, j_n is equal to 0 and therefore $\mathbf{F}$=$\mathbf{0}$ for all i. We see that the elimination can also be written in matrix notation. Hence, this approach is analogous to solving the system by Gaussian elimination. This approach can be used to solve PDE's in an efficient way.

After the elimination process the matrix is not in the form $\mathbf{G}+s\mathbf{C}$ anymore, but is a fraction of polynomials in s. To get an RC-circuit representation an approximation is needed. Given the approximation method that is applied, removing one node leads to a larger error than removing the other.

Many others have investigated methods which are strongly related to the approach described here, for instance a more symbolic approach. The strong attributes of the methods described above is that an RC circuit is the direct result. The error made with the reduction is controllable, but can be rather large. A disadvantage is that reducing an RLC-circuit in this way is more difficult and it is hard to get an RLC-circuit back after reduction.

3.8 Proper Orthogonal Decomposition

Apart from Krylov subspace methods and Truncation methods, there is Proper Orthogonal Decomposition (POD), also known as Karhunen-Loeve decomposition. This method is developed within the area of Computational Fluid Dynamics and nowadays used frequently in many CFD problems. The method is so common there, that it should at least be mentioned here as an option to reduce models derived in an electronic setting. The strong point of POD is that it can be applied to non-linear partial differential equations and is at the moment state-of-the-art for many of such problems.

The idea underlying this method is that the time response of a system given a certain input, contains the essential behavior of the system. The most important aspects of this output in time are retrieved to describe the system. Therefore, the set of outputs serves as a starting-point for POD. The outputs, which are called ‘snapshots’, must be given or else be computed first.

A snapshot consists of a column vector describing the state at a certain moment. Let $W \in \mathbb{R}^{N\times K}$ be the matrix consisting of the snapshots. N is the number of snapshots, K is the number of elements in every snapshot, say the number of state variables. Usually we have that $N < K$.

Let $\mathcal{X}$ be a separable Hilbert space with inner product $(.,.)$, and with an orthonormal basis $\{\varphi_i\}_{i\in I}$. Then, any element $T(x,t) \in \mathcal{X}$ can be written as:

$$T(x,t) = \sum_i a_i(t)\varphi_i(x) = \sum_i (T(x,t),\varphi_i(x))\varphi_i(x). \tag{16}$$

The time dependent coefficients a_i are called Fourier coefficients. We are looking for an orthonormal basis $\{\varphi_i\}_{i\in I}$ such that the averages of the Fourier-coefficients are ordered:

$$\langle a_1^2(t)\rangle \geq \langle a_2^2(t)\rangle \geq \dots, \tag{17}$$

where $\langle . \rangle$ is an averaging operator. In many practical applications the first few elements represent 99% of the content. Incorporating these elements in the approximation gives a good approximation. The *misfit*, the part to which the remaining elements contribute to, is small.

It can be shown that this basis can be found in the first eigenvectors of this operator:

$$C = \langle (T(t),\varphi)T(t)\rangle. \tag{18}$$

In case we consider a finite dimensional problem, in a discrete and finite set of time points, this definition of C comes down to:

$$C = \frac{1}{N} W W^{\mathrm{T}}. \tag{19}$$

Because C is self-adjoint, the eigenvectors are real and can be ordered, such that:

$$\lambda_1 \geq \lambda_2 \geq \dots \tag{20}$$

A basis consisting of the first, say q eigenvectors of this matrix form the optimal basis for POD of size q.

This leads to the following POD algorithm:

1. Input: the data in the matrix W consisting of the snapshots.
2. Define the correlation matrix C as:
$$C = \frac{1}{N} W W^{\mathrm{T}}.$$
3. Compute the eigenvalue decomposition $C\Phi = \Phi\Lambda$.
4. Output: The basis to project the system on, Φ.

This algorithm contains an eigenvalue problem of size $K \times K$, which can be computationally expensive. Therefore the 'method of snapshots' is designed. The reason underlying this different approach is that the eigenvalues of C are the same as the eigenvalues of $K = \frac{1}{N} W^{\mathrm{T}} W$ but K is smaller, namely $N \times N$.

Let the eigenvalues of K be Ψ_i, then the eigenvectors of C are defined as:

$$\varphi_i = \frac{1}{\|W_{snap}\Psi_i\|} W_{snap}\Psi_i. \tag{21}$$

However, also the singular value decomposition of the snapshot matrix W is a straightforward way to obtain the eigenvectors and eigenvalues of C.

$$W = \Phi \Sigma \Psi^{\mathrm{T}}, \tag{22}$$

$$C = \frac{1}{N} W W^{\mathrm{T}} = \frac{1}{N} \Phi \Sigma \Psi^{\mathrm{T}} \Psi \Sigma \Phi^{\mathrm{T}} = \frac{1}{N} \Phi \Sigma^2 \Phi^{\mathrm{T}}. \tag{23}$$

The eigenvectors of C are in Φ:

$$C\Phi = \frac{1}{N} \Phi \Sigma^2 \Phi^{\mathrm{T}} \Phi = \Phi \frac{1}{N} \Sigma^2. \tag{24}$$

From which it can be seen that the eigenvalues of C are $\frac{1}{N}\Sigma^2$.

Once the optimal orthonormal basis is found, the system is projected onto it. For this, we will focus on the following formulation of a possibly non-linear model:

$$\begin{aligned} C(\mathbf{x}) \frac{d}{dt}\mathbf{x} &= \mathbf{f}(\mathbf{x}, \mathbf{u}) \\ \mathbf{y} &= \mathbf{h}(\mathbf{x}, \mathbf{u}). \end{aligned}$$

$\mathbf{x}$ consists of a part in the space spanned by this basis and a residual:

$$\mathbf{x} = \hat{\mathbf{x}} + \mathbf{r}, \tag{25}$$

where $\hat{\mathbf{x}} = \sum_{k=1}^{Q} a_k(t)\mathbf{w}_k$. When $\hat{\mathbf{x}}$ is taken as state space in (25) an error is made:

$$C(\hat{\mathbf{x}}) \frac{d}{dt}\hat{\mathbf{x}} - \mathbf{f}(\hat{\mathbf{x}}, \mathbf{u}) = \rho \neq 0. \tag{26}$$

This error is forced to be perpendicular to the basis W. Forcing this defines the projection fully. In the following derivation we use that $\frac{d}{dt}\hat{\mathbf{x}} = \sum_{k=1}^{Q} \frac{d}{dt} a_k(t)\mathbf{w}_k$:

$$\begin{aligned} 0 = \left(C(\hat{\mathbf{x}}) \frac{d}{dt}\hat{\mathbf{x}} - \mathbf{f}(\hat{\mathbf{x}}, \mathbf{u}), \mathbf{w}_k \right) &= \left(C(\hat{\mathbf{x}}) \sum_{k=1}^{Q} \frac{d}{dt} a_k(t)\mathbf{w}_k - \mathbf{f}(\hat{\mathbf{x}}, \mathbf{u}), \mathbf{w}_k \right) \\ &= \sum_{k=1}^{Q} \frac{d}{dt} a_k(t) \left(C(\hat{\mathbf{x}})\mathbf{w}_k, \mathbf{w}_k \right) - \left(\mathbf{f}(\hat{\mathbf{x}}, \mathbf{u}), \mathbf{w}_k \right), \end{aligned} \tag{27}$$

for $j = 1, \ldots, Q$. Therefore the reduced model of order Q can be formulated as:

$$A(\mathbf{a})\frac{d}{dt}\mathbf{a} = \mathbf{g}(\mathbf{a}, \mathbf{u})$$
$$\mathbf{y} = \mathbf{h}\left(\sum_{k=1}^{Q} a_k \mathbf{w}_k, \mathbf{u}\right),$$

where:

$$\begin{aligned} A_{ij} &= \left(C(\textstyle\sum_{k=1}^{Q} a_k(t)\mathbf{w}_k)\mathbf{w}_i, \mathbf{w}_j\right) \\ \mathbf{a}_j &= a_j(t) \\ \mathbf{g}(\mathbf{a}(t), \mathbf{u}(t)) &= \left(\mathbf{f}(\textstyle\sum_{k=1}^{Q} a_k(t)\mathbf{w}_k, \mathbf{u}(t)), \mathbf{w}_j\right) \end{aligned}$$

Obviously, if the time domain output of a system has yet to be calculated, this method is far too expensive. Fortunately, the much cheaper to obtain frequency response can be used. Consider therefore the following linear system:

$$\begin{aligned} (G + j\omega C)\mathbf{x} &= B\mathbf{u} \\ \mathbf{y} &= L^{\mathrm{T}}\mathbf{x}. \end{aligned}$$

Calculate a set of frequency states, for certain choices of ω:

$$\mathbf{x}_{\omega_j} = [j\omega_j C + G]^{-1} B, \tag{29}$$

where $\mathbf{x}_{\omega_j} \in \mathbb{C}^{n \times 1}$. We can take the real and imaginary part, or linear combinations of both, for the POD process. We immediately see that the correlation matrix is an approximation of the controllability Gramian:

$$K = \frac{1}{M}\sum_{j=1}^{M} [j\omega_j C + G]^{-1} B B^* [-j\omega_j C^* + G^*]^{-1}. \tag{30}$$

This approach solves the problem of chosing which time-simulation is the most appropriate.

3.9 Other Methods

In the foregoing sections, we have reviewed a number of the most important methods for model order reduction. The discussion is certainly not exhaustive, alternative methods have been published. For example, we have not mentioned the method of *vector fitting*. This method builds rational approximations of the transfer function in a very clever and efficient way, and can be used to adaptively build reduced order models. The chapter by Deschrijver and Dhaene contains an account of recent developments in this area.

As model order reduction is a very active area of research, progress in this very active area may lead to an entirely new class of methods. The development of such new methods is often sparked by an industrial need. For example, right now there is a demand for reducing problems in the electronics industry that contain many inputs and outputs. It has already become clear that current methods cannot cope with such problems, as the Krylov spaces very quickly become inhibitively large, even after a few iterations. Hence, new ways of constructing reduced order models must be developed.

References

1. J.I. Aliaga, D.L. Boley, R.W. Freund and V. Hernández. A Lanczos-type method for multiple starting vectors. *Math. Comp.*, 69(232):1577-1601, May 1998.
2. A.C. Antoulas book. *Approximation of Large-Scale Dynamical Systems.* SIAM series on Advances in Design and Control, 2005.
3. P. Benner, V. Mehrmann and D.C. Sorensen. *Dimension Reduction of Large-Scale Systems.* Lecture Notes in Computational Science and Engineering, vol. 45, Springer-Verlag, June 2005.
4. A. Brandt. Multilevel adaptive solutions to boundary value problems. *Math. Comp.*, 31:333-390, 1977.
5. W.L. Briggs, Van Emden Henson and S.F. McCormick. *A multigrid tutorial.* SIAM, 2000.
6. *BSIM3 and BSIM4 Compact MOSFET Model Summary.* Online: http://www-device.eecs.berkeley.edu/ 3/get.html.
7. E. Chiprout and M.S. Nakhla. *Asymptotic Waveform Evaluation and moment matching for interconnect analysis.* Kluwer Academic Publishers, 1994.
8. N.S. Ellner and E.L. Wachspress. Alternating direction implicit iteration for systems with complex spectra. *SIAM J. Numer. Anal.*, 28(3):859.870, June 1991.
9. P. Feldmann and R. Freund. Efficient linear circuit analysis by Padé approximation via the Lanczos process. *IEEE Trans. Computer-Aided Design*, 14:137-158, 1993.
10. P. Feldmann and R. Freund. Reduced-order modeling of large linear subcircuits via a block Lanczos algorithm. *Proc. 32nd ACM/IEEE Design Automation Conf.*, June 1995.
11. R.W. Freund and P. Feldmann. *Reduced-order modeling of large passive linear circuits by means of the SyPVL algorithm.* Numerical Analysis Manuscript 96-13, Bell Laboratories, Murray Hill, N.J., May 1996.
12. R.W. Freund and P. Feldmann. Interconnect-Delay Computation and Signal-Integrity Verification Using the SyMPVL Algorithm. Proc. 1997 European Conf. Circuit Theory and Design, 408-413, 1997.
13. R.W. Freund and P. Feldmann. The SyMPVL algorithm and its application to interconnect simulation. *Proc. 1997 Int. Conf. Simulation of Semiconductor Processes and Devices*, 113-116, 1997.
14. R.W. Freund and P. Feldmann. Reduced-order modeling of large linear passive multi-terminal circuits using Matrix-Padé approximation. *Proc. DATE Conf. 1998*, 530-537, 1998.
15. G. Gildenblat, X. Li, W. Wu, H. Wang, A. Jha, R. van Langevelde, G.D.J. Smit, A.J. Scholten and D.B.M. Klaassen. PSP: An Advanced Surface-Potential-Based MOSFET Model for Circuit Simulation. *IEEE Trans. Electron Dev.*, 53(9):1979-1993, September 2006.
16. E.J. Grimme. *Krylov Projection Methods for model reduction.* PhD thesis, Univ. Illinois, Urbana-Champaign, 1997.

17. G. H. Golub and C. F. Van Loan. *Matrix Computations*. Johns Hopkins University Press, Baltimore, Maryland, 3rd edition, 1996.
18. K. Glover. Optimal Hankel-norm approximations of linear multivariable systems and their l_∞-error bounds. *Int. J. Control*, 39(6):115-193, 1984.
19. H.C. de Graaff and F.M. Klaassen. *Compact Transistor Modelling for Circuit Design*. Springer-Verlag, Wien, New York, 1990.
20. Heres2001
21. P.J. Heres. *Robust and efficient Krylov subspace methods for Model Order Reduction*. PhD Thesis, TU Eindhoven, The Netherlands, 2005.
22. M.R. Hestenes and E. Stiefel. Methods of Conjugate Gradients for the solution of linear systems. *J. Res. Res. Natl. Bur. Stand.*, 49:409-436, 1952.
23. L. Knockaert and D. De Zutter. Passive Reduced Order Multiport Modeling: The Padé-Arnoldi-SVD Connection. *Int. J. Electron. Commun. (AEU)*, 53:254-260, 1999.
24. L. Knockaert and D. De Zutter. Laguerre-SVD Reduced-Order Modelling. *IEEE Trans. Microwave Theory and Techn.*, 48(9):1469-1475, September 2000.
25. J.A. Meijerink and H.A. van der Vorst. An iterative solution method for linear systems of which the coefficient matrix is a symmetric M-matrix. *Math. Comp.*, 31:148-162, 1977.
26. A. Odabasioglu and M. Celik. PRIMA: passive reduced-order interconnect macromodeling algorithm. *IEEE Trans. Computer-Aided Design*, 17(8):645-654, August 1998.
27. J.R. Phillips, L. Daniel, and L.M. Silveira. Guaranteed passive balancing transformations for model order reduction. *Proc. 39th cConf. Design Automation*, 52-57, 2002.
28. J.R. Phillips and L.M. Silveira. Poor Man's TBR: A simple model reduction scheme. *IEEE. Trans. Comp. Aided Design ICS*, 24(1):43-55, January 2005.
29. L.T. Pillage and R.A. Rohrer. Asymptotic Waveform Evaluation for Timing Analysis. *IEEE Trans. Computer-Aided Design Int. Circ. and Syst.*, 9(4): 352-366, April 1990.
30. A. Quarteroni and A. Veneziani. Analysis of a geometrical multiscale model based on the coupling of PDE's and ODE's for blood flow simulations. *SIAM J. on MMS*, 1(2): 173-195, 2003.
31. Y. Saad. *Iterative Methods for Sparse Linear Systems*. SIAM, Philadelphia, 2nd edition, 2003.
32. W.H.A. Schilders and E.J.W. ter Maten. *Numerical methods in electromagnetics*. Handbook of Numerical Analysis, volume XIII, Elsevier, 2005.
33. G. Starke. Optimal alternating direction implicit parameters for nonsymmetric systems of linear equations. *SIAM J. Numer. Anal.*, 28(5):1431-1445, October 1991.
34. H.A. van der Vorst. Bi-CGSTAB: A fast and smoothly converging variant of Bi-CG for the solution of nonsymmetric linear systems. *SIAM J. Sci. Stat. Comput.*, 13(2):631-644, 1992.
35. T. Wittig. *Zur Reduktion der Modellordnung in elektromagnetischen Feldsimulationen*. PhD thesis, TU Darmstadt, 2003.
36. T. Wittig, I. Munteanu, R. Schuhmann, and T. Weiland. Two-step Lanczos algorithm for model order reduction. *IEEE Trans. Magnetics*, 38(2):673-676, March 2001.

Linear Systems, Eigenvalues, and Projection

Henk van der Vorst

Mathematical Institute, Utrecht University

1 Introduction

The development of Model Order Reduction techniques for various problems was triggered by the success of subspace projection methods for the solution of large linear systems and for the solution of matrix eigenvalue problems.

The idea behind a subspace projection method for the solution of a large linear system $Ax = b$, with $A \in \mathcal{R}^{n \times n}$, $x, b \in \mathcal{R}^n$, is to identify a subspace $\mathcal{R}^m$, with $m \ll n$. The system $Ax = b$ is projected onto $\mathcal{R}^m$, which yields a (much) smaller system $Hy - c$, with $y \in \mathcal{R}^m$. The latter system can be solved conveniently by standard solution techniques. From its solution y it is straightforward to generate an approximation for the vector x. The problem is now to generate a subspace such that we obtain acceptable approximations for x for relatively low values of m, in order to keep computational costs (storage and CPU-time) affordable.

The most well-known approaches in the subspace projection arena are based on the construction of Krylov subspaces. These subspaces were proposed in 1931 by Krylov for the explicit construction of the characteristic polynomial of a matrix, so that the eigenvalues could be computed as the roots of that polynomial. This initial technique proved to be unpractical for matrices of order larger than, say, 6 or 7. It failed because of the poor quality of the standard basis vectors for the Krylov subspace. An orthogonal basis for this subspace appeared to be an essential factor, as well as the way in which this orthogonal basis is generated. These were breakthroughs initiated by Lanczos [4] and Arnoldi [1], both in the early 1950's. In this chapter we will first discuss briefly some standard techniques for solving linear systems and for matrix eigenvalue problems. We will mention some relevant properties, but we refer the reader for background and more references to the standard text by Golub and van Loan [3].

We will then focus our attention on subspace techniques and highlight ideas that are relevant and can be carried over to Model Order Reduction approaches for other sorts of problems.

1.1 Some Basic Properties

We will consider linear systems $Ax = b$, where A is usually an n by n matrix:

$$A \in \mathcal{R}^{n \times n}.$$

The elements of A will be denoted as $a_{i,j}$. The vectors $x = (x_1, x_2, \ldots, x_n)^T$ and b belong to the linear space $\mathcal{R}^n$. Sometimes we will admit complex matrices $A \in \mathcal{C}^{n \times n}$ and vectors $x, b \in \mathcal{C}^n$, but that will be explicitly mentioned.

Over the space $\mathcal{R}^n$ we will use the Euclidean inner product between two vectors x and y:

$$x^T y = \sum_{i=1}^{n} x_i y_i,$$

and for $v, w \in \mathcal{C}^n$ we use the standard complex inner product:

$$v^H w = \sum_{i=1}^{n} \bar{v}_i w_i.$$

These inner products lead to the 2-norm or Euclidean length of a vector

$$\|x\|_2 = \sqrt{x^T x} \;\text{ for }\; x \in \mathcal{R}^n,$$

$$\|v\|_2 = \sqrt{v^H v} \;\text{ for }\; v \in \mathcal{C}^n.$$

With these norms we associate a 2-norm for matrices: for $A \in \mathcal{R}^{n \times n}$, its associated 2-norm $\|A\|_2$ is defined as

$$\|A\|_2 = \sup_{y \in \mathcal{R}^n, y \neq 0} \frac{\|Ay\|_2}{\|y\|_2},$$

and similarly in the complex case, using the complex inner product.

The associated matrix norms are convenient, because they can be used to bound products. For $A \in \mathcal{R}^{n \times k}$, $B \in \mathcal{R}^{k \times m}$, we have that

$$\|AB\|_2 \leq \|A\|_2 \|B\|_2,$$

in particular

$$\|Ax\|_2 \leq \|A\|_2 \|x\|_2.$$

The inverse of a nonsingular matrix A is denoted as A^{-1}. Particularly useful is the condition number of a square nonsingular matrix A, defined as

$$\kappa_2(A) = \|A\|_2 \|A^{-1}\|_2.$$

The condition number is used to characterize the sensitivity of the solution x of $Ax = b$ with respect to perturbations in b and A. For perturbed systems we have the following theorem.

Theorem 1. *[3, Theorem 2.7.2] Suppose*

$$Ax = b \quad A \in \mathcal{R}^{n\times n}, 0 \neq b \in \mathcal{R}^n$$

$$(A + \Delta A)y = b + \Delta b \quad \Delta A \in \mathcal{R}^{n\times n}, \Delta b \in \mathcal{R}^n,$$

with $\|\Delta A\|_2 \leq \epsilon\|A\|_2$ *and* $\|\Delta b\|_2 \leq \epsilon\|b\|_2$.

If $\epsilon\kappa_2(A) = r < 1$, *then* $A + \Delta A$ *is nonsingular and*

$$\frac{\|y - x\|_2}{\|x\|_2} \leq \frac{\epsilon}{1-r}\kappa_2(A).$$

With the superscript ^{T} we denote the transpose of a matrix (or vector): for $A \in \mathcal{R}^{n\times k}$, the matrix $B = A^T \in \mathcal{R}^{k\times n}$ is defined by

$$b_{i,j} = a_{j,i}.$$

If $E \in \mathcal{C}^{n\times k}$ then the superscript ^{H} is used to denote its complex conjugate $F = E^H$, defined as

$$f_{i,j} = \bar{e}_{j,i}.$$

Sometimes the superscript ^{T} is used for complex matrices in order to denote the transpose of a complex matrix.

The matrix A is symmetric if $A = A^T$, and $B \in \mathcal{C}^{n\times n}$ is Hermitian if $B = B^H$. Hermitian matrices have the attractive property that their spectrum is real. In particular, Hermitian (or symmetric real) matrices that are positive definite are attractive, because linear systems with such matrices can be solved rather easily by proper iterative methods (the CG method).

A Hermitian matrix $A \in \mathcal{C}^{n\times n}$ is positive definite if $x^H Ax > 0$ for all $0 \neq x \in \mathcal{C}^n$. A positive definite Hermitian matrix has only positive real eigenvalues.

We will encounter some special matrix forms, in particular tridiagonal matrices and (upper) Hessenberg matrices. The matrix $T = (t_{i,j}) \in \mathcal{R}^{n\times m}$ will be called tridiagonal if all elements for which $|i-j| > 1$ are zero. It is called upper Hessenberg if all elements for which $i > j+1$ are zero. In the context of Krylov subspaces, these matrices are often $k + 1$ by k and they will then be denoted as $T_{k+1,k}$.

For purposes of analysis it is often helpful or instructive to transform a given matrix to an easier form, for instance, diagonal or upper triangular form.

The easiest situation is the symmetric case: for a real symmetric matrix, there exists an orthogonal matrix $Q \in \mathcal{R}^{n\times n}$, so that $Q^T AQ = D$, where $D \in \mathcal{R}^{n\times n}$ is a diagonal matrix. The diagonal elements of D are the eigenvalues of A, the columns of Q are the corresponding eigenvectors of A. Note that the eigenvalues and eigenvectors of A are all real.

If $A \in \mathcal{C}^{n\times n}$ is Hermitian ($A = A^H$) then there exist $Q \in \mathcal{C}^{n\times n}$ and a diagonal matrix $D \in \mathcal{R}^{n\times n}$ so that $Q^H Q = I$ and $Q^H AQ = D$. This means that the eigenvalues of a Hermitian matrix are all real, but its eigenvectors may be complex.

Unsymmetric matrices do not in general have an orthonormal set of eigenvectors, and may not have a complete set of eigenvectors, but they can be transformed unitarily to Schur form:

$$Q^H AQ = R,$$

in which R is upper triangular. In fact, the symmetric case is a special case of this Schur decomposition, since a symmetric triangular matrix is clearly diagonal. Apart from the ordering of the eigenvalues along the diagonal of R and a factor -1 for each column of Q, the matrix Q is unique.

If the matrix A is complex, then the matrices Q and R may also be complex. However, they may be complex even when A is real unsymmetric. It may then be advantageous to work in real arithmetic. This can be realized because of the existence of the *real Schur decomposition*. If $A \in \mathcal{R}^{n\times n}$ then it can be transformed with an orthonormal $Q \in \mathcal{R}^{n\times n}$ as

$$Q^T AQ = \widetilde{R},$$

with

$$\widetilde{R} = \begin{bmatrix} \widetilde{R}_{1,1} & \widetilde{R}_{1,2} & \cdots & \widetilde{R}_{1,k} \\ 0 & \widetilde{R}_{2,2} & \cdots & \widetilde{R}_{2,k} \\ \vdots & \vdots & \ddots & \vdots \\ 0 & 0 & \cdots & \widetilde{R}_{k,k} \end{bmatrix} \in \mathcal{R}^{n\times n}.$$

Each $\widetilde{R}_{i,i}$ is either 1 by 1 or a 2 by 2 (real) matrix having complex conjugate eigenvalues. For a proof of this see [3, Chapter 7.4.1]. This form of $\widetilde{R}$ is referred to as an upper *quasi-triangular matrix*.

If all eigenvalues are distinct then there exists a nonsingular matrix X (in general not orthogonal) that transforms A to diagonal form:

$$X^{-1}AX = D.$$

A general matrix can be transformed to Jordan form with a nonsingular X:

$$X^{-1}AX = \mathrm{diag}(J_1, J_2, \ldots, J_k),$$

where

$$J_i = \begin{bmatrix} \lambda_i & 1 & 0 & \cdots & 0 \\ 0 & \lambda_i & \ddots & & \vdots \\ & \ddots & \ddots & \ddots & \\ \vdots & & \ddots & \ddots & 1 \\ 0 & \cdots & & 0 & \lambda_i \end{bmatrix}.$$

If there is a J_i with dimension greater than 1 then the matrix A is called defective. In this case A does not have a complete set of independent eigenvectors. In numerical computations we may argue that small perturbations lead to different eigenvalues and hence that it will be unlikely that A has a true Jordan form in actual computation. However, if A is close to a matrix with a nontrivial Jordan block, then this is reflected by a (severely) ill-conditioned eigenvector matrix X.

Matrices $A \in \mathcal{C}^{n \times n}$ that satisfy the property $A^H A = AA^H$ are called *normal.* Normal matrices also have a complete orthonormal eigensystem. For such matrices the distribution of the eigenvalues can help to explain (local) phenomena in the convergence behaviour of some methods. For unsymmetric matrices that are not normal, the eigenvalues are often insufficient for a detailed analysis. We refer to, e.g., Trefethen et al [10] for better alternatives, including the notion of pseudospectra.

We will also encounter eigenvalues that are called *Ritz values*. For simplicity, we will introduce them here for the real case. The subspace methods that will be discussed in this book are based on the identification of good solutions from certain low-dimensional subspaces $\mathcal{V}^k \subset \mathcal{R}^n$, where $k \ll n$ denotes the dimension of the subspace. If $V_k \in \mathcal{R}^{n \times k}$ denotes an orthogonal basis of $\mathcal{V}^k$ then the operator $H_k = V_k^T A V_k \in \mathcal{R}^{k \times k}$ represents the projection of A onto V_k. Assume that the eigenvalues and eigenvectors of H_k are represented as

$$H_k s_j^{(k)} = \theta_j^{(k)} s_j^{(k)},$$

then $\theta_j^{(k)}$ is called a *Ritz value* of A with respect to $\mathcal{V}^k$ and $V_k s_j^{(k)}$ is its corresponding *Ritz vector*. For a thorough discussion of Ritz values and Ritz vectors see, for instance, [6, 7, 9, 11].

The importance of the Ritz values is that they can be viewed as approximations for eigenvalues of A. Often they represent, even for modest values of k, very accurate approximations for some eigenvalues.

2 Linear Systems

Inverses of matrices play a role in formulas for important model order reduction problems. However, in numerical computations the explicit inversion of a matrix has to be avoided if possible. In most situations, the inverse of a matrix is used to denote an operator acting on a vector or another matrix. We consider the evaluation of the vector y defined as

$$y = A^{-1} b.$$

Determining the explicit inverse of A is expensive. The classical way is to compute an LU factorization first and then solve linear systems with L and U for all canonical basis vectors. After A^{-1} has been obtained, it has to be multiplied with b in order to obtain y.

The vector y can be obtained in a much cheaper way. We see that y has to be solved from

$$Ay = b.$$

Again, we use the LU factorization and first solve z from $Lz = b$, followed by the solution from $Uy = z$. These two solution steps require as much computation as the sole multiplication of A^{-1} and b, and we have avoided to solve the n systems with the canonical basis vectors as right hand sides. Likewise, always when the matrix A^{-1} occurs as an operator acting on some vector or some other matrix, then the expression can be numerically evaluated as we have demonstrated for y.

For sparse matrices, computational differences may be even much more dramatic. In relevant cases, A^{-1} may be dense, while L and U are sparse. For instance, if A is a positive definite tridiagonal matrix, then solving $Ax = b$ via LU decomposition requires only in the order of n arithmetic computations. The bottom line is: *avoid explicit inversion.*

Direct computation of y via LU requires, for dense matrices, in total in the order of n^3 floating point operations, which may pose practical limits for large values of n. This is a major motivation for Model Order Reduction: to save computational costs and computer memory storage.

Eigenvalue computations lead to similar observations. The standard way to compute non-trivial λ and x, satisfying

$$Ax = \lambda x,$$

requires the diagonalization of the matrix A. The state of the art technique for this is to first transform A to a more efficient to handle form by orthogonal transformations. In particular, A is transformed by a finite number of Householder or Givens transformations to upper Hessenberg form H: $Q^T AQ = H$. The eigenvalues of A are equal to those of A and the eigenvalues of H are computed with the QR-method. The operations in QR can efficiently be done on Hessenberg matrices. The eigenvalue computation is essentially an iterative process, but QR converges so fast that in practice the complete method (reduction to upper Hessenberg plus QR) is viewed as a direct technique. The whole process requires a few times n^3 arithmetic operations, and again this poses practical problems if n is large.

3 Subspace Methods

3.1 The Krylov Subspace

The main problem in Model Order Reduction is to identify a suitable subspace to which the given problem can be restricted. For a linear system $Ax = b$ it has been proven a fruitful idea to start from some convenient iteration method and to collect successive approximation vectors or residuals as the basis vectors for a subspace. We will explain this idea in a little more detail.

Suppose that we have an initial guess x_0 for x. Obviously, we want the correction z that satisfies

$$A(x_0 + z) = b,$$

which leads to the system

$$Az = b - Ax_0 \equiv r_0.$$

This system is just as difficult to solve as the given system, but we are satisfied for the moment with an approximation z_0 for the correction z, which may be obtained from an easier to solve nearby system $Kz_0 = r_0$. In the context of iteration processes, the operator K is usually referred to as the preconditioner. In order to keep the explanation simple, we select a very simple approximation K for A, namely

$K = I$. The inclusion of more practical K is straightforward, for more on this see, e.g. [12].

With this very simple K, we obtain $z_0 = r_0$, which leads to the updated approximation

$$x_1 = x_0 + r_0.$$

Carrying on in this fashion, we obtain the basic iteration method

$$\begin{aligned} x_{i+1} &= x_i + z_i \\ &= x_i + r_i \\ &= x_i + b - Ax_i. \end{aligned} \tag{1}$$

It is straightforward that with $x_0 = 0$, we have that $x_1 = b$, $x_2 = 2b - Ab$, etcetera, and in genaral x_i for the standard iteration (with $x_0 = 0$) can be written as a special combination of the vectors $b, Ab, \ldots, A^{i-1}b$. For $x_0 \neq 0$, we obtain a simple shift of these combinations of vectors (again, see [12] for more details). Since we have to carry out a matrix vector product with A in each iteration (to compute r_i) it seems to be attractive to forget about the iteration matrix and to generate the vectors $A^j b$ directly.

These vectors define the so-called Krylov subspace $\mathcal{K}^i(A; b)$ of dimension i and generated by A and b:

$$\mathcal{K}^i(A; b) \equiv span\{b, Ab, \ldots, A^{i-1}b\}. \tag{2}$$

The idea is to exploit this subspace for the computation of suitable approximations for x. The question is how to do this in a numerically stable way. The generation of the basis vectors $A^j b$ and orthogonalize them afterwards is not a good idea, because these vectors point more and more in the direction of a dominant eigenvector and, hence, will form an ill-conditioned set of vectors. Orthogonalization of such a set of ill-conditioned set of vectors may lead to a correct projection process, but most often it leads to a loss of information and loss of efficiency. Using the iteration vectors x_i or r_i is not a good alternative, because they also may suffer from near dependency. It is much better to generate an orthogonal basis for the Krylov subspace (or any other appropriate subspace) right from the start. We will explain later how to do that for the Krylov subspace.

For standard eigenproblems $Ax = \lambda x$, the subspace approach is even more obvious. Behind almost all effective eigenproblem solvers there is the Power Method in some disguise. With a convenient initial guess v for a dominating eigenvector, the Power Method generates the vectors Av, A^2v, ..., and one observes that $A^j v$ converges to the dominant eigenvector for increasing values of j. From ratios of (norms of) successive vectors one derives approximations for the corresponding eigenvalue. We see that the iteration vectors for the Power Method are just the defining vectors for the Krylov subspace, and again, as above, the idea is to compute a better basis for this subspace for the reduction of the given problem $Ax = \lambda x$.

Note that the construction of a better basis was not so obvious as it may seem now. Krylov used the defining basis vectors and it took almost 20 years before

Lanczos and Arnoldi, independently, came with the construction of a better basis (instead of orthogonalizing a set of Krylov vectors). It then took another 20 years before numerical errors in the generation of the orthogonal basis were well understood.

Before we turn our attention to the computation of an effective orthogonal basis for the Krylov subspace, we give an overview of some popular reduction techniques, once the orthogonal basis is available.

3.2 Reduction of the Problem

Reduction of the described linear problems comes down to the selection of a proper subspace and then solve a related problem over that subspace. To that end we have to construct an orthonormal basis $Y_k = \{y_1, ..., y_k\}$ for the k-dimensional subspace. We have to identify a suitable vector x_k in this subspace that approximates y. The most popular, and in some sense optimal, ways are:

1. The *Ritz-Galerkin approach*: Construct the x_k for which the residual is orthogonal to the current subspace: $b - Ax_k \perp Y_k$.
2. The *minimum norm residual approach*: Identify the x_k for which the Euclidean norm $\|b - Ax_k\|_2$ is minimal over Y_k.
3. The *Petrov-Galerkin approach*: Find an x_k so that the residual $b - Ax_k$ is orthogonal to some other suitable k-dimensional subspace.
4. The *minimum norm error approach*: Determine x_k in $A^T Y_k$ for which the Euclidean norm $||x_k - x||_2$ is minimal.

We will now focus our attention to the important aspect of the construction of an efficient orthogonal basis. This is the key element in any Model Order Reduction approach. After that we will give an example on how to use the orthogonal basis in order to obtain a smaller to solve system that replaces the given system $Ax = b$.

3.3 The Krylov Subspace

We now derive an orthogonal basis that, in exact arithmetic, spans the Krylov subspace. For this we follow ideas from [9, Chapter 4.3]. We start with the generic basis for $\mathcal{K}^{i+1}(A; r_0)$ and we denote the basis vectors by u_j:

$$u_j = A^{j-1} r_0.$$

We define the n by j matrix U_j as the matrix with columns $u_1, \ldots, u_j$. The connection between A and U_i is as follows:

$$AU_i = U_i B_i + u_{i+1} e_i^T, \tag{3}$$

with e_i the i-th canonical basis vector in $\mathcal{R}^i$, and B_i an i by i matrix with $b_{j+1,j} = 1$ and all other elements zero.

The next step is to decompose U_i, still in exact arithmetic, as

$$U_i = Q_i R_i,$$

with $Q_i^T Q_i = I$ and R_i upper triangular. Then, with (3), it follows that

$$AQ_iR_i = Q_iR_iB_i + u_{i+1}e_i^T,$$

or

$$\begin{aligned} AQ_i &= Q_iR_iB_iR_i^{-1} + u_{i+1}e_i^T R_i^{-1} \\ &= Q_i\widetilde{H}_i + u_{i+1}e_i^T R_i^{-1} \qquad (4) \\ &= Q_i\widetilde{H}_i + \frac{1}{r_{i,i}}u_{i+1}e_i^T. \qquad (5) \end{aligned}$$

The matrix $\widetilde{H}_i$ is an upper Hessenberg matrix.

We can also decompose U_{i+1} as $U_{i+1} = Q_{i+1}R_{i+1}$, and if we write the last column of R_{i+1} as $(\widetilde{r}, r_{i+1,i+1})^T$, that is

$$R_{i+1} = \begin{pmatrix} R_i & \widetilde{r} \\ 0 & r_{i+1,i+1} \end{pmatrix},$$

then it follows that

$$u_{i+1} = Q_i\widetilde{r} + r_{i+1,i+1}q_{i+1}.$$

In combination with (4), this gives

$$\begin{aligned} AQ_i &= Q_i(\widetilde{H}_i + \frac{1}{r_{i,i}}\widetilde{r}e_i^T) + \frac{r_{i+1,i+1}}{r_{i,i}}q_{i+1}e_i^T \\ &= Q_iH_i + \alpha q_{i+1}e_i^T. \qquad (6) \end{aligned}$$

From this expression we learn at least two things: first

$$Q_i^T AQ_i = H_i, \qquad (7)$$

with H_i upper Hessenberg, and second

$$q_{i+1}^T Aq_i = \alpha,$$

which, with $Q_{i+1}^T AQ_{i+1} = H_{i+1}$, leads to $\alpha = h_{i+1,i}$.

The implicit Q theorem [3, Theorem 7.4.2] states that the orthogonal Q that reduces A to upper Hessenberg form is uniquely determined by $q_1 = \frac{1}{\|r_0\|}r_0$, except for signs (that is, q_j may be multiplied by -1). The orthogonality of the q_j basis gives us excellent opportunities to compute this basis in finite precision arithmetic.

Arnoldi [1] proposed to compute the orthogonal basis as follows (in fact, with Arnoldi's procedure we compute in a straightforward manner the columns of Q_i and the elements of H_i). Start with $v_1 \equiv r_0/\|r_0\|_2$. Then compute Av_1, make it orthogonal to v_1 and normalize the result, which gives v_2. The general procedure is as follows. Assuming that we already have an orthonormal basis $v_1, \ldots, v_j$ for $K^j(A; r_0)$, this basis is expanded by computing $t = Av_j$ and by orthonormalizing this vector t with respect to $v_1, \ldots, v_j$. In principle, the orthonormalization process

$v_1 = r_0/\|r_0\|_2;$
for $j = 1, \ldots, m-1$
  $t = Av_j;$
  for $i = 1, \ldots, j$
    $h_{i,j} = v_i^T t;$
    $t = t - h_{i,j} v_i;$
  end;
  $h_{j+1,j} = \|t\|_2;$
  $v_{j+1} = t/h_{j+1,j};$
end

Fig. 1. Arnoldi's method with modified Gram–Schmidt orthogonalization

can be carried out in different ways, but the most commonly used approach is then the modified Gram–Schmidt procedure [3].

This leads to an algorithm for the creation of an orthonormal basis for $\mathcal{K}^m(A; r_0)$, as in Figure 1. It is easily verified that $v_1, \ldots, v_m$ form an orthonormal basis for $\mathcal{K}^m(A; r_0)$ (that is, if the construction does not terminate at a vector $t = 0$). The orthogonalization leads, in exact arithmetic, to the relation that we have seen before (cf. (6), but now expressed in terms of the v_j. Let V_j denote the matrix with columns v_1 up to v_j, then it follows that

$$AV_{m-1} = V_m H_{m,m-1}. \tag{8}$$

The m by $m-1$ matrix $H_{m,m-1}$ is upper Hessenberg, and its elements $h_{i,j}$ are defined by the Arnoldi algorithm.

From a computational point of view, this construction is composed of three basic elements: a matrix vector product with A, inner products, and vector updates. We see that this orthogonalization becomes increasingly expensive for increasing dimension of the subspace, since the computation of each $h_{i,j}$ requires an inner product and a vector update.

Note that if A is symmetric, then so is $H_{m-1,m-1} = V_{m-1}^T A V_{m-1}$, so that in this situation $H_{m-1,m-1}$ is tridiagonal. This means that in the orthogonalization process, each new vector has to be orthogonalized with respect to the previous two vectors only, since all other inner products vanish. The resulting three-term recurrence relation for the basis vectors of $K_m(A; r_0)$ is known as the *Lanczos method* [4] and some very elegant methods are derived from it. In this symmetric case the orthogonalization process involves constant arithmetical costs per iteration step: one matrix vector product, two inner products, and two vector updates.

A more accurate basis for the Krylov subspace

A more accurate implementation for the construction of an orthonormal basis, useful for ill-conditioned matrices A, was suggested by Walker [13]. He suggested employing Householder reflections instead of the modified Gram–Schmidt orthogonalization procedure.

An alternative is to do two iterations with (modified) Gram–Schmidt if necessary. This works as follows. If we want to have a set of orthogonal vectors to almost working precision then we have to check, after the orthogonalization of a new vector with respect to the existing set, whether the resulting unnormalized vector is significantly smaller in norm than the new vector at the start of the orthogonalization step, say more than $\kappa < 1$ smaller. In that case we may have had cancellation effects, and once again we apply modified Gram–Schmidt. This is the basis for the refinement technique suggested in [2]. It leads to a set of vectors for which the mutual loss of orthogonality is limited to $1/\kappa$, in a relative sense. In the template in Figure 2, we incorporate this technique into the Arnoldi algorithm.

Note that, in exact arithmetic, the constants ρ in Figure 2 are equal to zero. It is easily verified that, in exact arithmetic, the $v_1, \ldots, v_m$ form an orthonormal basis for $\mathcal{K}(A; v)$ (that is, if the construction does not terminate at a vector $t = 0$).

v is a convenient starting vector
Select a value for κ, e.g., $\kappa = .25$
$v_1 = v/\|v\|_2$
for $j = 1, \ldots, m-1$
    $t = Av_j$
    $\tau_{in} = ||t||_2$
    for $i = 1, \ldots, j$
        $h_{i,j} = v_i^* t$
        $t = t - h_{i,j} v_i$
    end
    if $\|t\|_2/\tau_{in} \le \kappa$
        for $i = 1, \ldots, j$
            $\rho = v_i^* t$
            $t = t - \rho v_i$
            $h_{i,j} = h_{i,j} + \rho$
        end
    endif
    $h_{j+1,j} = \|t\|_2$
    $v_{j+1} = t/h_{j+1,j}$
end

Fig. 2. The Arnoldi Method with refined modified Gram–Schmidt

3.4 Example of the Construction of Approximate Solutions

In this section we give one example of the reduction of a problem, in this case $Ax{=}b$, to a smaller problem over the Krylov subspace. For other approaches and more examples, see, e.g., [12].

The Ritz–Galerkin conditions imply that $r_k \perp \mathcal{K}^k(A; r_0)$, and this is equivalent to

$$V_k^T(b - Ax_k) = 0.$$

Since $b = r_0 = \|r_0\|_2 v_1$, it follows that $V_k^T b = \|r_0\|_2 e_1$ with e_1 the first canonical unit vector in $\mathcal{R}^k$. With $x_k = V_k y$ we obtain

$$V_k^T A V_k y = \|r_0\|_2 e_1.$$

This system can be interpreted as the system $Ax = b$ projected onto the subspace $\mathcal{K}^k(A; r_0)$.

Obviously we have to construct the $k \times k$ matrix $V_k^T A V_k$, but this is, as we have seen, readily available from the orthogonalization process:

$$V_k^T A V_k = H_{k,k},$$

so that the x_k for which $r_k \perp \mathcal{K}^k(A; r_0)$ can be easily computed by first solving $H_{k,k} y = \|r_0\|_2 e_1$, and then forming $x_k = V_k y$. This algorithm is known as FOM or GENCG [8].

When A is symmetric, then $H_{k,k}$ reduces to a tridiagonal matrix $T_{k,k}$, and the resulting method is known as the *Lanczos* method [5]. When A is in addition positive definite then we obtain, at least formally, the *Conjugate Gradient* method. In commonly used implementations of this method, an LU factorization for $T_{k,k}$ is implicitly formed without generating $T_{k,k}$ itself, and this leads to very elegant short recurrences for the x_j and the corresponding r_j.

For eigenvalues and eigenvectors, the Krylov subspace approach can be used in a similar fashion, for details we refer to [6].

References

1. W. E. Arnoldi. The principle of minimized iteration in the solution of the matrix eigenproblem. *Quart. Appl. Math.*, 9:17–29, 1951.
2. J. W. Daniel, W. B. Gragg, L. Kaufmann, and G. W. Stewart. Reorthogonalization and stable algorithms for updating the Gram-Schmidt QR factorization. *Math. Comp.*, 30:772–795, 1976.
3. G. H. Golub and C. F. Van Loan. *Matrix Computations*. The Johns Hopkins University Press, Baltimore, 1996.
4. C. Lanczos. An iteration method for the solution of the eigenvalue problem of linear differential and integral operators. *J. Res. Natl. Bur. Stand*, 45:225–280, 1950.
5. C. Lanczos. Solution of systems of linear equations by minimized iterations. *J. Res. Natl. Bur. Stand*, 49:33–53, 1952.

6. B. N. Parlett. *The Symmetric Eigenvalue Problem.* Prentice-Hall, Englewood Cliffs, N.J., 1980.
7. Y. Saad. *Numerical methods for large eigenvalue problems.* Manchester University Press, Manchester, UK, 1992.
8. Y. Saad and M. H. Schultz. GMRES: a generalized minimal residual algorithm for solving nonsymmetric linear systems. *SIAM J. Sci. Statist. Comput.*, 7:856–869, 1986.
9. G. W. Stewart. *Matrix Algorithms, Vol. II: Eigensystems.* SIAM, Philadelphia, 2001.
10. L. N. Trefethen. Computation of pseudospectra. *Acta Numerica*, 8:247–295, 1999.
11. H. A. van der Vorst. Computational methods for large eigenvalue problems. In P. G. Ciarlet and J. L. Lions, editors, *Handbook of Numerical Analysis*, volume VIII, pages 3–179. North-Holland, Amsterdam, 2002.
12. H. A. van der Vorst. *Iterative Krylov Methods for Large Linear Systems.* Cambridge University Press, Cambridge, UK, 2003.
13. H. F. Walker. Implementation of the GMRES method using Householder transformations. *SIAM J. Sci. Statist. Comput.*, 9:152–163, 1988.

Part II

Theory

Structure-Preserving Model Order Reduction of RCL Circuit Equations

Roland W. Freund

Department of Mathematics, University of California at Davis, One Shields Avenue, Davis, CA 95616, U.S.A.
freund@math.ucdavis.edu

Summary. In recent years, order-reduction techniques based on Krylov subspaces have become the methods of choice for generating macromodels of large multi-port RCL circuits. Despite the success of these techniques and the extensive research efforts in this area, for general RCL circuits, the existing Krylov subspace-based reduction algorithms do not fully preserve all essential structures of the given large RCL circuit. In this paper, we describe the problem of structure-preserving model order reduction of general RCL circuits, and we discuss two state-of-the-art algorithms, PRIMA and SPRIM, for the solution of this problem. Numerical results are reported that illustrate the higher accuracy of SPRIM vs. PRIMA. We also mention some open problems.

1 Introduction

Electronic circuits often contain large linear subnetworks of passive components. Such subnetworks may represent interconnect (IC) automatically extracted from layout as large RCL networks, models of IC packages, or models of wireless propagation channels. Often these subnetworks are so large that they need to be replaced by much smaller reduced-order models, before any numerical simulation becomes feasible. Ideally, these models would produce a good approximation of the input-output behavior of the original subnetwork, at least in a limited domain of interest, e.g., a frequency range.

In recent years, reduced-order modeling techniques based on Padé or Padé-type approximation have been recognized to be powerful tools for various circuit simulation tasks. The first such technique was asymptotic waveform evaluation (AWE) [31], which uses explicit moment matching. More recently, the attention has moved to reduced-order models generated by means of Krylov-subspace algorithms, which avoid the typical numerical instabilities of explicit moment matching; see, e.g., the survey papers [14–16].

PVL [9, 10] and its multi-port version MPVL [11] use variants of the Lanczos process [26] to stably compute reduced-order models that represent Padé or matrix-Padé approximations [5] of the circuit transfer function. SyPVL [21] and its multi-port version SyMPVL [12, 23, 24] are versions of PVL and MPVL, respectively, that are tailored to RCL circuits. By exploiting the symmetry of RCL transfer functions, the computational costs of SyPVL and SyMPVL are only half of those of general PVL and MPVL.

Reduced-order modeling techniques based on the Arnoldi process [3], which is another popular Krylov-subspace algorithm, were first proposed in [8, 28–30, 33]. Arnoldi-based reduced-order models are defined by a certain Padé-type approximation property, rather than Padé approximation, and as a result, in general, they are not as accurate as a Padé-based model of the same size. In fact, Arnoldi-based models are known to match only half as many moments as Lanczos-based models; see [15, 28, 29, 33].

In many applications, in particular those related to VLSI interconnect, the reduced-order model is used as a substitute for the full-blown original model in higher-level simulations. In such applications, it is very important for the reduced-order model to preserve the passivity properties of the original circuit. In [4, 23, 24], it is shown that SyMPVL is passive for RC, RL, and LC circuits. However, the Padé-based reduced-order models that characterize SyMPVL cannot be guaranteed to be passive for general RCL circuits. On the other hand, in [28–30], it was proved that the Arnoldi-based reduction technique PRIMA produces passive reduced-order for general RCL circuits. PRIMA employs a block version of the Arnoldi process and then obtains reduced-order models by projecting the matrices defining the RCL transfer function onto the Arnoldi basis vectors. While PRIMA generates provably passive reduced-order models, it does not preserve other structures, such as reciprocity or the block structure of the circuit matrices, inherent to RCL circuits. This has motivated the development of the reduction technique SPRIM [17, 18], which overcomes these disadvantages of PRIMA. In particular, SPRIM generates provably passive and reciprocal macromodels of multi-port RCL circuits. Furthermore, SPRIM models match twice as many moments as the corresponding PRIMA models obtained with identical computational work. In this paper, we describe the problem of structure-preserving model order reduction of general RCL circuits, and we discuss the PRIMA and SPRIM algorithms for the solution of this problem.

The remainder of this article is organized as follows. In Section 2, we review the formulation of general RCL circuits as systems of integro-differential-algebraic equations (integro-DAEs). In Section 3, we describe the problem of structure-preserving model order reduction of systems of integro-DAEs. In Section 4, we present an equivalent formulation of such systems as time-invariant linear dynamical systems. In Section 5, we review oder reduction based on projection onto Krylov subspaces and the PRIMA algorithm. In Section 6, we describe the SPRIM algorithm for order reduction of general RCL circuits, and in Section 7, we present some theoretical properties of SPRIM. In Section 8, we report the results of some numerical experiments with SPRIM and PRIMA. Finally, in Section 9, we mention some open problems and make some concluding remarks.

Throughout this article the following notation is used. The set of real and complex numbers is denoted by $\mathbb{R}$ and $\mathbb{C}$, respectively. Unless stated otherwise, all vectors and matrices are allowed to have real or complex entries. For (real or complex) matrices $M = \left[m_{jk}\right]$, we denote by $M^T = \left[m_{kj}\right]$ the *transpose* of M, and by

$M^* := [\overline{m_{kj}}]$ the *Hermitian* (or *complex conjugate*) of M. The identity matrix is denoted by I and the zero matrix by 0; the actual dimensions of I and 0 will always be apparent from the context. The notation $M \succeq 0$ ($M \succ 0$) is used to indicate that a real or complex square matrix M is *Hermitian positive semidefinite* (*positive definite*). If all entries of the matrix $M \succeq 0$ ($M \succ 0$) are real, then M is said to be *symmetric positive semidefinite* (*positive definite*). The kernel (or null space) of a matrix M is denoted by ker M.

2 Formulation of General RCL Circuits as Integro-DAEs

In this section, we review the formulation of general RCL circuits as systems of integro-DAEs.

2.1 Electronic Circuits as Directed Graphs

We use the standard lumped-element approach that models general electronic circuits as directed graphs; see, e.g., [7, 34]. More precisely, a given circuit is described as a directed graph $\mathcal{G} = (\mathcal{N}, \mathcal{E})$ whose edges $e \in \mathcal{E}$ correspond to the circuit elements and whose nodes $n \in \mathcal{N}$ correspond to the interconnections of the circuit elements. For each element for which the direction of the current flow through the element is known beforehand, the corresponding edge is oriented in the direction of the current flow; for example, current sources and voltage sources are elements with known direction of current flow. For all other elements, arbitrary directions are assigned to the edges corresponding to these elements. Each edge $e \in \mathcal{E}$ can be written as an ordered pair of nodes, $e = (n_1, n_2)$, where the direction of e is from node n_1 to node n_2. We say that the edge $e = (n_1, n_2)$ *leaves* node n_1 and *enters* node n_2.

The directed graph $\mathcal{G} = (\mathcal{N}, \mathcal{E})$ can be described by its *incidence matrix* $\mathcal{A} = [a_{jk}]$. The rows and columns of $\mathcal{A}$ correspond to the nodes and edges of the directed graph, respectively, and the entries a_{jk} of $\mathcal{A}$ are defined as follows:

$$a_{jk} = \begin{cases} 1 & \text{if edge } e_k \text{ leaves node } n_j, \\ -1 & \text{if edge } e_k \text{ enters node } n_j, \\ 0 & \text{otherwise.} \end{cases}$$

In order to avoid redundancy, any one of the nodes can be selected as the *datum* (or *ground*) node of the circuit. We denote by n_0 the datum node, by $\mathcal{N}_0 = \mathcal{N} \setminus \{ n_0 \}$ the remaining non-datum nodes, and by $\mathcal{A}_0$ the matrix obtained by deleting from $\mathcal{A}$ the row corresponding to n_0. Note that $\mathcal{A}_0$ is called the *reduced incidence matrix* of the directed graph $\mathcal{G}$. We remark that $\mathcal{A}_0$ has full row rank, i.e.,

$$\operatorname{rank} \mathcal{A}_0 = |\mathcal{N}_0|,$$

provided the graph $\mathcal{G}$ is connected; see, e.g., [7, Theorem 9-6].

We denote by $v = v(t)$ the vector of nodal voltages at the non-datum nodes $\mathcal{N}_0$, i.e., the k-th entry of v is the voltage at node $n_k \in \mathcal{N}_0$. We denote by $v_{\mathcal{E}} = v_{\mathcal{E}}(t)$ and $i_{\mathcal{E}} = i_{\mathcal{E}}(t)$ the vectors of edge voltages and currents, respectively, i.e., the j-th entry of $v_{\mathcal{E}}$ is the voltage across the circuit element corresponding to edge $e_j \in \mathcal{E}$, and the j-th entry of $i_{\mathcal{E}}$ is the current through the circuit element corresponding to edge $e_j \in \mathcal{E}$.

Any general electronic circuit is described completely by three types of equations: *Kirchhoff's current laws* (KCLs), *Kirchhoff's voltage laws* (KVLs), and the *branch constitutive relations* (BCRs); see, e.g., [34]. The KCLs state that at each node $n \in \mathcal{N}$, the currents along all edges leaving and entering the node n sum up to zero. In terms of the reduced incidence matrix $\mathcal{A}_0$ of $\mathcal{G}$ and the vector $i_{\mathcal{E}}$, the KCLs can be expressed in the following compact form:

$$\mathcal{A}_0\, i_{\mathcal{E}} = 0. \tag{1}$$

Similarly, the KCVs state that for any closed (undirected) loop in the graph $\mathcal{G}$, the voltages along the edges of the loop sum up to zero. The KCLs can be expressed in the following compact form:

$$\mathcal{A}_0^T\, v = v_{\mathcal{E}}. \tag{2}$$

The BCRs are the equations that describe the physical behavior of the circuit elements.

2.2 RCL Circuit Equations

We now restrict ourselves to general linear RCL circuits. The possible element types of such circuits are resistors, capacitors, inductors, independent voltage sources, and independent current sources. We use subscripts r, c, l, v, and i to denote edge quantities corresponding to resistors, capacitors, inductors, voltage sources, and current sources of the RCL circuit, respectively. Moreover, we assume that the edges $\mathcal{E}$ are ordered such that we have the following partitionings of the reduced incidence matrix and the vectors of edge voltages and currents:

$$\mathcal{A}_0 = \begin{bmatrix} \mathcal{A}_r \ \mathcal{A}_c \ \mathcal{A}_l \ \mathcal{A}_v \ \mathcal{A}_i \end{bmatrix}, \quad v_{\mathcal{E}} = \begin{bmatrix} v_r \\ v_c \\ v_l \\ v_v \\ v_i \end{bmatrix}, \quad i_{\mathcal{E}} = \begin{bmatrix} i_r \\ i_c \\ i_l \\ i_v \\ i_i \end{bmatrix}. \tag{3}$$

The BCRs for the resistors, capacitors, and inductors can be expressed in the following compact form:

$$v_r(t) = R\, i_r(t), \quad i_c(t) = C\,\frac{d}{dt} v_c(t), \quad v_l(t) = L\,\frac{d}{dt} i_l(t). \tag{4}$$

Here,

$$R \succ 0, \quad C \succ 0, \quad \text{and} \quad L \succ 0 \tag{5}$$

are symmetric positive definite matrices. Furthermore, R and C are diagonal matrices whose diagonal entries are the resistances and capacitances of the resistors and capacitors, respectively. The diagonal entries of L are the inductances of the inductors. Often L is also diagonal, but in general, when mutual inductances are included, L is not diagonal. The BCRs for the voltage sources simply state that $v_v(t)$ is a given input vector, the entries of which are the voltages provided by the voltages sources. Similarly, the BCRs for the current sources state that $i_i(t)$ is a given input vector, the entries of which are the currents provided by the current sources.

The KCLs (1), the KVLs (2), and the BCRs (4), together with initial conditions for the nodal voltages $v(t_0)$ at some initial time t_0, describe the behavior of a given RCL circuit. Without loss of generality, we set $t_0 = 0$. The initial condition then reads

$$v(0) = v^{(0)}, \tag{6}$$

where $v^{(0)}$ is a given vector. Moreover, for simplicity, we also assume that

$$i_l(0) = 0.$$

Then, the BCRs for the inductors in (4) can be equivalently stated as follows:

$$i_l(t) = L^{-1} \int_0^t v_l(\tau)\, d\tau. \tag{7}$$

The resulting set of equations describing a given RCL circuit can be simplified considerably by eliminating the edge quantities corresponding to the resistors, capacitors, and inductors. To this end, we first use the partitionings (3) to rewrite the KCLs (1) as follows:

$$\mathcal{A}_r\, i_r + \mathcal{A}_c\, i_c + \mathcal{A}_l\, i_l + \mathcal{A}_v\, i_v + \mathcal{A}_i\, i_i = 0. \tag{8}$$

Similarly, the KCVs (2) can be expressed as follows:

$$\mathcal{A}_r^T\, v = v_r, \quad \mathcal{A}_c^T\, v = v_c, \quad \mathcal{A}_l^T\, v = v_l, \quad \mathcal{A}_v^T\, v = v_v, \quad \mathcal{A}_i^T\, v = v_i. \tag{9}$$

From (4), (7), and (9), it follows that

$$\begin{aligned} i_r(t) &= R^{-1}\mathcal{A}_r^T\, v(t), \quad i_c(t) = C\,\mathcal{A}_c^T\, \frac{d}{dt} v(t), \\ i_l(t) &= L^{-1}\mathcal{A}_l^T \int_0^t v(\tau)\, d\tau. \end{aligned} \tag{10}$$

Inserting (10) into (8), and using (9), we obtain

$$\begin{aligned} M_{11}\, \frac{d}{dt} v(t) + D_{11}\, v(t) + \mathcal{A}_v\, i_v(t) + K_{11} \int_0^t v(\tau)\, d\tau &= -\mathcal{A}_i\, i_i(t), \\ -\mathcal{A}_v^T\, v(t) &= -v_v(t), \\ v_i(t) &= \mathcal{A}_i^T\, v(t), \end{aligned} \tag{11}$$

where

$$M_{11} := \mathcal{A}_c\, C\, \mathcal{A}_c^T, \quad D_{11} := \mathcal{A}_r\, R^{-1} \mathcal{A}_r^T, \quad K_{11} := \mathcal{A}_l\, L^{-1} \mathcal{A}_l^T. \tag{12}$$

The equations (11) can be viewed as a *linear dynamical system* for the unknown *state-space vector*

$$z(t) := \begin{bmatrix} v(t) \\ i_v(t) \end{bmatrix}, \tag{13}$$

with given *input vector* and unknown *output vector*

$$u(t) := \begin{bmatrix} -i_i(t) \\ v_v(t) \end{bmatrix} \quad \text{and} \quad y(t) := \begin{bmatrix} v_i(t) \\ -i_v(t) \end{bmatrix}, \tag{14}$$

respectively. Indeed, setting

$$\begin{aligned} M &:= \begin{bmatrix} M_{11} & 0 \\ 0 & 0 \end{bmatrix}, \quad D := \begin{bmatrix} D_{11} & \mathcal{A}_v \\ -\mathcal{A}_v^T & 0 \end{bmatrix}, \\ K &:= \begin{bmatrix} K_{11} & 0 \\ 0 & 0 \end{bmatrix}, \quad F := \begin{bmatrix} \mathcal{A}_i & 0 \\ 0 & -I \end{bmatrix}, \quad z^{(0)} := \begin{bmatrix} v^{(0)} \\ i_v(0) \end{bmatrix}, \end{aligned} \tag{15}$$

and using (13), (14), and (9), the equations (11) can be rewritten in the form

$$\begin{aligned} M \frac{d}{dt} z(t) + D\, z(t) + K \int_0^t z(\tau)\, d\tau &= F\, u(t), \\ y(t) &= F^T z(t), \end{aligned} \tag{16}$$

and the initial conditions (6) can be stated in the form

$$z(0) = z^{(0)}.$$

Note that, in (16), M, D, and K are $N_0 \times N_0$ matrices and F is an $N_0 \times m$ matrix. Here, N_0 is the sum of the number of non-datum nodes in the circuit and the number of voltage sources, and m denotes the number of all voltage and current sources. We remark that N_0 is the *state-space dimension* of the linear dynamical system (16), and m is the number of inputs (and outputs) of (16). In general, the matrix M is singular, and thus the first equation of (16) is a system of *integro-differential-algebraic equations* (integro-DAEs). Finally, note that the matrices (15), M, D, K, and F, exhibit certain structures. In particular, from (5), (12), and (15), it follows that

$$M = \begin{bmatrix} M_{11} & 0 \\ 0 & 0 \end{bmatrix} \succeq 0, \quad D + D^T = \begin{bmatrix} 2D_{11} & 0 \\ 0 & 0 \end{bmatrix} \succeq 0, \quad \text{and} \quad K \succeq 0. \tag{17}$$

3 Structure-Preserving Model Order Reduction

In this section, we formulate the problems of model order reduction and structure preservation.

3.1 Model Order Reduction

A *reduced-order model* of the linear dynamical system (16) is a system of the same form as (16), but with smaller state-space dimension n_0 ($<N_0$). More precisely, a reduced-order model of (16) with state-space dimension n_0 is a system of the form

$$\begin{aligned} \widetilde{M}\,\frac{d}{dt}\widetilde{z}(t) + \widetilde{D}\,\widetilde{z}(t) + \widetilde{K}\int_0^t \widetilde{z}(\tau)\,d\tau &= \widetilde{F}\,u(t), \\ \widetilde{y}(t) &= \widetilde{F}^T\widetilde{z}(t), \end{aligned} \tag{18}$$

with initial conditions

$$\widetilde{z}(0) = \widetilde{z}^{(0)},$$

where

$$\widetilde{M},\ \widetilde{D},\ \widetilde{K} \in \mathbb{R}^{n_0 \times n_0},\quad \widetilde{F} \in \mathbb{R}^{n_0 \times m},\quad \text{and}\quad \widetilde{z}^{(0)} \in \mathbb{R}^{n_0}. \tag{19}$$

The general problem of order reduction of a given linear dynamical system (16) is to determine a reduced state-space dimension n_0 and data (19) such that the corresponding reduced-order model (18) is a 'sufficiently accurate' approximation of (16).

A practical way of assessing the accuracy of reduced-order models is based on the concept of Laplace-domain transfer functions of linear dynamical systems. The *transfer function* of the original linear dynamical system (16) is given by

$$H(s) = F^T\Big(s\,M + D + \frac{1}{s}\,K\Big)^{-1} F. \tag{20}$$

Here, we assume that the matrix $s\,M + D + \frac{1}{s}\,K$ is singular only for finitely many values of $s \in \mathbb{C}$. Conditions that guarantee this assumption are given in Section 4 below.

In analogy to (20), the transfer function of a reduced-order model (18) of (16) is given by

$$\widetilde{H}(s) = \widetilde{F}^T\Big(s\,\widetilde{M} + \widetilde{D} + \frac{1}{s}\,\widetilde{K}\Big)^{-1} \widetilde{F}. \tag{21}$$

Note that both

$$H : \mathbb{C} \mapsto (\mathbb{C} \cup \infty)^{m \times m} \quad \text{and} \quad \widetilde{H} : \mathbb{C} \mapsto (\mathbb{C} \cup \infty)^{m \times m}$$

are $m \times m$-matrix-valued rational functions.

In terms of transfer functions, the problem of order reduction of the original system (16) is equivalent to the problem of constructing the matrices $\widetilde{M}$, $\widetilde{D}$, $\widetilde{K}$, and $\widetilde{F}$ in (18) such that the transfer function (21), $\widetilde{H}(s)$, is a 'sufficiently accurate' approximation to the original transfer function (20), $H(s)$.

3.2 Structure Preservation

Recall that the linear dynamical system (16) with data matrices given in (15) describes the behavior of a given RCL circuit. Therefore, the reduced-order model (18) should be constructed such that it corresponds to an actual RCL circuit. This is the problem of *structure-preserving* model order reduction of RCL circuits: Generate matrices $\widetilde{M}$, $\widetilde{D}$, $\widetilde{K}$, and $\widetilde{F}$ such that the reduced-order model (18) can be synthesized as an RCL circuit. Obviously, (18) corresponds to an actual RCL circuit if the matrices $\widetilde{M}$, $\widetilde{D}$, $\widetilde{K}$, and $\widetilde{F}$ are constructed such that they have analogous structures as the matrices M, D, K, and F of the original given RCL circuit. Unfortunately, for general RCL circuits, no order reduction method is known that is guaranteed to preserve all these 'RCL structures'. However, the SPRIM algorithm described in Section 6 below does generate reduced-order models (18) with matrices $\widetilde{M}$, $\widetilde{D}$, $\widetilde{K}$, and $\widetilde{F}$ that preserve the block structure (15) of the original matrices M, D, K, and F, as well as the semidefiniteness properties (17) of M, D, and K.

For the special cases of RC, RL, and LC circuits, there are variants of the general MPVL (Matrix-Padeé Via Lanczos) method [11] that do preserve the RC, RL, and LC structures, respectively. In particular, the SyPVL and SyMPVL algorithms are procedures for generating reduced-order models that can be synthesized as RC, RL, and LC circuits, respectively; see [22–24].

3.3 Passivity

An important property of general RCL circuits is passivity. Roughly speaking, a system is *passive* if it does not generate energy. In particular, any RCL circuit is passive. For linear dynamical systems of the form (16), passivity is equivalent to *positive realness* of the associated transfer function (20), $H(s)$; see, e.g., [2,30]. The general definition of positive realness is as follows.

Definition 1. *An $m \times m$-matrix-valued function $H : \mathbb{C} \mapsto (\mathbb{C} \cup \infty)^{m \times m}$ is called* positive real *if the following three conditions are satisfied:*

(i) *H is analytic in $\mathbb{C}_+ := \{ s \in \mathbb{C} \mid \operatorname{Re} s > 0 \}$;*
(ii) *$H(\overline{s}) = \overline{H(s)}$ for all $s \in \mathbb{C}$;*
(iii) *$H(s) + (H(s))^* \succeq 0$ for all $s \in \mathbb{C}_+$.*

Since any RCL circuit is passive, positive realness of the reduced-order transfer function (21), $\widetilde{H}(s)$, is a necessary condition for the associated reduced-order model (18) to be synthesizable as an actual RCL circuit. However, in general, positive realness of $\widetilde{H}(s)$ is not a sufficient condition. Nevertheless, any reduced-order model (18) with a positive real transfer function (21) can be synthesized as an actual physical electronic circuit, but it may contain other electronic devices besides resistors, capacitors, and inductors. We refer the reader to [2] for a discussion of the problem of synthesis of positive real transfer functions.

4 Equivalent First-Order Form of Integro-DAEs

The system of integro-DAEs (16) can also be formulated as an equivalent first-order system. In this section, we discuss such a first-order formulation and some of its properties.

4.1 First-Order Formulation

Consider equations (11) and their equivalent statement (16) as a system of integro-DAEs. It turns out that (11) (and thus (16)) can be rewritten as a *first-order time-invariant linear dynamical system* of the form

$$\begin{aligned} E\,\frac{d}{dt}x(t) &= A\,x(t) + B\,u(t), \\ y(t) &= B^T x(t), \end{aligned} \tag{22}$$

with initial conditions

$$x(0) = x^{(0)}.$$

Indeed, by adding the vector of inductance currents, $i_l(t)$, to the original state-space vector (13), $z(t)$, and using the last relation of (10), one readily verifies that the equations (11) can be stated in the form (22) with data matrices, state-space vector, and initial vector given by

$$\begin{gathered} A := -\begin{bmatrix} D_{11} & \mathcal{A}_l & \mathcal{A}_v \\ -\mathcal{A}_l^T & 0 & 0 \\ -\mathcal{A}_v^T & 0 & 0 \end{bmatrix}, \quad E := \begin{bmatrix} M_{11} & 0 & 0 \\ 0 & L & 0 \\ 0 & 0 & 0 \end{bmatrix}, \\ B := \begin{bmatrix} \mathcal{A}_i & 0 \\ 0 & 0 \\ 0 & -I \end{bmatrix}, \quad x(t) := \begin{bmatrix} v(t) \\ i_l(t) \\ i_v(t) \end{bmatrix}, \quad \text{and} \quad x^{(0)} := \begin{bmatrix} v^{(0)} \\ 0 \\ i_v(0) \end{bmatrix}. \end{gathered} \tag{23}$$

Here, M_{11} and D_{11} are the matrices defined in (12). Moreover, $A,\ E \in \mathbb{R}^{N\times N}$, $B \in \mathbb{R}^{N\times m}$, and $x^{(0)} \in \mathbb{R}^{N\times m}$, where N denotes the *state-space dimension* of the system (22). We remark that N is the sum of the state-space dimension N_0 of the equivalent system of integro-DAEs (16) and the number of inductors of the RCL circuit. Note that, in (22), the input vector $u(t)$ and the output vector $y(t)$ are the same as in (16), namely the vectors defined in (14). In particular, both systems (16) and (22) have m inputs and m outputs.

4.2 Regularity of the First-Order Matrix Pencil

Next, we consider the matrix pencil

$$s\,E - A, \quad s \in \mathbb{C}, \tag{24}$$

where A and E are the matrices defined in (23). The pencil (24) is said to be *regular* if the matrix $sE - A$ is singular only for finitely many values of $s \in \mathbb{C}$. In this subsection, we present conditions for regularity of (24).

In view of the definitions of A and E in (23), we have

$$s\,E - A = \begin{bmatrix} s\,M_{11} + D_{11} & \mathcal{A}_l & \mathcal{A}_v \\ -\mathcal{A}_l^T & s\,L & 0 \\ -\mathcal{A}_v^T & 0 & 0 \end{bmatrix} \quad \text{for all} \quad s \in \mathbb{C}. \tag{25}$$

Now assume that $s \neq 0$, and set

$$U_1(s) = \begin{bmatrix} I & -\frac{1}{s}\,\mathcal{A}_l L^{-1} & 0 \\ 0 & I & 0 \\ 0 & 0 & I \end{bmatrix} \quad \text{and} \quad U_2(s) = \begin{bmatrix} I & 0 & 0 \\ \frac{1}{s}\,L^{-1}\mathcal{A}_l^T & I & 0 \\ 0 & 0 & I \end{bmatrix}. \tag{26}$$

Then, one readily verifies that, for all $s \neq 0$,

$$U_1(s)\,\bigl(s\,E - A\bigr)\,U_2(s) = \begin{bmatrix} s\,M_{11} + D_{11} + \frac{1}{s}\,K_{11} & 0 & \mathcal{A}_v \\ 0 & s\,L & 0 \\ -\mathcal{A}_v^T & 0 & 0 \end{bmatrix}, \tag{27}$$

where K_{11} is the matrix defined in (12).

We now use the relation (27) to establish a necessary and sufficient condition for regularity of (24). Recall from (3) that $\mathcal{A}_r$, $\mathcal{A}_c$, $\mathcal{A}_l$, and $\mathcal{A}_v$ are the submatrices of the reduced incidence matrix $\mathcal{A}_0$ corresponding to the resistors, capacitors, inductors, and voltage sources of the RCL circuit, respectively.

Theorem 1. (Regularity of the matrix pencil (24).)

(a) *The pencil* (24) *is regular if, and only if, the matrix-valued function*

$$F(s) := \begin{bmatrix} F_{11}(s) & \mathcal{A}_v \\ -\mathcal{A}_v^T & 0 \end{bmatrix}, \quad \textit{where} \quad F_{11}(s) := s\,M_{11} + D_{11} + \frac{1}{s}\,K_{11}, \tag{28}$$

is regular, i.e., the matrix $F(s)$ is singular only for finitely many values of $s \in \mathbb{C}$, $s \neq 0$.

(b) *The pencil* (24) *is regular if, and only if, the matrix $\mathcal{A}_v$ has full column rank and the matrix*

$$\mathcal{A}_1 := \bigl[\mathcal{A}_r\ \mathcal{A}_c\ \mathcal{A}_l\ \mathcal{A}_v\bigr] \tag{29}$$

has full row rank.

Proof. Part (a) readily follows from (27) and the fact that, in view of (5), the matrix $L \succ 0$ is nonsingular. Indeed, since the matrices (26), $U_1(s)$ and $U_2(s)$, are nonsingular for all $s \neq 0$, it follows from (27) that the pencil (24) is regular if, and only if, the matrix-valued function on the right-hand side of (27) is regular. Since L is nonsingular, it follows that for the matrix-valued function (27) is regular if, and only if, $F(s)$, is regular.

To prove part (b), we make use of part (a). and we show that $F(s)$ is regular if, and only if $\mathcal{A}_v$ has full column rank and the matrix $\mathcal{A}_1$ has full row rank. Suppose $\mathcal{A}_v$ does not have full column rank, and let $c \neq 0$ a nontrivial vector in $\ker \mathcal{A}_v$. Then

$$F(s)\begin{bmatrix} 0 \\ c \end{bmatrix} = 0, \quad \begin{bmatrix} 0 \\ c \end{bmatrix} \neq 0,$$

and thus $F(s)$ is singular for all s. Therefore, we can assume that

$$\ker \mathcal{A}_v = \{\,0\,\}. \tag{30}$$

Next, note that the function $s\det F(s)$ is a polynomial in s, and thus $F(s)$ is regular unless $\det F(s) = 0$ for all s. Therefore, it is sufficient to consider the matrix (28), $F(s)$, for $s > 0$ only. Using (12) and the definition (29) of $\mathcal{A}_1$, the submatrix $F_{11}(s)$ of $F(s)$ can be expressed as follows:

$$F_{11}(s) = \begin{bmatrix} \mathcal{A}_r & \mathcal{A}_c & \mathcal{A}_l \end{bmatrix} \begin{bmatrix} sC & 0 & 0 \\ 0 & R^{-1} & 0 \\ 0 & 0 & \frac{1}{s}L^{-1} \end{bmatrix} \begin{bmatrix} \mathcal{A}_r & \mathcal{A}_c & \mathcal{A}_l \end{bmatrix}^T. \tag{31}$$

In view of (5), the 3×3 block diagonal matrix in (31) is symmetric positive definite for $s > 0$. It follows that for all $s > 0$, we have

$$F_{11}(s) \succeq 0 \quad \text{and} \quad \ker\big(F_{11}(s)\big) = \ker\left(\begin{bmatrix} \mathcal{A}_r & \mathcal{A}_c & \mathcal{A}_l \end{bmatrix}^T\right). \tag{32}$$

Finally, we apply Theorem 3.2 from [6], which gives a necessary and sufficient condition for the nonsingularity of 2×2 block matrices of the form (28) with subblocks satisfying (30) and the first condition in (32). By [6, Theorem 3.2], it follows that for $s > 0$, the matrix $F(s)$ is nonsingular if, and only if,

$$\ker\big(F_{11}(s)\big) \cap \ker\big(\mathcal{A}_v^T\big) = \{\,0\,\}. \tag{33}$$

Using the second relation in (32), we can rewrite (33) as follows:

$$\left(\ker\big(\mathcal{A}_1^T\big) =\right) \ker\left(\begin{bmatrix} \mathcal{A}_r & \mathcal{A}_c & \mathcal{A}_l & \mathcal{A}_v \end{bmatrix}^T\right) = \{\,0\,\}.$$

This condition is equivalent to the matrix (29), $\mathcal{A}_1$, having full row rank, and thus the proof of part (b) is complete. □

Remark 1. In terms of the given RCL circuit, the rank conditions in part (b) of Theorem 1 have the following meaning. In view of (3), the matrix (29), $\mathcal{A}_1$, is the reduced incidence matrix of the subcircuit obtained from the given RCL circuit by deleting all independent current sources. This matrix has full row rank if this subcircuit is connected; see, e.g., [7, Theorem 9-6]. The matrix $\mathcal{A}_v$ is the reduced incidence matrix of the subcircuit consisting of only the independent voltage sources. his matrix has full column rank if this subcircuit does not contain any closed (undirected) loop; see, e.g., [7, Section 9-8].

Since the two circuit conditions in Remark 1 are satisfied for any practical RCL circuit, from now on, we assume that the matrix pencil (24) is regular.

4.3 First-Order Transfer Function

In analogy to (20), the *transfer function* of the first-order formulation (22) is the matrix-valued rational function given by

$$H : \mathbb{C} \mapsto (\mathbb{C} \cup \infty)^{m \times m}, \quad H(s) = B^T \left(s\,E - A\right)^{-1} B. \tag{34}$$

We remark that (34) is a well-defined rational function since the matrix pencil $s\,E - A$ is assumed to be regular. Recall that the system of integro-DAEs (16) and its first-order formulation (22) have the same input and output vectors. Since transfer functions only depend on the input-output behavior of the system, it follows that the transfer functions (20) and (34) are identical, i.e.,

$$\begin{aligned} H(s) &= B^T \left(s\,E - A\right)^{-1} B \\ &= F^T \left(s\,M + D + \frac{1}{s}\,K\right)^{-1} F \quad \text{for all} \quad s \in \mathbb{C}. \end{aligned} \tag{35}$$

Here, A, E, B and M, D, K, F are the matrices given in (23) and (15), respectively. In particular, the regularity of the matrix pencil $s\,E - A$ also guarantees the existence of the transfer function (20) of the system of integro-DAEs (16).

Remark 2. The relation (35) can also be verified directly using the identity (27), (26), the definition of the matrix B in (23), and the definitions of the matrices M, D, K, and F in (15).

Remark 3. The definitions of A and E in (23), together with (5) and (17), imply that

$$-A - A^* \succeq 0 \quad \text{and} \quad E \succeq 0. \tag{36}$$

The matrix properties (36) in turn guarantee that the transfer function (34), H satisfies all conditions of Definition 1, and thus H is positive real.

4.4 Reduced-Order Models

A *reduced-order model* of the linear dynamical system (22) is a system of the same form as (22), but with smaller state-space dimension n $(< N)$. More precisely, a reduced-order model of (22) with state-space dimension n is a system of the form

$$\begin{aligned} E_n \frac{d}{dt}\widetilde{x}(t) &= A_n \widetilde{x}(t) + B_n u(t), \\ \widetilde{y}(t) &= B_n^T \widetilde{x}(t), \end{aligned} \tag{37}$$

with initial conditions

$$\widetilde{x}(0) = \widetilde{x}^{(0)}, \tag{38}$$

where A_n and E_n are $n \times n$ matrices, B_n is an $n \times m$ matrix, and $\widetilde{x}^{(0)}$ is a vector of length n.

Provided that the reduced-order matrix pencil

$$s\,E_n - A_n, \quad s \in \mathbb{C}, \tag{39}$$

is regular, the transfer function of the reduced-order model (37) is given by

$$H_n : \mathbb{C} \mapsto (\mathbb{C} \cup \infty)^{m \times m}, \quad H_n(s) = B_n^T \left(s\,E_n - A_n\right)^{-1} B_n. \tag{40}$$

5 Krylov-Subspace Projection and PRIMA

In this section, we review the generation of reduced-order models (37) via projection, in particular onto block Krylov subspaces.

5.1 Order Reduction Via Projection

A simple approach to model order reduction is to use projection. Let

$$V_n \in \mathbb{C}^{N \times n}, \quad \text{rank}\, V_n = n, \tag{41}$$

be any given matrix with full column rank. Then, by setting

$$A_n := V_n^* A V_n, \quad E_n := V_n^* E V_n, \quad B_n := V_n^* B \tag{42}$$

one obtains reduced data matrices that define a reduced-order model (37). From (36) and (42), it readily follow that

$$-A_n - A_n^* \succeq 0 \quad \text{and} \quad E_n \succeq 0. \tag{43}$$

If in addition, the matrix V_n is chosen as a real matrix and the matrix pencil (39) is assumed to be regular, then the reduced-order transfer function (40), H_n, satisfies all conditions of Definition 1, and thus H_n is positive real; see [15, Theorem 13].

5.2 Block Krylov Subspaces

The simple projection approach (42) yields powerful model-order reduction techniques when the columns of the matrix (41), V_n, are chosen as basis vectors of certain block Krylov subspaces.

To this end, let $s_0 \in \mathbb{C}$ be a suitably chosen expansion point such that the matrix $s_0\,E - A$ is nonsingular. Note that, in view of the regularity of the matrix pencil (24), there are only finitely many values of s_0 for which $s_0\,E - A$ is singular. We can then rewrite the transfer function (34), H, as follows:

$$\begin{aligned} H(s) &= B^T\big(s_0\,E - A + (s-s_0)E\big)^{-1}B \\ &= B^T\big(I + (s-s_0)A\big)^{-1}R, \end{aligned} \tag{44}$$

where

$$M := \big(s_0\,E - A\big)^{-1}E, \quad R := \big(s_0\,E - A\big)^{-1}B. \tag{45}$$

We will use block Krylov subspaces induced by the matrices M and R in (45) to generate reduced-order models.

Next, we briefly review the notion of block Krylov subspaces; see [1] for a more detailed discussion. The matrix sequence

$$R, MR, M^2R, \ldots, M^{j-1}R, \ldots$$

is called a *block Krylov sequence*. The columns of the matrices in this sequence are vectors of length N, and thus at most N of these columns are linearly independent. By scanning the columns of the matrices in the block Krylov sequence from left to right and deleting each column that is linearly dependent on earlier columns, we obtain the *deflated* block Krylov sequence

$$R_1, MR_2, M^2R_3, \ldots, M^{j-1}R, \ldots, M^{j_{\max}-1}R_{j_{\max}}. \tag{46}$$

This process of deleting linearly dependent vectors is called *deflation*. In (46), each R_j is a submatrix of R_{j-1}. Denoting by m_j the number of columns of R_j, we thus have

$$m \geq m_1 \geq m_2 \geq \cdots \geq m_j \geq \cdots \geq m_{j_{\max}} \geq 1. \tag{47}$$

By construction, the columns of the matrices (46) are linearly independent, and for each n, the subspace spanned by the first n of these columns is called the *n-th block Krylov subspace* (induced by M and R) and denoted by $\mathcal{K}_n(M,R)$ in the sequel.

For $j = 1, 2, \ldots, j_{\max}$, we set

$$n(j) := m_1 + m_2 + \cdots + m_j. \tag{48}$$

For $n = n(j)$, the n-th block Krylov subspace is given by

$$\mathcal{K}_n(M,R) = \text{colspan}\ \big[R_1\ MR_2\ M^2R_3\ \cdots\ M^jR_j\big].$$

Here and in the sequel, we use colspan V to denote the subspace spanned by the columns of the matrix V. Finally, we remark that, by (47), $n(j) \leq m \cdot j$ with $n(j) = m \cdot j$ if no deflation has occurred.

5.3 Projection Onto Block Krylov Subspaces and PRIMA

PRIMA [28–30] combines projection with block Krylov subspaces. More precisely, the n-th PRIMA reduced-order model is defined by (37) and (42), where the matrix (41), V_n, is chosen such that its columns span the n-th block Krylov subspace $\mathcal{K}_n(M,R)$, i.e., colspan $V_n = \mathcal{K}_n(M,R)$. We refer to any such matrix V_n as a *basis matrix* of the n-th Krylov subspace $\mathcal{K}_n(M,R)$.

Although the PRIMA reduced-order models are defined by simple projection, the combination with block Krylov subspaces guarantees that the PRIMA reduced-order models satisfy a Padé-type approximation property. For the special case $s_0 = 0$ and basis vectors generated by a block Arnoldi process without deflation, this Padé-type approximation property was first observed in [28]. In [13], this result was extended to the most general case where possibly nonzero expansion points s_0 are allowed and where the underlying block Krylov method allows the necessary deflation of linearly dependent vectors. The result can be stated as follows; for a proof, we refer the reader to [15, Theorem 7].

Theorem 2. *Let* $n = n(j)$ *be of the form* (48) *for some* $1 \leq j \leq j_{\max}$, *and let* $V_n \in \mathbb{C}^{N \times n}$ *be any matrix such that*

$$\operatorname{colspan} V_n = \mathcal{K}_n(M, R). \tag{49}$$

Then the transfer function (40), H_n, *of the reduced-order model* (37) *defined by the projected data matrices* (42) *satisfies:*

$$H_n(s) = H(s) + \mathcal{O}\big((s - s_0)^j\big). \tag{50}$$

If in addition, the expansion point s_0 is chosen to be real,

$$s_0 \in \mathbb{R}, \tag{51}$$

then the matrices (45), M and R, are real and the basis matrix V_n in (49) can be constructed to be real. In fact, any of the usual Krylov subspace algorithms for constructing basis vectors for $\mathcal{K}_n(M, R)$, such as the band Lanczos method or the band Arnoldi process [16], generate a real basis matrix V_n. In this case, as mentioned at the end of Section 5.1, the transfer function H_n is positive real, and thus the PRIMA reduced-order models are passive.

On the other hand, the data matrices (42) of the PRIMA reduced-order models are full in general, and thus, PRIMA does not preserve the special block structure of the original data matrices (23).

6 The SPRIM Algorithm

In this section, we describe the SPRIM algorithm [17, 20], which unlike PRIMA, preserves the block structure of the data matrices (23).

6.1 The Projection Theorem

It turns out that in order to guarantee a Padé-type property (50) of the reduced-order transfer function, the condition (49) on the matrix V_n can be relaxed. In fact, let $\widehat{V} \in \mathbb{C}^{N \times \widehat{n}}$ be any matrix with the property

$$\mathcal{K}_n(M, R) \subseteq \operatorname{colspan} \widehat{V}. \tag{52}$$

Then the statement of Theorem 2 remains correct when (49) is replaced by the weaker condition (52). This result, which is sometimes referred to as the *projection theorem*, was derived by Grimme in [25]. A different proof of the projection theorem is given in [18, Theorem 8.6.1]. Note that, in view of (52), $\widehat{V}$ must have at least as many columns as any matrix V_n satisfying (49).

The projection theorem can be used to devise an order reduction algorithm that in the Padé-type sense (50), is at least as accurate as PRIMA, but unlike PRIMA preserves the block structure of the original data matrices (23). Indeed, let V_n be any basis matrix of the n-th Krylov subspace $\mathcal{K}_n(M, R)$. Let

$$V_n = \begin{bmatrix} \widehat{V}_1 \\ \widehat{V}_2 \\ \widehat{V}_3 \end{bmatrix}$$

be the partitioning of V_n corresponding to the block sizes of the matrices A and E in (23), and formally set

$$\widehat{V} = \begin{bmatrix} \widehat{V}_1 & 0 & 0 \\ 0 & \widehat{V}_2 & 0 \\ 0 & 0 & \widehat{V}_3 \end{bmatrix}. \tag{53}$$

Since V_n is a basis matrix of $\mathcal{K}_n(M, R)$, the matrix (53) satisfies (52). Thus, we can replace V_n by $\widehat{V}$ in (42) and still obtain a reduced-order model (37) that satisfies a Padé-type property (50). In view of the block structures of the original data matrices (23) and of the matrix (53), the reduced-order matrices are of the form

$$A_n = -\begin{bmatrix} \widetilde{D}_{11} & \widetilde{A}_l & \widetilde{A}_v \\ -\widetilde{A}_l^T & 0 & 0 \\ -\widetilde{A}_v^T & 0 & 0 \end{bmatrix}, \quad E_n = \begin{bmatrix} \widetilde{M}_{11} & 0 & 0 \\ 0 & \widetilde{L} & 0 \\ 0 & 0 & 0 \end{bmatrix}, \quad B_n = \begin{bmatrix} \widetilde{A}_i & 0 \\ 0 & 0 \\ 0 & -\widehat{V}_3^T \end{bmatrix},$$

and thus the block structure of the original data matrices (23) is now preserved. The resulting order reduction procedure is the most basic form of the SPRIM algorithm.

We remark that in this most basic form of SPRIM, the relative sizes of the blocks in (23) are not preserved. Recall that the sizes of the three diagonal blocks of A and E in (23) are the number of interconnections, the number of inductors, and the number of voltage sources, respectively, of the given RCL circuit. These numbers are very different in general. Typically, there are only very few voltage sources. Similarly, the number of inductors is typically significantly smaller than the number of interconnections. Consequently, unless n is smaller than the number of voltage sources, the subblock $\widehat{V}_3$ does not have full column rank. The sizes of the subblocks in the third block rows and columns of the reduced-order data matrices can thus be reduced further by replacing $\widehat{V}_3$ with a matrix whose columns span the same subspace as $\widehat{V}_3$, but which has full column rank, before the projection is performed. Similar size reductions are possible if $\widehat{V}_2$ or $\widehat{V}_1$ do not have full column rank.

6.2 SPRIM

The basic form of SPRIM, with possible size reduction of the subblocks $\widehat{V}_l$, $l = 1, 2, 3$, as an optional step, can be summarized as follows.

Algorithm 1. (SPRIM for general RCL circuits)

- *Input: matrices of the form*

$$A = -\begin{bmatrix} D_{11} & \mathcal{A}_l & \mathcal{A}_v \\ -\mathcal{A}_l^T & 0 & 0 \\ -\mathcal{A}_v^T & 0 & 0 \end{bmatrix}, \quad E = \begin{bmatrix} M_{11} & 0 & 0 \\ 0 & L & 0 \\ 0 & 0 & 0 \end{bmatrix}, \quad B = \begin{bmatrix} \mathcal{A}_i & 0 \\ 0 & 0 \\ 0 & -I \end{bmatrix},$$

  where $D_{11}, M_{11} \succeq 0$*;*
  an expansion point $s_0 \in \mathbb{R}$*.*
- *Formally set*

$$M = (s_0\, E - A)^{-1}\, C, \quad R = (s_0\, E - A)^{-1}\, B.$$

- *Until* n *is large enough, run your favorite block Krylov subspace method (applied to* M *and* R*) to construct the columns of the basis matrix*

$$V_n = \begin{bmatrix} v_1 \; v_2 \; \cdots \; v_n \end{bmatrix}$$

  of the n*-th block Krylov subspace* $\mathcal{K}_n(M, R)$*, i.e.,*

$$\operatorname{colspan} V_n = \mathcal{K}_n(M, R).$$

- *Let*

$$V_n = \begin{bmatrix} \widehat{V}_1 \\ \widehat{V}_2 \\ \widehat{V}_3 \end{bmatrix}$$

  be the partitioning of V_n *corresponding to the block sizes of* A *and* E*.*
- (Optional step) *For* $l = 1, 2, 3$ *do:*
  If $r_l := \operatorname{rank} \widehat{V}_l < n$*, determine an* $N \times r_l$ *matrix* $\widetilde{V}_l$ *with*

$$\operatorname{colspan} \widetilde{V}_l = \operatorname{colspan} \widehat{V}_l, \quad \operatorname{rank} \widetilde{V}_l = r_l,$$

  and set $\widehat{V}_l := \widetilde{V}_l$*.*
- *Set*

$$\widetilde{D}_{11} = \widehat{V}_1^* D_{11} \widehat{V}_1, \quad \widetilde{\mathcal{A}}_l = \widehat{V}_1^* \mathcal{A}_l \widehat{V}_2, \quad \widetilde{\mathcal{A}}_v = \widehat{V}_1^* \mathcal{A}_v \widehat{V}_3,$$
$$\widetilde{M}_{11} = \widehat{V}_1^* M_{11} \widehat{V}_1, \quad \widetilde{L} = \widehat{V}_2^* L \widehat{V}_2, \quad \widetilde{\mathcal{A}}_i = \widehat{V}_1^* \mathcal{A}_i.$$

- *Output: the data matrices*

$$A_n = -\begin{bmatrix} \widetilde{D}_{11} & \widetilde{\mathcal{A}}_l & \widetilde{\mathcal{A}}_v \\ -\widetilde{\mathcal{A}}_l^T & 0 & 0 \\ -\widetilde{\mathcal{A}}_v^T & 0 & 0 \end{bmatrix}, \quad E_n = \begin{bmatrix} \widetilde{M}_{11} & 0 & 0 \\ 0 & \widetilde{L} & 0 \\ 0 & 0 & 0 \end{bmatrix},$$

$$B_n = \begin{bmatrix} \widetilde{\mathcal{A}}_i & 0 \\ 0 & 0 \\ 0 & -\widehat{V}_3^T \end{bmatrix} \tag{54}$$

of the SPRIM reduced-order model

$$\begin{aligned} E_n \frac{d}{dt}\widetilde{x}(t) &= A_n \widetilde{x}(t) + B_n u(t), \\ \widetilde{y}(t) &= B_n^T \widetilde{x}(t). \end{aligned}$$

We remark that the main computational cost of the SPRIM algorithm is running the block Krylov subspace method to obtain $\widehat{V}_n$. This is the same as for PRIMA. Thus generating the PRIMA reduced-order model and the SPRIM reduced-order model H_n involves the same computational costs. Implementation details of the SPRIM algorithm can be found in [20].

7 Padé-Type Approximation Property of SPRIM

While PRIMA and SPRIM generate different reduced-order models, the projection theorem suggests that both models have comparable accuracy in the sense of the Padé-type approximation property (50). However, as long as the expansion point s_0 is chosen to be real, cf. (51), numerical experiments show that SPRIM is significantly more accurate than PRIMA; see the numerical results in Section 8. This higher accuracy is a consequence of the structure preservation of the SPRIM reduced-order data matrices (54). We stress that the restriction (51) of the expansion point s_0 to real values is needed anyway for both PRIMA and SPRIM, in order to guarantee that the PRIMA and SPRIM reduced-order models are passive.

For the special case of RCL circuits with current sources only, which means that the third block rows and columns in (54) are not present, it was proven in [18, Theorem 8.7.2] that the SPRIM reduced-order transfer function satisfies (50) with j replaced by $2j$.

A recent result [19] shows that the higher accuracy of SPRIM holds true in the more general context of Padé-type model order reduction of J-Hermitian linear dynamical systems. A square matrix A is said to be *J-Hermitian* with respect to a given nonsingular matrix J of the same size as A if

$$JA = A^* J.$$

Clearly, for RCL circuits, in view of (23), the original data matrices A and E are J-Hermitian with respect to the indefinite matrix

$$J = \begin{bmatrix} I & 0 & 0 \\ 0 & -I & 0 \\ 0 & 0 & -I \end{bmatrix}.$$

Furthermore, due to the structure preservation of SPRIM, the reduced-order data matrices A_n and E_n in (54) are J_n-Hermitian with respect to a matrix J_n of the same form as J, but with correspondingly smaller blocks. Finally, the matrix (53),$\widehat{V}$, which is used to generate the SPRIM models, satisfies the *compatibility condition*

$$J\widehat{V} = \widehat{V} J_n.$$

The result in [19] shows that for J-Hermitian data matrices and J_n-Hermitian reduced-order data matrices, the compatibility condition implies the higher accuracy of Padé-type reduced-order models. In particular, as a special case of this more general result, we have the following theorem.

Theorem 3. *Let $n = n(j)$ be of the form (48) for some $1 \leq j \leq j_{\max}$, and assume that $s_0 \in \mathbb{R}$. Then the transfer function (40), H_n, of the* SPRIM *reduced-order model (37) defined by the projected data matrices (54) satisfies:*

$$H_n(s) = H(s) + \mathcal{O}\big((s - s_0)^{2j}\big).$$

8 Numerical Examples

In this section, we present results of some numerical experiments with the SPRIM algorithm. These results were first reported in [17]. The results in this section illustrate the higher accuracy of the SPRIM reduced-order models vs. the PRIMA reduced-order models.

8.1 A PEEC Circuit

The first example is a circuit resulting from the so-called PEEC discretization [32] of an electromagnetic problem. The circuit is an RCL network consisting of 2100 capacitors, 172 inductors, 6990 inductive couplings, and a single resistive source that drives the circuit. The circuit is formulated as a 2-port. We compare the PRIMA and SPRIM models corresponding to the same dimension n of the underlying block Krylov subspace. The expansion point $s_0 = 2\pi \times 10^9$ was used. In Figure 1, we plot the absolute value of the $(2,1)$ component of the 2×2-matrix-valued transfer function over the frequency range of interest. The dimension $n = 120$ was sufficient for SPRIM to match the exact transfer function. The corresponding PRIMA model of the same dimension, however, has not yet converged to the exact transfer function in large parts of the frequency range of interest. Figure 1 clearly illustrates the better approximation properties of SPRIM due to matching of twice as many moments as PRIMA.

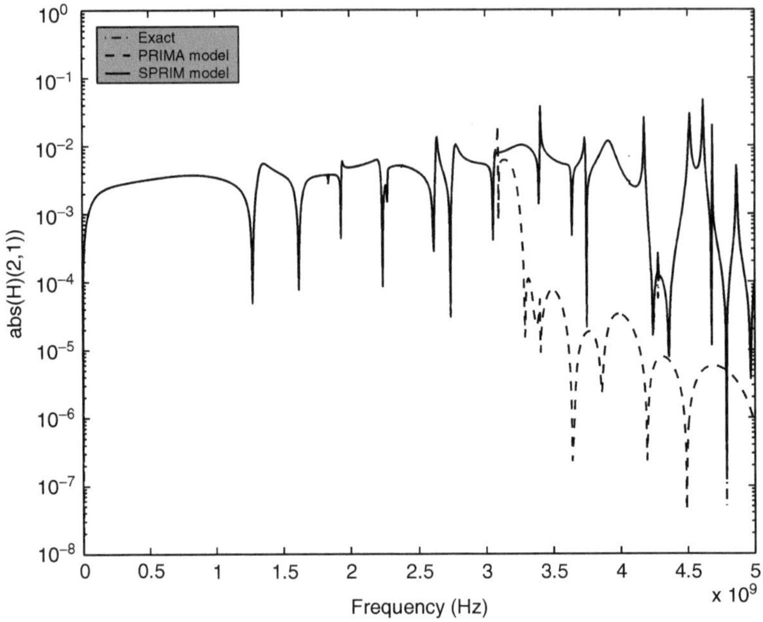

Fig. 1. $|H_{2,1}|$ for PEEC circuit

8.2 A Package Model

The second example is a 64-pin package model used for an RF integrated circuit. Only eight of the package pins carry signals, the rest being either unused or carrying supply voltages. The package is characterized as a 16-port component (8 exterior and 8 interior terminals). The package model is described by approximately 4000 circuit elements, resistors, capacitors, inductors, and inductive couplings. We again compare the PRIMA and SPRIM models corresponding to the same dimension n of the underlying block Krylov subspace. The expansion point $s_0 = 5\pi \times 10^9$ was used. In Figure 2, we plot the absolute value of one of the components of the 16×16-matrix-valued transfer function over the frequency range of interest. The state-space dimension $n = 80$ was sufficient for SPRIM to match the exact transfer function. The corresponding PRIMA model of the same dimension, however, does not match the exact transfer function very well near the high frequencies; see Figure 3.

8.3 A Mechanical System

Exploiting the equivalence (see, e.g., [27]) between RCL circuits and mechanical systems, both PRIMA and SPRIM can also be applied to reduced-order modeling of mechanical systems. Such systems arise for example in the modeling and simulation of MEMS devices. In Figure 4, we show a comparison of PRIMA and SPRIM for a finite-element model of a shaft. The expansion point $s_0 = \pi \times 10^3$ was used. The

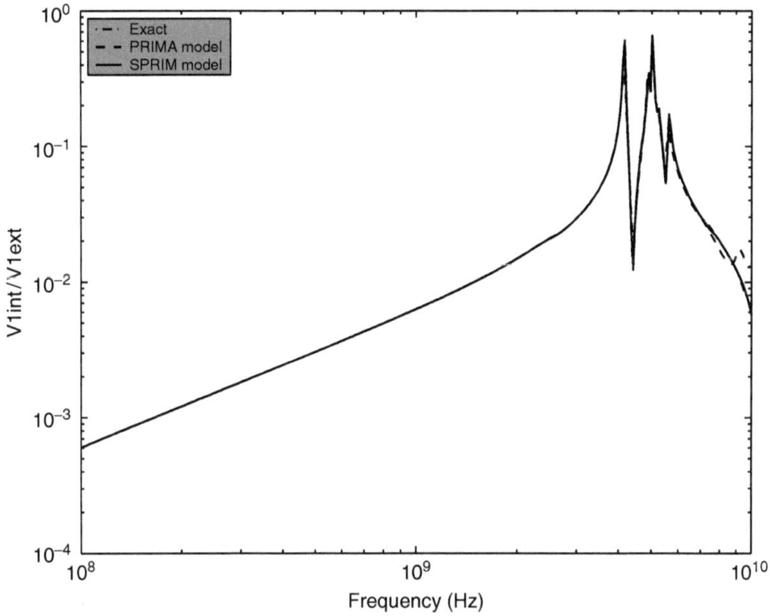

Fig. 2. The package model

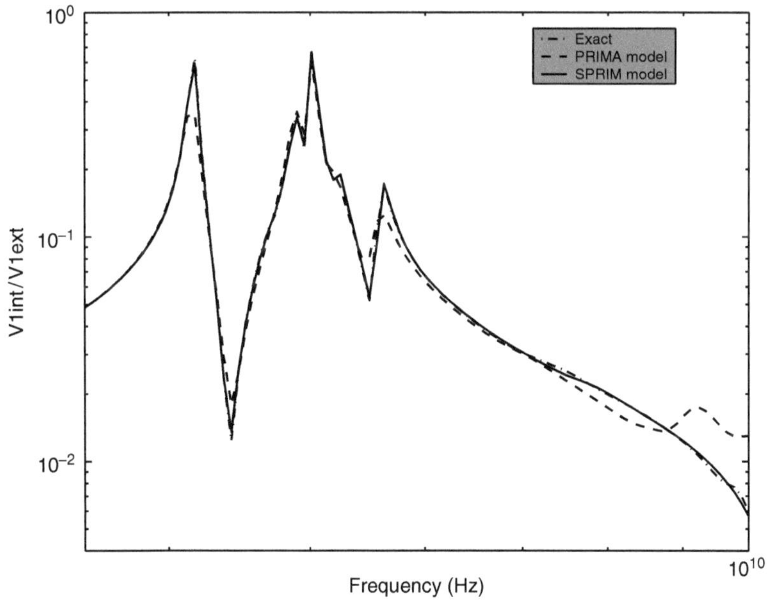

Fig. 3. The package model, high frequencies

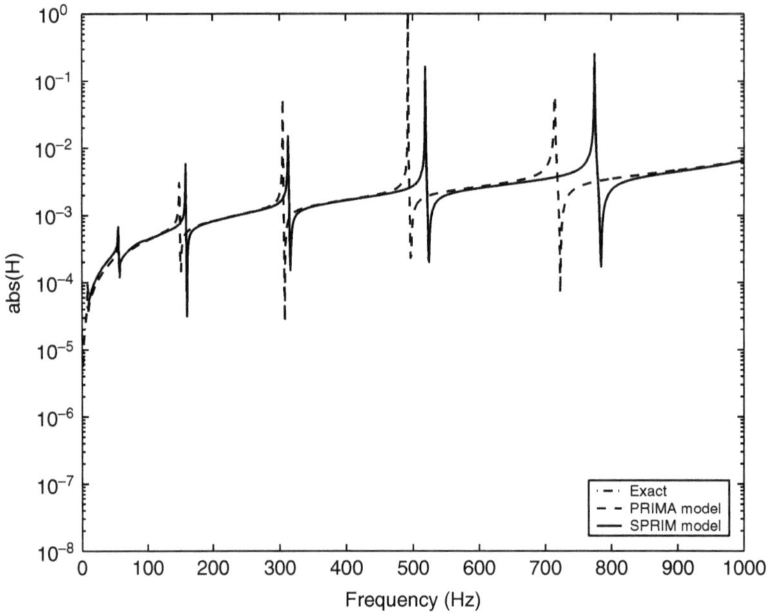

Fig. 4. A mechanical system

dimension $n = 15$ was sufficient for SPRIM to match the exact transfer function in the frequency range of interest. The corresponding PRIMA model of the same dimension, however, has not converged to the exact transfer function in large parts of the frequency range of interest. Figure 4 again illustrates the better approximation properties of SPRIM due to the matching of twice as many moments as PRIMA.

9 Concluding Remarks

In this paper, we reviewed the formulation of general RCL circuits as linear dynamical systems and discussed the problem of structure-preserving model reduction of such systems. We described the general framework of order reduction via projection and discussed two state-of-the-art projection algorithms, namely PRIMA and SPRIM.

While there has been a lot of progress in Krylov subspace-based structure-preserving model reduction of large-scale linear dynamical systems in recent years, there are still many open problems. All state-of-the-art structure-preserving methods, such as SPRIM, first generate a basis matrix of the underlying Krylov subspace and then employ explicit projection using some suitable partitioning of the basis matrix to obtain a structure-preserving reduced-order model. In particular, there are two major problems with the use of such explicit projections. First, it requires the storage of the basis matrix, which becomes prohibitive in the case of truly large-scale linear dynamical systems. Second, the approximation properties of the resulting structure-preserving reduced-order models are far from optimal, and they show that the available degrees of freedom are not fully used. It would be highly desirable to have structure-preserving reduction method that do no involve explicit projection and would thus be applicable in the truly large-scale case. Other unresolved issues include the automatic and adaptive choice of suitable expansion points s_0 and robust and reliable stopping criteria and error bounds.

Acknowledgement

This work was supported in part by the National Science Foundation through Grant DMS-0613032.

References

1. J. I. Aliaga, D. L. Boley, R. W. Freund, and V. Hernández. A Lanczos-type method for multiple starting vectors. *Math. Comp.*, 69:1577–1601, 2000.
2. B. D. O. Anderson and S. Vongpanitlerd. *Network Analysis and Synthesis*. Prentice-Hall, Englewood Cliffs, New Jersey, 1973.
3. W. E. Arnoldi. The principle of minimized iterations in the solution of the matrix eigenvalue problem. *Quart. Appl. Math.*, 9:17–29, 1951.
4. Z. Bai, P. Feldmann, and R. W. Freund. How to make theoretically passive reduced-order models passive in practice. In *Proc. IEEE 1998 Custom Integrated Circuits Conference*, pages 207–210, Piscataway, New Jersey, 1998. IEEE.
5. G. A. Baker, Jr. and P. Graves-Morris. *Padé Approximants*. Cambridge University Press, New York, New York, Second edition, 1996.
6. M. Benzi, G. H. Golub, and J. Liesen. Numerical solution of saddle point problems. *Acta Numerica*, 14:1–137, 2005.
7. N. Deo. *Graph Theory with Applications to Engineering and Computer Science*. Prentice-Hall, Englewood Cliffs, New Jersey, 1974.
8. I. M. Elfadel and D. D. Ling. Zeros and passivity of Arnoldi-reduced-order models for interconnect networks. In *Proc. 34nd ACM/IEEE Design Automation Conference*, pages 28–33, New York, New York, 1997. ACM.
9. P. Feldmann and R. W. Freund. Efficient linear circuit analysis by Padé approximation via the Lanczos process. In *Proceedings of EURO-DAC '94 with EURO-VHDL '94*, pages 170–175, Los Alamitos, California, 1994. IEEE Computer Society Press.
10. P. Feldmann and R. W. Freund. Efficient linear circuit analysis by Padé approximation via the Lanczos process. *IEEE Trans. Computer-Aided Design*, 14:639–649, 1995.
11. P. Feldmann and R. W. Freund. Reduced-order modeling of large linear subcircuits via a block Lanczos algorithm. In *Proc. 32nd ACM/IEEE Design Automation Conference*, pages 474–479, New York, New York, 1995. ACM.
12. P. Feldmann and R. W. Freund. Interconnect-delay computation and signal-integrity verification using the SyMPVL algorithm. In *Proc. 1997 European Conference on Circuit Theory and Design*, pages 132–138, Los Alamitos, California, 1997. IEEE Computer Society Press.

13. R. W. Freund. Passive reduced-order models for interconnect simulation and their computation via Krylov-subspace algorithms. In *Proc. 36th ACM/IEEE Design Automation Conference*, pages 195–200, New York, New York, 1999. ACM.
14. R. W. Freund. Reduced-order modeling techniques based on Krylov subspaces and their use in circuit simulation. In B. N. Datta, editor, *Applied and Computational Control, Signals, and Circuits*, Vol. 1, pages 435–498. Birkhäuser, Boston, 1999.
15. R. W. Freund. Krylov-subspace methods for reduced-order modeling in circuit simulation. *J. Comput. Appl. Math.*, 123(1–2):395–421, 2000.
16. R. W. Freund. Model reduction methods based on Krylov subspaces. *Acta Numerica*, 12: 267–319, 2003.
17. R. W. Freund. SPRIM: structure-preserving reduced-order interconnect macromodeling. In *Tech. Dig. 2004 IEEE/ACM International Conference on Computer-Aided Design*, pages 80–87, Los Alamitos, California, 2004. IEEE Computer Society Press.
18. R. W. Freund. Padé-type model reduction of second-order and higher-order linear dynamical systems. In P. Benner, V. Mehrmann, and D. C. Sorensen, editors, *Dimension Reduction of Large-Scale Systems*, Lecture Notes in Computational Science and Engineering, Vol. 45, pages 191–223, Berlin/Heidelberg, 2005. Springer-Verlag.
19. R. W. Freund. On Padé-type model order reduction of J-Hermitian linear dynamical systems. Technical Report, Department of Mathematics, University of California, Davis, California, 2007. Available online at `http://www.math.ucdavis.edu/~freund/reprints.html`.
20. R. W. Freund. The SPRIM algorithm for structure-preserving order reduction of general RCL circuits. Technical Report, Department of Mathematics, University of California, Davis, California, 2008. In preparation.
21. R. W. Freund and P. Feldmann. Reduced-order modeling of large passive linear circuits by means of the SyPVL algorithm. In *Tech. Dig. 1996 IEEE/ACM International Conference on Computer-Aided Design*, pages 280–287, Los Alamitos, California, 1996. IEEE Computer Society Press.
22. R. W. Freund and P. Feldmann. Small-signal circuit analysis and sensitivity computations with the PVL algorithm. *IEEE Trans. Circuits and Systems—II: Analog and Digital Signal Processing*, 43:577–585, 1996.
23. R. W. Freund and P. Feldmann. The SyMPVL algorithm and its applications to interconnect simulation. In *Proc. 1997 International Conference on Simulation of Semiconductor Processes and Devices*, pages 113–116, Piscataway, New Jersey, 1997. IEEE.
24. R. W. Freund and P. Feldmann. Reduced-order modeling of large linear passive multi-terminal circuits using matrix-Padé approximation. In *Proc. Design, Automation and Test in Europe Conference 1998*, pages 530–537, Los Alamitos, California, 1998. IEEE Computer Society Press.
25. E. J. Grimme. *Krylov projection methods for model reduction*. PhD Thesis, Department of Electrical Engineering, University of Illinois at Urbana-Champaign, Urbana-Champaign, Illinois, 1997.
26. C. Lanczos. An iteration method for the solution of the eigenvalue problem of linear differential and integral operators. *J. Res. Nat. Bur. Standards*, 45:255–282, 1950.
27. R. Lozano, B. Brogliato, O. Egeland, and B. Maschke. *Dissipative Systems Analysis and Control*. Springer-Verlag, London, 2000.
28. A. Odabasioglu. Provably passive RLC circuit reduction. M.S. Thesis, Department of Electrical and Computer Engineering, Carnegie Mellon University, 1996.
29. A. Odabasioglu, M. Celik, and L. T. Pileggi. PRIMA: passive reduced-order interconnect macromodeling algorithm. In *Tech. Dig. 1997 IEEE/ACM International Conference on*

Computer-Aided Design, pages 58–65, Los Alamitos, California, 1997. IEEE Computer Society Press.

30. A. Odabasioglu, M. Celik, and L. T. Pileggi. PRIMA: passive reduced-order interconnect macromodeling algorithm. *IEEE Trans. Computer-Aided Design*, 17(8):645–654, 1998.
31. L. T. Pillage and R. A. Rohrer. Asymptotic waveform evaluation for timing analysis. *IEEE Trans. Computer-Aided Design*, 9:352–366, 1990.
32. A. E. Ruehli. Equivalent circuit models for three-dimensional multiconductor systems. *IEEE Trans. Microwave Theory Tech.*, 22:216–221, 1974.
33. L. M. Silveira, M. Kamon, I. Elfadel, and J. White. A coordinate-transformed Arnoldi algorithm for generating guaranteed stable reduced-order models of RLC circuits. In *Tech. Dig. 1996 IEEE/ACM International Conference on Computer-Aided Design*, pages 288–294, Los Alamitos, California, 1996. IEEE Computer Society Press.
34. J. Vlach and K. Singhal. *Computer Methods for Circuit Analysis and Design*. Van Nostrand Reinhold, New York, New York, Second edition, 1994.

A Unified Krylov Projection Framework for Structure-Preserving Model Reduction

Zhaojun Bai[1], Ren-cang Li[2], and Yangfeng Su[3]

[1] Department of Computer Science and Department of Mathematics, University of California, Davis, CA 95616, USA
bai@cs.ucdavis.edu

[2] Department of Mathematics, University of Texas, Arlington, TX 76019
rcli@uta.edu

[3] School of Mathematical Science, Fudan University, Shanghai 200433, China
yfsu@fudan.edu.cn

1 Introduction

Physical systems often have certain characteristics that are critical in determining the system behavior. Often these characteristics appear in the form of system matrices that are naturally blocked with each sub-block having its own physical relevance. For example, the system matrices from linearizing a second order dynamical system admit a natural 2-by-2 block partitioning. General purpose subspace projection techniques for model order reduction usually destroy any block structure and thus the reduced systems may not be of the same type as the original system. For similar reasons we would like to preserve the block structure and hence some of the important characteristics so that the reduced systems are much like the original system but only at a much smaller scale.

Structure-preserving Krylov subspace projection methods have received much attention in recent years. In this chapter, we discuss the advance of the structure-preserving methods under a unified Krylov projection formulation. We shall start by building a mathematical foundation and a general paradigm to preserve important block structures under subspace projections. The general paradigm provides a unified projection formulation. When necessary, the technique can be used to preserve certain blocks in the system matrices. We then go on to study in detail model order reductions of RCL and RCS systems.

The remainder of this chapter is organized as follows. In Section 2, we discuss a unified Krylov subspace projection formulation for model order reduction with properties of structure-preserving and moment-matching, and present a generic algorithm for constructing structure-preserving projection matrices. The inherent structural properties of Krylov subspaces for certain block matrices are presented in Section 3. Section 4 examines structure-preserving model order reduction of RCL and RCS equations including the objective to develop synthesized RCL and RCS equations.

Throughout the chapter, $\mathbb{R}^{k\times\ell}$ is the set of $k \times \ell$ real matrices. I is the identity matrix and its dimension will be clear from the context. Unless otherwise explicitly stated, capital letters denote matrices, while lower case letters denote vectors or scalars. X^{T} is the transpose of the matrix X, span$\{X\}$ is the subspace spanned by the columns of X.

Let A be $N \times N$, and let B be $N \times p$. The *kth Krylov subspace* generated by A on B is defined to be

$$\mathcal{K}_k(A,B) = \text{span}\{B, AB, \ldots, A^{k-1}B\}. \tag{1}$$

For convenience, when $k = 0$, define $\mathcal{K}_0(A,B) = \{0\}$, the subspace of the zero vector.

2 A unified Krylov Projection Structure-Preserving Model Order Reduction Framework

Consider the *matrix-valued transfer function* of the first-order multi-input multi-output (MIMO) linear dynamical system

$$H(s) = \mathbf{L}^{\mathrm{T}}(s\mathbf{C} + \mathbf{G})^{-1}\mathbf{B}, \tag{2}$$

where $\mathbf{C}$ and $\mathbf{G}$ are $N \times N$, $\mathbf{B}$ is $N \times m$ and $\mathbf{L}$ is $N \times p$. Often $p \ll N$ and $m \ll N$.

Assume that $\mathbf{G}$ is nonsingular. The transfer function can be expanded around $s = 0$ as

$$\begin{aligned} H(s) &= \sum_{\ell=0}^{\infty}(-1)^{\ell}s^{\ell}\mathbf{L}^{\mathrm{T}}(\mathbf{G}^{-1}\mathbf{C})^{\ell}\mathbf{G}^{-1}\mathbf{B} \\ &\equiv \sum_{\ell=0}^{\infty}(-1)^{\ell}s^{\ell}M_{\ell}, \end{aligned}$$

where the matrices M_ℓ, defined by

$$M_\ell = \mathbf{L}^{\mathrm{T}}(\mathbf{G}^{-1}\mathbf{C})^{\ell}\mathbf{G}^{-1}\mathbf{B}$$

are referred to as the *moments* at $s = 0$. In the case when $\mathbf{G}$ is singular or approximations to $H(s)$ around a selected point $s_0 \neq 0$ are sought[1], a shift

$$s = (s - s_0) + s_0 \equiv \sigma + s_0 \tag{3}$$

can be carried out and then

[1] It is assumed that the matrix pencil $s\mathbf{C} + \mathbf{G}$ is regular, meaning that there are at most N values of s at which $s\mathbf{C} + \mathbf{G}$ is singular.

$$sC + G = (s - s_0)C + s_0C + G \equiv \sigma C + (s_0C + G).$$

Upon substitutions (i.e., renaming)

$$\mathbf{G} \leftarrow s_0\mathbf{C} + \mathbf{G}, \quad s \leftarrow \sigma,$$

the problem of approximating $H(s)$ around $s = s_0$ becomes equivalent to approximating the substituted $H(\sigma)$ around $\sigma = 0$. For this reason, without loss of generality we shall focus mostly on approximations around $s = 0$ in this chapter, unless some care has to be taken for computational efficiency, for instance, when a shift like (3) has to be performed.

Many transfer functions appearing in different forms can be reformulated in first order form (2).

Example 1. Consider a system of *integro-differential-algebraic equations* (Integro-DAEs) arising from the MNA formulation of circuits, such as the ones described in the chapters by Freund and by Gad, Nakhla, and Achar in this book:

$$\begin{cases} C\frac{d}{dt}z(t) + Gz(t) + \Gamma\int_0^t z(\tau)d\tau = Bu(t), \\ \qquad\qquad\qquad\qquad\qquad y(t) = B^{\mathrm{T}}z(\tau). \end{cases}$$

The transfer function of the Integro-DAEs is given by

$$H(s) = B^{\mathrm{T}}\left(sC + G + \frac{1}{s}\Gamma\right)^{-1}B. \tag{4}$$

By defining

$$\mathbf{C} = \begin{bmatrix} C & 0 \\ 0 & -W \end{bmatrix}, \ \mathbf{G} = \begin{bmatrix} G & \Gamma \\ W & 0 \end{bmatrix}, \ \mathbf{L} = \mathbf{B} = \begin{bmatrix} B \\ 0 \end{bmatrix} \tag{5}$$

for any *nonsingular* matrix W, the transfer function is of the form (2), namely

$$H(s) = \mathbf{B}^{\mathrm{T}}(s\mathbf{C} + \mathbf{G})^{-1}\mathbf{B}. \tag{6}$$

In (5), the matrix W is usually taken to be Γ (if it is nonsingular) or simply the identity matrix.

Alternatively, if one defines

$$\mathbf{C} = \begin{bmatrix} G & C \\ W & 0 \end{bmatrix}, \ \mathbf{G} = \begin{bmatrix} \Gamma & 0 \\ 0 & -W \end{bmatrix}, \ \mathbf{L} = \mathbf{B} = \begin{bmatrix} B \\ 0 \end{bmatrix} \tag{7}$$

again for any *nonsingular* matrix W (usually taken to be C if it is nonsingular, or simply the identity matrix), then the transfer function is turned into the form

$$H(s) = s\,\mathbf{B}^{\mathrm{T}}(s\mathbf{C} + \mathbf{G})^{-1}\mathbf{B}. \tag{8}$$

Leaving out the front factor s, (8) is in the form of (2). In the second linearization (8), matrix-vector products with the matrices $\mathbf{G}^{-1}\mathbf{C}$ and $\mathbf{G}^{-\mathrm{T}}\mathbf{C}^{\mathrm{T}}$ are much easier to handle than the first linearization (5) and (6). These two types of matrix-vector products are needed for forming Krylov subspaces for calculating approximations to $H(s)$ around $s = 0$. In this respect, the first linearization (5) and (6) favors approximations around $s = \infty$. In the case when approximations near a finite point $s_0 \neq 0$ are sought, a shift like (3) must be carried out and then neither linearization has cost advantage over the other because $s_0\mathbf{C} + \mathbf{G}$ matrix is no longer block diagonal for both (5) and (7). However, we point out that if the shift is carried out before linearization, then the same advantage as the second linearization over the first one for approximations near $s = 0$ is retained. Details for this shift-*before*-linearization are discussed in Section 4. ◇

Example 2. The interconnected (coupled) system described in [12] and [18] (see also the chapters by Reis and Stykel and by Vandendorpe and Van Dooren in this book) gives rise to the following transfer function

$$H(s) = L_0^{\mathrm{T}}\,(I - W(s)\mathcal{E})^{-1}\,W(s)B_0,$$

where $\mathcal{E}$ is the subsystem incidence matrix as a glue for connecting all subsystems $H_1(s), \ldots, H_k(s)$ together, and

$$\begin{aligned} W(s) &= \mathrm{diag}(\ H_1(s),\ \ldots,\ H_k(s)\ ) \\ &= \mathrm{diag}(\ L_1^{\mathrm{T}}(sI - A_1)^{-1}B_1,\ \ldots, L_k^{\mathrm{T}}(sI - A_k)^{-1}B_k\ ). \end{aligned}$$

Let $A = \mathrm{diag}(A_1, \ldots, A_k)$, $B = \mathrm{diag}(B_1, \ldots, B_k)$, and $L = \mathrm{diag}(L_1, \ldots, L_k)$, then the transfer function $H(s)$ can be formulated in the form (2), namely

$$H(s) = \mathbf{L}^{\mathrm{T}}(s\mathbf{C} + \mathbf{G})^{-1}\mathbf{B},$$

where $\mathbf{C} = I$, $\mathbf{G} = -A - B\mathcal{E}L$, $\mathbf{B} = BB_0$ and $\mathbf{L} = LL_0$. ◇

Model order reduction of the transfer function $H(s)$ defined by (2) via subspace projection starts by computing matrices

$$\mathcal{X}, \mathcal{Y} \in \mathbb{R}^{N\times n} \quad \text{such that} \quad \mathcal{Y}^{\mathrm{T}}\mathbf{G}\mathcal{X} \text{ is nonsingular,}$$

which then leads to a *reduced-order transfer function*

$$H_{\mathrm{r}}(s) = \mathbf{L}_{\mathrm{r}}^{\mathrm{T}}(s\mathbf{C}_{\mathrm{r}} + \mathbf{G}_{\mathrm{r}})^{-1}\mathbf{B}_{\mathrm{r}}, \tag{9}$$

where

$$\mathbf{C}_{\mathrm{r}} = \mathcal{Y}^{\mathrm{T}}\mathbf{C}\mathcal{X}, \quad \mathbf{G}_{\mathrm{r}} = \mathcal{Y}^{\mathrm{T}}\mathbf{G}\mathcal{X}, \quad \mathbf{B}_{\mathrm{r}} = \mathcal{Y}^{\mathrm{T}}\mathbf{B}, \quad \mathbf{L}_{\mathrm{r}} = \mathcal{X}^{\mathrm{T}}\mathbf{L}. \tag{10}$$

Similarly, the reduced transfer function $H_{\mathrm{r}}(s)$ can be expanded around $s = 0$:

$$\begin{aligned} H_{\mathrm{r}}(s) &= \sum_{\ell=0}^{\infty}(-1)^{\ell}s^{\ell}\mathbf{L}_{\mathrm{r}}^{\mathrm{T}}(\mathbf{G}_{\mathrm{r}}^{-1}\mathbf{C}_{\mathrm{r}})^{\ell}\mathbf{G}_{\mathrm{r}}^{-1}\mathbf{B}_{\mathrm{r}} \\ &= \sum_{\ell=0}^{\infty}(-1)^{\ell}s^{\ell}M_{\mathrm{r},\ell}, \end{aligned}$$

where the matrices

$$M_{\mathrm{r},\ell} = \mathbf{L}_\mathrm{r}^\mathrm{T}(\mathbf{G}_\mathrm{r}^{-1}\mathbf{C}_\mathrm{r})^\ell \mathbf{G}_\mathrm{r}^{-1}\mathbf{B}_\mathrm{r}$$

are referred to as the *moments* of the reduced system.

In practice it is often the case that $n \ll N$. This makes the reduced system matrices $\mathbf{G}_\mathrm{r}$, $\mathbf{C}_\mathrm{r}$, $\mathbf{L}_\mathrm{r}$, and $\mathbf{B}_\mathrm{r}$ much smaller. By choosing $\mathcal{X}$ and $\mathcal{Y}$ properly, the reduced system associated with the reduced transfer function can be forced to resemble the original system sufficiently well to have practical relevance.

The following theorem dictates how well a reduced transfer function approximates the original transfer function. For the case when $\mathbf{G}$ is the identity matrix, the result is due to [19]. The general form as stated in the following theorem was proved by [6]; a new proof in a projection context was given later in [9]. Its implication for structure-preserving model reduction was also first realized in [9].

Theorem 1. *Suppose that* $\mathbf{G}$ *and* $\mathbf{G}_\mathrm{r}$ *are nonsingular. If*

$$\mathcal{K}_k(\mathbf{G}^{-1}\mathbf{C}, \mathbf{G}^{-1}\mathbf{B}) \subseteq \mathrm{span}\{\mathcal{X}\}$$

and

$$\mathcal{K}_j(\mathbf{G}^{-\mathrm{T}}\mathbf{C}^\mathrm{T}, \mathbf{G}^{-\mathrm{T}}\mathbf{L}) \subseteq \mathrm{span}\{\mathcal{Y}\},$$

then the moments of $H(s)$ *and of its reduced function* $H_\mathrm{r}(s)$ *satisfy*

$$M_\ell = M_{\mathrm{r},\ell} \quad \textit{for} \quad 0 \le \ell \le k+j-1,$$

which imply

$$H_\mathrm{r}(s) = H(s) + \mathcal{O}(s^{k+j}).$$

Remark 1. The conditions suggest that by enforcing $\mathrm{span}\{\mathcal{X}\}$ and/or $\mathrm{span}\{\mathcal{Y}\}$ to contain more appropriate Krylov subspaces associated with multiple points, $H_\mathrm{r}(s)$ can be constructed to approximate $H(s)$ sufficiently well near all those points. See [6] and [13, 14] for more details.

Let us now discuss the objectives of structure-preserving model order reduction. For simplicity of exposition, consider system matrices $\mathbf{G}$, $\mathbf{C}$, $\mathbf{B}$, and $\mathbf{L}$ with the following 2×2 block structure

$$\begin{aligned}
&\mathbf{C} = \begin{array}{c} \scriptstyle N_1' \\ \scriptstyle N_2' \end{array}\!\!\overset{\begin{array}{cc}\scriptstyle N_1 & \scriptstyle N_2\end{array}}{\begin{bmatrix} C_{11} & 0 \\ 0 & C_{22} \end{bmatrix}}, \quad \mathbf{G} = \begin{array}{c} \scriptstyle N_1' \\ \scriptstyle N_2' \end{array}\!\!\overset{\begin{array}{cc}\scriptstyle N_1 & \scriptstyle N_2\end{array}}{\begin{bmatrix} G_{11} & G_{12} \\ G_{21} & 0 \end{bmatrix}}, \\
&\mathbf{B} = \begin{array}{c} \scriptstyle N_1' \\ \scriptstyle N_2' \end{array}\!\!\overset{\scriptstyle p}{\begin{bmatrix} B_1 \\ 0 \end{bmatrix}}, \quad \mathbf{L} = \begin{array}{c} \scriptstyle N_1 \\ \scriptstyle N_2 \end{array}\!\!\overset{\scriptstyle m}{\begin{bmatrix} L_1 \\ 0 \end{bmatrix}},
\end{aligned} \tag{11}$$

where $N_1 + N_2 = N_1' + N_2' = N$. System matrices from the *time-domain modified nodal analysis* (MNA) circuit equations take such forms (see Section 4).

A structure-preserving model order reduction technique generates a reduced-order system that structurally preserves the block structure with

$$\mathbf{C}_\mathrm{r} = \begin{matrix} & \begin{matrix} n_1 & n_2 \end{matrix} \\ \begin{matrix} n_1' \\ n_2' \end{matrix} & \begin{bmatrix} C_{\mathrm{r},11} & 0 \\ 0 & C_{\mathrm{r},22} \end{bmatrix} \end{matrix}, \quad \mathbf{G}_\mathrm{r} = \begin{matrix} & \begin{matrix} n_1 & n_2 \end{matrix} \\ \begin{matrix} n_1' \\ n_2' \end{matrix} & \begin{bmatrix} G_{\mathrm{r},11} & G_{\mathrm{r},12} \\ G_{\mathrm{r},21} & 0 \end{bmatrix} \end{matrix},$$
$$\mathbf{B}_\mathrm{r} = \begin{matrix} & p \\ \begin{matrix} n_1' \\ n_2' \end{matrix} & \begin{bmatrix} B_{\mathrm{r},1} \\ 0 \end{bmatrix} \end{matrix}, \quad \mathbf{L}_\mathrm{r} = \begin{matrix} & m \\ \begin{matrix} n_1 \\ n_2 \end{matrix} & \begin{bmatrix} L_{\mathrm{r},1} \\ 0 \end{bmatrix} \end{matrix}, \tag{12}$$

where $n_1 + n_2 = n_1' + n_2' = n$. Furthermore, each sub-block is a direct reduction of the corresponding sub-block in the original system.

In the formulation of subspace projection, this objective of structure-preserving model order reduction can be accomplished by picking the projection matrices

$$\mathcal{X} = \begin{matrix} & \begin{matrix} n_1 & n_2 \end{matrix} \\ \begin{matrix} N_1 \\ N_2 \end{matrix} & \begin{bmatrix} X_1 & \\ & X_2 \end{bmatrix} \end{matrix}, \quad \mathcal{Y} = \begin{matrix} & \begin{matrix} n_1' & n_2' \end{matrix} \\ \begin{matrix} N_1' \\ N_2' \end{matrix} & \begin{bmatrix} Y_1 & \\ & Y_2 \end{bmatrix} \end{matrix}. \tag{13}$$

Then

$$\mathcal{Y}^\mathrm{T}\mathbf{C}\mathcal{X} = \begin{bmatrix} Y_1^\mathrm{T} & \\ & Y_2^\mathrm{T} \end{bmatrix} \begin{bmatrix} C_{11} & 0 \\ 0 & C_{22} \end{bmatrix} \begin{bmatrix} X_1 & \\ & X_2 \end{bmatrix} = \begin{bmatrix} C_{\mathrm{r},11} & 0 \\ 0 & C_{\mathrm{r},22} \end{bmatrix} = \mathbf{C}_\mathrm{r},$$

$$\mathcal{Y}^\mathrm{T}\mathbf{G}\mathcal{X} = \begin{bmatrix} Y_1^\mathrm{T} & \\ & Y_2^\mathrm{T} \end{bmatrix} \begin{bmatrix} G_{11} & G_{12} \\ G_{21} & 0 \end{bmatrix} \begin{bmatrix} X_1 & \\ & X_2 \end{bmatrix} = \begin{bmatrix} G_{\mathrm{r},11} & G_{\mathrm{r},12} \\ G_{\mathrm{r},21} & 0 \end{bmatrix} = \mathbf{G}_\mathrm{r},$$

$$\mathcal{Y}^\mathrm{T}\mathbf{B} = \begin{bmatrix} Y_1^\mathrm{T} & \\ & Y_2^\mathrm{T} \end{bmatrix} \begin{bmatrix} B_1 \\ 0 \end{bmatrix} = \begin{bmatrix} B_{\mathrm{r},1} \\ 0 \end{bmatrix} = \mathbf{B}_\mathrm{r},$$

$$\mathcal{X}^\mathrm{T}\mathbf{L} = \begin{bmatrix} X_1^\mathrm{T} & \\ & X_2^\mathrm{T} \end{bmatrix} \begin{bmatrix} L_1 \\ 0 \end{bmatrix} = \begin{bmatrix} L_{\mathrm{r},1} \\ 0 \end{bmatrix} = \mathbf{L}_\mathrm{r}.$$

For the case when $\mathcal{Y}$ is taken to be the same as $\mathcal{X}$, this idea is exactly the so-called "*split congruence transformations*" in [8]. A discussion of this idea in a general framework was described in [9].

We now discuss a generic algorithm to generate the desired projection matrices $\mathcal{X}$ and $\mathcal{Y}$ as in (13). Suppose that we have computed the basis matrices $\widetilde{X}$ and $\widetilde{Y}$ by, e.g., a block Arnoldi procedure [15], such that

$$\mathcal{K}_k(\mathbf{G}^{-1}\mathbf{C}, \mathbf{G}^{-1}\mathbf{B}) \subseteq \text{span}\left\{\widetilde{X}\right\}$$

and

$$\mathcal{K}_j(\mathbf{G}^{-\mathrm{T}}\mathbf{C}^\mathrm{T}, \mathbf{G}^{-\mathrm{T}}\mathbf{L}) \subseteq \text{span}\left\{\widetilde{Y}\right\}.$$

In general, $\widetilde{X}$ and $\widetilde{Y}$ generated by an Arnoldi process do not have the form as the desired $\mathcal{X}$ and $\mathcal{Y}$ in, e.g., (13). Hence, taking $\mathcal{X} = \widetilde{X}$ and $\mathcal{Y} = \widetilde{Y}$ will not preserve the 2-by-2 block structure presented in the matrices as in (11). So instead of simply taking $\mathcal{X} = \widetilde{X}$ and $\mathcal{Y} = \widetilde{Y}$, we need to seek $\mathcal{X}$ and $\mathcal{Y}$ of the form as in (13) and meanwhile satisfying

$$\operatorname{span}\left\{\widetilde{X}\right\} \subseteq \operatorname{span}\left\{\mathcal{X}\right\} \text{ and } \operatorname{span}\left\{\widetilde{Y}\right\} \subseteq \operatorname{span}\left\{\mathcal{Y}\right\} \tag{14}$$

so that the first $k + j$ moments of $H(s)$ and its reduced function $H_{\mathrm{r}}(s)$ match as claimed by Theorem 1.

This task can be accomplished by the following algorithm that for a given $\widetilde{Z} = \begin{bmatrix} \widetilde{Z}_1 \\ \widetilde{Z}_2 \end{bmatrix}$, computes $\mathcal{Z} = \begin{bmatrix} Z_1 & \\ & Z_2 \end{bmatrix}$ satisfying

$$\operatorname{span}\{\widetilde{Z}\} \subseteq \operatorname{span}\{\mathcal{Z}\}.$$

Algorithm 1

1. Compute Z_i having full column rank such that $\operatorname{span}\{\widetilde{Z}_i\} \subseteq \operatorname{span}\{Z_i\}$;
2. Output $\mathcal{Z} = \begin{bmatrix} Z_1 & \\ & Z_2 \end{bmatrix}$.

Remark 2. There are various ways to realize Step 1: Rank revealing QR decompositions, modified Gram-Schmidt process, or singular value decompositions [3, 4, 7]. For maximum efficiency, one should construct Z_i with as few columns as possible. Note that the smallest possible number is $\operatorname{rank}(\widetilde{Z}_i)$, but one may have to add a few additional columns to ensure the total number of columns in all X_i and those in all Y_i are the same when constructing $\mathcal{X}$ and $\mathcal{Y}$ below in (15).

For convenience, we introduce the notation $\rightsquigarrow$ that transforms $\widetilde{Z}$ to $\mathcal{Z}$, i.e.,

$$\widetilde{Z} = \begin{bmatrix} \widetilde{Z}_1 \\ \widetilde{Z}_2 \end{bmatrix} \rightsquigarrow \mathcal{Z} = \begin{bmatrix} Z_1 & \\ & Z_2 \end{bmatrix} \text{ satisfying } \operatorname{span}\{\widetilde{Z}\} \subseteq \operatorname{span}\{\mathcal{Z}\}.$$

Returning to the subspace embedding objective (14), by Algorithm 1, we partition $\widetilde{X}$ and $\widetilde{Y}$ as

$$\widetilde{X} = \begin{bmatrix} \widetilde{X}_1 \\ \widetilde{X}_2 \end{bmatrix} \text{ and } \widetilde{Y} = \begin{bmatrix} \widetilde{Y}_1 \\ \widetilde{Y}_2 \end{bmatrix},$$

consistently with the block structures in $\mathbf{G}$, $\mathbf{C}$, $\mathbf{L}$, and $\mathbf{B}$, and then transform

$$\begin{bmatrix} \widetilde{X}_1 \\ \widetilde{X}_2 \end{bmatrix} \rightsquigarrow \mathcal{X} = \begin{bmatrix} X_1 & \\ & X_2 \end{bmatrix} \text{ and } \begin{bmatrix} \widetilde{Y}_1 \\ \widetilde{Y}_2 \end{bmatrix} \rightsquigarrow \mathcal{Y} = \begin{bmatrix} Y_1 & \\ & Y_2 \end{bmatrix}. \tag{15}$$

There are numerical more efficient alternatives when further characteristics in the sub-blocks in $\mathbf{G}$ and $\mathbf{C}$ are known. For example, when $\mathbf{G}$ and $\mathbf{C}$ are as in (7) from linearizing a transfer function like (4), X_1 and Y_1 can be computed directly via the Second-Order Arnoldi process (SOAR) [1,2]. More details are given in the next section.

3 Structure of Krylov Subspace and Arnoldi Process

The generic Algorithm 1 presents a way to preserve the sub-block structure in the reduced systems by first computing the bases of the related Krylov subspaces and then splitting the basis matrices. In this section, we discuss a situation where this first-computing-then-splitting technique can be combined into one technique to generate the desired $\mathcal{X}$ and $\mathcal{Y}$ directly. This is made possible by taking advantage of a structural property of Krylov subspaces for certain block matrix. The next theorem was implicitly in [2,16] (see also [9]).

Theorem 2. *Suppose that A and B admit the following partitioning*

$$A = \begin{array}{c} \\ {\scriptstyle N} \\ {\scriptstyle N} \end{array}\!\!\begin{array}{c} \begin{array}{cc} {\scriptstyle N} & {\scriptstyle N} \end{array} \\ \begin{bmatrix} A_{11} & A_{12} \\ \alpha I & 0 \end{bmatrix} \end{array}, \quad B = \begin{array}{c} \\ {\scriptstyle N} \\ {\scriptstyle N} \end{array}\!\!\begin{array}{c} {\scriptstyle p} \\ \begin{bmatrix} B_1 \\ B_2 \end{bmatrix} \end{array}, \tag{16}$$

where α is a scalar. Let a basis matrix $\widetilde{X}$ of the Krylov subspace $\mathcal{K}_k(A, B)$ be partitioned as

$$\widetilde{X} = \begin{array}{c} {\scriptstyle N} \\ {\scriptstyle N} \end{array}\!\!\begin{bmatrix} \widetilde{X}_1 \\ \widetilde{X}_2 \end{bmatrix}.$$

Then

$$\operatorname{span}\{\widetilde{X}_2\} \subseteq \operatorname{span}\{B_2, \widetilde{X}_1\}.$$

In particular, if also $B_2 = 0$, then $\operatorname{span}\{\widetilde{X}_2\} \subseteq \operatorname{span}\{\widetilde{X}_1\}$.

This theorem provides a theoretical foundation to simply compute $\widetilde{X}_1$, then to expand $\widetilde{X}_1$ to X_1 so that $\operatorname{span}\{X_1\} = \operatorname{span}\{B_2, \widetilde{X}_1\}$ (by orthogonalizing the columns ofB_2 against those of $\widetilde{X}_1$), and finally to obtain

$$\mathcal{X} = \begin{bmatrix} X_1 & \\ & X_1 \end{bmatrix}.$$

In practice, X_1 can be computed directly by a structured Arnoldi process, referred to as the *second-order Arnoldi process* (SOAR) in [1,2,16], as given below.

Algorithm 2 Structured Arnoldi process (framework)

Input: A **and** B **as in (16).**
Output: $\widetilde{X}_1$ **as in Theorem 2 and** X_1 **with** $\text{span}\{X_1\} = \text{span}\{B_2, \widetilde{X}_1\}$.

$B_1 = Q_1 R$ **(QR decomposition)**
$P_1 = \alpha B_2 R_2^{-1}$
for $j = 1, 2, \ldots, k$ **do**
  $T = A_{11} Q_j + A_{12} P_j$
  $S = \alpha Q_j$
  for $i = 1, 2, \ldots, j$ **do**
    $Z = Q_i^{\mathrm{T}} T$
    $T = T - Q_i Z$
    $S = S - P_i Z$
  enddo
  $T = Q_j R$ **(QR decomposition)**
  $P_j = S R^{-1}$
enddo
$\widetilde{X}_1 = [Q_1, Q_2, \ldots, Q_k]$
$T = B_2;$
for $j = 1, 2, \ldots, k$ **do**
  $Z = Q_j^{\mathrm{T}} T$
  $T = T - Q_j Z$
enddo
$T = QR$ **(QR decomposition)**
$X_1 = [\widetilde{X}_1, Q]$

Remark 3. Algorithm 2 is a simplified version to illustrate the key ingredients. Practical implementation will have to incorporate the possibility when various QR decompositions produce (nearly) singular upper triangular matrices R.

4 RCL and RCS Systems

4.1 Basic Equations

The MNA (*modified nodal analysis*) formulation [20] of an RCL circuit network in the frequency domain is of the form

$$\begin{cases} \left(s \begin{bmatrix} C & 0 \\ 0 & L \end{bmatrix} + \begin{bmatrix} G & E \\ -E^{\mathrm{T}} & 0 \end{bmatrix} \right) \begin{bmatrix} v(s) \\ i(s) \end{bmatrix} = \begin{bmatrix} B_v \\ 0 \end{bmatrix} u(s), \\ \qquad\qquad\qquad\qquad y(s) = \begin{bmatrix} D_v^{\mathrm{T}} & 0 \end{bmatrix} \begin{bmatrix} v(s) \\ i(s) \end{bmatrix}, \end{cases} \tag{17}$$

where $v(s)$ and $i(s)$ denote the N_1 nodal voltages and the N_2 auxiliary branch currents, respectively; u and y are the input current sources and output voltages; B_v and D_v denote the incidence matrices for the input current sources and output node voltages; C, L and G represent the contributions of the capacitors, inductors and resistors, respectively; and E is the incidence matrix for the inductances.

When an RCL network is modeled with a 3-D extraction method for interconnection analysis, the resulted inductance matrix L is usually very large and dense [10]. This may cause major difficulties for the subsequent simulation process. As an alternative approach, we can use the susceptance matrix $S = L^{-1}$, which is sparse after dropping small entries [5, 22]. The resulting equations are called *the RCS equations*:

$$\begin{cases} \left(s \begin{bmatrix} C & 0 \\ 0 & I \end{bmatrix} + \begin{bmatrix} G & E \\ -SE^{\mathrm{T}} & 0 \end{bmatrix} \right) \begin{bmatrix} v(s) \\ i(s) \end{bmatrix} = \begin{bmatrix} B_v \\ 0 \end{bmatrix} u(s), \\ \qquad\qquad y(s) = \begin{bmatrix} D_v^{\mathrm{T}} & 0 \end{bmatrix} \begin{bmatrix} v(s) \\ i(s) \end{bmatrix}. \end{cases} \tag{18}$$

Accordingly, the equations in (17) are called *the RCL equations*.

Eliminating the branch current variable $i(s)$ of the RCL and RCS equations in (17) and (18), we have the so-called *second-order form*

$$\begin{cases} \left(sC + G + \frac{1}{s}\Gamma \right) v(s) = B_v u(s), \\ \qquad\qquad y(s) = D_v^{\mathrm{T}} v(s), \end{cases} \tag{19}$$

with

$$\Gamma = EL^{-1}E^{\mathrm{T}} = ESE^{\mathrm{T}}.$$

The transfer function $H(s)$ of the RCL and RCS equations in (17) and (18) can thus be rewritten as

$$H(s) = D_v^{\mathrm{T}} \left(sC + G + \frac{1}{s}\Gamma \right)^{-1} B_v. \tag{20}$$

Perform the shift (3) to obtain

$$\begin{aligned} H(s) &= s\, D_v^{\mathrm{T}} (s^2 C + sG + \Gamma)^{-1} B_v \\ &= (s_0 + \sigma)\, D_v^{\mathrm{T}} [\sigma^2 C + \sigma(2s_0 C + G) + (s_0^2 C + s_0 G + \Gamma)]^{-1} B_v \\ &= (s_0 + \sigma)\, \mathbf{L}^{\mathrm{T}} (\sigma \mathbf{C} + \mathbf{G})^{-1} \mathbf{B}, \end{aligned}$$

where

$$\mathbf{C} = \begin{bmatrix} G_0 & C \\ W & 0 \end{bmatrix}, \quad \mathbf{G} = \begin{bmatrix} \Gamma_0 & 0 \\ 0 & -W \end{bmatrix}, \quad \mathbf{L} = \begin{bmatrix} D_v \\ 0 \end{bmatrix}, \quad \mathbf{B} = \begin{bmatrix} B_v \\ 0 \end{bmatrix}, \tag{21}$$

and $G_0 = 2s_0 C + G$, $\Gamma_0 = s_0^2 C + s_0 G + \Gamma$ and W is a free to chose *nonsingular* matrix.

4.2 Model Order Reduction

The SPRIM method described in the chapter by Freund in this book provides a structure-preserving model order reduction method for the RCL equations in (17).

In this section, we discuss an alternative structure-preserving method for the RCL equations in (17) and the the RCS equations in (18) using the framework presented in Sections 2 and 3. The method is referred to as *the SAPOR method*, initially published in [11, 17]. The SAPOR method exploits the second-order form (19).

For the system matrices $\mathbf{C}$, $\mathbf{G}$ and $\mathbf{B}$ in (21), we have[2]

$$\mathbf{G}^{-1}\mathbf{C} = \begin{bmatrix} \Gamma_0^{-1}G_0 & \Gamma_0^{-1}C \\ -I & 0 \end{bmatrix}, \ \mathbf{G}^{-1}\mathbf{B} = \begin{bmatrix} \Gamma_0^{-1}B_v \\ 0 \end{bmatrix}.$$

They have the exact block structures required by Theorem 2. Apply Algorithm 2 to compute X_r with orthonormal columns such that

$$\mathcal{K}_k(\mathbf{G}^{-1}\mathbf{C}, \mathbf{G}^{-1}\mathbf{B}) \subset \operatorname{span}\left\{ \begin{bmatrix} X_\mathrm{r} & \\ & X_\mathrm{r} \end{bmatrix} \right\}, \tag{22}$$

which is required by Theorem 1 for matching the first k moments. The framework of the projection technique in Section 2 can also be viewed as a change-of-variables

$$v(s) \approx X_\mathrm{r} v_\mathrm{r}(s), \tag{23}$$

where $v_\mathrm{r}(s)$ is a vector of dimension n. Substituting (23) into (19), and multiplying the first equation in (19) by X_r^T from the left yields the reduced-order model of the second-order equations in (19):

$$\begin{cases} \left(sC_\mathrm{r} + G_\mathrm{r} + \dfrac{1}{s}\Gamma_\mathrm{r} \right) v_\mathrm{r}(s) = B_{\mathrm{r},v} u(s), \\ \qquad\qquad\qquad\qquad\quad \widetilde{y}(s) = D_{\mathrm{r},v}^\mathrm{T} v_\mathrm{r}(s), \end{cases} \tag{24}$$

where

$$C_\mathrm{r} = X_\mathrm{r}^\mathrm{T} C X_\mathrm{r}, \ G_\mathrm{r} = X_\mathrm{r}^\mathrm{T} G X_\mathrm{r}, \ \Gamma_\mathrm{r} = E_\mathrm{r}^\mathrm{T} \Gamma E_\mathrm{r}, \ E_\mathrm{r} = X_\mathrm{r}^\mathrm{T} E, \tag{25}$$

and

$$B_{\mathrm{r},v} = X_\mathrm{r}^\mathrm{T} B_v, \ D_{\mathrm{r},v} = X_\mathrm{r}^\mathrm{T} D_v.$$

The transfer function of the reduced system (24) is given by

$$H_\mathrm{r}(s) = D_{\mathrm{r},v}^\mathrm{T} \left(sC_\mathrm{r} + G_\mathrm{r} + \frac{1}{s}\Gamma_\mathrm{r} \right)^{-1} B_{\mathrm{r},v}. \tag{26}$$

The reduced second-order form (24) corresponds to a reduced order system of the original RCS equations in (18). This can be seen by setting

$$\mathcal{X} = \mathcal{Y} = \begin{array}{c} \\ {\scriptstyle N_1} \\ {\scriptstyle N_2} \end{array} \!\!\begin{array}{c} \begin{array}{cc} {\scriptstyle n} & {\scriptstyle N_2} \end{array} \\ \begin{bmatrix} X_\mathrm{r} & \\ & I \end{bmatrix} \end{array},$$

[2] To preserve the symmetry in C, G, and Γ as by (25), we do not need a Krylov subspace of $\mathbf{G}^{-\mathrm{T}}\mathbf{C}^\mathrm{T}$ on $\mathbf{G}^{-\mathrm{T}}\mathbf{L}$.

and then projecting the original RCS equations in (18) as in Section 2 to obtain the reduced order equations

$$\begin{cases} \left(s\begin{bmatrix} C_{\rm r} & 0 \\ 0 & I \end{bmatrix} + \begin{bmatrix} G_{\rm r} & E_{\rm r} \\ -SE_{\rm r}^{\rm T} & 0 \end{bmatrix}\right)\begin{bmatrix} v_{\rm r}(s) \\ \widetilde{i}(s) \end{bmatrix} = \begin{bmatrix} B_{{\rm r},v} \\ 0 \end{bmatrix} u(s), \\ \qquad\qquad\qquad\qquad \widetilde{y}(s) = \begin{bmatrix} D_{{\rm r},v}^{\rm T} & 0 \end{bmatrix}\begin{bmatrix} v_{\rm r}(s) \\ \widetilde{i}(s) \end{bmatrix}. \end{cases} \tag{27}$$

Note that $\widetilde{i}(s)$ is a vector with N_2 components, corresponding with the number of original auxiliary branch currents $i(s)$.

4.3 Towards a Synthesized System

The reduced system (27) preserves the block structures and the symmetry of system data matrices of the original RCS system (18). However, the matrix $E_{\rm r}$ in the reduced-order RCS system (27) cannot be interpreted as an incidence matrix. Towards the objective of synthesis based on the reduced-order model, we shall reformulate the projection (23) and the reduced-order system (24). This work was first published in [21]. Here we derive a rigorous mathematical formulation of the approach.

We begin with the original RCS equations in (18). Let

$$\widehat{i}(s) = E\, i(s). \tag{28}$$

Then the RCS equations in (18) can be written as

$$\begin{cases} \left(s\begin{bmatrix} C & 0 \\ 0 & I \end{bmatrix} + \begin{bmatrix} G & I \\ -\Gamma & 0 \end{bmatrix}\right)\begin{bmatrix} v(s) \\ \widehat{i}(s) \end{bmatrix} = \begin{bmatrix} B_v \\ 0 \end{bmatrix} u(s), \\ \qquad\qquad\qquad\qquad y(s) = \begin{bmatrix} D_v^{\rm T} & 0 \end{bmatrix}\begin{bmatrix} v(s) \\ \widehat{i}(s) \end{bmatrix}. \end{cases} \tag{29}$$

With the change-of-variables (28), the incidence matrix E in the original RCS equations in (18) is now the identity matrix I in (29). The matrix Γ plays the role of the susceptance matrix. An identity incidence matrix can be interpreted as "*self-inductance*", although the susceptance matrix Γ is not diagonal yet. We will discuss how to do so later in this subsection.

Note that the new current vector $\widehat{i}(s)$ is of the size N_1, typically $N_1 \geq N_2$. The order of the new RCS equations in (29) is $2N_1$. The equations in (18) and (29) have the same voltage variables and the same output. However, they are not equivalent since the current variables $i(s)$ cannot be recovered from $\widehat{i}(s)$. The reformulated equations in (29) are referred to as *the expanded RCS equations*, or RCSe for short.

In first-order form, the transfer function $H(s)$ of the RCSe equations in (29) is given by

$$H(s) = \mathbf{L}^{\rm T}(s\mathbf{C} + \mathbf{G})^{-1}\mathbf{B}, \tag{30}$$

where $\mathbf{G}$ and $\mathbf{C}$ are $2N_1 \times 2N_1$:

$$\mathbf{C} = \begin{bmatrix} C & 0 \\ 0 & I \end{bmatrix}, \quad \mathbf{G} = \begin{bmatrix} G & I \\ -\Gamma & 0 \end{bmatrix},$$

and

$$\mathbf{B} = \begin{bmatrix} B_v \\ 0 \end{bmatrix}, \quad \mathbf{L} = \begin{bmatrix} D_v \\ 0 \end{bmatrix}.$$

For the reduced-order model of the RCSe equations in (29), we define

$$\mathcal{X} = \mathcal{Y} = \begin{array}{c} \\ N_1 \\ N_1 \end{array} \overset{\begin{array}{cc} n & n \end{array}}{\begin{bmatrix} X_{\mathrm{r}} & \\ & X_{\mathrm{r}} \end{bmatrix}}. \tag{31}$$

Then by the change-of-variables

$$v(s) \approx X_{\mathrm{r}} v_{\mathrm{r}}(s) \quad \text{and} \quad \widehat{i}(s) \approx X_{\mathrm{r}} i_{\mathrm{r}}(s), \tag{32}$$

and using the projection procedure in Section 2, we have the reduced-order RCSe equations

$$\begin{cases} \left(s \begin{bmatrix} C_{\mathrm{r}} & 0 \\ 0 & I \end{bmatrix} + \begin{bmatrix} G_{\mathrm{r}} & I \\ -\Gamma_{\mathrm{r}} & 0 \end{bmatrix} \right) \begin{bmatrix} v_{\mathrm{r}}(s) \\ i_{\mathrm{r}}(s) \end{bmatrix} = \begin{bmatrix} B_{\mathrm{r},v} \\ 0 \end{bmatrix} u(s), \\ \qquad\qquad\qquad \widetilde{y}(s) = \begin{bmatrix} D_{\mathrm{r},v}^{\mathrm{T}} & 0 \end{bmatrix} \begin{bmatrix} v_{\mathrm{r}}(s) \\ i_{\mathrm{r}}(s) \end{bmatrix}. \end{cases} \tag{33}$$

Compared with the RCSe equations in (29), the reduced equations in (33) not only preserve the 2-by-2 block structure of the system data matrices $\mathbf{G}$ and $\mathbf{C}$, but also preserve the identity of the incidence matrix.

For the objective of synthesis of the original RCL and RCS equations in (17) and (18), let us further consider the structures of the input and output matrices and the incidence matrix. Without loss of generality, we assume that the sub-blocks B_v and D_v in the input and output of the RCS equations in (18) are of the form:

$$B_v = \begin{array}{c} \\ p_1 \\ N_1 - p_1 \end{array} \overset{p}{\begin{bmatrix} B_{v1} \\ 0 \end{bmatrix}}, \quad D_v = \begin{array}{c} \\ p_1 \\ N_1 - p_1 \end{array} \overset{m}{\begin{bmatrix} D_{v1} \\ 0 \end{bmatrix}}. \tag{34}$$

This indicates that there are totally p_1 different input and output nodes. Alternatively, we can reorder the nodes in the RLC/RCS circuit network such that B_v and D_v are in the desired forms.

Furthermore, we assume that the incidence matrix E in (18) has the zero block on the top, conformal with the partition of the input and output matrices in (34):

$$E = \begin{array}{c} \\ p_1 \\ N_1 - p_1 \end{array} \overset{N_2}{\begin{bmatrix} 0 \\ \widetilde{E} \end{bmatrix}}. \tag{35}$$

This assumption means that there is no susceptance (inductor) directly connecting to the input and output nodes [21].

With the assumptions in (34) and (35), let X_{r} be an orthonormal basis for the projection subspace in (22). Using partitioning-and-embedding steps in Algorithm 1 of Section 2, we have that

$$X_{\mathrm{r}} = \begin{matrix} & n \\ {\scriptstyle p_1} & \\ {\scriptstyle N_1-p_1} & \end{matrix}\!\!\begin{bmatrix} X_{\mathrm{r}}^{(1)} \\ X_{\mathrm{r}}^{(2)} \end{bmatrix} \rightsquigarrow \widehat{X}_{\mathrm{r}} = \begin{matrix} {\scriptstyle p_1} \\ {\scriptstyle N_1-p_1} \end{matrix}\!\!\overset{\begin{matrix} p_1 & n \end{matrix}}{\begin{bmatrix} I & \\ & X_2 \end{bmatrix}},$$

where the columns of X_2 form an orthonormal basis for the range of $X_{\mathrm{r}}^{(2)}$. For simplicity, we assume that there is no deflation, namely, $\mathrm{rank}(X_{\mathrm{r}}^{(2)}) = \mathrm{rank}(X_2) = n$.

Similarly to (31) and (32), using the subspace projection with

$$\mathcal{X} = \mathcal{Y} = \begin{matrix} {\scriptstyle N_1} \\ {\scriptstyle N_1} \end{matrix}\!\!\overset{\begin{matrix} p_1+n & p_1+n \end{matrix}}{\begin{bmatrix} \widehat{X}_{\mathrm{r}} & \\ & \widehat{X}_{\mathrm{r}} \end{bmatrix}},$$

we have the reduced-order RCSe equations

$$\begin{cases} \left(s \begin{bmatrix} C_{\mathrm{r}} & 0 \\ 0 & I \end{bmatrix} + \begin{bmatrix} G_{\mathrm{r}} & I \\ -\Gamma_{\mathrm{r}} & 0 \end{bmatrix} \right) \begin{bmatrix} v_{\mathrm{r}}(s) \\ i_{\mathrm{r}}(s) \end{bmatrix} = \begin{bmatrix} B_{\mathrm{r},v} \\ 0 \end{bmatrix} u(s), \\ \qquad\qquad\qquad\qquad \widetilde{y}(s) = \begin{bmatrix} D_{\mathrm{r},v}^{\mathrm{T}} & 0 \end{bmatrix} \begin{bmatrix} v_{\mathrm{r}}(s) \\ i_{\mathrm{r}}(s) \end{bmatrix}, \end{cases} \tag{36}$$

where C_{r}, G_{r}, and Γ_{r} are $(p_1+n) \times (p_1+n)$ matrices:

$$C_{\mathrm{r}} = \widehat{X}_{\mathrm{r}}^{\mathrm{T}} C \widehat{X}_{\mathrm{r}}, \; G_{\mathrm{r}} = \widehat{X}_{\mathrm{r}}^{\mathrm{T}} G \widehat{X}_{\mathrm{r}}, \; \Gamma_{\mathrm{r}} = \widehat{X}_{\mathrm{r}}^{\mathrm{T}} \Gamma \widehat{X}_{\mathrm{r}}.$$

The input and output sub-block matrices $B_{\mathrm{r},v}$ and $D_{\mathrm{r},v}$ preserve the structure in (34):

$$B_{\mathrm{r},v} = \widehat{X}_{\mathrm{r}}^{\mathrm{T}} \begin{bmatrix} B_{v1} \\ 0 \end{bmatrix} = \begin{matrix} {\scriptstyle p_1} \\ {\scriptstyle n} \end{matrix}\!\!\overset{p}{\begin{bmatrix} B_{v1} \\ 0 \end{bmatrix}}, \quad D_{\mathrm{r},v} = \widehat{X}_{\mathrm{r}}^{\mathrm{T}} \begin{bmatrix} D_{v1} \\ 0 \end{bmatrix} = \begin{matrix} {\scriptstyle p_1} \\ {\scriptstyle n} \end{matrix}\!\!\overset{m}{\begin{bmatrix} D_{v1} \\ 0 \end{bmatrix}}.$$

Note that

$$\mathrm{span}\left\{ \begin{bmatrix} X_{\mathrm{r}} & \\ & X_{\mathrm{r}} \end{bmatrix} \right\} \subseteq \mathrm{span}\left\{ \begin{bmatrix} \widehat{X}_{\mathrm{r}} & \\ & \widehat{X}_{\mathrm{r}} \end{bmatrix} \right\}.$$

The reduced RCSe system (36) also preserves the moment-matching property for the system (33) by Theorem 1.

Now we turn to the diagonalization of Γ in the RCSe equations in (29) for the objective of synthesis. The assumption (35) for the incidence matrix E implies that Γ is of the form

$$\Gamma = EL^{-1}E^{\mathrm{T}} = \begin{matrix} & p_1 & N_1-p_1 \\ p_1 & 0 & 0 \\ N_1-p_1 & 0 & \widetilde{\Gamma} \end{matrix}.$$

It can be seen that in the reduced RCSe equations in (36), Γ_{r} has the same form

$$\Gamma_{\mathrm{r}} = \begin{matrix} & p_1 & n \\ p_1 & 0 & 0 \\ n & 0 & \widetilde{\Gamma}_{\mathrm{r}} \end{matrix},$$

where $\widetilde{\Gamma}_{\mathrm{r}} = X_2^{\mathrm{T}}\widetilde{\Gamma}X_2$. Note that $\widetilde{\Gamma}$ is symmetric semi-positive definite, and so is $\widetilde{\Gamma}_{\mathrm{r}}$. Let

$$\widetilde{\Gamma}_{\mathrm{r}} = \widetilde{V}\Lambda\widetilde{V}^{\mathrm{T}}$$

be the eigen-decomposition of $\widetilde{\Gamma}_{\mathrm{r}}$, where V is orthogonal and Λ is diagonal. Define

$$V = \begin{matrix} & p_1+n & p_1+n \\ p_1+n & \widehat{V} & \\ p_1+n & & \widehat{V} \end{matrix},$$

where

$$\widehat{V} = \begin{matrix} & p_1 & n \\ p_1 & I & \\ n & & \widetilde{V} \end{matrix}.$$

Then by a congruence transformation using the matrix V, the reduced-order RCSe equations in (36) is equivalent to the equations

$$\begin{cases} \left(s\begin{bmatrix} \widehat{C}_{\mathrm{r}} & 0 \\ 0 & I \end{bmatrix} + \begin{bmatrix} \widehat{G}_{\mathrm{r}} & I \\ -\widehat{\Gamma}_{\mathrm{r}} & 0 \end{bmatrix}\right)\begin{bmatrix} \widehat{v}_{\mathrm{r}}(s) \\ \widehat{i}_{\mathrm{r}}(s) \end{bmatrix} = \begin{bmatrix} \widehat{B}_{\mathrm{r},v} \\ 0 \end{bmatrix} u(s), \\ \widehat{y}(s) = \begin{bmatrix} \widehat{D}_{\mathrm{r},v}^{\mathrm{T}} & 0 \end{bmatrix}\begin{bmatrix} \widehat{v}_{\mathrm{r}}(s) \\ \widehat{i}_{\mathrm{r}}(s) \end{bmatrix}, \end{cases} \tag{37}$$

where $\widehat{v}_{\mathrm{r}}(s) = \widehat{V}^{\mathrm{T}}v_{\mathrm{r}}(s)$ and $\widehat{i}_{\mathrm{r}}(s) = \widehat{V}^{\mathrm{T}}i_{\mathrm{r}}(s)$. $\widehat{C}_{\mathrm{r}}$, $\widehat{G}_{\mathrm{r}}$, and $\widehat{\Gamma}_{\mathrm{r}}$ are $(p_1+n)\times(p_1+n)$ matrices:

$$\widehat{C}_{\mathrm{r}} = \widehat{V}^{\mathrm{T}}C_{\mathrm{r}}\widehat{V}, \quad \widehat{G}_{\mathrm{r}} = \widehat{V}^{\mathrm{T}}G_{\mathrm{r}}\widehat{V}, \quad \widehat{\Gamma}_{\mathrm{r}} = \widehat{V}^{\mathrm{T}}\Gamma_{\mathrm{r}}\widehat{V}.$$

Moreover, with V block diagonal, the input and output structures are preserved too:

$$\widehat{B}_{\mathrm{r},v} = \widehat{V}^{\mathrm{T}}B_{\mathrm{r},v} = \begin{matrix} & p \\ p_1 & B_{v1} \\ n & 0 \end{matrix}, \quad \widehat{D}_{\mathrm{r},v} = \widehat{V}^{\mathrm{T}}D_{\mathrm{r},v} = \begin{matrix} & p \\ p_1 & D_{v1} \\ n & 0 \end{matrix}.$$

We note that after the congruence transformation, $\widehat{\Gamma}_r$ is diagonal

$$\widehat{\Gamma}_{\rm r} = \begin{array}{c} \\ p_1 \\ n \end{array} \overset{\begin{array}{cc} p_1 & n \end{array}}{\begin{bmatrix} 0 & 0 \\ 0 & \Lambda \end{bmatrix}}$$

Therefore, to avoid large entries in the synthesized inductors for synthesized RCL equations, we partition the eigenvalue matrix Λ of $\widetilde{\Gamma}_{\rm r}$ into

$$\Lambda = \begin{array}{c} \ell \\ n-\ell \end{array} \overset{\begin{array}{cc} \ell & n-\ell \end{array}}{\begin{bmatrix} \Lambda_1 & \\ & \Lambda_2 \end{bmatrix}},$$

where Λ_2 contains the $n-\ell$ smallest eigenvalues that are smaller than a given threshold ϵ in magnitude. Setting $\Lambda_2 = 0$, we derive reduced RCSe equations of the same form as in (37) with the "susceptance" matrix

$$\widehat{\Gamma}_{\rm r} = \begin{array}{c} p_1 \\ \ell \\ n-\ell \end{array} \overset{\begin{array}{ccc} p_1 & \ell & n-\ell \end{array}}{\begin{bmatrix} 0 & & \\ & \Lambda_1 & \\ & & 0 \end{bmatrix}}.$$

Subsequently, we define reduced-order equations to resemble the RCL form (17):

$$\begin{cases} \left(s \begin{bmatrix} \widehat{C}_{\rm r} & 0 \\ 0 & \widehat{L}_{\rm r} \end{bmatrix} + \begin{bmatrix} \widehat{G}_{\rm r} & I \\ -I & 0 \end{bmatrix} \right) \begin{bmatrix} \widehat{v}_{\rm r}(s) \\ \widehat{i}_{\rm r}(s) \end{bmatrix} = \begin{bmatrix} B_{{\rm r},v} \\ 0 \end{bmatrix} u(s), \\ \widetilde{y}(s) = \begin{bmatrix} D_{{\rm r},v}^{\rm T} & 0 \end{bmatrix} \begin{bmatrix} \widehat{v}_{\rm r}(s) \\ \widehat{i}_{\rm r}(s) \end{bmatrix}, \end{cases} \tag{38}$$

where the inductance matrix $\widehat{L}_{\rm r}$ is given by

$$\widehat{L}_{\rm r} = \begin{array}{c} p_1 \\ \ell \\ n-\ell \end{array} \overset{\begin{array}{ccc} p_1 & \ell & n-\ell \end{array}}{\begin{bmatrix} 0 & & \\ & \Lambda_1^{-1} & \\ & & 0 \end{bmatrix}}.$$

Since $\widehat{L}_{\rm r}$ is diagonal, there is no inductance loop in the synthesized network. We refer to the equations in (38) as *the synthesized RCL equations* of the original RCL equations in (17). The use of the equation (38) for equivalent RLC circuit synthesis is discussed in [21].

Example 3. We consider a 64-bit bus circuit network with 8 inputs and 8 outputs. The order N of the corresponding RCL model is $N = 16963$. By the structure-preserving model order reduction described in this section, we obtain a reduced-order RLC equations of the form (38), with order $n = 640$.

SPICE transient analysis is performed on the original RLC circuit and on the synthesized circuit (38) with excitations of pulse current sources at eight inputs.

The transient simulation results are shown in Figure 1. The transient response of the synthesized circuit is visually indistinguishable from that of the original RLC circuit. SPICE AC analysis was also performed on the original RLC circuit and on the synthesized RLC circuit with current excitation at the near end of the first line. The voltage at the far end of the first line is considered as the observing point. The AC simulation results are shown in Figure 2. We see that two curves are visually

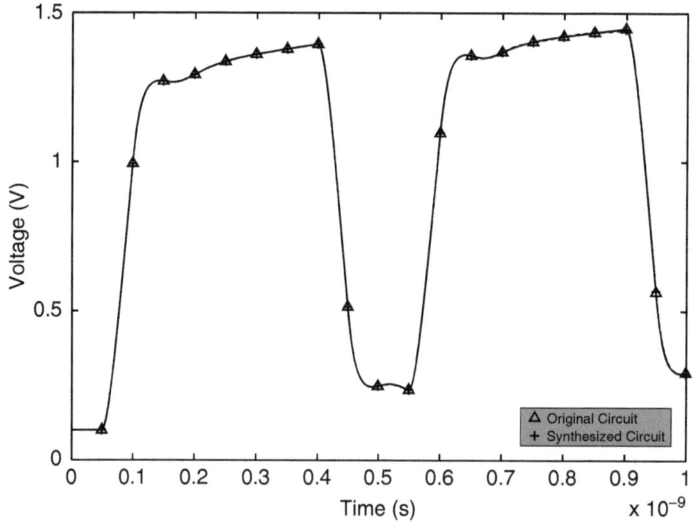

Fig. 1. Transient analysis of the bus circuit

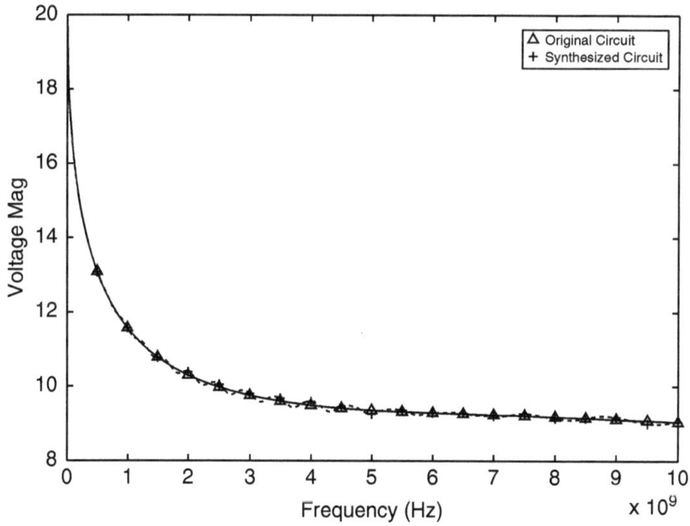

Fig. 2. AC analysis of the bus circuit

indistinguishable. The CPU elapsed time for the transient and AC analysis are shown in the following table

	Full RCL	Synthesized RCL
Dimensionality	16963	640
Transient analysis	5007.59 (sec.)	90.16 (sec.)
AC analysis	29693.02 (sec.)	739.29 (sec.)

From the table, we see that with the reduced RCL equations a factor of 50 of speedup for the transient analysis and a factor of 40 of speedup for the AC analysis has been achieved. ◇

Acknowledgment

Bai is supported in part by NSF grant DMS-0611548. Li is supported in part by NSF Grant DMS-0702335. Su is supported in part by Shanghai Dawn Project 200601 and China NSF project 90307017.

References

1. Z. Bai and Y. Su. Dimension reduction of large-scale second-order dynamical systems via a second-order Arnoldi method. *SIAM J. Sci. Comput.*, 25(5):1692–1709, 2005.
2. Z. Bai and Y. Su. SOAR: A second-order Arnoldi method for the solution of the quadratic eigenvalue problem. *SIAM J. Matrix Anal. Appl.*, 26(3):640–659, Mar. 2005.
3. Ake Björck. *Numerical Methods for Least Squares Problems*. SIAM, Philadelphia, 1996.
4. J. Demmel. *Applied Numerical Linear Algebra*. SIAM, Philadelphia, 1997.
5. A. Devgan, H. Ji, and W. Dai. How to efficiently capture on-chip inductance effects: Introducing a new circuit element K. In *Proceedings of IEEE/ACM ICCAD 2000*, pages 150–155, 2000.
6. E.J. Grimme. *Krylov Projection Methods for model reduction*. PhD thesis, Univ. Illinois, Urbana-Champaign, 1997.
7. G. H. Golub and C. F. Van Loan. *Matrix Computations*. Johns Hopkins University Press, Baltimore, Maryland, 3rd edition, 1996.
8. Kevin J. Kerns and Andrew T. Yang. Preservation of passivity during RLC network reduction via split congruence transformations. *IEEE Trans. Computer-Aided Design of Integrated Circuits and Systems*, 17(7):582–591, July 1998.
9. Ren-Cang Li and Zhaojun Bai. Structure-preserving model reduction. *Comm. Math. Sci.*, 3(2):179–199, 2005.
10. D. Ling and A. Ruehli. *Circuit Analysis, Simulation and Design - Advances in CAD for VLSI*, volume 3, chapter 11. Elsevier Science Publisher, 1987.
11. B. Liu, X. Zeng, and Y. Su. Block SAPOR: Block second-order arnoldi method for passive order reduction of multi-input multi-output RCS interconnect circuits. In *Proc. IEEE/ACM ASP-DAC*, pages 244–249, 2005.
12. T. Reis and T. Stykel. Stability analysis and model order reduction of coupled systems. *Math. Comput. Model. Dyn. Syst.*, 13:413–436, 2007.
13. Axel Ruhe. Rational Krylov sequence methods for eigenvalue computation. *Linear Algebra Appl.*, 58:391–405, 1984.

14. Axel Ruhe. Rational Krylov algorithms for nonsymmetric eigenvalue problems. II. matrix pairs. *Linear Algebra Appl.*, 197-198:283–295, 1994.
15. Yousef Saad. *Iterative Methods for Sparse Linear Systems*. SIAM, Philadelphia, 2nd edition, 2003.
16. T.-J. Su and R. R. Craig. Model reduction and control of flexible structures using Krylov vectors. *J. Guidance, Control, and Dynamics*, 14(2):260–267, 1991.
17. Y. Su, J. Wang, X. Zeng, Z. Bai, C. Chiang, and D. Zhou. SAPOR: Second-order arnoldi method for passive order reduction of RCS circuits. In *Proc. IEEE/ACM Int. Conf. Computer Aided Design*, pages 74–79, Nov. 2004.
18. A. Vandendorpe and P. Van Dooren. On model reduction of interconnected systems. Proceedings of the 16th International Symposium of Mathematical Theory of Networks and Systems, Leuven, Belgium, 2004.
19. C. E. Villemagne and R. E. Skelton. Model reduction using a projection formulation. *Internat. J. Control*, 46(6):2141–2169, 1987.
20. J. Vlach and K. Singhal. *Computer Methods for Circuit Analysis and Design*. Van Nostrand Reinhold, New York, second edition, 1994.
21. F. Yang, X. Zeng, Y. Su, and D. Zhou. RLC equivalent circuit synthesis method for structure-preserved reduced-order model of interconnect in VLSI. *Commun. Comput. Phys.*, 3:376–396, February 2008.
22. H. Zheng, B. Krauter, M. Beattie, and L. Pileggi. Window-based susceptance models for large-scale RLC circuit analyses. In *Proceedings of IEEE/ACM DATE 2002*, pages 628–633, 2002.

Model Reduction via Proper Orthogonal Decomposition

René Pinnau

Fachbereich Mathematik, Technische Universität Kaiserslautern, D-67663 Kaiserslautern, Germany
pinnau@mathematik.uni-kl.de

1 Introduction

In many fields of science and engineering, like fluid or structural mechanics and electric circuit design, large–scale dynamical systems need to be simulated, optimized or controlled. They are often given by discretizations of systems of nonlinear partial differential equations yielding high–dimensional discrete phase spaces. For this reason during the last decades research was mainly focused on the development of sophisticated analytical and numerical tools to understand the overall system behavior. Not surprisingly, the number of degrees of freedom for simulations kept pace with the increasing computing power. But when it comes to optimal design or control the problems are in general to large to be tackled with standard techniques. Hence, there is a strong need for *model reduction techniques* to reduce the computational costs and storage requirements. They should yield low–dimensional approximations for the full high–dimensional dynamical system, which reproduce the characteristic dynamics of the system.

In this work, we present a method known as *proper orthogonal decomposition* (POD), which is widely discussed in literature during the last decades. The original concept goes back to *Pearson* [28]. The method is also known as Karhunen–Loève decomposition [15, 22] or principal component analysis [13]. Further names are factor analysis or total least–squares estimation. It provides an optimally ordered, orthonormal basis in the least–squares sense for a given set of theoretical, experimental or computational data [2]. Reduced order models or surrogate models are then obtained by truncating this optimal basis. Clearly, the choice of the data set plays a crucial role and relies either on guesswork, intuition or simulations. Most prominent is the *method of snapshots* introduced by Sirovich [36]. Here, the data set is chosen as time snapshots containing the spatial distribution of a numerical simulation at certain time instances reflecting the system dynamics.

As an a posteriori, data dependent method it does not need a priori knowledge of the system behavior and can also be used to analyze patterns in data. Due to this fact, it was intensively used to study turbulence phenomena and coherent structures in fluid dynamics [4, 14, 32, 35, 36] as well as in signal analysis and pattern recognition [9, 33]. More recently, it has been used in optimal control of partial differential equations [1, 8, 10, 12, 16–19, 31], inverse problems in structural dynamics [6] and controller design for real–time control [3, 27, 37].

POD falls into the general category of projection methods where the dynamical system is projected onto a subspace of the original phase space. In combination with Galerkin projection [12, 17, 18] it provides a powerful tool to derive surrogate models for high–dimensional or even infinite dimensional dynamical systems, since the subspace is composed of basis functions inheriting already special characteristics of the overall solution. This is in contrast to standard finite element discretizations where the choice of the basis functions is in general independent of the system dynamics.

The main advantage of POD lies in the fact that it requires only standard matrix computations, despite of its application to nonlinear problems. Although projecting only onto linear or affine subspaces the overall nonlinear dynamics is preserved, since the surrogate model will still be nonlinear. Nevertheless, it is computationally more convenient than to reduce the dynamics onto a curved manifold [30], like it is done in the methods of intrinsic lower–dimensional manifolds (ILDM) for reducing chemical kinetics [23].

In Section 2 we present the construction of the POD basis which is either based on data sets or on the method of snapshots. Further, we use Galerkin projections to reduce the system dimensionality and discuss the connection of POD and singular value decomposition. Section 3 is dedicated to an numerical test of the POD method in radiative heat transfer. Finally, we give in Section 4 some conclusions and future research perspectives.

2 Proper Orthogonal Decomposition

POD can be seen as a model reduction technique or as a method for data representation. Being a projection method the latter point of view can be translated into the question [29, 30]:

> *Find a subspace approximating a given set of data in an optimal least–squares sense.*

This is related to model reduction of dynamical systems by the choice of the data points, which are either given by samplings from experiments or by trajectories of the physical system extracted from simulations of the full model.

2.1 Construction of the POD Basis

To put all this into a mathematical framework (see also [30] for a more detailed discussion) we start with a vector space V of finite or infinite dimension and a given set of data in V. Considering a dynamical system described by partial differential equations this resembles the phase space of an ordinary differential system, which one gets after a spatial discretization via a method of lines, or to the infinite dimensional state space in which the solution lies. In the following we will restrict ourself to finite dimensions and set $V = \mathbb{R}^n$. Then, a set of sampled data $Y = \{y_1(t), \ldots, y_m(t)\}$ is given by trajectories $y_i(t) \in \mathbb{R}^n$, $i = 1, \ldots, m$ and $t \in [0, T]$.

Next, we use a *principal component analysis* of this data to find a d–dimensional subspace $V_d \subset V$ approximating the data in some optimal least–squares sense, i.e. we seek an orthogonal projection $\Pi_d : V \to V_d$ of fixed rank d that minimizes the total least–squares distance

$$\|Y - \Pi_d Y\|^2 := \sum_{i=1}^{m} \int_0^T \|y_i(t) - \Pi_d y_i(t)\|^2 \, dt.$$

The solution of this problem relies on the introduction of the *correlation matrix* $K \in \mathbb{R}^{n \times n}$ defined by

$$K = \sum_{i=1}^{m} \int_0^T y_i(t) y_i(t)^* \, dt, \tag{1}$$

where the star stands for the transpose (with additional complex conjugation in case of $V = \mathbb{C}^n$) of a vector or a matrix. By definition, K is a symmetric positive semi-definite matrix with real, nonnegative ordered eigenvalues $\lambda_1 \geq \cdots \geq \lambda_n \geq 0$. Let u_j denote the corresponding eigenvectors given by

$$K u_j = \lambda_j u_j, \qquad j = 1, \ldots, n.$$

Due to the special structure of the matrix K we can choose them in fact as an orthonormal basis of V.

The main result of POD is that the optimal subspace V_d of dimension d representing the data is given by $V_d = \text{span}\,\{\mathrm{u}_1, \ldots, \mathrm{u}_\mathrm{d}\}$. The vectors u_j, $j = 1, \ldots, d$ are then called *POD modes*. More precisely, we have the following result [20]:

Theorem 1. *Let K be the correlation matrix of the data defined by :=correlation and let $\lambda_1 \geq \cdots \geq \lambda_n \geq 0$ be the ordered eigenvalues of K. Then it holds*

$$\min_{V_d} \|Y - \Pi_d Y\| = \sum_{j=n-d+1}^{n} \lambda_j,$$

where the minimum is taken over all subspaces V_d of dimension d. Further, the optimal orthogonal projection $\Pi_d : V \to V_d$, with $\Pi_d \Pi_d^* = I$, is given by

$$\Pi_d = \sum_{j=1}^{d} u_j u_j^*.$$

Each data vector $y_i(t) \in V$ can be written as

$$y_i(t) = \sum_{j=1}^{n} y_{ij}(t) u_j,$$

where $y_{ij}(t) = \langle y_i(t), u_j \rangle$. Then it holds

$$\Pi_d y_i(t) = \sum_{j=1}^{d} u_j u_j^* \left(\sum_{l=1}^{n} y_{il}(t) u_l \right) = \sum_{j=1}^{d} y_{ij}(t) u_j,$$

since $\langle u_i, u_j \rangle = \delta_{ij}$.

Remark 1. Often, one is interested in finding rather an approximating affine subspace than a linear subspace [30]. Consider for example the flow around a cylinder, where one can observe Karman's vortex street [12]. Physically speaking, we have then the superposition of the mean flow, in which we are not interested, and the vortex structures on which our main focus lies. So, we construct first the mean value of the data given by

$$\bar{y} := \frac{1}{mT} \sum_{i=1}^{m} \int_0^T y_i(t)\, dt$$

and then build–up the *covariance matrix* $\bar{K}$ defined by

$$\bar{K} := \sum_{i=1}^{m} \int_0^T (y_i(t) - \bar{y})(y_i(t) - \bar{y})^*\, dt.$$

Now, we can proceed in analogy. Let $\lambda_1 \geq \cdots \geq \lambda_n \geq 0$ be the ordered eigenvalues of $\bar{K}$ and u_j the corresponding eigenvectors. We define $V_d = \mathrm{span}\{\mathrm{u}_1, \ldots, \mathrm{u}_\mathrm{d}\}$. Then the optimal affine subspace fixed in $\bar{y}$ is given by $V_{d,\bar{y}} = \bar{y} + V_d$ and the optimal orthogonal projection is given by

$$\Pi_d y := \Pi_d(y - \bar{y}) + \bar{y}.$$

2.2 Choosing the Dimension

Finally, we have to answer the question how to choose the dimension d of the subspace V_d such that we get a *good* approximation of our data set. Here, Theorem 1 can guide us, since it provides the overall least–squares error. Hence, we only have to study the eigenvalues of K. In terms of a dynamical system, large eigenvalues correspond to main characteristics of the system, while small eigenvalues give only small perturbations of the overall dynamics. The goal is to choose d small enough while the *relative information content* [1] of the basis for V_d, defined by

$$I(d) = \frac{\sum_{j=1}^{d} \lambda_j}{\sum_{j=1}^{n} \lambda_j},$$

is near to one. I.e. if the subspace V_d should contain a percentage p of the information in V, then one should choose d such that

$$d = \mathrm{argmin}\left\{\mathrm{I(d)} : \quad \mathrm{I(d)} \geq \frac{\mathrm{p}}{100}\right\}.$$

Remark 2. If one wants to significantly reduce the dimension of the problem, i.e. $d \ll n$, one needs clearly that the eigenvalues decrease sufficiently fast. In many applications like fluid dynamics or heat transfer one observes an exponential decrease of the eigenvalues, such that one has indeed a good chance to derive low–order approximate models (see also Section 3).

Remark 3. Note, that the POD modes are optimally approximating a given data set in the least–squares sense, but they are not constructed to be the modes approximating the dynamics generating the given the data set. For example, consider a low Mach number flow where acoustic effects play a crucial role [34]. Due to their small energy compared to the high energy hydrodynamic pressure fluctuations, they would be neglected in our reduced model, although being a relevant feature of the full dynamical system. Further, although an increase of the number of POD modes leads to a decrease of the least squares error, it might happen that more POD modes lead to a worse approximation of the full dynamics. New approaches exploiting the relation between POD and balanced truncation [21, 34, 37] or dual techniques [24, 25] yield a way out of this problem.

2.3 POD and Galerkin Projection

To get reduced order models for dynamical systems one uses a *Galerkin projection* onto the subspace V_d. This is the standard technique to reduce partial differential equations with a method of lines to a system of ordinary differential equations. Standard finite element approaches for the spatial discretization are using basis functions which are in general not correlated with the overall system dynamics [12, 17]. This approach holds for any subspace V_d, but having now the POD modes at hand one can use naturally this optimal approximating subspace. So, let $f : V \to V$ be a vector field and consider the solution $y(t) : [0, T] \to V$ of the dynamical system

$$\dot{y}(t) = f(y(t)),$$

which we e.g. get from a discretization of a partial differential equation via finite elements or finite differences. Further, we construct an approximating d–dimensional subspace $V_d = \operatorname{span}\{\mathrm{u}_1, \ldots, \mathrm{u}_\mathrm{d}\}$ via POD. The reduced order model is then given by

$$\dot{y}_d(t) = \Pi_d f(y_d(t)) \tag{2}$$

with solution $y_d(t) : [0, T] \to V_d$. Here, $\Pi_d f(y_d(t))$ is just the projection of the original vector field f onto the subspace V_d. To rewrite :=reduced component wise we use

$$y_d(t) = \sum_{j=1}^{d} \chi_j(t) u_j$$

and substitute this into :=reduced. Then, a multiplication with u_j^* yields

$$\dot{\chi}_j(t) = u_j^* f(y_d(t)) = u_j^* f\left(\sum_{j=1}^{d} \chi_j(t) u_j\right), \quad j = 1, \ldots, k,$$

i.e. we are left with a coupled system of d ordinary differential equations for the evolution of $y_d(t)$. Clearly, also the initial condition has to be projected, i.e.

$$y_d(0) = \Pi_d y(0).$$

Remark 4. For an affine subspace $V_{d,\bar{y}}$ the reduced model for the dynamics of $y_d : [0,T] \to V_{d,\bar{y}}$ can be derived in analogy.

2.4 POD and Snapshots

Concerning real world applications like flow problems, one has to encounter many degrees of freedom, such that the dimension n of the phase space might be very large. In practical computations this can lead to a phase space with dimension $n = 10^6$–10^{10}. Hence, the calculation of the POD modes would require the solution of a large eigenvalue problem for a full matrix $K \in \mathbb{R}^{n\times n}$, which might be infeasible. To overcome this problem Sirovich [36] proposed the *method of snapshots*, which proves to a powerful tool for the computation of the eigenfunctions. Instead of solving the eigensystem for the matrix $K \in \mathbb{R}^{n\times n}$ one only needs to consider a matrix in $\mathbb{R}^{m\times m}$, where m is the number of snapshots considered (for a more detailed discussion we refer to [12, 17] and the references therein).

Snapshots are constructed from the trajectories of the dynamical system by evaluating them at certain discrete time instances $t_1, \dots, t_m \in [0,T]$, i.e. they are given by $y_i = y(t_i) \in \mathbb{R}^n$. Then, we get a new correlation matrix K defined by

$$K = \sum_{i=1}^{m} y(t_i) y(t_i)^*. \tag{3}$$

Remark 5. But how many snapshots should one choose? An educated guess would be to choose less than n, since we cannot expect to get more than n linearly independent vectors. On the other hand, snapshots should be always taken whenever the dynamics of the system is changing. Hence, it might happen that $m > n$. Be aware that the chosen snapshot vectors might be linearly dependent, such that it is not clear, if one can reconstruct a suitable basis. Further note that the snapshots depend clearly on the initial datum and on a given input.

We build the matrix $Y = (y(t_1), \dots y(t_m)) \in \mathbb{R}^{n\times m}$ consisting in the columns of the snapshots. Hence, in each row we find the trajectories of the dynamical system at discrete time events. Then, the sum (3) can be written as $K = YY^*$. In the method of snapshots one considers now instead the matrix $Y^*Y \in \mathbb{R}^{m\times m}$ and solves the eigenvalue problem

$$Y^*Y v_j = \lambda_j v_j, \quad j = 1, \dots, m, \quad v_j \in \mathbb{R}^m.$$

In the following we will see that the eigenvalues are indeed the same. Again, we can choose an orthonormal basis of eigenvectors $\{v_1, \dots, v_m\}$ and the corresponding POD modes are given then given by

$$u_j = \frac{1}{\sqrt{\lambda_j}} Y v_j, \quad j = 1, \dots, m.$$

2.5 POD and SVD

The above discussion suggest that there is indeed a strong connection of POD and *singular value decomposition* (SVD) for rectangular matrices (for an excellent overview see [12, 16, 17]). Consider a matrix $Y \in \mathbb{R}^{n\times m}$ with rank d. From standard SVD we know that there exist real numbers $\sigma_1 \geq \ldots \geq \sigma_d > 0$ and unitary matrices $U \in \mathbb{R}^{n\times n}$ and $V \in \mathbb{R}^{m\times m}$ such that

$$U^*YV = \begin{pmatrix} \Sigma_d & 0 \\ 0 & 0 \end{pmatrix} = \Sigma \in \mathbb{R}^{n\times m}, \tag{4}$$

where $\Sigma_d = \text{diag}(\sigma_1, \ldots, \sigma_\text{d}) \in \mathbb{R}^{\text{d}\times\text{d}}$.

The positive numbers σ_i are called singular values of Y. For $U = (u_1, \ldots, u_n)$ and $V = (v_1, \ldots, v_m)$ we call $u_i \in \mathbb{R}^n$ the left singular vectors and $v_i \in \mathbb{R}^m$ the right singular vectors, which satisfy

$$Yv_i = \sigma_i u_i \quad \text{and} \quad Y^*u_i = \sigma_i v_i, \quad i = 1, \ldots, d.$$

These are eigenvectors of YY^* and Y^*Y with eigenvalues σ_i^2, $i = 1, \ldots, d$.

The link between POD and SVD lies in the fact that the approximating POD basis should contain as much information or energy as possible. Mathematically, we can write the problem of approximating the snapshot vectors y_i by a single vector u as the constrained optimization problem

$$\max \sum_{j=1}^{m} |\langle y_j, u\rangle|^2 \quad \text{s.t.} \quad |u| = 1. \tag{5}$$

Using the Lagrangian formalism we derive that a necessary condition for this problem is given by the eigenvalue problem

$$YY^*u = \sigma^2 u.$$

The singular value analysis yields that u_1 solves this eigenvalue problem and the functional value is indeed σ_1^2. Now, we iterate this procedure and derive that u_i, $i = 1, \ldots, d$ solves

$$\max \sum_{j=1}^{m} |\langle y_j, u\rangle|^2 \quad \text{s.t.} \quad |u| = 1 \text{and} \, \langle u, u_j\rangle = 0, \; j = 1 \ldots, i-1 \tag{6}$$

and the value of the functional is given by σ_i^2.

By construction it is clear that for every $d \leq m$ the approximation of the columns $Y = (y_1, \ldots, y_m)$ by the first d singular vectors $\{u_i\}_{i=1}^d$ is optimal in the least–squares sense among all rank d approximations to the columns of Y.

Altogether, this leads the way to the practical determination of a POD basis of rank d. If $m < n$ holds, then one can compute the m eigenvectors v_i corresponding to the largest eigenvalues of $Y^*Y \subset \mathbb{R}^{n\times n}$. These relate to the POD basis as follows

$$u_i = \frac{1}{\sigma_i} Y v_i, \quad i = 1, \ldots d.$$

3 POD in Radiative Heat Transfer

In the following we want to test the POD approach for model reduction in a radiative heat transfer problem given by the so–called Rosseland model [26], which is given by the nonlinear parabolic partial differential equation

$$\partial_t y(x,t) = \mathrm{div}\left(\left(\mathrm{k_h} + \mathrm{k_r y^3(x,t)}\right) \nabla \mathrm{y(x,t)}\right) \tag{7}$$

for the temperature distribution $y(x,t)$. Here, (x,t) is in the space–time cylinder $Q = \Omega \times (0,T)$, where Ω is a bounded domain in $\mathbb{R}^2$. The coefficients k_h and k_r are positive constants related to the conductive and radiative heat transfer. This model has to be supplemented with boundary data

$$n \cdot \nabla y(x,t) = h(b(x,t) - y(x,t)) + \alpha\left(b^4(x,t) - y^4(x,t)\right) \tag{8}$$

for $(x,t) \in \partial\Omega \times (0,T)$ and an initial datum

$$y(x,0) = y_0(x) \tag{9}$$

for $x \in \Omega$. Here, $y_0(x)$ is the initial temperature, $b(x,t)$ a specified boundary temperature, and h and α measure the conductive and radiative heat loss over the boundary, respectively. For the forthcoming simulations we choose

$$k_h = 1,\ k_r = 10^{-7},\ h = 1,\ \alpha = 5 \cdot 10^{-7},\ y_0(x) = 500,\ b(x,t) = 300.$$

Remark 6. Note, that this model has two nonlinearities: One in the heat conductivity which models volume radiation and one in the boundary condition which adds additional surface radiation to the standard Newton cooling law, where the temperature flux is proportional to the temperature difference. Hence, we can expect that boundary layers will appear in the solution such that the POD modes will significantly differ from the eigenfunctions of the Laplacian.

This nonlinear partial differential equation is solved using the finite element package FEMLAB. The computational domain $\Omega \subset \mathbb{R}^2$ is an ellipse depicted in Figure 1 with center zero and an aspect ratio of two. There, one also finds the triangular mesh consisting of 1769 degrees of freedom. For the discretization we use linear finite elements with the nodal basis $\{\varphi_i(x)\}_{i=1}^n$ and write the approximate solution as

$$y_h(x,t) = \sum_{i=1}^{n} y_i(t)\varphi_i(x).$$

The finite element ansatz for the equation :=RHT yields a dynamical system for $Y(t) = (y_1(t), \ldots, y_n(t))$

$$\dot{Y}(t) = f(Y(t)),$$

where the right hand side is computed via the Galerkin projection onto the finite element space.

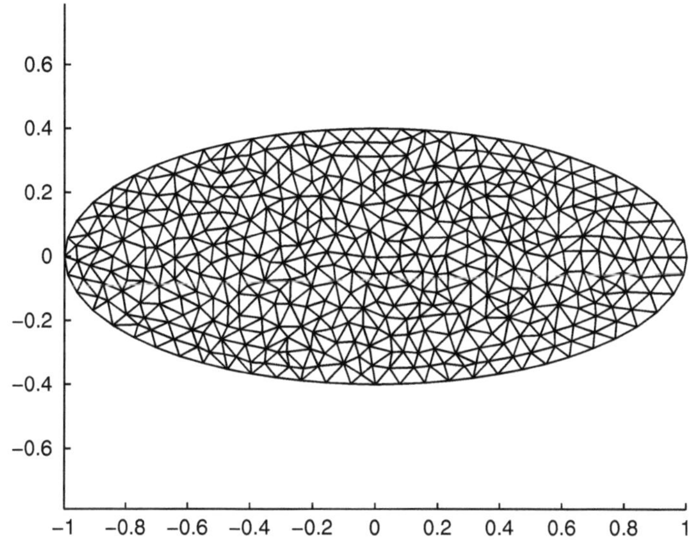

Fig. 1. Computational Domain and Mesh

For the POD analysis we take $m = 81$ equidistantly distributed snapshots $\{Y(t_j)\}_{j=1}^m$ in the interval $[0, T]$, build up the snapshot matrix

$$Y = \begin{pmatrix} y_1(t_1) & \cdots & y_1(t_m) \\ \vdots & & \vdots \\ y_n((t_1) & \cdots & y_n(t_m) \end{pmatrix} \in \mathbb{R}^{n \times m}$$

and introduce the correlation matrix $K = Y^*Y \in \mathbb{R}^{m \times m}$. Using the MATLAB routine eigs.m one can easily compute the eigenvalues λ_j and the eigenvectors u_j, $j = 1, \ldots m$. Despite of the high nonlinearity of our problem, we get exponential decay of the eigenvalues, which can be seen in Figure 2.

From the eigenvectors we can compute the POD basis $\{u_i\}_{i=1}^d$ of rank d as follows

$$u_i = \frac{1}{\sqrt{\lambda_i}} \sum_{j=1}^m v_i^j y_j, \quad i = 1, \ldots, d,$$

where v_i^j is the j–th component of the eigenvector v_i. In the following we will use the normalized POD basis functions $\psi_i = u_i / \|u_i\|$. The first eight of these basis functions can be found in Figure 4. Computing the relative information content for the first mode yields already $I(1) = 99.96\%$.

To get a reduced POD model we use the POD basis $\{\psi_i\}_{i=1}^d$ in a Galerkin ansatz

$$y_d(x, t) = \sum_{i=1}^d \chi_i(t) \psi_i(x),$$

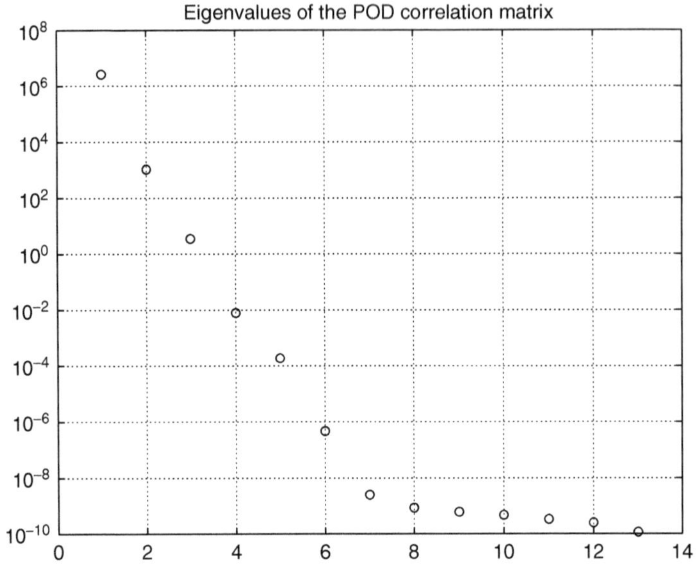

Fig. 2. Eigenvalues of the correlation matrix

plug this into the equation and test with χ_j, which yields

$$\begin{aligned}\partial_t \chi_i(t) = &-\int_\Omega \left(k_h + k_r \left(\sum_{k=1}^d \chi_k(t)\psi_k(x) \right)^3 \right) \sum_{i=1}^d \chi_i(t) \nabla\psi_i(x) \cdot \nabla\psi_j(x)\, dx \\ &+ \int_{\partial\Omega} \left(k_h + k_r \left(\sum_{k=1}^d \chi_k(t)\psi_k(x) \right)^3 \right) \sum_{i=1}^d \chi_i(t)\, n \cdot \nabla\psi_i(x)\, \psi_j \, ds \\ &=: g(\chi(t)), \quad i = 1, \dots, d, \quad \chi = (\chi_1, \dots, \chi_d).\end{aligned}$$

This gives the ordinary differential system $\dot\chi(t) = g(\chi(t))$ of size $d \times d$, which can be solved with an implicit Euler method for example.

Remark 7. Note that the ordinary differential system can be solved quite fast due to its small size. Nevertheless, one should be aware that building up the system might need some time since we have to compute the inner products for *global basis functions*, in contrast to the finite element basis which has compact support. Alternatively, one can compute the projections of the finite element matrices onto the reduced space.

To get an idea how well our reduced model approximates the full model, we measure the difference of $y_h(x,t)$ and $y_d(x,t)$ in the norm of the space $L^2(0,T;L^2(\Omega))$, i.e.

$$\|y_h - y_d\|^2_{L^2(0,T;L^2(\Omega))} = \int_0^T \int_\Omega \left(y_h(x,t) - y_d(x,t) \right)^2 \, dx\, dt.$$

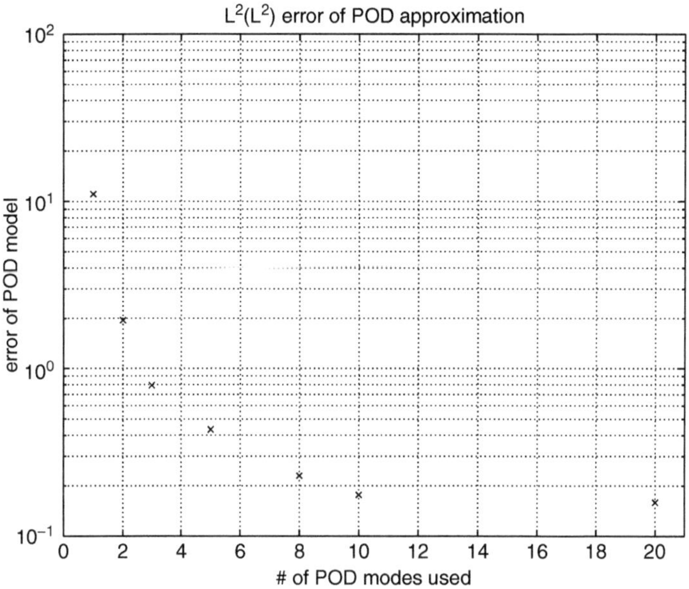

Fig. 3. Error between the full and the reduced model

This error is plotted in Figure 3 for different sizes d of the reduced model. Most remarkable is that already three POD modes yield a mean error of 1 in the temperature, which yields a relative mean error of less than 1%.

Remark 8. It is worth noting that in the context of partial differential equations, it is also possible to build up the correlation matrix using a different inner product, which is more related to the Galerkin ansatz, i.e. here one could also use the inner product of $H^1(\Omega)$. I.e. one replaces the correlation matrix by $K = YMY^*$, where $M \in \mathbb{R}^{n\times n}$ is a symmetric, problem dependent matrix. This yields a different POD basis which might give even better results (for the linear case see e.g. [11]). Further, it was pointed out that one may increase the accuracy of the reduced model by adding finite differences of the snapshots to the snapshot set, i.e. also considering time derivatives of the snapshots [12, 17]. Clearly, this does not change the space spanned by the snapshots since we are only adding linearly dependent vectors. Nevertheless, we get different weights in the correlation and thus again different modes.

4 Conclusions and Future Perspectives

Being a powerful tool for model reduction of large–scale dynamical systems, POD is acquiring increasing attention in the mathematics and engineering community. Presently, there is the tendency to test its performance in more and more fields of application, like fluid and structural dynamics, reducing models based on partial

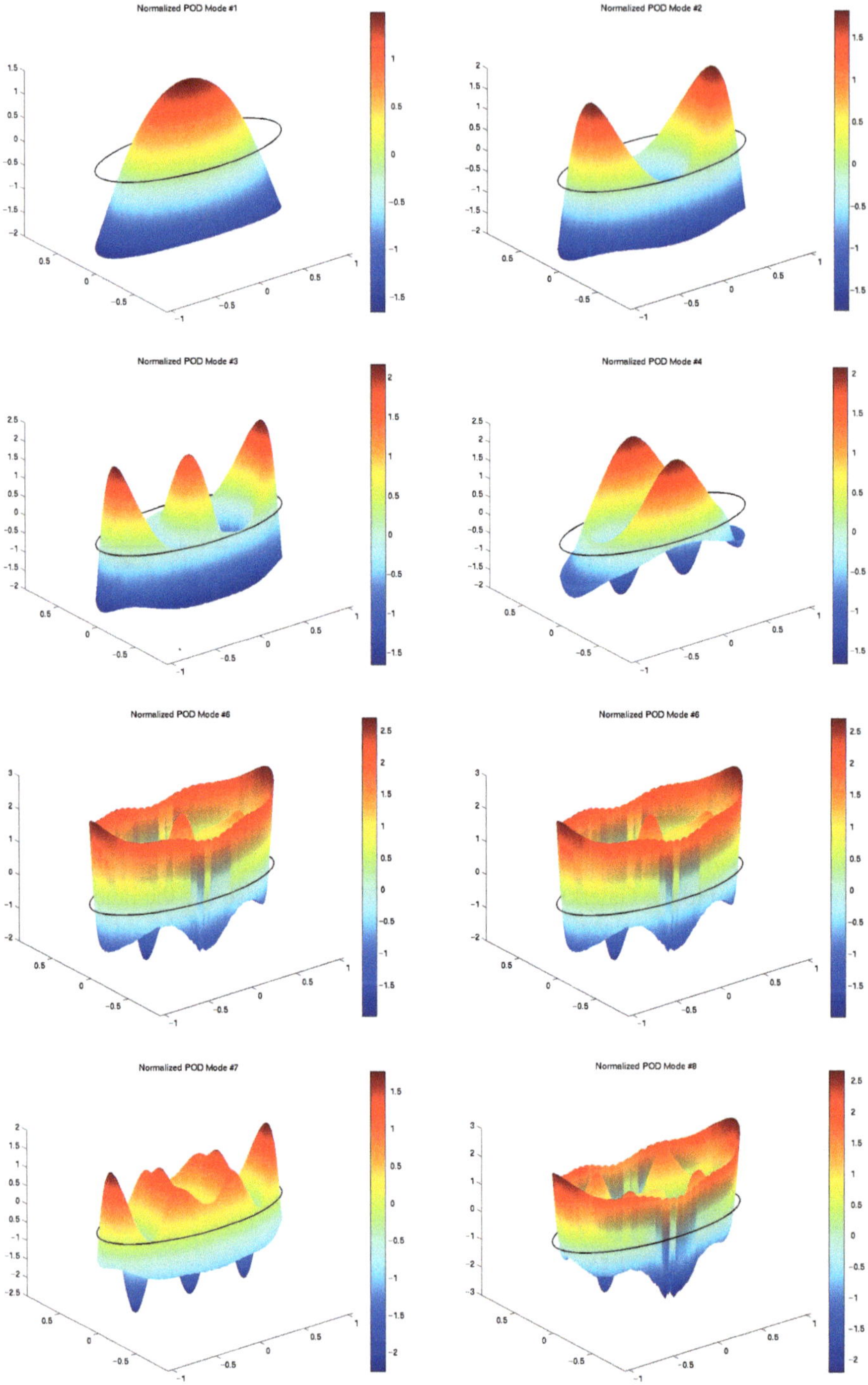

Fig. 4. The first 8 POD modes

differential equations for devices in electric circuits, frequency averaging in radiative heat transfer, or even tire modeling. These different fields clearly will have different requirements on the POD method. Either, one wants to have structure preserving reduced models [5, 7, 20] or estimates on the quality of the surrogate model [11, 24, 25, 29, 30]. In particular, the derivation of error estimates for POD models is a field of intensive research which follows two lines: First, the combination of POD and balanced truncation (c.f. [34, 37] and the references therein) and second exploiting the special structure of Galerkin approximations to partial differential equations (c.f. [12, 17] and the references therein).

Acknowledgments

The author would like to thank Michael Hinze for many helpful discussion on proper orthogonal decomposition. Further, he is indebted to Alexander Schulze who provided the numerical examples.

References

1. Konstantin Afanasiev and Michael Hinze. Adaptive control of a wake flow using proper orthogonal decomposition. In *Shape optimization and optimal design (Cambridge, 1999)*, volume 216 of *Lecture Notes in Pure and Appl. Math.*, pages 317–332. Dekker, New York, 2001.
2. Vidal R. Algazi and David J. Sakrison. On the optimality of the Karhunen-Loève expansion. *IEEE Trans. Information Theory*, IT-15:319–320, 1969.
3. Jeanne A. Atwell and Belinda B. King. Reduced order controllers for spatially distributed systems via proper orthogonal decomposition. *SIAM J. Sci. Comput.*, 26(1):128–151, 2004.
4. Nadine Aubry, Philip Holmes, John L. Lumley, and Emily Stone. The dynamics of coherent structures in the wall region of a turbulent boundary layer. *J. Fluid Mech.*, 192:115–173, 1988.
5. Nadine Aubry, Wen Yu Lian, and Edriss S. Titi. Preserving symmetries in the proper orthogonal decomposition. *SIAM J. Sci. Comput.*, 14(2):483–505, 1993.
6. H. T. Banks, Michele L. Joyner, Buzz Wincheski, and William P. Winfree. Nondestructive evaluation using a reduced-order computational methodology. *Inverse Problems*, 16(4):929–945, 2000.
7. Gal Berkooz and Edriss S. Titi. Galerkin projections and the proper orthogonal decomposition for equivariant equations. *Phys. Lett. A*, 174(1-2):94–102, 1993.
8. Marco Fahl and Ekkehard W. Sachs. Reduced order modelling approaches to PDE-constrained optimization based on proper orthogonal decomposition. In *Large-scale PDE-constrained optimization (Santa Fe, NM, 2001)*, volume 30 of *Lect. Notes Comput. Sci. Eng.*, pages 268–280. Springer, Berlin, 2003.
9. Keinosuke Fukunaga. *Introduction to statistical pattern recognition*. Computer Science and Scientific Computing. Academic Press Inc., Boston, MA, second edition, 1990.
10. M. Hinze and K. Kunisch. Three control methods for time - dependent fluid flow. *Flow, Turbulence and Combustion*, 65:273–298, 2000.

11. M. Hinze and S. Volkwein. Error estimates for abstract linear-quadratic optimal control problems using proper orthogonal decomposition. *Technical Report IMA02-05, KFU Graz*, 2005.
12. M. Hinze and S. Volkwein. Proper orthogonal decomposition surrogate models for nonlinear dynamical systems: Error estimates and suboptimal control. In D. Sorensen, P. Benner, V. Mehrmann, editor, *Dimension Reduction of Large-Scale Systems, Lecture Notes in Computational and Applied Mathematics*, pages 261–306. 2005.
13. H. Hoetelling. Simplified calculation of principal component analysis. *Psychometrica*, 1:27–35, 1935.
14. Philip Holmes, John L. Lumley, and Gal Berkooz. *Turbulence, coherent structures, dynamical systems and symmetry*. Cambridge Monographs on Mechanics. Cambridge University Press, Cambridge, 1996.
15. Kari Karhunen. Zur Spektraltheorie stochastischer Prozesse. *Ann. Acad. Sci. Fennicae. Ser. A. I. Math.-Phys.*, 1946(34):7, 1946.
16. K. Kunisch and S. Volkwein. Control of the Burgers equation by a reduced-order approach using proper orthogonal decomposition. *J. Optim. Theory Appl.*, 102(2):345–371, 1999.
17. K. Kunisch and S. Volkwein. Galerkin proper orthogonal decomposition methods for parabolic problems. *Numer. Math.*, 90(1):117–148, 2001.
18. K. Kunisch and S. Volkwein. Galerkin proper orthogonal decomposition methods for a general equation in fluid dynamics. *SIAM J. Numer. Anal.*, 40(2):492–515, 2002.
19. K. Kunisch, S. Volkwein, and L. Xie. HJB-POD-based feedback design for the optimal control of evolution problems. *SIAM J. Appl. Dyn. Syst.*, 3(4):701–722, 2004.
20. Sanjay Lall, Petr Krysl, and Jerrold E. Marsden. Structure-preserving model reduction for mechanical systems. *Phys. D*, 184(1-4):304–318, 2003.
21. Sanjay Lall, Jerrold E. Marsden, and Sonja Glavaški. A subspace approach to balanced truncation for model reduction of nonlinear control systems. *Internat. J. Robust Nonlinear Control*, 12(6):519–535, 2002.
22. M. Loeve. Fonctions aleatoire de second ordre. *Revue*, 48:195–206, 1946.
23. U. Maas and S. B. Pope. Simplifying chemical kinetics: Intrinsic low-dimensional manifolds in compositional space. *Combust. Flame*, 88:239–264, 1992.
24. H.G. Matthies and M. Meyer. Nonlinear galerkin methods for the model reduction of nonlinear dynamical systems. *Computers & Structures*, 81(12), 2003.
25. M. Meyer and H.G. Matthies. Efficient model reduction in nonlinear dynamics using the karhunen-love expansion and dual-weighted-residual methods. *Computational Mechanics*, 31(1+2), 2003.
26. M.F. Modest. *Radiative Heat Transfer*. McGraw-Hill, 1993.
27. B.C. Moore. Principal component analysis in linear systems: Controlability, observability and model reduction. *IEEE Trans. Automat. Contr.*, 26(1), 1981.
28. K. Pearson. On lines and planes of closest to points in space. *Philosophical Magazine*, 2:609–629, 1901.
29. Muruhan Rathinam and Linda R. Petzold. Dynamic iteration using reduced order models: a method for simulation of large scale modular systems. *SIAM J. Numer. Anal.*, 40(4):1446–1474, 2002.
30. Muruhan Rathinam and Linda R. Petzold. A new look at proper orthogonal decomposition. *SIAM J. Numer. Anal.*, 41(5):1893–1925, 2003.
31. S.S. Ravindran. Adaptive reduced-order controllers for a thermal flow system using proper orthogonal decomposition. *SIAM J. Sci. Comput.*, 23(6):1924–1942, 2002.
32. D. Rempfer. On dynamical systems obtained via Galerkin projections onto low-dimensional bases of eigenfunctions. In *Fundamental problematic issues in turbulence (Monte Verita, 1998)*, Trends Math., pages 233–245. Birkhäuser, Basel, 1999.

33. Azriel Rosenfeld and Avinash C. Kak. *Digital picture processing. Vol. 2*. Computer Science and Applied Mathematics. Academic Press Inc., New York, second edition, 1982.
34. C.W. Rowley. Model reduction for fluids, using balanced proper orthogonal decomposition. *Internat. J. Bifur. Chaos Appl. Sci. Engrg.*, 15(3):997–1013, 2005.
35. Clarence W. Rowley, Tim Colonius, and Richard M. Murray. Model reduction for compressible flows using POD and Galerkin projection. *Phys. D*, 189(1-2):115–129, 2004.
36. Lawrence Sirovich. Turbulence and the dynamics of coherent structures. I—III. *Quart. Appl. Math.*, 45(3):561–590, 1987.
37. Karen Willcox and J. Peraire. Balanced model reduction via the proper orthogonal decomposition. *AIAA*, 40(11):2323–2330, 2002.

PMTBR: A Family of Approximate Principal-components-like Reduction Algorithms

Joel R. Phillips[1], Zhenhai Zhu[1], and L. Miguel Silveira[2]

[1] Cadence Berkeley Laboratories
Berkeley, CA 94704, U.S.A.
{zhzhu, jrp}@cadence.com
[2] Cadence Laboratories / INESC ID
IST / Tech. U. Lisbon
Lisbon, Portugal
lms@inesc-id.pt

Summary. In this chapter we present a family of algorithms that can be considered intermediate between frequency domain projection methods and approximation of truncated balanced realizations. The methods discussed are computationally simple to implement, have good error properties, and possess simple error estimation and order control procedures. By tailoring the method to take into account a statistical representation of individual problem characteristics, more efficient, improved results have been obtained in several situations, meaning models of small order that retain acceptable accuracy, on problems for which many other methods struggle. Examples are shown to illustrate the algorithms in the contexts of frequency weighting, circuit simulation with parasitics networks having large numbers of input/output ports, and interconnect modeling in the presence of parameter change due to process variation.

1 Introduction

Model order reduction methods are now commonly used in the modeling, simulation, and analysis of integrated circuit (IC) components, particularly circuit interconnect and packaging. The dominant characteristics of IC and other problems in electronic design automation (EDA) is the large scale of the problems encountered. In a reduction context, both the size of the systems encountered, as well as the number of systems to be reduced, can be quite large. This necessitates reduction methods which are efficient in producing very small models, are effective on fairly large systems, and are very fast in reduction time.

Development of reduction methods in IC EDA has historically proceeded in four main stages. The first methods to be proposed were more or less geometric in nature, and based on heuristics supported by reasonings from circuit theory. These methods still exist today, under the guise of "realizable" or graph-based reduction techniques [1], and can be quite effective in certain applications, and for limited, but important, classes of systems like pure resistor-capacitor networks. They are however hard to generalize and do not perform well on many other relevant problems. The second reduction family appeared in the late 1980s with the popularization of "moment-matching", i.e. Padé approximation [2]. While still used even today, the numerical disadvantages of these methods were fairly quickly recognized. Next, the connection to numerically stable Krylov-subspace algorithms [3, 4] and the advent of general projection techniques [5] opened the era of modern reduction approaches. Finally, it was inevitable that approaches developed earlier in the systems and control literature, such as balanced truncation [6–8] and optimal Hankel norm approximation [9], would be examined in light of this renew interest. Due to the high computational cost of these techniques, they were usually applied either in conjunction with the more scalable Krylov-projection techniques, or on very limited problem sets, though recent work on iterative solution of Lyapunov or Ricatti equations may make these methods more widely applicable [10–14].

Eventually various questions arose in practical application of the competing techniques and their relation. It is widely expected that, on average, the balanced truncation type methods should give better results than the more-or-less heuristic moment-matching-via-projection approaches. At the same time, many procedures for computing the controllability/observability Gramians involve computations remarkably similar to moment-matching, implying a connection between the methods, and we argue below that it is in fact fairly straightforward to connect the two families, and interpolate between them. More interestingly, we have empirically observed that one can often do better (in the sense of smaller model for a given allowable error) with the seemingly ad-hoc techniques than with e.g. balanced truncation for which global error bounds are known to exist. While this appears somewhat counternature, we will argue that this is not the result of pure luck and furthermore, with the insights gained, reduction algorithms can be constructed that are effective on systems (e.g. some systems with large numbers of inputs and outputs) that appear to be "non-reducible" by more conventional approaches.

In this chapter we review a simple family of algorithms that are based on cheap approximations to a principal components analysis, implemented via random sampling [15]. These techniques are easy to implement, illustrate rapid convergence on industrial problems, and have several practical advantages. They can be tailored to gain efficiency on some hard problems that are difficult to treat by other means [16]. In practical implementations, error/order control is very important, and in the proposed approaches it can be done in an 'on-the-fly' manner, enabling incremental implentation of the algorithms which is very important for practical implementations. The family of algorithms we discuss in this chapter also generalizes nicely to some related problems, such as reduction of affinely parameterized systems [17].

The algorithms we have been working with are relatives of the 'principal orthogonal decomposition' (POD) methods used for some time in [18–20]. However, the motivation and algorithms in this chapter are somewhat different, in particular as regards selection of sampling strategies, which is somewhat ad-hoc in other work we are aware of, but plays a critical role in our analysis.

The outline of the chapter is as follows. In Section 2 we describe the basics of the algorithms and discuss some of its advantages over competing techniques. In Section 3 we present some variants of the algorithm which have proven sucessfull in specific settings. Then in Section 4 we provide a thourough analysis of the method and comparisons to existing alternatives. Finally in Section 5 we provide examples from applications of the techniques to various problems and settings. Conclusions are drawn in Section 6.

2 Basic Algorithm

2.1 Sampling Approximations

Consider the state-space model

$$\frac{dx}{dt} = Ax + Bu(t) \tag{1}$$

$$y(t) = Cx + Du(t). \tag{2}$$

with input $u(t)$ and output $y(t)$, that are described by the matrices $A \in \mathbb{R}^{n\times n}, B \in \mathbb{R}^{n\times p}, C \in \mathbb{R}^{p\times n}$, where p is the number of inputs and outputs. The goal of Model Order Reduction is to produce a reduced model

$$\frac{dz}{dt} = \widehat{A}z + \widehat{B}u, \quad y = \widehat{C}z \tag{3}$$

where $\widehat{A} \in \mathbb{R}^{q\times q}, \widehat{B} \in \mathbb{R}^{q\times p}, \widehat{C} \in \mathbb{R}^{p\times q}$, $q \ll n$. As was discussed in previous chapters, there are many different ways to achieve this goal, with projection schemes perhaps the more common in IC EDA applications. In that context, a pair of matrices W and V can be constructed, whose columns span a "useful" subspace, and the original equations can then be projected into those spaces as

$$\widehat{A} \equiv W^T AV \quad \widehat{B} \equiv W^T B \quad \widehat{C} \equiv CV. \tag{4}$$

Most common choices are based on picking the columns of W, V to span a Krylov subspace [4, 21], and a very common scheme is to choose $W = V$. Of course, different choices will lead to different algorithms with slightly different properties but an overall similar "flavour".

Let us assume that the system (2) is stable which implies the eigenvalues of A to have negative real part. The controllability Gramian, which can be computed as the solution of the Lyapunov equation

$$AX + XA^T = -BB^T, \tag{5}$$

can be computed in the time domain as

$$X = \int_0^\infty e^{At}BB^Te^{A^Tt}dt. \tag{6}$$

or in the frequency domain as

$$X = \int_{-\infty}^{\infty} (j\omega I - A)^{-1} B B^T (j\omega I - A)^{-H} d\omega. \tag{7}$$

where superscript H denotes Hermitian transpose. Neglecting for the moment the issue of balancing, a model order reduction procedure can be obtained from the eigendecomposition

$$X = V_L \Sigma V_L^T. \tag{8}$$

where $V_L^T V_L = I$ since X is real symmetric. An obvious candidate for reduction would be to pick a projection matrix formed from the columns of V_L corresponding to the dominant eigenvalues of X.

As a first step to a simple procedure, consider evaluating X via applying numerical quadrature to (7). Given a quadrature scheme with nodes ω_k and weights w_k, and defining

$$z_k = (j\omega_k I - A)^{-1} B, \tag{9}$$

an approximation $\widehat{X}$ to X can be computed as

$$\widehat{X} = \sum_k w_k z_k z_k^H. \tag{10}$$

Letting Z be a matrix whose columns are z_k, and W a diagonal matrix with diagonal entries $W_{kk} = \sqrt{w_k}$, Eqn. (10) can be written as

$$\widehat{X} = Z W^2 Z^H = (ZW)(ZW)^H. \tag{11}$$

Given the singular value decomposition of ZW.

$$ZW = V_Z S_Z U_Z \tag{12}$$

with S_Z real diagonal, V_Z and U_Z unitary matrices, clearly

$$\widehat{X} = V_Z S_Z^2 V_Z^T. \tag{13}$$

In an appropriately chosen limit, the dominant eigenvectors of $\widehat{X}$, the dominant singular vectors in V_Z, will converge to the dominant eigenspace of X. As an engineering approximation, it seems a reasonable strategy to search for some computationally cheaper approximation to such an eigenspace. With something of a leap of faith, we thus posit Algorithm 1 for solving the slightly generalized set of parametric model equations

$$L(\Lambda)x = Bu \tag{14}$$

$$y = Cx + Du(t). \tag{15}$$

where $\Lambda = (s, \lambda_1, \ldots, \lambda_P)$ is a vector of parameters, in which we include a frequency variable, that parameterizes the matrix operator L.

Algorithm **1** : **PMTBR: Poor Man's TBR**

1. *Do until satisfied:*
2. *Select a frequency/parameter point* Λ_i.
3. *Compute* $z_i = [L(\Lambda_i)]^{-1}B$.
4. *Form the matrix of columns*
5. *[real* s_i*]:* $Z_{(i)} = [z_1, z_2, \ldots, z_i]$.
   [complex s_i*]:* $Z_{(i)} = [z_1, z_1^*, z_2, z_2^*, \ldots, z_i, z_i^*]$.
6. *Construct the SVD of* $Z_{(i)}$.
   If the error is satisfactory, set $Z = Z_{(i)}$*, go to Step 7.*
   Otherwise, go to Step 2.
7. *Construct the projection space V from the orthogonalized column span of Z, dropping columns whose associated singular values fall below a desired tolerance.*

Algorithm 1 is based on the idea that appropriately chosen points in the multidimensional frequency and parameter spaces are good quadrature points for approximating the dominant eigenspace of X. Our empirical experience seems to indicate that furthermore, on a wide variety of problems of practical relevance, very good models can be obtained with a fairly small number of sample points. We denote our method "Poor Man's" TBR (PMTBR), since the quantities computed are, from a certain perspective, cheap approximations to those required to perform a full truncated balanced realizations (TBR) procedure.

Consider first a most restricted case: that the system in study is LTI and in balanced coordinates, and that the samples are chosen via quadrature points placed in a sufficiently dense manner over the entire imaginary axis as to accurately compute the true Gramian X. Then, the singular values of Z are simply the Hankel singular values, and provide error bounds through the usual connection to TBR [6,9]. If samples Λ_i are drawn i.i.d from a uniform distribution over a sufficiently wide bandwidth, the procedure becomes a Monte Carlo estimate for the integral determining the Gramian, and again in some limit we expect the singular values to provide error estimates. Of course, it is impractical to take such a large number of samples as to converge these integrals. Nor is it desirable to sample from a distribution uniform over a wide frequency range. We therefore propose two modifications to the procedure that have proved effective in engineering applications.

First, we sample with a distribution that is restricted to match the characteristics of the expect problem inputs as closely as possible. By restricting the allowed range of inputs, the dimensionality of the associated Gramians, and therefore the size of the reduced model that is needed to achieve a given error tolerance, is likewise reduced. Exactly how the weighting is done in fact determines the different possible concrete algorithms. Section 3 is devoted to a discussion of some possible choices. Obviously, this changes the nature of the error control strategy since we no longer compute, even asymptotically, system Gramians. But it is a standard result that for a large number of samples, the singular values of Z give a measure of the average error in reconstructing the space spanned by the z_i. However, unlike in TBR, this error estimate applies only to the specific class of system inputs defined in the probabilistic model.

Second, we strive to take a fairly limited number of samples. In particular, we propose to guide the sampling process by analyzing the singular values of the matrix Z at each iteration, as constructed by the samples themselves. This usually leads us to terminate the sampling process after a very few samples. Thus, we never converge, in fact even come close to converging, the relevant Gramian integrals. However, we usually obtain sufficiently good estimates of the important subspaces early on in the process. Section 4.3 discusses error analysis in more detail.

2.2 Advantages

The Algorithm proposed in the previous section prescribes the usage of sampling points in a multidimensional space in order to generate the projection matrix for reduction. At first sight this introduced a new problem, i.e. where to place the samples in order to get a good, i.e. small and accurate, model. In fact there are several practical advantages that come from using the sampling viewpoint.

First, it is simple to incorporate weighting functions that are expressed in a "non-analytic" manner. For example, it is common in IC interconnect analysis to know only a "maximum" or "cutoff" frequency of interest. In an attempt to improve the model in the frequency range of interest, sometimes weighting functions are used to bias the transfer function. In our algorithm, this can be done in a straightforward way that fits readily into our computation: samples above the cutoff frequency should be weighted with zero probability, samples below with equal weight. The weighted versions of, e.g., balanced truncation, with which we are familiar require analytic weighting, which introduces additional complexity to the procedure and can interfere with error estimates. In the worst case, complicated weighting functions can conflict with demands for very compact (e.g. order 2 or 3) models.

Second, and more importantly, we can build information about input structure (or other available system structure) into the PMTBR method. Any specific information about some particular aspect of the structure of the system, or its inputs, can be used as a guideline when sampling. In some instances this can result in orders of magnitude improvement in performance. Another way of expressing our claim is to say that the traditional methods contain *implicit assumptions* about the relationships between input data or system structure that can often interfere with effective reduction procedures.

As a result, in many *practical* applications, PMTBR performs *better* than TBR in the sense of giving more accurate models for a given model size or amount of effort. By connecting our method to multipoint moment matching, we believe we can explain the instances where these methods also have superior performance. In essence this links back to the empirical observation previously mentioned that in many applications, ad-hoc techniques outperformed TBR, a seemingly puzzling fact in view of the purported close-to-optimal reduction available through TBR-like methods.

Here we summarize our basic claims surrounding the PMTBR family of algorithms.

- Efficiency in reduction algorithms is obtained through restricted assumptions on system inputs.
- Only approximations of relevant subspaces are needed for engineering accuracy.
- Random sampling is a cheap but effective way of estimating the important subspaces.
- With proper sampling, error control comes as a by-product.

The rest of the chapter is devoted to substantiating these claims. In many cases we have only empirical evidence to offer.

3 Algorithmic Variants

We stated in Section 2.2 that an important advantage of working from a sampling viewpoint was that it allowed us to incorporate knowledge about the system structure or environment into the reduction process, in order to improve the quality of the model (i.e. to generate more accurate models, or a smaller models of the same accuracy). In this Section we describe in detail some particular contexts in which we have applied this philosophy with success [15].

Our philosophy is similar to that in a Bayesian view of statistics: we believe assumptions in the modeling process, in this case about the system inputs, should be stated clearly, and that these assumptions should, first, be stated clearly, but second, taken into account in the modeling procedure, possibly leading to better results in the process. One way of quantifying uncertainty is via probabilistic models that we embed in our sampling procedure.

We believe this perspective presents subtle distinction, both with many principal-components/POD procedures as well as with TBR. In many POD procedures, the sampling is somewhat ad-hoc, which can lead to pathologies in the algorithms. For example, as time-domain simulation is expensive, on occasion approaches will be proposed that consider only a single input trajectory, in other words, a single input sample, which is too coarse a sampling even by our spartan standards. Likewise, we believe that the TBR procedure can be interpreted as a special case of our approach, where all possible inputs are considered equally important. In no practical context are all possible inputs equally important, so this formulation unnecessarily over-constraints the model reduction procedure. These assumptions are not typically explicitly stated, but they are (implicitly) present nonetheless.

Now we will discuss three contexts in which we have exercised the PMTBR approach: restricted frequency ranges, estimating information about restricted or relations between circuit inputs, and parameter-varying problems.

3.1 Frequency Selectivity

Consider defining a "frequency-weighted" Gramian as

$$X_{FW} = \int_{-\infty}^{\infty} (j\omega E - A)^{-1} BB^T (j\omega E - A)^{-H} w(\omega) d\omega. \tag{16}$$

where $w(\omega)$ is the "weighting" function. The more appropriate the weighting function to our problem at hand, the better we expect the performance of the reduction algorithm to be. Seen from this viewpoint, TBR is a generic, somewhat naive, algorithm as it presumes complete ignorance of frequency content. The weighting function in "standard" TBR is most appropriate for white noise inputs where nothing is really known about frequency content. However, it is quite common to have some knowledge of the actual frequency distribution of the inputs. Often, the inputs are limited to a known band, for example we may approximate them as zero above a maximum frequency.[1] Therefore, we propose to select the weighting such that we can truncate the integral in Eqn. (16) to a finite interval, and use the resulting "finite-bandwidth" Gramian for model reduction. Since the resulting Gramian places more emphasis on frequencies of relevance, we expect to achieve better performance, for a given model order, on problems with finite bandwidth inputs.

In a practical implementation, with a finite number of frequency samples, weighting can be accomplished by adjusting the weights w_k and/or location of samples ω_k. In fact, every ZW-matrix implicitly defines a frequency weighting scheme. For this reason, it is better to choose points/weights in PMTBR (perhaps adaptively) according to the expected frequency profile of the system and the inputs, than to try to achieve convergence to the TBR singular values themselves.

This analysis provides another explanation for the empirically observed fact that multipoint projection can sometimes exhibit better relative error performance than generic TBR. Unlike TBR, which spreads out the effort in a uniform way, multipoint projection concentrates its effort on points where the projection is performed, which in essence is a weighting procedure similar to the one prescribed in (16). By placing higher weight on specific points and areas, multipoint projection leads to better relative performance in those areas. In the search for good global error performance, TBR can over-emphasize areas of the transfer function that are large in magnitude. When such regions are of interest to the problem at hand, TBR is a nearly optimal method. However, when such regions lie out of the frequency band of interest, or lead to excessive sacrifice of relative error for absolute error, TBR may not do as well as multi-point projection. PMTBR, on the other hand, can always be tailored to the problem at hand.

The Frequency Selective TBR procedure is shown as Algorithm 2. The similarity with Algorithm 1 should be fairly obvious, the main distinction being in the point selection algorithm.

[1] No causal system has zero frequency response over an finite interval, but this is a common engineering approximation.

Algorithm **2** : **Frequency Selective TBR**

/* *assuming M bands of interest are previously defined* */
1. *Define the range corresponding to the frequency bands of interest,* $S = \bigcup_{k=1}^{M} S_k$
2. *Do until satisfied:*
3. *Select a frequency point* s_i **within** S.
4. *Compute* $z_i = [s_i I - A]^{-1} B$.
5. *Form the matrix of columns* $Z_{(i)}^{fs} = [z_1, z_2, \ldots, z_i]$.
6. *Construct the SVD of* $Z_{(i)}^{fs}$.
   If the error is satisfactory, set $Z = Z_{(i)}^{fs}$, *go to Step 7.*
   Otherwise, go to Step 3.
7. *Construct the projection space V from the orthogonalized column span of Z, dropping columns whose associated singular values fall below a desired tolerance.*

3.2 Nested Bandlimited Models

In many contexts it is reasonable to *enforce* a frequency-range limit. The modeling system might provide a series of models, M_1, M_2, M_3 valid to frequencies $f_1 < f_2 < f_3$ respectively. The parent application may choose M_1, presumably the most compact model, if it is known in a given context that inputs are bandlimited to below f_1. If the most complex models are needed less often, or with less probability, than the simple models, then the computational savings in the parent analysis can be substantial.

Note that by properly ordering the sampling computations, it is possible to compute a single "parent" model from which all the child models M_1, M_2, M_3 can be simply obtained. There is therefore little overhead introduced into the reduction procedure, compared to a worst-case analysis, by using this strategy. This idea of incremental model re-construction was exploited in [22]; implementation details are discussed therein. In fact, there is considerable advantage to performing *all* the reduction computations in such an incremental fashion. In contrast, as far as we know, it is not easy to build TBR-like models incrementally, nor is it useful to consider relations among different models implicit in TBR.

3.3 Correlated Inputs

In the previous section, we argued that exploiting knowledge of the frequency profiles of the inputs can lead to more efficient reduction procedures. This is the simplest case to illustrate, although the computational gains, though real, are limited on practical problems. In this section we present one approach for exploiting the relation *between* the inputs of a multi-input system.

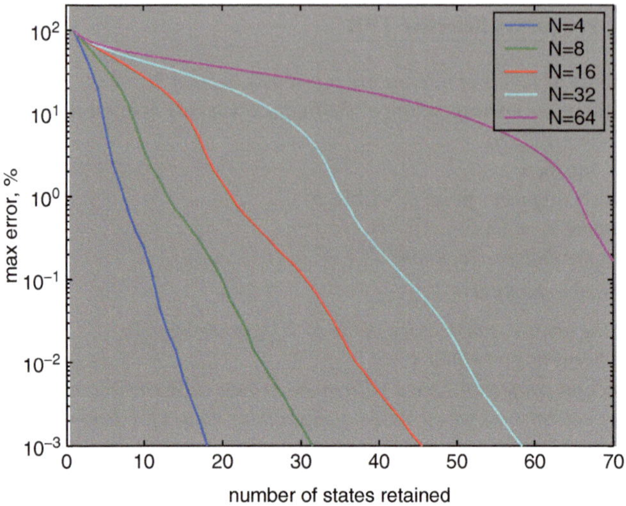

Fig. 1. TBR error bounds for 12×12 RC mesh as function of number of inputs.

Circuit networks with a large numbers of input/output ports, that is, networks with many columns in the matrices defining the inputs, are not very "reducible" by most projection methods. However, such "massively coupled" parasitic networks occur in many important cases, such as substrate analysis, package modeling, and on large digital interconnect networks and any technique that is able to compress such networks is of eminent practical use. To motivate our analysis for this particular setting, we will use as an example system a simple RC mesh. To gain better understanding of such systems, we vary the number of inputs attached to a mesh of fixed size. Figure 1 shows the TBR error bound obtained from the Hankel singular values as a function of the number of inputs. We can conclude that the order needed for good accuracy grows with the number of inputs. Even in this simple RC circuit case, for the 64-input case, low accuracy (20% error bound) requires at least 40 states in the reduced model. Other available procedures are likewise impractical. The moment-matching (Krylov-subspace) family of algorithms, such as PRIMA [4] and PVL [21], lead to models whose size is the number ports multiplied by the number of moments matched. However, while often it is necessary to retain all the input ports if the full impact of parasitic effects is to be correctly estimated [23], often there are relations between the inputs (or outputs) at different network ports that can be exploited to give a smaller model.

One way to obtain a useful relation between different model inputs is by estimating a relation between representative time-domain waveforms. Consider taking a set of N samples of input waveforms, u_k for input k, $k = 1 \ldots p$ [16]. A correlation matrix [24] K for the inputs can be estimated as

$$K_{ij} = \frac{1}{N} \sum_{l=1}^{N} u_i^l u_j^l \tag{17}$$

Algorithm **3** : **Input Correlated TBR**

1. *Construct the SVD of inputs* $\mathcal{U} = V_K S_K U_K^T$
2. *Do until bored:*
3. *Draw a vector* $r \in \mathbb{R}^p$ *by taking* p *draws from a normal distribution, variances given by* S_K.
4. *Select a frequency point* s_i.
5. *Compute* $z_i = [s_i I - A]^{-1} B U_K r$.
6. *Form the matrix of columns* $Z_{(i)}^{ic} = [z_1, z_2, \ldots, z_i]$.
7. *Construct the SVD of* $Z_{(i)}^{ic}$.
   If the error is satisfactory, set $Z = Z_{(i)}^{ic}$, *go to Step 8.*
   Otherwise, go to Step 3.
8. *Construct the projection space* V *from the orthogonalized column span of* Z, *dropping columns whose associated singular values fall below a desired tolerance.*

As is the usual case, the actual correlation matrix need not be formed. Instead, we can take the SVD of the matrix $\mathcal{U}$ whose columns are the input samples u_k, i.e. $\mathcal{U} = V_K S_K U_K^T$ with U_K, V_K orthonormal.

The key insight is, the closer K is to being a low-rank matrix, the faster the eigenvalues of the approximation Gramian (the singular values of Z) decay, and the smaller the model, for a given expected error, that can be tolerated. This will be the case if the inputs exhibit some correlated behavior. Under the hypothesis that K is a suitably representative model of the possible inputs, no accuracy will be lost. In practical problems such fidelity can be guaranteed if we are somewhat conservative in the specification of the correlation matrix which we can do by being conservative in the choice of samples to be sure we "over-approximate" K. $K = I$ corresponds to the ultimate degree of safety, total ignorance about the inputs, but it also usually leads to overly large models. As in the frequency selective case, though the physical interpretation as an absolute error bound no longer applies, the eigenvalues of the Gramian can still be used for error control, as they can be given an interpretation associated with likelihood of error in the probabilistic input model. The final algorithm is shown as Algorithm 3.

3.4 Statistical Parameter Variation

Parametric model order reduction has become an area of intense research in the last few years, owing to increasing variability in process control as technologies continue to shrink. For the purposes of this chapter we will lump all such variations into a set of parameters $\lambda_1, \ldots, \lambda_M$ through which the matrices of the interconnect model vary. We assume the equations are written in such a way that the variations are confined to the G, C matrices, so that the parametric model is

$$C(\lambda_1,\ldots,\lambda_M)\frac{dx}{dt}+G(\lambda_1,\ldots,\lambda_M)x=Bu,\quad y=L^Tx. \tag{18}$$

The fundamental difficulty in generating reduced models from (18) is that the number of parameters M (the dimension of the parameter variable space) may be large. Interconnect modeling in a parameter variation context consists of two main steps. First, the variation of the unreduced circuit matrices must be modeled in a tractable form. A typical assumption is of small variation, such that the parameter-varying matrices can be expressed as linear deviations from some nominal value. This leads to the affine model [25, 26]

$$G(\lambda_1,\ldots,\lambda_M)=G_0+\sum_k \Delta G_k\delta\lambda_k \tag{19}$$
$$C(\lambda_1,\ldots,\lambda_M)=C_0+\sum_k \Delta C_k\delta\lambda_k.$$

Linearity is often a good assumption for process variation in IC analysis. Our analysis does not require affine models, in fact the algorithm can be applied to any model form in which the projected matrices (with possibly nonlinear parameter dependence) can be computed efficiently. However, affine models are a simple form in which to illustrate the overall strategy, so we restrict the discussion to this form here.

The second question to be addressed is the choice of the projection matrix. Assuming that a probability density for λ exists on some domain S_λ, a "variational Gramian" may be defined as

$$X_\lambda=\int_{S_\lambda}\int_{-\infty}^{\infty}(j\omega C_\lambda+G_\lambda)^{-1}BB^T(j\omega C_\lambda+G_\lambda)^{-H}p(\omega,\lambda)\,d\omega d\lambda. \tag{20}$$

Any sort of explicit manipulation of this quantity is impractical due to the exponential growth in complexity with the number of parameters, each parameter leading to a new dimension in the integral. For example, it is not practical to compute the integral via products of 1D quadrature rules. Multi-dimensional moment-matching algorithms are likewise impractical. However, one could imagine that a Monte Carlo like approach to integration might be useful. The PMTBR procedure for parameter-varying systems, shown as Algorithm 4, is precisely such an approach [17].

It is important to understand that the accuracy of the model reduction algorithm is generally much better than would be expected merely by examining the the convergence of the integral in Eqn. (20). Monte Carlo integration generally incurs an error that decreases slowly, as (at best) $N^{-1/2}$ where N is the number of samples, and irregularly. On examples occuring in integrated circuit analysis, the performance of the PMTBR approach is vastly better. This is because the accuracy of the reduced order model depends on the projection subspaces, not on the accuracy of the quadratures. For example, a two-state circuit is exactly represented by a model obtained via *any* projection matrix V with two columns. More generally, the topology of practical circuits puts strong constraints on the possible interesting subspaces. For example, RC circuits do not exhibit irregular or oscillatory behavior. Even with parameter variation, the space of responses "smooth" on a given topology is effectively

Algorithm **4** : **Poor Man's Variational TBR**

1. *Select a quadrature rule of* N *nodes in an* $M+1$ *dimensional space* $(s, \lambda_1, \ldots, \lambda_M)$ *with weights* w.
2. *For each node* i, *compute*
$$z_i = [s_i C(\lambda_1^i, \ldots, \lambda_M^i) + G(\lambda_1^i, \ldots, \lambda_M^i)]^{-1} B.$$
3. *Form the matrix of columns* $Z = [z_1, z_2, \ldots, z_N]$ *and weights* $W_{ii} = w_i$.
4. *Construct the singular value decomposition of* ZW.
5. *Construct the projection space* V *from the orthogonalized column span of* Z, *dropping columns whose associated singular values fall below a desired tolerance.*

of small dimension. Thus, even if the number of statistically varying parameters is large, the projection matrix V may capture the dominant behavior with a small number of columns, if the columns are chosen wisely. Moment matching schemes cannot do this because they are keyed to parameter counts.

4 Analysis and Comparisons

4.1 Comparison to Moment-Matching Schemes

Many methods prominent in the integrated circuit analysis literature are projection-based rational interpolation methods [3–5]. Most popular are "moment-matching" methods that contruct the projection matrix V from a Krylov subspace thus matching the transfer function and its derivatives ("moments", when expressed in time-domain) at a specified point, typically zero frequency. There are good practical efficiency reasons for this choice. For a given model order the multipoint approximants tend to be more accurate, but are usually more expensive to construct. Given M complex frequency points s_k, a projection matrix may be constructed whose kth column is

$$z_k = (s_k I - A)^{-1} B. \tag{21}$$

Clearly, in the limit where all the sampling vectors are retained, the previously discussed methods degenerate to multipoint rational interpolation.

4.2 Comparison to Balanced Truncation

Model reduction via balanced truncation [6, 27] is based on analysis of the controllability and observability Gramians X, Y respectively computed from the Lyapunov equations

$$AX + XA^T = -BB^T, \tag{22}$$
$$A^T Y + YA = -C^T C. \tag{23}$$

An obvious tactic for improvement of multipoint rational interpolation is to perform an SVD on the vectors z_k computed as in Eqn. (21). So, in the opposite limit, with many samples suitably spaced, the PMTBR algorithm, which constructs projection matrices by multipoint frequency sampling, as in (21), followed by an SVD, in fact produces the same information as in X, Y. The singular values obtained from such a procedure approximate the Hankel singular values, and can thus be used for order and error control. We will argue below that even with the approximations obtained from small number of samples, these values can be used for reasonably effective error control.

Two potentially problematic differences arise in our approach when compared with full balanced truncation. First, in TBR, reduction is performed by projection onto the invariant subspaces associated with the dominant eigenvalues of the product of Gramians XY. In special "balanced" coordinates, $X = Y$, and it is sufficient to use "one-sided" approximations, i.e., those computed from the matrix pair (A, B). In integrated circuit analysis, there are important special cases where $A = A^T, C^T = B$, and so $X = Y$. Therefore we can ignore balancing considerations. It is possible to make our approach 'two-sided by also working with samples of $(sI - A^T)^{-1}C^T$ [8]. However, in fact it is fairly unusual to encounter IC systems that are "very" unbalanced. The second difference concerns stability. In general, since the PMTBR procedure does not construct the true system Gramians, the connection to stability analysis, and stability of models obtained via truncation, is lost. However, many integrated circuit problems are naturally posed in a coordinate system [4] such that *any* system obtained from congruence transform and/or truncation also satisfies a Lyapunov inequality, which ensures stability of the reduced models.

4.3 Error Analysis

In general in the PMTBR procedure, due to both the selective sampling and the aggressive approximation process, the singular values of the Z-matrix can no longer be considered estimates of the Hankel singular values, and therefore precise statements about error bounds are lost. However, good models might be obtained well before convergence to the integral form of the Gramians. A good example is a very high-Q two-state RLC circuit. Only two sample vectors are required to obtain the exact model, which PMTBR will correctly predict, but obtaining the exact Gramian by numerical integration could require a very large number of quadrature points, especially if the points are placed in a naive manner such as a uniform distribution. We argue that the singular values from truncated modes can still be interpreted as errors on the "filtered" system, i.e. finite-bandwidth or weighted errors. The matrix ZZ^T is an estimator for the correlation matrix K_{xx} of the random process obtained from the mapping of the inputs u to the state x. The eigenanalysis of K_{xx} reveals which subspaces are necessary to capture the dominant portion of the variance in x. So we have found that, again assuming a sampling density consistent with the weighting $w(\omega)$, the singular values usually give a fairly good guide to model order well before convergence is achieved. Our experiments indicate that when, for a number of samples in excess (e.g. twice) of the model order, the singular value distribution exhibits a small "tail" (that is, for a "small" ϵ, $\exists k : \epsilon > \sum_{i=k+1}^{\infty} \sigma_i$), then sufficient order and point placement has been achieved.

To minimize computational effort, it is desirable to run the algorithm until a desired accuracy level is obtained, and at that point cease to add points into the sample space of PMTBR. We would make this decision by looking at a small number of the trailing singular values of the Z matrix, and stopping when the sum drops below a given threshold. This means that estimates of the singular values of ZW must be available each time a sample becomes available. In the previous development we utilized the SVD decomposition because of its direct connection to eigenvalue analysis. However, the SVD is not the most appropriate tool for such an adaptive order control procedure, since no fast update procedure is known. Since in our case, we do not need the actual singular values themselves, only the ability to estimate the magnitude of a trailing few, and to obtain a basis for the dominant subspaces, other rank revealing factorizations that possess better updating properties may be more appropriate, for example rank-revealing QR factorizations [28, 29]. For additional work on point selection, see [22].

5 Experimental Results

5.1 Error Control

In the first set of computational examples we examine the empirical properties of the error control strategy. Several thousand resistor-capacitor (RC) networks were obtained from parasitic extraction of industrial circuits. We set the PMTBR algorithm to sample in the interval from DC to 10GHz, and computed the maximum error between the input admittances of the reduced models and the original models at a set of discrete frequency points densely spaced in the same interval. Our target accuracy was two digits.

We will present two sets of statistics collected from a large set (over 25,000 cases) of RC networks representative of those encountered in timing and signal integrity analysis. Figure 2 shows a scatterplot of the measured maximum error in the multiport driving admittance transfer functions plotted vs. the estimated error from the leading singular value of the sample matrix Z. Figure 3 shows a histogram (number of examples achieving a given worst-case error value) of the relative error results.

These results are typical in our experience for this problem class. We observe from Figure 2 that the measured errors are indeed highly correlated to the predicted errors. This enables an effective error control strategy, as motivated by Figure 3. Note the "wall" at about two digits of accuracy in the histogram that indicates correct operation of the strategy.

We need to present several caveats and clarifications about these results. First, in applications it is typically the maximum error that needs to be controlled, whereas SVD estimates we might expect to be more closely tied to average errors. We introduce some error margin to account for this. Second, both relative and absolute error metrics need to be considered, especially for multi-port networks. Often one error can be artificially inflated. For example, small transfer function entries lead to

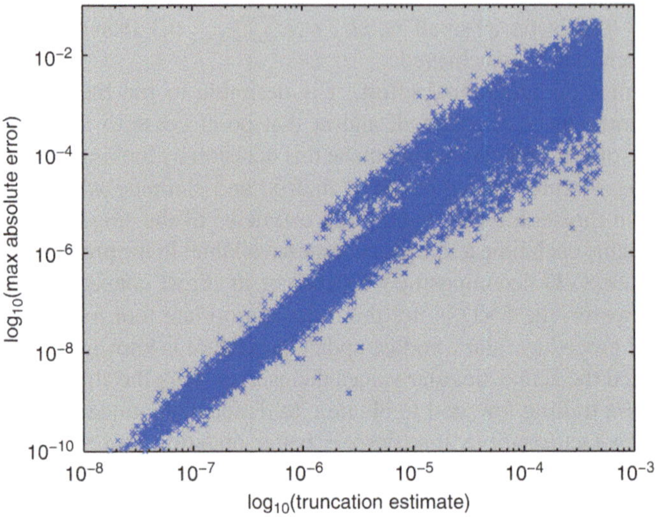

Fig. 2. Error correlation statistics for RC networks. Note log scales on axes.

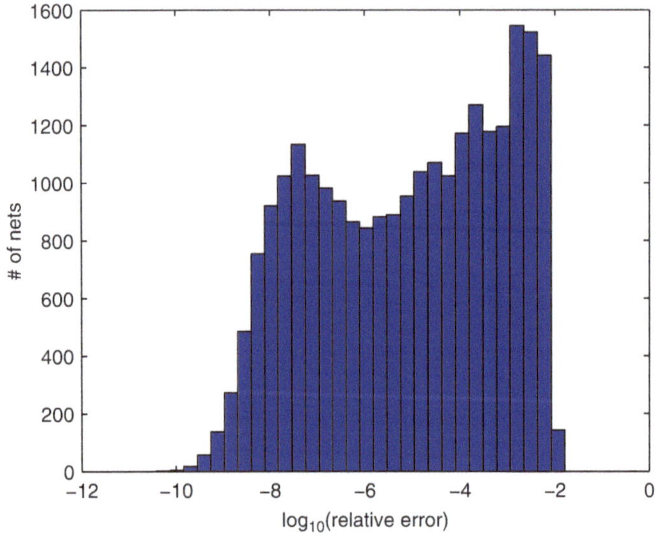

Fig. 3. Error statistics for RC networks. Note log scale on x-axis.

artificially inflated relative error metric that can lead to anomalous out-of-spec entries in plots like the above, for samples which on an absolute basis have perfectly satisfactory error. Finally, estimates based purely on the singular values, while workable on many problems, are relatively crude metrics on which we hope to improve in the future.

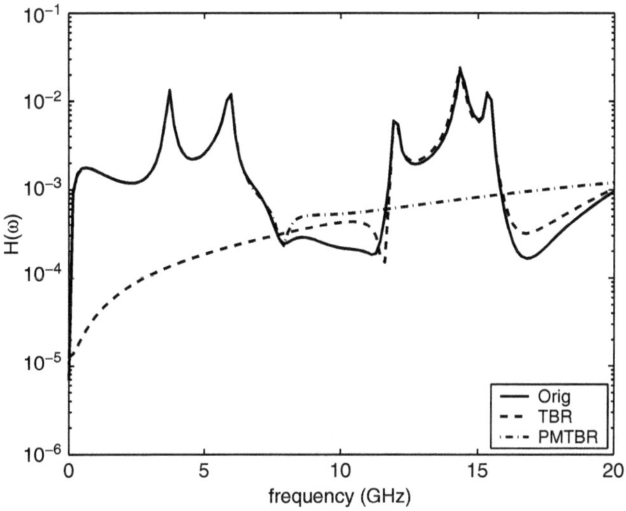

Fig. 4. Transfer function approximations for the connector example. Note PMTBR has better finite-bandwidth performance than TBR.

5.2 Frequency Selectivity

Now we begin to investigate the behavior of the different algorithmic variants of PMTBR. The first algorithm variant we demonstrate illustrates frequency selectivity. Figure 4 shows a plot of the exact transfer function of an 18 pin shielded connector structure [30] as well as approximations obtained with TBR and PMTBR. We are interested in testing the ability of the PMTBR algorithm to produce approximations on a finite bandwidth, in this case the frequency range of of zero to 8 GHz. In PMTBR, we set the sampling selection mechanism to generate samples uniformly distributed in the the frequency range from DC to $8GHZ$. These samples were used, after SVD, to generate an order 18 PMTBR approximation. For the TBR model, we found that 30 was the minimum order required for TBR to provide reasonable representation of *any* features in the 0-8GHz range, so those results are shown. PMTBR concentrates its effort in the desired interval, 0-8GHz, and shows good accuracy there. On the other hand, TBR concentrates effort around 15GHz because of the relative amplitude of the transfer function. Even with the higher order approximation, the TBR model is not accurate in the frequency range of interest. PMTBR is easily focused on the 8GHz and below range merely by selection of sampling points, and does not waste effort with approximation at higher frequencies.

5.3 Input Compression

The next algorithmic variant demonstrates how a two-stage approach to system approximation can result in dramatic increases in reduction efficiency by first approximately learning, then exploiting in reduction, particular system features. The test system is an analog circuit (a data converter) with parasitics from an extracted substrate network. The substrate network considered by itself is for the most part unreducible with standard projection methods.

In the first stage of analysis we perform a simulation of the data converter circuit without the substrate parasitics. From the MOS transistor bulk current signals we obtain an estimate of the correlation matrix for the inputs to the substrate network. In the second stage, we use this correlation matrix to drive sampling in the input-correlated PMTBR procedure. For a 1000-port segment of the substrate network, Figure 5 shows the error estimate data obtained from the singular value analysis in PMTBR. In this case acceptable accuracy can be achieved with a 30-state model. This is a 30X reduction in model size from the original system, which translates to a more than 1000X reduction in simulation run time and memory consumption. In fact it was not practical to complete detailed circuit simulations with the original model.

Due to these constraints on model size, to show that the final models do indeed produce results, we have examined only a section of the full parasitics network. The approach described above was applied to a 150-port subsection of the substrate network. Figure 6 shows the results of comparing time-domain simulations with the full model to simulations with a four- and eight- state reduced models. With the 20X reduction of the eight-state model, comparable to the results shown for the full 1000-port network above, very good match is obtained. We emphasize these results are dependent on the specific interactions between the substrate network and the circuit activity. Without the estimates of circuit activity, we could not obtain models of small size. With poor estimates – in particular, using this model for an entirely different circuit – the results would likely be inaccurate.

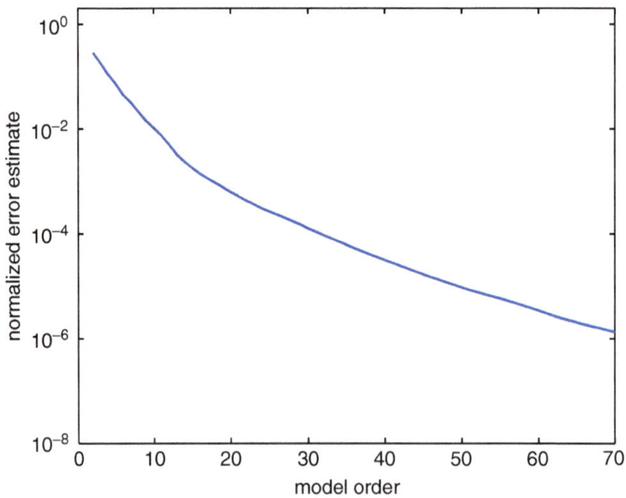

Fig. 5. Error estimate based on singular value analysis of Z-matrix from input-correlated TBR, for 1000-port substrate network with inputs from data converter example.

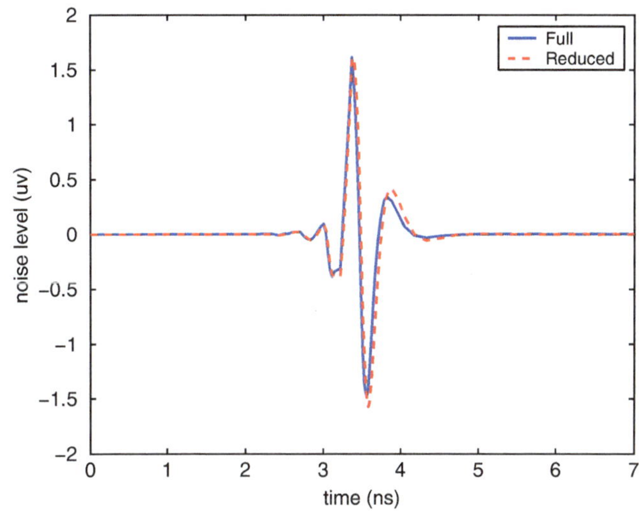

Fig. 6. Simulation results for data converter example, 150 port substrate models, full vs. 4-state reduced model.

5.4 Parametric Networks

The final example comes from analysis of parameter-varying systems. Two metrics are of concern here. First, we are interested in whether the sampling and error control strategies produce satisfactory worst-case errors over a large collection of networks. We have performed similar computations as in Figure 3, computing error statistics for large collections of industrial examples. The results are quite similar to the nominal case, so due to space considerations we omit the detailed results here. A more interesting question to investigate is the tradeoff of error versus the number of samples. Therefore, to gain more insight into the behavior of the algorithm, we show the convergence behavior for a single, moderately large, network, with a large number of parameters. Due to either the Monte Carlo nature of the sampling, or the large dimensionality of the parameter space, we might expect convergence to slow considerably when variation is introduced. Figure 7 shows the error of the reduced network as a function of the model order. As generally the model order is equal to the number of samples, or one less, the sample number is a good measure of computational effort. This particular network had about 600 nodes and thirty parameters in the network. The parameter range of variation was typical for a contemporary (65nm) IC fabrication process. Despite the large number of parameters in the network, the convergence is seen to be very rapid. In our experience this is a fairly typical result for RC networks with process variation – actually, this example is relatively difficult among signal net reduction problems. Due to the interaction of network topology and parameter variation, the subspaces computed by PMTBR are low-dimensional, and in fact the addition of variation does not affect the dominant subspaces to a great degree. This is an important experimental observation that is not expected from direct analogies to interpolation-based methods in multiple dimensions.

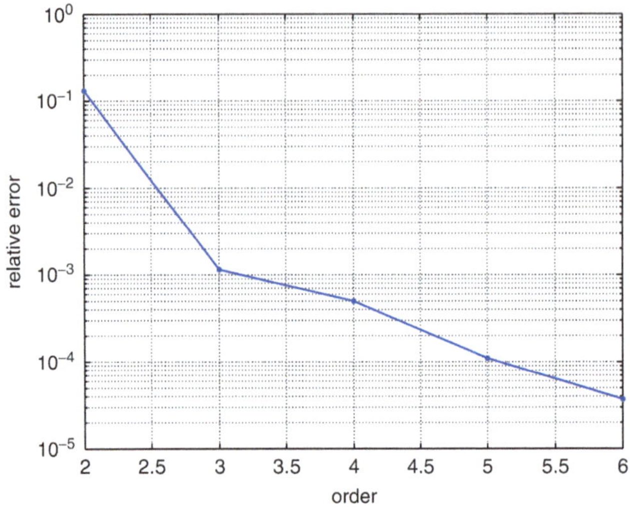

Fig. 7. Error vs. order for parameter-varying RC network.

6 Conclusions

In this work we discussed a connection between truncated balanced realization (TBR) model reduction methods and multipoint rational approximation/projection techniques. Building on this connection we presented a family of algorithms, generically entitled PMTBR, that can be viewed as bridging both type of methods and leading to useful new techniques that have been shown to have some advantages, particularly in generating smaller reduced models, and possibly in order control and error estimation. TBR is a principal components analysis of the functionals defined by the state-space model, and naturally arises from time-domain theory of state-space systems. PMTBR arises naturally from a numerical approximation viewpoint of frequency domain data, with the principal components analysis manifested purely in the SVD of sampled data.

A potentially more important observation is that the existing model order reduction algorithms contain implicit assumptions about the inputs to the systems being modeled. To each set of assumptions corresponds an implicit model of the inputs themselves. When correlated with actual information available from application domains, these "input models" seem unduly restrictive, implying that the assumptions implicit in the standard model order reduction schemes may be somewhat naive. Our hope is that increased care in modeling the system inputs themselves can lead to more powerful modeling schemes.

We have presented technique for model reduction of systems with band-limited inputs, large number of inputs, as well as systems in the presence of random model parameter perturbations, and demonstrated their effectiveness on industrial examples. The proposed methods are computationally simple to implement, have good accuracy and error properties, and possess simple error estimation and order control procedures.

The studies in this chapter also suggest some interesting conclusions about general problems of nonlinear modeling. While principal-components type analyses, such as the TBR and PMTBR [15] algorithms can generate somewhat more compact models than moment- and point- matching schemes, their utility is much enhanced in the higher-dimensional spaces induced by problems with multiple parameters. The performance gap between the "best" linear approximation spaces and the moment-matching spaces can be huge in high dimensions.

References

1. Bernard N. Sheehan. TICER: Realizable reduction of extracted RC circuits. In *International Conference on Computer Aided-Design*, pages 200–203, Santa Clara, CA, November 1999.
2. Lawrence T. Pillage and Ronald A. Rohrer. Asymptotic Waveform Evaluation for Timing Analysis. *IEEE Trans. Computer-Aided Design*, 9(4):352–366, April 1990.
3. Peter Feldmann and Roland W. Freund. Efficient linear circuit analysis by Padé approximation via the Lanczos process. In *EURO-DAC'94 with EURO-VHDL'94*, September 1994.
4. A. Odabasioglu, M. Celik, and L. T. Pileggi. PRIMA: passive reduced-order interconnect macromodeling algorithm. *IEEE Trans. Computer-Aided Design*, 17(8):645–654, August 1998.
5. Eric Grimme. *Krylov Projection Methods for Model Reduction*. PhD thesis, Coordinated-Science Laboratory, University of Illinois at Urbana-Champaign, Urbana-Champaign, IL, 1997.
6. Bruce Moore. Principal Component Analysis in Linear Systems: Controllability, Observability, and Model Reduction. *IEEE Transactions on Automatic Control*, AC-26(1):17–32, February 1981.
7. M. Green. Balanced stochastic realizations. *Linear Algebra and its Applications*, 98:211–247, January 1988.
8. Athanasios C. Antoulas. *Approximation of Large-Scale Dynamical Systems*. Society for Industrial and Applied Mathematics, Philadelphia, PA, USA, 2005.
9. Keith Glover. All optimal Hankel-norm approximations of linear multivariable systems and their l^{∞}-error bounds. *International Journal on Control*, 39(6):1115–1193, June 1984.
10. I. M. Jaimoukha and E. M. Kasenally. Krylov subspace methods for solving large Lyapunov equations. *SIAM Journal on Numerical Analysis*, 31:227–251, 1994.
11. I. M. Jaimoukha and E. M. Kasenally. Oblique projection methods for large scale model reduction. *SIAM J. Matrix Anal. Appl.*, 16:602–627, 1995.
12. Jing-Rebecca Li, F. Wang, and J. White. An efficient Lyapunov equation-based approach for generating reduced-order models of interconnect. In 36th *ACM/IEEE Design Automation Conference*, pages 1–6, New Orleans, Louisiana, June 1999.
13. Jing-Rebecca Li, F. Wang, and J. White. Efficient model reduction of interconnect via approximate system grammians. In *International Conference on Computer Aided-Design*, pages 380–383, SAN JOSE, CA, November 1999.

14. D. C. Sorensen and A. C. Antoulas. Projection methods for balanced model reduction. Technical Report TR01-03, Rice University, Houston, TX, March 2001.
15. Joel R. Phillips and L. Miguel Silveira. Poor Man's TBR: A simple model reduction scheme. *IEEE Trans. Computer-Aided Design*, 24(1):43–55, Jan. 2005.
16. L. Miguel Silveira and Joel Phillips. Exploiting input information in a model reduction algorithm for massively coupled parasitic networks. In 41st *ACM/IEEE Design Automation Conference*, pages 385–388, San Diego, CA, USA, June 2004.
17. Joel Phillips. Variational interconnect analysis via PMTBR. In *International Conference on Computer Aided-Design*, pages 872–879, San Jose, CA, USA, November 2004.
18. M. Kirby and L. Sirovich. Application of the Karhunen Loeve procedure for the characterization of human faces. *IEEE Transactions on Pattern Analysis and Machine*, 12(1):103–108, 1990.
19. Sanjay Lall, Jerrold E. Marsden, and Sonja Glavaski. A subspace approach to balanced truncation for model reduction of nonlinear control systems. *Int. J. of Robust and Nonlinear Control*, 12(6):519–535, 2002.
20. K. Willcox and J. Peraire. Balanced model reduction via the proper orthogonal decomposition. *American Institute of Aeronautics and Astronautics Journal*, 40(11):2323–2330, 2002.
21. Peter Feldmann and Roland W. Freund. Reduced-order modeling of large linear subcircuits via a block Lanczos algorithm. In 32nd *ACM/IEEE Design Automation Conference*, pages 474–479, San Francisco, CA, June 1995.
22. L. Miguel Silveira and Joel R. Phillips. Resampling plans for sample point selection in multipoint model order reduction. *IEEE Trans. Computer-Aided Design*, 25(12):2775–2783, December 2006.
23. Joel R. Phillips and L. Miguel Silveira. Simulation approaches for strongly coupled interconnect systems. In *International Conference on Computer Aided-Design*, pages 430–437, November 2001.
24. A. Papoulis. *Probability, random variables, and stochastic processes*. McGraw Hill, New York, 1991.
25. Y. Liu, L. T. Pileggi, and A. J. Strojwas. Model order reduction of RC(L) interconnect including variational analysis. In 36th *ACM/IEEE Design Automation Conference*, pages 201–206, June 1999.
26. L. Daniel, O. C. Siong, S. C. Low, K. H. Lee, and J. K. White. A multiparameter moment-matching model-reduction approach for generating geometrically parametrized interconnect performance models. *IEEE Trans. Computer-Aided Design*, 23:678–693, May 2004.
27. M. G. Safonov and R. Y. Chiang. A Schur method for balanced truncation model reduction. *IEEE Transactions on Automatic Control*, 34(7):729–733, July 1989.
28. T. F. Chan. Rank revealing qr factorizations. *Linear Algebra and its Applications*, 88/89:67–82, 1987.
29. Christian H. Bischof and G. Quintana-Ortí. Computing rank-revealing QR factorizations of dense matrices. *ACM Trans. Math. Softw.*, 24(2):226–253, 1998.
30. Nuno Marques, Mattan Kamon, Jacob White, and L. Miguel Silveira. A mixed nodal-mesh formulation for efficient extraction and passive reduced-order modeling of 3d interconnects. In 35th *ACM/IEEE Design Automation Conference*, pages 297–302, San Francisco, CA, USA, June 1998.

A Survey on Model Reduction of Coupled Systems*

Timo Reis[1] and Tatjana Stykel[2]

[1] Institut für Mathematik, MA 4-5, Technische Universität Berlin, Straße des 17. Juni 136, 10623 Berlin, Germany
reis@math.tu-berlin.de.

[2] Institut für Mathematik, MA 4-5, Technische Universität Berlin, Straße des 17. Juni 136, 10623 Berlin, Germany
stykel@math.tu-berlin.de.

Summary. In this paper we give an overview of model order reduction techniques for coupled systems. We consider linear time-invariant control systems that are coupled through input-output relations and discuss model reduction of such systems using moment matching approximation and balanced truncation. Structure-preserving approaches to model order reduction of coupled systems are also presented. Numerical examples are given.

1 Introduction

Modelling and simulation of complex physical and technical processes yield coupled systems that consist of ordinary differential equations, differential-algebraic equations and partial differential equations. Such systems arise in many practical applications including very large system integrated (VLSI) chip design and micro-electro-mechanical systems (MEMS), e.g. [10, 14, 21, 52, 58]. As the number and density of components on a single chip increase and feature sizes decrease, different physical effects such as thermal interaction, electromagnetic radiation, substrate noise and crosstalk cannot be ignored anymore. Furthermore, the design of micro- and nano-structures requires the development of new multi-physical models describing their complex internal behavior. Another application area of coupled systems is in subdomain decomposition. Partial differential equations on complicated spatial geometries may be represented as a system of partial differential equations on simpler domains coupled, for example, through boundary conditions.

As the mathematical models get more detailed and different coupling effects have to be included, the development of efficient simulation and optimization tools for large-scale coupled systems is a challenging task. Such systems consist of several subsystems whose inputs and outputs are coupled via additional algebraic relations. The subsystems usually have a high number of internal variables that leads to large

* Supported by the DFG Research Center MATHEON "Mathematics for key technologies" in Berlin.

memory requirements and computational complexity. To handle such large systems in simulation, control and optimization, their *model order reduction* (or *reduced-order modelling*) is indispensable. A general idea of model order reduction is to approximate a large-scale system by a reduced model of lower state space dimension that has the same behavior as the original system.

In the last years, many different model reduction methods have been developed in computational fluid dynamics, control design and electrical and mechanical engineering, see [4,11,47] for books on this topic. In this paper we review recent progress in dimension reduction of coupled systems. In structural dynamics, model reduction methods based on subsystem structuring have been of interest already for a long time [16,35,41]. Here, we will not consider these methods, but will rather focus on general concepts of model reduction of coupled systems developed in [42,53,60].

This paper is organized as follows. In Section 2 we introduce linear time-invariant coupled systems and give their closed-loop formulation. Section 3 deals with model order reduction of coupled systems. To make the paper self-contained, we briefly review model reduction techniques of balanced truncation and moment matching approximation. Furthermore, we report two approaches for reduced-order modelling of coupled systems based on the reduction of closed-loop systems (Section 3.1) and on structure-preserving model reduction (Section 3.2). The discussion of the advantages and disadvantages of these approaches is presented in Section 3.3. Finally, in Section 4 we consider some numerical examples.

2 Coupled Systems

Consider a system of k coupled linear time-invariant generalized state space subsystems in the first-order form

$$\begin{aligned} E_j\dot{x}_j(t) &= A_jx_j(t) + B_ju_j(t), \\ y_j(t) &= C_jx_j(t), \end{aligned} \tag{1}$$

or in the second-order form

$$\begin{aligned} M_j\ddot{x}_j(t) \,+\, D_j\,\dot{x}_j(t) \,+\, S_j\,x_j(t) &= B_ju_j(t), \\ C_{j2}\dot{x}_j(t) + C_{j1}x_j(t) &= y_j(t), \end{aligned} \tag{2}$$

that are coupled through the relations

$$u_j(t) = K_{j1}y_1(t) + \ldots + K_{jk}y_k(t) + H_ju(t), \quad j = 1,\ldots,k, \tag{3}$$

$$y(t) = R_1\,y_1(t) \,+ \ldots +\, R_k\,y_k(t). \tag{4}$$

Here E_j, A_j, M_j, D_j, $S_j \in \mathbb{R}^{n_j,n_j}$, $B_j \in \mathbb{R}^{n_j,m_j}$, C_j, C_{j1}, $C_{j2} \in \mathbb{R}^{p_j,n_j}$, $x_j(t)\in\mathbb{R}^{n_j}$ are internal state vectors, $u_j(t) \in \mathbb{R}^{m_j}$ are internal inputs and $y_j(t) \in \mathbb{R}^{p_j}$ are internal outputs. Furthermore, $K_{jl} \in \mathbb{R}^{m_j,p_l}$, $H_j \in \mathbb{R}^{m_j,m}$, $R_j \in \mathbb{R}^{p,p_j}$, $u(t) \in \mathbb{R}^m$ is an external input and $y(t) \in \mathbb{R}^p$ is an external output. Coupled systems of the form (1)–(4) are also known as *interconnected* or *composite systems*. The first-order systems of the form (1) arise in simulation of linear RLC circuits that consist of resistors, capacitors, inductors, voltage and current sources only [2, 34, 62]. In this case the components of the state vector $x_j(t)$ are the nodal voltages, the inductor currents and the currents through the voltage sources, $u_j(t)$ contains the currents and voltages of the current and voltage sources, respectively, and $y_j(t)$ consists of the voltages across the current sources and the currents through the voltage sources. The linear RLC circuits are often used to model the interconnections of VLSI networks. They can also be described by the second-order systems (2), where $x_j(t)$ consists of the nodal voltages only. Systems of the form (2) appear also in mechanical and structural dynamics. In this case, $x_j(t)$ is the displacement vector and $u_j(t)$ is the acting force. Furthermore, systems (1) and (2) arise from spatial discretization of instationary linear partial differential equations that describe, for example, heat transfer, vibrations, electromagnetic radiation or fluid flow.

Since the second-order system (2) can be rewritten as an equivalent first-order system of the form (1), in the following we will consider the coupled system (1), (4), (4) only. The matrices E_j in (1) may be singular, but we will assume that the pencils $\lambda E_j - A_j$ are *regular*, i.e., $\det(\lambda E_j - A_j) \not\equiv 0$ for $j = 1, \ldots, k$. In this case we can consider the transfer function of (1) given by $\boldsymbol{G}_j(s) = C_j(sE_j - A_j)^{-1}B_j$. If $E_j x_j(0) = 0$, then applying the Laplace transform to (1), we find that $\boldsymbol{y}_j(s) = \boldsymbol{G}_j(s)\boldsymbol{u}_j(s)$, where $\boldsymbol{y}_j(s)$ and $\boldsymbol{u}_j(s)$ are the Laplace transforms of $y_j(t)$ and $u_j(t)$, respectively. Thus, $\boldsymbol{G}_j(s)$ describes the input-output relation of system (1) in the frequency domain.

The transfer function $\boldsymbol{G}_j(s)$ is called *proper* if $\lim_{s\to\infty} \boldsymbol{G}_j(s) < \infty$, and *improper*, otherwise. System (1) is *asymptotically stable* if the pencil $\lambda E_j - A_j$ is *stable*, i.e., all its finite eigenvalues have negative real part. The transfer function $\boldsymbol{G}_j(s)$ of (1) is called *stable* if it has no poles in the closed right half-plane. Clearly, the asymptotically stable system (1) has the stable transfer function $\boldsymbol{G}_j(s)$. Note that the stability of $\boldsymbol{G}_j(s)$ does not, in general, imply that the pencil $\lambda E_j - A_j$ is stable. However, for any stable transfer function $\boldsymbol{G}_j(s)$ one can find a generalized state space representation (1) such that $\boldsymbol{G}_j(s) = C_j(sE_j - A_j)^{-1}B_j$ and $\lambda E_j - A_j$ is stable, see [36]. Let $\mathbb{H}_\infty$ be the space of all proper and stable rational transfer functions. We provide this space with the $\mathbb{H}_\infty$-*norm* defined for $\boldsymbol{G} \in \mathbb{H}_\infty$ by

$$\|\boldsymbol{G}\|_{\mathbb{H}_\infty} := \sup_{\mathbb{R}\mathrm{ee}(s)>0} \|\boldsymbol{G}(s)\|_2 = \sup_{\omega\in\mathbb{R}} \|\boldsymbol{G}(\mathrm{i}\omega)\|_2, \tag{5}$$

where $\|\cdot\|_2$ denotes the matrix spectral norm.

Let $n = n_1 + \ldots + n_k$, $p_0 = p_1 + \ldots + p_k$ and $m_0 = m_1 + \ldots + m_k$. Consider the coupling block matrices

$$R = [\,R_1, \;\ldots,\; R_k\,] \in \mathbb{R}^{p,p_0}, \qquad H = [\,H_1^T, \;\ldots,\; H_k^T\,]^T \in \mathbb{R}^{m_0,m}, \tag{6}$$

and $K = [K_{j,l}]_{j,l=1}^k \in \mathbb{R}^{m_0,p_0}$ together with the block diagonal matrices

$$\begin{aligned} E &= \mathrm{diag}(E_1, \ldots, E_k) \in \mathbb{R}^{n,n}, & A &= \mathrm{diag}(A_1, \ldots, A_k) \in \mathbb{R}^{n,n}, \\ B &= \mathrm{diag}(B_1, \ldots, B_k) \in \mathbb{R}^{n,m_0}, & C &= \mathrm{diag}(C_1, \ldots, C_k) \in \mathbb{R}^{p_0,n}. \end{aligned} \tag{7}$$

Let $\boldsymbol{G}(s) = C(sE-A)^{-1}B = \operatorname{diag}\big(\boldsymbol{G}_1(s), \ldots, \boldsymbol{G}_k(s)\big)$. If $I-\boldsymbol{G}(s)K$ is invertible, then the input-output relation of the coupled system (1), (4), (4) can be written as $\boldsymbol{y}(s) = \boldsymbol{\mathcal{G}}(s)\boldsymbol{u}(s)$, where $\boldsymbol{y}(s)$ and $\boldsymbol{u}(s)$ are the Laplace transforms of the external output $y(t)$ and the external input $u(t)$, respectively, and the closed-loop transfer function $\boldsymbol{\mathcal{G}}(s)$ has the form

$$\boldsymbol{\mathcal{G}}(s) = R\big(I - \boldsymbol{G}(s)K\big)^{-1}\boldsymbol{G}(s)H = R\,\boldsymbol{G}(s)\big(I - K\boldsymbol{G}(s)\big)^{-1}H. \qquad (8)$$

A generalized state space realization of $\boldsymbol{\mathcal{G}}(s)$ is given by

$$\begin{aligned} \mathcal{E}\,\dot{x}(t) &= \mathcal{A}\,x(t) + \mathcal{B}\,u(t), \\ y(t) &= \mathcal{C}\,x(t), \end{aligned} \qquad (9)$$

where

$$\begin{aligned} \mathcal{E} &= E \in \mathbb{R}^{n,n}, & \mathcal{A} &= A + BKC \in \mathbb{R}^{n,n}, \\ \mathcal{B} &= BH \in \mathbb{R}^{n,m}, & \mathcal{C} &= RC \in \mathbb{R}^{p,n}. \end{aligned} \qquad (10)$$

Note that $I-\boldsymbol{G}(s)K$ is invertible if and only if the pencil $\lambda E - A - BKC$ is regular. Moreover, if $\boldsymbol{G}(s)$ and $(I - \boldsymbol{G}(s)K)^{-1}$ are proper, then the coupled system (1), (4), (4) is well-posed in the sense that the closed-loop transfer function $\boldsymbol{\mathcal{G}}(s)$ exists and it is proper. In a schematic way, an example of a coupled system is shown in Fig. 1.

The model reduction problem for the coupled system (1), (4), (4) consists in an approximation of the global mapping from the external input $u(t)$ to the external output $y(t)$. In other words, we want to find a reduced-order model

$$\begin{aligned} \widetilde{\mathcal{E}}\,\dot{\widetilde{x}}(t) &= \widetilde{\mathcal{A}}\,\widetilde{x}(t) + \widetilde{\mathcal{B}}\,u(t), \\ \widetilde{y}(t) &= \widetilde{\mathcal{C}}\,\widetilde{x}(t), \end{aligned} \qquad (11)$$

with $\widetilde{\mathcal{E}}$, $\widetilde{\mathcal{A}} \in \mathbb{R}^{\ell,\ell}$, $\widetilde{\mathcal{B}} \in \mathbb{R}^{\ell,m}$, $\widetilde{\mathcal{C}} \in \mathbb{R}^{\ell,n}$ and $\ell \ll n$ that approximates the closed-loop system (9). In the frequency domain, the model reduction problem can be reformulated as follows: for given $\boldsymbol{\mathcal{G}}(s) = \mathcal{C}(s\,\mathcal{E} - \mathcal{A})^{-1}\mathcal{B}$, find an approximation

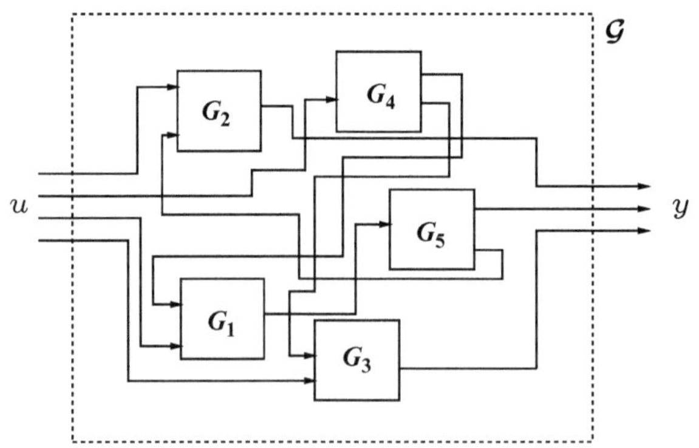

Fig. 1. Coupled system.

$\widetilde{\mathcal{G}}(s) = \widetilde{\mathcal{C}}(s\widetilde{\mathcal{E}} - \widetilde{\mathcal{A}})^{-1}\widetilde{\mathcal{B}}$ such that $\widetilde{\mathcal{E}}, \widetilde{\mathcal{A}} \in \mathbb{R}^{\ell,\ell}$ and $\| \widetilde{\mathcal{G}} - \mathcal{G} \|$ is small in some system norm. For instance, the approximation error can be estimated in the $\mathbb{H}_\infty$-norm. Apart from having a small state space dimension ℓ, it is also required that the reduced-order system (11) preserves essential properties of (9) like stability and passivity. Note that passivity, in general, means that the system does not produce energy and it is an important system property, especially in circuit design [2].

3 Model Reduction Approaches for Coupled Systems

There exist two main approaches for model order reduction of coupled systems. The first approach is to consider all subsystems together in the closed-loop form (9) and to compute the reduced-order system (11) by applying any model reduction method to (9). The second approach consists in replacing subsystems (1) by reduced-order models that are coupled then through the same interconnection relations. In this section we discuss these two approaches in more detail and mention their advantages and disadvantages.

3.1 Model Reduction of the Closed-Loop System

Most of the model reduction methods for linear time-invariant dynamical systems are based on the projection of the system onto lower dimensional subspaces. Using these methods for the closed-loop system (9), we can compute the reduced-order model (11) by projection

$$\widetilde{\mathcal{E}} = \mathcal{W}^T\mathcal{E}\mathcal{T}, \qquad \widetilde{\mathcal{A}} = \mathcal{W}^T\mathcal{A}\mathcal{T}, \qquad \widetilde{\mathcal{B}} = \mathcal{W}^T\mathcal{B}, \qquad \widetilde{\mathcal{C}} = \mathcal{C}\mathcal{T}, \tag{12}$$

where the projection matrices $\mathcal{W}, \mathcal{T} \in \mathbb{R}^{n,\ell}$ determine the subspaces of interest. For example, in modal model reduction the columns of $\mathcal{W}$ and $\mathcal{T}$ span, respectively, the left and right deflating subspaces of the pencil $\lambda\mathcal{E} - \mathcal{A}$ corresponding to the dominant eigenvalues [18, 44]. Balanced truncation model reduction is based on the projection of system (9) onto the subspaces corresponding to the dominant Hankel singular values of (9), see [46, 54]. In the moment matching approximation, one chooses the projection matrices $\mathcal{W}$ and $\mathcal{T}$ whose columns form the bases of certain Krylov subspaces associated with (9), e.g. [5, 22]. In the next subsections we briefly describe balanced truncation and moment matching methods.

Balanced Truncation

One of the most studied model reduction techniques is *balanced truncation*, an approach first proposed for standard state space systems in [19, 27, 46, 54] and then extended to generalized state space systems in [45, 50, 56]. An important property of balanced truncation model reduction methods is that stability is preserved in the reduced-order system. Moreover, the existence of computable error bounds allows an adaptive choice of the state space dimension ℓ of the approximate model. A disadvantage of these methods is that (generalized) Lyapunov equations have to be solved. However, recent results on low rank approximations to the solutions of matrix equations [9, 13, 29, 33, 43, 49] make the balanced truncation model reduction approach attractive for large-scale problems.

Consider the closed-loop system (9) with the stable pencil $\lambda\mathcal{E} - \mathcal{A}$. For simplicity, we will assume that the matrix $\mathcal{E}$ is nonsingular. However, all results of this subsection can also be extended for systems with singular $\mathcal{E}$, see [45, 56] for details. The balanced truncation model reduction method is closely related to the *controllability Gramian* $\mathcal{P}$ and the *observability Gramian* $\mathcal{Q}$ that are unique symmetric, positive semidefinite solutions of the generalized Lyapunov equations

$$\mathcal{E}\mathcal{P}\mathcal{A}^T + \mathcal{A}\mathcal{P}\mathcal{E}^T = -\mathcal{B}\mathcal{B}^T, \tag{13}$$

$$\mathcal{E}^T\mathcal{Q}\mathcal{A} + \mathcal{A}^T\mathcal{Q}\mathcal{E} = -\mathcal{C}^T\mathcal{C}. \tag{14}$$

The matrix $\mathcal{P}\mathcal{E}^T\mathcal{Q}\mathcal{E}$ has nonnegative eigenvalues, and the square roots of these eigenvalues $\sigma_j = \sqrt{\lambda_j(\mathcal{P}\mathcal{E}^T\mathcal{Q}\mathcal{E})}$ define the *Hankel singular values* of system (9). We will assume that σ_j are ordered decreasingly. System (9) is called *balanced* if $\mathcal{P} = \mathcal{Q} = \operatorname{diag}(\sigma_1, \ldots, \sigma_n)$. The Hankel singular values characterize the 'importance' of state variables in (9). States of the balanced system corresponding to the small Hankel singular values are difficult to reach and to observe at the same time. Such states are less involved in the energy transfer from inputs to outputs, and, therefore, they can be truncated without changing the system properties significantly [46]. Thus, a general idea of balanced truncation is to transform system (9) into a balanced form and to truncate the states that correspond to the small Hankel singular values. In practice, balancing and truncation can be combined by projecting system (9) onto the dominant subspaces of the matrix $\mathcal{P}\mathcal{E}^T\mathcal{Q}\mathcal{E}$. This can be done in a numerically efficient way using the following algorithm that is an obvious generalization of the square root method [39, 59].

Algorithm 1. Generalized square root balanced truncation method.

Given system (9) with the transfer function $\mathcal{G}(s) = \mathcal{C}(s\mathcal{E} - \mathcal{A})^{-1}\mathcal{B}$, compute the reduced-order system (11).

1. Compute the Cholesky factors $L_\mathcal{P}$ and $L_\mathcal{Q}$ of the Gramians $\mathcal{P} = L_\mathcal{P}L_\mathcal{P}^T$ and $\mathcal{Q} = L_\mathcal{Q}L_\mathcal{Q}^T$ that satisfy the Lyapunov equations (13) and (14).
2. Compute the singular value decomposition

$$L_\mathcal{P}^T\mathcal{E}^T L_\mathcal{Q} = [\,U_1,\ U_2\,]\begin{bmatrix}\Sigma_1 & 0\\ 0 & \Sigma_2\end{bmatrix}[\,V_1,\ V_2\,]^T, \tag{15}$$

   where the matrices $[\,U_1,\ \ U_2\,]$ and $[\,V_1,\ \ V_2\,]$ have orthonormal columns, $\Sigma_1 = \operatorname{diag}(\sigma_1, \ldots, \sigma_\ell)$, $\Sigma_2 = \operatorname{diag}(\sigma_{\ell+1}, \ldots, \sigma_r)$ with $r = \operatorname{rank}(L_\mathcal{P}^T\mathcal{E}^T L_\mathcal{Q})$.
3. Compute the reduced system (11) with

$$\widetilde{\mathcal{E}} = \mathcal{W}^T\mathcal{E}\mathcal{T}, \qquad \widetilde{\mathcal{A}} = \mathcal{W}^T\mathcal{A}\mathcal{T}, \qquad \widetilde{\mathcal{B}} = \mathcal{W}^T\mathcal{B}, \qquad \widetilde{\mathcal{C}} = \mathcal{C}\mathcal{T},$$

   where $\mathcal{W} = L_\mathcal{Q}V_1\Sigma_1^{-1/2}$ and $\mathcal{T} = L_\mathcal{P}U_1\Sigma_1^{-1/2}$.

One can show that the reduced-order system $\widetilde{\mathcal{G}}(s) = \widetilde{\mathcal{C}}(s\widetilde{\mathcal{E}} - \widetilde{\mathcal{A}})^{-1}\widetilde{\mathcal{B}}$ computed by this algorithm is stable and the $\mathbb{H}_\infty$-norm error bound

$$\| \widetilde{\mathcal{G}} - \mathcal{G} \|_{\mathbb{H}_\infty} \leq 2(\sigma_{\ell+1} + \ldots + \sigma_r) \tag{16}$$

holds, where $\sigma_{\ell+1}, \ldots, \sigma_r$ are the truncated Hankel singular values of (9), see [19, 27,51]. To solve the large-scale generalized Lyapunov equations (13) and (14) for the Cholesky factors without forming the Gramians $\mathcal{P}$ and $\mathcal{Q}$ explicitly, we can use the ADI method [43, 49], the cyclic Smith method [33, 49] or the sign function method [9, 13].

Apart from the balanced truncation method considered here, other balancing-based model reduction techniques have been developed, see [32, 47]. These include LQG balancing, stochastic balancing, positive real balancing and bounded real balancing. All these techniques are related to algebraic Riccati equations and aim to capture specific system properties like closed-loop performance, minimum phase property, passivity and $\mathbb{H}_\infty$-gain.

Moment Matching Approximation

An alternative model reduction approach for linear time-invariant systems is a *moment matching approximation* based on Krylov subspace methods, see [4, 5, 22] for recent surveys on these methods. Suppose that $s_0 \in \mathbb{C}$ is not an eigenvalue of the pencil $\lambda\mathcal{E} - \mathcal{A}$. Then the matrix $\mathcal{A} - s_0\mathcal{E}$ is nonsingular, and the transfer function $\mathcal{G}(s) = \mathcal{C}(s\mathcal{E} - \mathcal{A})^{-1}\mathcal{B}$ of the closed-loop system (9) can be expanded into a Taylor series at s_0 as

$$\begin{aligned}\mathcal{G}(s) &= -\mathcal{C}\big(I - (s - s_0)(\mathcal{A} - s_0\mathcal{E})^{-1}\mathcal{E}\big)^{-1}(\mathcal{A} - s_0\mathcal{E})^{-1}\mathcal{B} \\ &= \mathcal{M}_0 + \mathcal{M}_1(s - s_0) + \mathcal{M}_2(s - s_0)^2 + \ldots,\end{aligned}$$

where the matrices

$$\mathcal{M}_j = -\mathcal{C}\left((\mathcal{A} - s_0\mathcal{E})^{-1}\mathcal{E}\right)^j(\mathcal{A} - s_0\mathcal{E})^{-1}\mathcal{B} \tag{17}$$

are called the *moments* of system (9) at the expansion point s_0. The moment matching approximation problem consists in determining a reduced-order system (11) whose transfer function $\widetilde{\mathcal{G}}(s) = \widetilde{\mathcal{C}}(s\widetilde{\mathcal{E}} - \widetilde{\mathcal{A}})^{-1}\widetilde{\mathcal{B}}$ has the Taylor series expansion at s_0 of the form

$$\widetilde{\mathcal{G}}(s) = \widetilde{\mathcal{M}}_0 + \widetilde{\mathcal{M}}_1(s - s_0) + \widetilde{\mathcal{M}}_2(s - s_0)^2 + \ldots, \tag{18}$$

where the moments $\widetilde{\mathcal{M}}_j$ satisfy the moment matching conditions

$$\mathcal{M}_j = \widetilde{\mathcal{M}}_j, \qquad j = 0, 1, \ldots, q. \tag{19}$$

For $s_0 = 0$, the approximation (18), (19) is the matrix *Padé approximation* of $\mathcal{G}(s)$, e.g. [8]. For an arbitrary complex number $s_0 \neq 0$, the moment matching approximation is the problem of rational interpolation [1]. Besides a single interpolation point, it is also possible to construct a reduced-order system with the transfer function $\widetilde{\mathcal{G}}(s)$ that matches $\mathcal{G}(s)$ at multiple points $\{s_0, s_1, \ldots, s_l\}$. Such an approximation is called a multi-point rational interpolant and has been studied in [25, 30]. Furthermore, one can consider the Laurent expansion of $\mathcal{G}(s)$ at $s_0 = \infty$ given by

$$\mathcal{G}(s) = \mathcal{M}_{-d}s^d + \ldots + \mathcal{M}_{-1}s + \mathcal{M}_0 + \mathcal{M}_1 s^{-1} + \mathcal{M}_2 s^{-2} + \ldots,$$

where the coefficients $\mathcal{M}_j$ are known as *Markov parameters* of system (9). In this case, computing the approximation

$$\widetilde{\mathcal{G}}(s) = \widetilde{\mathcal{M}}_{-d}s^d + \ldots + \widetilde{\mathcal{M}}_{-1}s + \widetilde{\mathcal{M}}_0 + \widetilde{\mathcal{M}}_1 s^{-1} + \widetilde{\mathcal{M}}_2 s^{-2} + \ldots$$

with $\mathcal{M}_j = \widetilde{\mathcal{M}}_j$ for $j = -d, \ldots, -1, 0, 1, \ldots, q$ reduces to the partial realization problem [12, 28].

In order to determine the reduced-order system (11) satisfying the moment matching condition (19), the explicit computation of the moments can be avoided by using the following connection between the Padé (or Padé-type) approximation and the right and left Krylov subspaces

$$\mathcal{K}_{q_r}(\mathcal{A}_0^{-1}\mathcal{E}, \mathcal{A}_0^{-1}\mathcal{B}) = \operatorname{Im}[\,\mathcal{A}_0^{-1}\mathcal{B},\ \mathcal{A}_0^{-1}\mathcal{E}\mathcal{A}_0^{-1}\mathcal{B},\ \ldots,\ (\mathcal{A}_0^{-1}\mathcal{E})^{q_r-1}\mathcal{A}_0^{-1}\mathcal{B}\,],$$
$$\mathcal{K}_{q_l}(\mathcal{A}_0^{-T}\mathcal{E}^T, \mathcal{A}_0^{-T}\mathcal{C}^T) = \operatorname{Im}[\,\mathcal{A}_0^{-T}\mathcal{C}^T,\ \mathcal{A}_0^{-T}\mathcal{E}^T\mathcal{A}_0^{-T}\mathcal{C}^T,\ \ldots,\ (\mathcal{A}_0^{-T}\mathcal{E}^T)^{q_l-1}\mathcal{A}_0^{-T}\mathcal{C}^T],$$

with $\mathcal{A}_0 = \mathcal{A} - s_0\mathcal{E}$ and $\mathcal{A}_0^{-T} = (\mathcal{A}_0^{-1})^T$.

Theorem 1. [26, 30] *Consider the closed-loop system* (9) *and the reduced-order system* (11), (12) *with some projection matrices* $\mathcal{W}, \mathcal{T} \in \mathbb{R}^{n,\ell}$. *Let* $s_0 \in \mathbb{C}$ *be not an eigenvalue of* $\lambda\mathcal{E} - \mathcal{A}$ *and* $\lambda\widetilde{\mathcal{E}} - \widetilde{\mathcal{A}}$, *and let* $\mathcal{M}_j$ *and* $\widetilde{\mathcal{M}}_j$ *be the moments of systems* (9) *and* (11), (12), *respectively.*

1. *If* $\mathcal{K}_{q_r}(\mathcal{A}_0^{-1}\mathcal{E}, \mathcal{A}_0^{-1}\mathcal{B}) \subseteq \operatorname{Im}\mathcal{T}$ *and* $\mathcal{W} = \mathcal{T}$, *then* $\mathcal{M}_j = \widetilde{\mathcal{M}}_j$ *for* $j = 0, \ldots, q_r - 1$.
2. *If* $\mathcal{K}_{q_r}(\mathcal{A}_0^{-1}\mathcal{E}, \mathcal{A}_0^{-1}\mathcal{B}) \subseteq \operatorname{Im}\mathcal{T}$ *and* $\mathcal{K}_{q_l}(\mathcal{A}_0^{-T}\mathcal{E}^T, \mathcal{A}_0^{-T}\mathcal{C}^T) \subseteq \operatorname{Im}\mathcal{W}$, *then* $\mathcal{M}_j = \widetilde{\mathcal{M}}_j$ *for* $j = 0, \ldots, q_r + q_l - 1$.

This theorem proposes to take the projection matrices $\mathcal{T}$ and $\mathcal{W}$ as the bases of the Krylov subspaces $\mathcal{K}_{q_r}(\mathcal{A}_0^{-1}\mathcal{E}, \mathcal{A}_0^{-1}\mathcal{B})$ and $\mathcal{K}_{q_l}(\mathcal{A}_0^{-T}\mathcal{E}^T, \mathcal{A}_0^{-T}\mathcal{C}^T)$, respectively. Such bases can be efficiently computed by a Lanczos or Arnoldi process [5, 20, 25] in the single-input single-output case and by Lanczos- or Arnoldi-type methods [22, 24, 30, 48] in the multi-input multi-output case.

While the Krylov-based moment matching methods are efficient for very large sparse problems, the reduced-order systems computed by these methods have only locally good approximation properties. So far, no global error bound is known, see [5, 6, 31] for recent contributions to this topic. The location of the interpolation points strongly influences the approximation quality. The optimal choice of these points remains an open problem. Another drawback of the moment matching methods is that stability and passivity are not necessarily preserved in the resulting reduced-order model, so that usually post-processing is needed to realize these properties. Recently, passivity-preserving model reduction methods based on Krylov subspaces have been developed for standard state space systems [3, 55] and also for structured generalized state space systems arising in circuit simulation [23, 24, 38, 48].

3.2 Structure-Preserving Model Reduction

Model order reduction of the closed-loop system (9) does not preserve the interconnection structure in the approximate system (11). Although many different model reduction methods have been developed for linear dynamical systems, *structure-preserving reduced-order modelling* of coupled systems has received only recently attention [41,42,53,60]. Instead of reduction of the entire system (9), one can replace each subsystem (1), or a suitable selection of them, by a reduced-order model

$$\begin{aligned} \widetilde{E}_j \dot{\widetilde{x}}_j(t) &= \widetilde{A}_j \widetilde{x}_j(t) + \widetilde{B}_j \widetilde{u}_j(t), \\ \widetilde{y}_j(t) &= \widetilde{C}_j \widetilde{x}_j(t), \end{aligned} \tag{20}$$

where $\widetilde{E}_j$, $\widetilde{A}_j \in \mathbb{R}^{\ell_j,\ell_j}$, $\widetilde{B}_j \in \mathbb{R}^{\ell_j,m_j}$, $\widetilde{C}_j \in \mathbb{R}^{p_j,\ell_j}$ with $\ell_j \ll n_j$, and then couple these subsystems through the same interconnection relations

$$\widetilde{u}_j(t) = K_{j1}\widetilde{y}_1(t) + \ldots + K_{jk}\widetilde{y}_k(t) + H_j u(t), \quad j = 1, \ldots, k, \tag{21}$$

$$\widetilde{y}(t) = R_1\,\widetilde{y}_1(t) \; + \; \ldots + \; R_k\,\widetilde{y}_k(t). \tag{22}$$

Note that since the internal outputs $y_j(t)$ are replaced by the approximate outputs $\widetilde{y}_j(t)$, due to (21), the internal inputs $u_j(t)$ in (20) should also be replaced by the approximate inputs $\widetilde{u}_j(t)$. Let

$$\begin{aligned} \widetilde{E} &= \mathrm{diag}(\widetilde{E}_1, \ldots, \widetilde{E}_k), & \widetilde{A} &= \mathrm{diag}(\widetilde{A}_1, \ldots, \widetilde{A}_k), \\ \widetilde{B} &= \mathrm{diag}(\widetilde{B}_1, \ldots, \widetilde{B}_k), & \widetilde{C} &= \mathrm{diag}(\widetilde{C}_1, \ldots, \widetilde{C}_k). \end{aligned} \tag{23}$$

If the reduced-order pencils $\lambda\widetilde{E} - \widetilde{A}$ and $\lambda\widetilde{E} - \widetilde{A} - \widetilde{B}K\widetilde{C}$ are regular, then the reduced-order closed-loop system has the form (11) with

$$\widetilde{\mathcal{E}} = \widetilde{E}, \qquad \widetilde{\mathcal{A}} = \widetilde{A} + \widetilde{B}K\widetilde{C}, \qquad \widetilde{\mathcal{B}} = \widetilde{B}H, \qquad \widetilde{\mathcal{C}} = R\widetilde{C}. \tag{24}$$

The transfer function of this system is given by

$$\widetilde{\mathcal{G}}(s) = R\big(I - \widetilde{\boldsymbol{G}}(s)K\big)^{-1}\widetilde{\boldsymbol{G}}(s)H = R\widetilde{\boldsymbol{G}}(s)\big(I - K\widetilde{\boldsymbol{G}}(s)\big)^{-1}H, \tag{25}$$

where $\widetilde{\boldsymbol{G}}(s) = \mathrm{diag}(\widetilde{\boldsymbol{G}}_1(s), \ldots, \widetilde{\boldsymbol{G}}_k(s))$ with $\widetilde{\boldsymbol{G}}_j(s) = \widetilde{C}_j(s\,\widetilde{E}_j - \widetilde{A}_j)^{-1}\widetilde{B}_j$.

The reduced-order subsystems (20) can be computed by projection

$$\widetilde{E}_j = W_j^T E_j T_j, \quad \widetilde{A}_j = W_j^T A_j T_j, \quad \widetilde{B}_j = W_j^T B_j, \quad \widetilde{C}_j = C_j T_j, \tag{26}$$

where the projection matrices $W_j, T_j \in \mathbb{R}^{n_j,\ell_j}$ are determined for every subsystem either independently or using interconnection structure as it was proposed in [42,60]. Note that in this case the matrix coefficients of the reduced-order system (11) have the form (12) with the block diagonal projection matrices

$$\mathcal{W} = \mathrm{diag}(W_1, \ldots, W_k), \qquad \mathcal{T} = \mathrm{diag}(T_1, \ldots, T_k). \tag{27}$$

The following theorem gives a bound on the $\mathbb{H}_\infty$-norm of the error $\widetilde{\mathcal{G}} - \mathcal{G}$. For the time being, we assume that all the subsystems are asymptotically stable.

Theorem 2. *Consider the coupled system* (1)–(4) *with asymptotically stable subsystems and consider the reduced-order coupled system* (20)–(22)*. Let*

$$\Pi_l = \operatorname{diag}(\xi_1 I_{p_1}, \ldots, \xi_k I_{p_k}), \qquad \Pi_r = \operatorname{diag}(\xi_1 I_{m_1}, \ldots, \xi_k I_{m_k}), \tag{28}$$

where $\xi_j = 1$ *if* $\widetilde{\boldsymbol{G}}_j \neq \boldsymbol{G}_j$ *and* $\xi_j = 0$*, otherwise. Let*

$$\begin{aligned} g_1 &= \|\Pi_r K(I - \boldsymbol{G}K)^{-1}\|_{\mathbb{H}_\infty}, & g_2 &= \|R(I - \boldsymbol{G}K)^{-1}\Pi_l\|_{\mathbb{H}_\infty}, \\ g_3 &= \|(I - K\boldsymbol{G})^{-1}K\Pi_l\|_{\mathbb{H}_\infty}, & g_4 &= \|\Pi_r(I - K\boldsymbol{G})^{-1}H\|_{\mathbb{H}_\infty}. \end{aligned} \tag{29}$$

If

$$2\max\{g_1, g_3\} \max_{1\le j\le k} \|\widetilde{\boldsymbol{G}}_j - \boldsymbol{G}_j\|_{\mathbb{H}_\infty} < 1, \tag{30}$$

then the absolute error $\widetilde{\mathcal{G}} - \mathcal{G}$ *is bounded as*

$$\|\widetilde{\mathcal{G}} - \mathcal{G}\|_{\mathbb{H}_\infty} \le \min\{c_1, c_2\} \max_{1\le j\le k} \|\widetilde{\boldsymbol{G}}_j - \boldsymbol{G}_j\|_{\mathbb{H}_\infty}, \tag{31}$$

where $c_1 = 2g_2(\|H\|_2 + g_1\|\boldsymbol{G}H\|_{\mathbb{H}_\infty})$ *and* $c_2 = 2g_4(\|R\|_2 + g_3\|R\,\boldsymbol{G}\|_{\mathbb{H}_\infty})$*.*

Proof. The result immediately follows from [53, Theorem 3.1]. □

Note that Theorem 2 provides not only the approximation error bounds but also gives sufficient criteria for the stability of the reduced-order system. Indeed, if $\mathcal{G}$ is stable, $\|\boldsymbol{G}H\|_{\mathbb{H}_\infty}$ or $\|R\,\boldsymbol{G}\|_{\mathbb{H}_\infty}$ is bounded and condition (30) holds, then Theorem 2 implies that $\widetilde{\mathcal{G}}$ is also stable. Further aspects of stability of coupled systems can be found in [37, 53].

Subsystem Model Reduction by Balanced Truncation

Now we apply the $\mathbb{H}_\infty$-norm estimates provided by balanced truncation to the coupled system (1)–(4), where all subsystems are asymptotically stable. As a consequence of Theorem 2 we obtain the following error bounds for the closed-loop system (11) computed by the balanced truncation model reduction method applied to the subsystems.

Corollary 1. *Consider the coupled system* (1)–(4) *with asymptotically stable subsystems and consider the reduced-order coupled system* (20)–(22)*, where subsystems* (20) *are computed by Algorithm* 1 *applied to* (1)*. Let*

$$\gamma = 2\max_{1\le j\le k} (\sigma^{(j)}_{\ell_j+1} + \ldots + \sigma^{(j)}_{n_j}),$$

where $\sigma^{(j)}_{\ell_j+1}, \ldots, \sigma^{(j)}_{n_j}$ *denote the truncated Hankel singular values of the* j*th subsystem* (1)*. Further, let* g_1*,* g_3*,* c_1 *and* c_2 *be as in Theorem* 2*. If* $2\gamma\max\{g_1, g_3\} < 1$*, then the* $\mathbb{H}_\infty$*-norm of the error* $\widetilde{\mathcal{G}} - \mathcal{G}$ *can be bounded as*

$$\|\widetilde{\mathcal{G}} - \mathcal{G}\|_{\mathbb{H}_\infty} \le \gamma\min\{c_1, c_2\}. \tag{32}$$

Note that the computation of the a priori error bounds (31) and (32) for large-scale systems is expensive, since we need to calculate the $\mathbb{H}_\infty$-norm of the transfer functions of the state space dimension $n_1 + \ldots + n_k$. Similar to Theorem 2 and Corollary 1, we can also obtain the a posteriori error bounds like (31) and (32), with $\boldsymbol{G}$ replaced by $\widetilde{\boldsymbol{G}}$ in the constants g_j and c_j.

An essential assumption in Theorem 2 and Corollary 1 was the asymptotic stability of the subsystems (1). However, the asymptotic stability of the involved subsystems is neither necessary nor sufficient for the asymptotic stability of the closed-loop system (9). Since unstable subsystems can be artificially represented as a coupling of stable subsystems, we are then in the situation of Theorem 2 and Corollary 1. A possibility for the representation of an unstable subsystem (1) as a coupling of stable ones is based on the coprime factorization.

Consider now the transfer function $\boldsymbol{G}_j(s) = C_j(sE_j - A_j)^{-1}B_j$ which is not necessarily in $\mathbb{H}_\infty$. Such a transfer function admits a representation $\boldsymbol{G}_j(s) = \boldsymbol{N}_j(s)\boldsymbol{D}_j(s)^{-1}$, where $\boldsymbol{D}_j \in \mathbb{H}_\infty$ is square and $\boldsymbol{N}_j \in \mathbb{H}_\infty$ has the same matrix dimensions as $\boldsymbol{G}_j$. If, additionally, there exist $\boldsymbol{X}_j, \boldsymbol{Y}_j \in \mathbb{H}_\infty$ such that $\boldsymbol{X}_j(s)\boldsymbol{D}_j(s)+\boldsymbol{Y}_j(s)\boldsymbol{N}_j(s) = I$, then $\boldsymbol{D}_j$ and $\boldsymbol{N}_j$ are called *right coprime factors* of $\boldsymbol{G}_j$. For system (1) with no unstable and coevally uncontrollable modes, the coprime factors can be determined via a state feedback matrix $F_j \in \mathbb{R}^{m_j,n_j}$ with the property that the pencil $sE_j - A_j - B_jF_j$ is stable and of index at most one [15,61]. In this case, $\boldsymbol{N}_j$ and $\boldsymbol{D}_j$ can be chosen as

$$
\begin{aligned}
\boldsymbol{N}_j(s) &= C_j(sE_j - A_j - B_jF_j)^{-1}B_j, \\
\boldsymbol{D}_j(s) &= F_j(sE_j - A_j - B_jF_j)^{-1}B_j + I.
\end{aligned}
\tag{33}
$$

Then the extended transfer function

$$
\boldsymbol{G}_{ext,j}(s) = \begin{bmatrix} \boldsymbol{N}_j(s) \\ \boldsymbol{D}_j(s) - I \end{bmatrix}
\tag{34}
$$

is stable and has the generalized state space representation

$$
\begin{aligned}
E_j\dot{x}_j(t) &= (A_j + B_jF_j)x_j(t) + B_jv_j(t), \\
\begin{bmatrix} y_{1j}(t) \\ y_{2j}(t) \end{bmatrix} &= \begin{bmatrix} C_j \\ F_j \end{bmatrix} x_j(t).
\end{aligned}
\tag{35}
$$

Coupling this system with itself by the relations

$$
\begin{aligned}
v_j(t) &= -y_{2j}(t) + u_j(t) = [\,0, \;\; -I\,] \begin{bmatrix} y_{1j}(t) \\ y_{2j}(t) \end{bmatrix} + u_j(t), \\
y_j(t) &= y_{1j}(t) = [\,I, \;\; 0\,] \begin{bmatrix} y_{1j}(t) \\ y_{2j}(t) \end{bmatrix},
\end{aligned}
\tag{36}
$$

we obtain the coupled system which has the same transfer function $\boldsymbol{G}_j(s)$ as system (1). Such a coupled system is shown in Fig. 2. Note that the state space dimension of (35) coincides with that of (1).

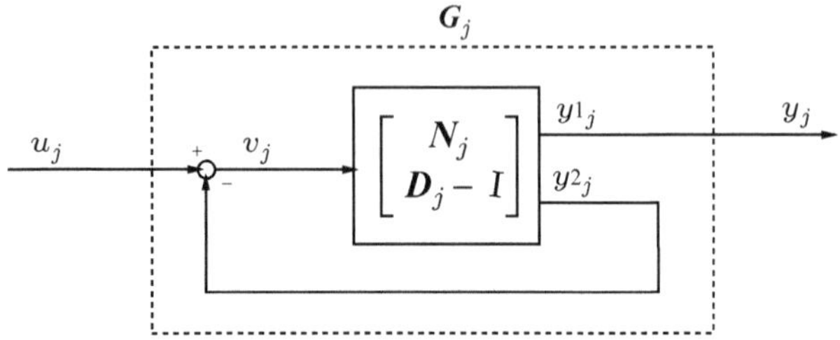

Fig. 2. Coprime factorization as a coupled system.

In the following, we discuss the benefits of the coprime factorization in the structure-preserving model reduction of the coupled system (1), (4), (4), where we now allow some unstable subsystems. Without loss of generality, we may assume that the first q subsystems are unstable and the corresponding right coprime factorizations are given by $\boldsymbol{G}_j(s) = \boldsymbol{N}_j(s)\boldsymbol{D}_j(s)^{-1}$ for $j = 1, \ldots, q$. The unstable subsystems can now be replaced by the asymptotically stable models (35) with the internal inputs and outputs satisfying (36). In this case, the coupling relations (4) and (4) take the form

$$\begin{aligned} v_j(t) &= -y_{2j}(t) + u_j(t) && 1 \le j \le q, \\ &= K_{j1}y_{11}(t) + \ldots + K_{jq}y_{1q}(t) - y_{2j}(t) \\ &\quad + K_{j,q+1}y_{q+1}(t) + \ldots + K_{jk}y_k(t) + H_j u(t), \\ u_j(t) &= K_{j1}y_{11}(t) + \ldots + K_{jq}y_{1q}(t) && q < j \le k, \\ &\quad + K_{j,q+1}y_{q+1}(t) + \ldots + K_{jk}y_k(t) + H_j u(t), \\ y(t) &= R_1 y_{11}(t) + \ldots + R_q y_{1q}(t) + R_{q+1}y_{q+1}(t) + \ldots + R_k y_k(t). \end{aligned} \tag{37}$$

The closed-loop transfer function of the new extended coupled system is given by $\mathcal{G}_{ext}(s) = R_{ext}(I - \boldsymbol{G}_{ext}(s)K_{ext})^{-1}\boldsymbol{G}_{ext}(s)H$, where

$$\begin{aligned} K_{ext} &= K\mathrm{diag}\big([I_{p_1},\ 0], \ldots, [I_{p_q},\ 0],\ I\big) - \mathrm{diag}\big([0,\ I_{m_1}], \ldots, [0,\ I_{m_q}],\ 0\big)\,, \\ R_{ext} &= \big[\, R_1, 0, \ldots, R_q, 0, R_{q+1}, R_{q+2}, \ldots, R_k \,\big] \end{aligned} \tag{38}$$

and

$$\boldsymbol{G}_{ext}(s) = \mathrm{diag}\big(\boldsymbol{G}_{ext,1}(s), \ldots, \boldsymbol{G}_{ext,q}(s), \boldsymbol{G}_{q+1}(s), \ldots, \boldsymbol{G}_k(s)\big) \tag{39}$$

with $\boldsymbol{G}_{ext,j}(s)$ as in (34). It has been shown in [53] that $\mathcal{G}_{ext}(s)$ coincides with the transfer function $\mathcal{G}(s)$ of the closed-loop system (9). This allows us to apply Theorem 2 and Corollary 1 to the extended coupled system with all subsystems being asymptotically stable in order to obtain the error bounds for the reduced-order system.

Another structure-preserving balancing-based model reduction method for coupled systems has been considered in [60]. There it has been proposed to project the subsystems (1) with $E_j = I$ onto the dominant eigenspaces of the matrices

$P_{jj}Q_{jj}$, where $P_{jj}, Q_{jj} \in \mathbb{R}^{n_j n_j}$ are the diagonal blocks of the controllability and observability Gramians $\mathcal{P} = [P_{jl}]_{j,l=1}^{k}$ and $\mathcal{Q} = [Q_{jl}]_{j,l=1}^{k}$ of the closed-loop system (9). Clearly, in the generalized state space case we should consider the matrices $P_{jj}E_j^T Q_{jj} E_j$. A drawback of this approach is that stability is not necessarily preserved in the reduced-order subsystems (20). Furthermore, we cannot make use of the error bound (31) since there are no global error estimates on $\widetilde{\boldsymbol{G}}_j - \boldsymbol{G}_j$.

Krylov Subspace Structure-Preserving Techniques

In this subsection we review structure-preserving model reduction methods based on Krylov subspaces. These methods have been previously proposed for second-order systems from structural dynamics, MEMS simulation and electronic circuit design [7, 23, 57] and then extended to coupled systems in [60]. A general framework for Krylov-based structure-preserving model reduction methods for partitioned systems can be found in [40, 42].

As mentioned above, for general projection matrices $\mathcal{W}$ and $\mathcal{T}$, the reduced-order system (11), (12) does not preserve the interconnection structure. This can be avoided if we take the block diagonal projection matrices $\mathcal{W}$ and $\mathcal{T}$ as in (27). However, in order to guarantee the moment matching conditions (19), the diagonal blocks in $\mathcal{W}$ and $\mathcal{T}$ have to satisfy certain subspace conditions as specified in the following theorem.

Theorem 3. *Let* $\widehat{\mathcal{W}} = [\,\widehat{W}_1^T, \ldots, \widehat{W}_k^T\,]^T$ *and* $\widehat{\mathcal{T}} = [\,\widehat{T}_1^T, \ldots, \widehat{T}_k^T\,]^T$ *with* $\widehat{W}_j, \widehat{T}_j \in \mathbb{R}^{n_j \ell}$. *Assume that the reduced-order systems* (20) *are computed by projection* (26), *where* $W_j, T_j \in \mathbb{R}^{n_j \ell_j}$ *have full column rank and satisfy*

$$\operatorname{Im}\widehat{W}_j \subseteq \operatorname{Im} W_j, \qquad \operatorname{Im}\widehat{T}_j \subseteq \operatorname{Im} T_j.$$

Let $\mathcal{M}_j$ *and* $\widetilde{\mathcal{M}}_j$ *be the moments of the closed-loop systems* (9), (10) *and* (11), (24), *respectively.*

1. *If* $\mathcal{K}_{q_r}(\mathcal{A}_0^{-1}\mathcal{E}, \mathcal{A}_0^{-1}\mathcal{B}) \subseteq \operatorname{Im}\widehat{\mathcal{T}}$ *and* $W_i = T_i$ *for* $i = 1, \ldots, k$, *then* $\mathcal{M}_j = \widetilde{\mathcal{M}}_j$ *for* $j = 0, \ldots, q_r - 1$.
2. *If* $\mathcal{K}_{q_r}(\mathcal{A}_0^{-1}\mathcal{E}, \mathcal{A}_0^{-1}\mathcal{B}) \subseteq \operatorname{Im}\widehat{\mathcal{T}}$ *and* $\mathcal{K}_{q_l}(\mathcal{A}_0^{-T}\mathcal{E}^T, \mathcal{A}_0^{-T}\mathcal{C}^T) \subseteq \operatorname{Im}\widehat{\mathcal{W}}$, *then* $\mathcal{M}_j = \widetilde{\mathcal{M}}_j$ *for* $j = 0, \ldots, q_r + q_l - 1$.

Proof. See [42, Theorem 4.1] and [60, Lemma 7]. □

A natural way to determine the projection matrices T_j and W_j is to compute the QR decomposition or the singular value decomposition of the matrices $\widehat{T}_j$ and $\widehat{W}_j$ such that the columns of $\widehat{\mathcal{T}} = [\,\widehat{T}_1^T, \ldots, \widehat{T}_k^T\,]^T$ and $\widehat{\mathcal{W}} = [\,\widehat{W}_1^T, \ldots, \widehat{W}_k^T\,]^T$ span the Krylov subspaces $\mathcal{K}_{q_r}(\mathcal{A}_0^{-1}\mathcal{E}, \mathcal{A}_0^{-1}\mathcal{B})$ and $\mathcal{K}_{q_l}(\mathcal{A}_0^{-T}\mathcal{E}^T, \mathcal{A}_0^{-T}\mathcal{C}^T)$, respectively. The matrices $\widehat{T}_j$ and $\widehat{W}_j$, in turn, can be computed simultaneously by applying a Lanczos- or Arnoldi-type method to the closed-loop system (9). The following theorem shows that $\widehat{T}_j$ and $\widehat{W}_j$ can also be generated separately by Krylov subspace methods applied to (1).

Theorem 4. *Consider the closed-loop system* (9), (10). *Let* $s_0 \in \mathbb{C}$ *be neither an eigenvalue of the pencil* $\lambda E - A$ *nor an eigenvalue of the pencil* $\lambda \mathcal{E} - \mathcal{A}$. *Then*

$$\mathcal{K}_{q_r}(\mathcal{A}_0^{-1}\mathcal{E},\, \mathcal{A}_0^{-1}\mathcal{B}) \subseteq \mathcal{K}_{q_r}((s_0E - A)^{-1}E,\, (s_0E - A)^{-1}B),$$
$$\mathcal{K}_{q_l}(\mathcal{A}_0^{-T}\mathcal{E}^T, \mathcal{A}_0^{-T}\mathcal{C}^T) \subseteq \mathcal{K}_{q_l}((s_0E - A)^{-T}E^T, (s_0E - A)^{-T}C^T).$$

Proof. These inclusions can be proved similarly to the case $E = I$, see [60, Lemma 6]. □

3.3 Comparison of Two Approaches for Model Reduction of Coupled Systems

The computation of the reduced-order model (11) by applying a model reduction method to the closed-loop system (9) has a couple of disadvantages. First of all note that the behavior of coupled systems is determined by different interconnected subsystems that are usually governed by entirely different physical laws and they often act in different spaces and time scales. There is no general model reduction technique, which can be considered as optimal, since the reliability, computation time and approximation quality of reduced-order models strongly depend on system properties. In model reduction of the closed-loop system (9), we ignore the special properties of the subsystems and destroy the coupling structure. Also in structure-preserving model reduction, where the projection matrices W_j and T_j are determined from the closed-loop system (9), we do not make use of subsystem properties. If we slightly change the coupled system, for example, by adding new subsystems, by replacing some of them by new ones or by changing the coupling configuration, we have to re-compute the reduced-order model again.

Subsystem model reduction, where the projection matrices W_j and T_j are computed separately from the subsystems (1), is free of these difficulties. In this approach, every subsystem can be reduced by a most suitable model reduction method that takes into consideration the structure and properties of the subsystem. If error estimates for subsystems are available, then using bound (31) we can evaluate how well the subsystems should be approximated to attain a prescribed accuracy in the reduced-order closed-loop system (11). Finally, subsystem model reduction is attractive for parallelization, since all k subsystems may be reduced simultaneously using k processors.

On the other hand, separate reduction of the subsystems usually yields the approximate model (11) of larger state space dimension than the system computed by projection of the closed-loop system (9). Furthermore, subsystem model reduction is often restricted to coupled systems whose subsystems have a small number of internal inputs and outputs.

4 Numerical Examples

In this section we present two numerical examples to demonstrate the properties of the discussed model reduction approaches for coupled systems. The computations were performed using MATLAB 7.

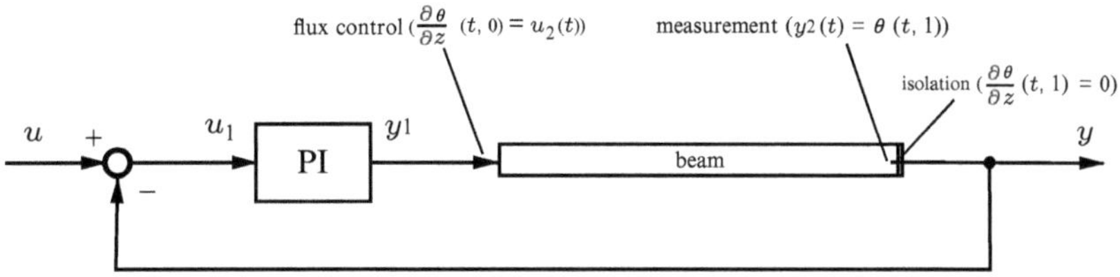

Fig. 3. A heated beam with a PI-controller.

Example 1. Consider a heated beam whose temperature is steered by a PI-controller as shown in Fig. 3. The transfer function of the PI-controller is given by $\boldsymbol{G}_1(s) = k_P + k_I s^{-1}$ and it is realized by the descriptor system

$$\begin{aligned} \begin{bmatrix} 1 & 0 \\ 0 & 0 \end{bmatrix} \dot{x}_1(t) &= \begin{bmatrix} 0 & 0 \\ 0 & 1 \end{bmatrix} x_1(t) + \begin{bmatrix} k_I \\ -k_P \end{bmatrix} u_1(t), \\ y_1(t) &= \begin{bmatrix} 1 & 1 \end{bmatrix} x_1(t). \end{aligned} \tag{40}$$

The heat transfer along the 1D beam of length 1 is described by

$$\frac{\partial\theta}{\partial t}(t,z) = \kappa \frac{\partial^2\theta}{\partial z^2}(t,z), \tag{41}$$

where $t > 0$ is the time, $z \in [0,1]$ is the position, $\theta(t,z)$ is the temperature distribution and κ is the heat conductivity of the material. On the left-hand side of the beam, the temperature flux is controlled by an input $u_2(t)$, whereas the beam is assumed to be perfectly isolated on the right-hand side. From this, we get the boundary conditions

$$\frac{\partial\theta}{\partial z}(t,0) = u_2(t), \qquad \frac{\partial\theta}{\partial z}(t,1) = 0. \tag{42}$$

The temperature is measured at $z = 1$ and it forms the output of the system, i.e., $y_2(t) = \theta(t,1)$ and $y(t) = y_2(t)$. By a spatial discretization of the heat equation with $n_2 + 1$ equidistant grid points, we obtain the system

$$\begin{aligned} E_2\dot{x}_2(t) &= A_2 x_2(t) + B_2 u_2(t), \\ y_2(t) &= C_2 x_2(t), \end{aligned} \tag{43}$$

where $E_2 = I_{n_2}$ and

$$A_2 = \kappa(n_2+1)^2 \begin{bmatrix} -1 & 1 & & & \\ 1 & -2 & 1 & & \\ & \ddots & \ddots & \ddots & \\ & & 1 & -2 & 1 \\ & & & 1 & -1 \end{bmatrix}, \; B_2 = \begin{bmatrix} \kappa(n_2+1) \\ 0 \\ \vdots \\ 0 \\ 0 \end{bmatrix}, \; C_2 = \begin{bmatrix} 0 \\ 0 \\ \vdots \\ 0 \\ 1 \end{bmatrix}^T. \tag{44}$$

The interconnection of the PI-controller and the beam is expressed by the relations

$$u_1(t) = u(t) - y_2(t), \quad u_2(t) = y_1(t). \tag{45}$$

Note that both the subsystems (40) and (43) are not asymptotically stable, since their transfer functions

$$\boldsymbol{G}_1(s) = C_1(sE_1 - A_1)^{-1}B_1 \quad \text{and} \quad \boldsymbol{G}_2(s) = C_2(sE_2 - A_2)^{-1}B_2$$

have a pole at the origin. The stabilizing state feedback matrices can be chosen as $F_1 = [\,0,\ -1\,]$ and $F_2 = [\,-n_2 - 1,\ 0,\ \cdots,\ 0\,]$. In this case, we obtain an extended coupled system with the stable subsystems $\boldsymbol{G}_{ext,1}(s)$ and $\boldsymbol{G}_{ext,2}(s)$ as in (34) and the interconnection matrices

$$K_{ext} = \begin{bmatrix} 0 & -1 & -1 & 0 \\ 1 & 0 & 0 & -1 \end{bmatrix}, \quad H = \begin{bmatrix} 1 \\ 0 \end{bmatrix}, \quad R_{ext} = [\,0,\ 0,\ 1,\ 0\,].$$

For our experiments, we chose the numerical values $k_P = k_I = \kappa = 1$ and $n_2 = 1000$. The second subsystem $\boldsymbol{G}_{ext,2}$ has been approximated by a reduced model $\widetilde{\boldsymbol{G}}_{ext,2}$ of order $\ell_2 = 21$ computed by balanced truncation. Figure 4 shows the absolute error $\|\widetilde{\boldsymbol{G}}_{ext,2}(\mathrm{i}\omega) - \boldsymbol{G}_{ext,2}(\mathrm{i}\omega)\|_2$ for the frequency range $\omega \in [\,10^{-1},\ 10^4\,]$ and the error bound γ that is twice the sum of the truncated Hankel singular values of $\boldsymbol{G}_{ext,2}$. We chose ℓ_2 such that $\gamma < 10^{-6}$. The resulting approximate closed-loop system with the transfer function $\widetilde{\boldsymbol{\mathcal{G}}}(s)$ has order $\ell = 23$.

Figure 5 shows the absolute error $\|\widetilde{\boldsymbol{\mathcal{G}}}(\mathrm{i}\omega) - \boldsymbol{\mathcal{G}}(\mathrm{i}\omega)\|_2$ and the a posteriori error bound $\gamma_{ext} = \gamma \min\{c_1, c_2\}$, where c_1 and c_2 are as in Theorem 2 with $\boldsymbol{G}$, K and R replaced by $\widetilde{\boldsymbol{G}}_{ext} = \mathrm{diag}(\boldsymbol{G}_{ext,1}, \widetilde{\boldsymbol{G}}_{ext,2})$, K_{ext} and R_{ext}, respectively. Comparing the approximation errors, we see that the error in the closed-loop system is larger

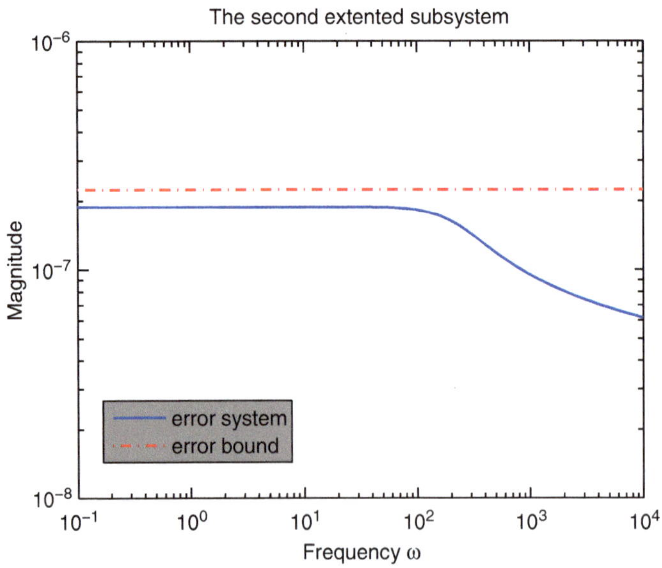

Fig. 4. Example 1: the absolute error $\|\widetilde{\boldsymbol{G}}_{ext,2}(\mathrm{i}\omega) - \boldsymbol{G}_{ext,2}(\mathrm{i}\omega)\|_2$ and the error bound γ.

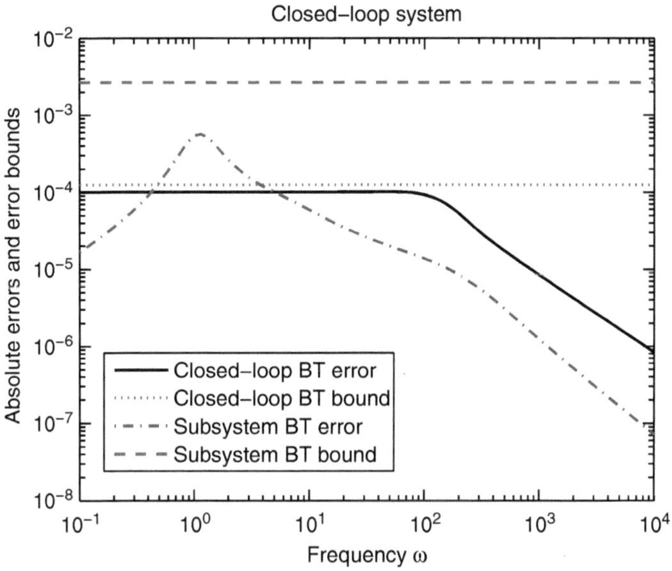

Fig. 5. Example 1: the absolute errors $\|\widetilde{\mathcal{G}}(\mathrm{i}\omega) - \mathcal{G}(\mathrm{i}\omega)\|_2$ and the error bounds γ_{ext} *(dashed line)* and γ_{cl} *(dotted line)*.

than the error in the subsystem due to the coupling. Furthermore, we applied the balanced truncation method to the closed-loop system and selected the order of the reduced model as a minimal integer ℓ such that the error bound $\gamma_{cl} = 2(\sigma_{\ell+1} + \ldots + \sigma_n)$ is smaller than γ_{ext}. We obtained the reduced model of order $\ell = 5$ with the approximation error comparable with the error in subsystem model reduction, see Fig. 5. Note, however, that if we change the parameters k_P, k_I and κ, then the closed-loop system is also changed, and we need to re-compute the reduced model. On the other hand, the reduced closed-loop system computed by subsystem model reduction can easily be modified by changing the first subsystem and by re-scaling the matrix coefficients in the reduced-order second subsystem.

Example 2. Consider the delay-differential system

$$\begin{aligned} \dot{x}(t) &= -x(t-1) + u(t), \\ y(t) &= x(t). \end{aligned} \tag{46}$$

This system can be represented as an interconnection of

$$\begin{aligned} \dot{x}_1(t) &= 0 \cdot x_1(t) + \begin{bmatrix} 1, & -1 \end{bmatrix} u_1(t), \\ y_1(t) &= x_1(t) \end{aligned} \tag{47}$$

with the system representing the pure unit delay

$$\widehat{y}_2(t) = u_2(t-1). \tag{48}$$

The coupling relations are

$$\begin{aligned} u_1(t) &= \begin{bmatrix} 0 \\ 1 \end{bmatrix} \widehat{y}_2(t) + \begin{bmatrix} 1 \\ 0 \end{bmatrix} u(t), \\ u_2(t) &= y_1(t), \qquad y(t) = y_1(t). \end{aligned} \tag{49}$$

The subsystems (47) and (48) have the transfer functions $\boldsymbol{G}_1(s) = [\,1/s,\ -1/s\,]$ and $\widehat{\boldsymbol{G}}_2(s) = \mathrm{e}^{-s}$, respectively. Due to the irrationality of $\widehat{\boldsymbol{G}}_2$, its system realizations have an infinite dimensional state space [17, 52]. The delay can be achieved by the following partial differential equation with boundary control and observation

$$\begin{aligned} \frac{\partial f}{\partial t}(t,z) &= -\frac{\partial f}{\partial z}(t,z), \\ f(t,0) &= u_2(t), \\ f(t,1) &= \widehat{y}_2(t). \end{aligned} \tag{50}$$

A spatial discretization of this equation with n_2 equidistant grid points leads to the subsystem

$$\begin{aligned} E_2\dot{x}_2(t) &= A_2x_2(t) + B_2u_2(t), \\ \widehat{y}_2(t) &= C_2x_2(t), \end{aligned} \tag{51}$$

with $E_2 = I_{n_2}$ and

$$A_2 = n_2 \begin{bmatrix} -1 & 1 & & & \\ & -1 & 1 & & \\ & & \ddots & \ddots & \\ & & & -1 & 1 \\ & & & & -1 \end{bmatrix}, \quad B_2 = \begin{bmatrix} 0 \\ 0 \\ \vdots \\ 0 \\ n_2 \end{bmatrix}, \quad C_2 = \begin{bmatrix} 1 \\ 0 \\ \vdots \\ 0 \\ 0 \end{bmatrix}^T. \tag{52}$$

The transfer function of this subsystem is given by

$$\boldsymbol{G}_2(s) = \frac{n_2}{(n_2+s)^{n_2}}. \tag{53}$$

Clearly, $\boldsymbol{G}_2 \in \mathbb{H}_\infty$. The coupled system (47), (49) and (51) is a finite dimensional approximant of the originally infinite dimensional delay-differential system (46). The estimation of the discretization error in the $\mathbb{H}_\infty$-norm is treated in [52].

The first subsystem (47) is not asymptotically stable since its transfer function $\boldsymbol{G}_1(s)$ has a pole at the origin. A stabilizing state feedback matrix can be taken as $F_1 = [0,\ 2]^T$. Thus, we obtain an extended coupled system with the stable subsystems $\boldsymbol{G}_{ext,1}(s)$ as in (34), $\boldsymbol{G}_2(s)$ and the interconnection matrices

$$K_{ext} = \begin{bmatrix} 0 & -1 & 0 & 0 \\ 0 & 0 & -1 & 1 \\ 1 & 0 & 0 & 0 \end{bmatrix}, \quad H = \begin{bmatrix} 1 \\ 0 \\ 0 \end{bmatrix}, \quad R_{ext} = [1,\ 0,\ 0,\ 0].$$

The second subsystem (51) of order $n_2 = 1000$ has been approximated by a reduced model of order $\ell_2 = 41$ computed by the balanced truncation method. The absolute values of the frequency responses $\widetilde{\boldsymbol{G}}_2(\mathrm{i}\omega)$ and $\boldsymbol{G}_2(\mathrm{i}\omega)$ of the original

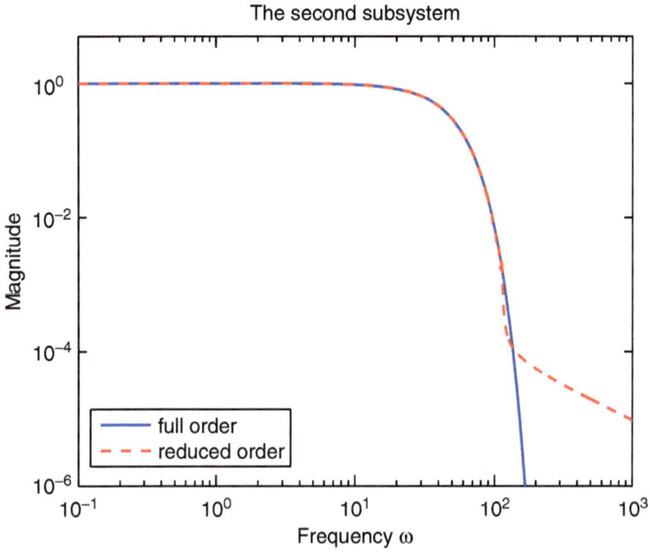

Fig. 6. Example 2: the absolute values of frequency responses $G_2(\mathrm{i}\omega)$ and $\widetilde{G}_2(\mathrm{i}\omega)$.

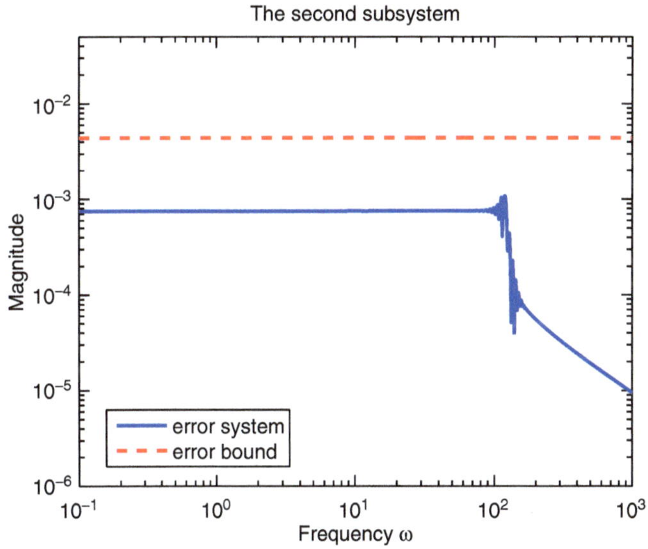

Fig. 7. Example 2: the absolute error $\|\widetilde{G}_2(\mathrm{i}\omega) - G_2(\mathrm{i}\omega)\|_2$ and the error bound γ.

and reduced-order subsystems are given in Fig. 6, whereas the absolute error $\|\widetilde{G}_2(\mathrm{i}\omega) - G_2(\mathrm{i}\omega)\|_2$ and the error bound

$$\gamma = 2(\sigma_{\ell_2+1}^{(2)} + \ldots + \sigma_{n_2}^{(2)})$$

are presented in Fig. 7. We see that the reduced-order subsystem approximates (51) satisfactorily.

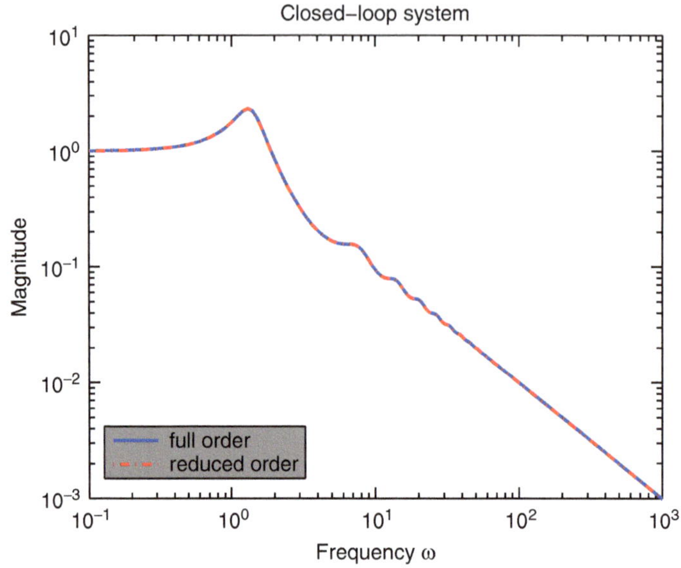

Fig. 8. Example 2: the absolute values of frequency responses $\boldsymbol{\mathcal{G}}(\mathrm{i}\omega)$ and $\widetilde{\boldsymbol{\mathcal{G}}}(\mathrm{i}\omega)$.

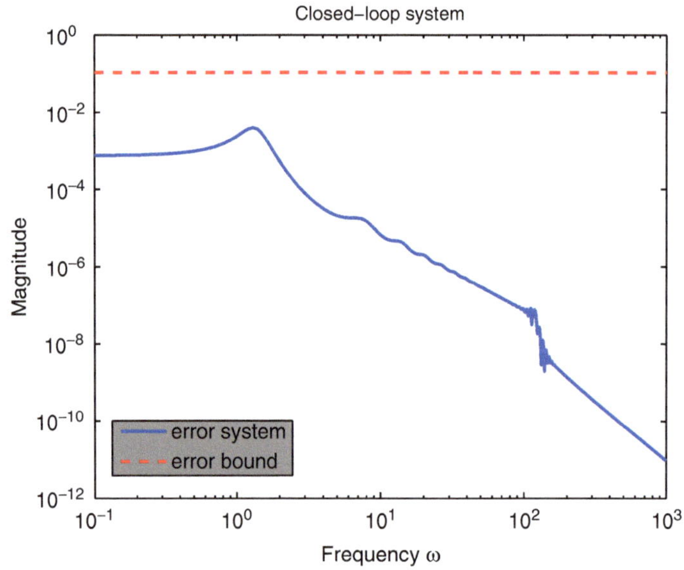

Fig. 9. Example 2: the absolute error $\|\widetilde{\boldsymbol{\mathcal{G}}}(\mathrm{i}\omega) - \boldsymbol{\mathcal{G}}(\mathrm{i}\omega)\|_2$ and the error bound γ_{ext}.

In Fig. 8 we plotted the absolute values of the frequency responses $\boldsymbol{\mathcal{G}}(\mathrm{i}\omega)$ and $\widetilde{\boldsymbol{\mathcal{G}}}(\mathrm{i}\omega)$ of the original and the reduced-order closed-loop systems. Figure 9 shows the error $\|\widetilde{\boldsymbol{\mathcal{G}}}(\mathrm{i}\omega) - \boldsymbol{\mathcal{G}}(\mathrm{i}\omega)\|_2$ and the a posteriori error bound

$$\gamma_{ext} = \gamma \min\{c_1, c_2\},$$

where the constants c_1 and c_2 are as in Theorem 2 with G, K and R replaced by $\widetilde{G}_{ext} = \mathrm{diag}(G_{ext,1}, \widetilde{G}_2)$, K_{ext} and R_{ext}, respectively. One can see that over the frequency range $[\,10^{-1},\ 10^{3}\,]$ there is no visible difference between the magnitude plots of $\widetilde{\mathcal{G}}$ and $\mathcal{G}$ and that the absolute error is smaller than 10^{-2}.

References

1. Anderson, B.D.O., Antoulas, A.C.: Rational interpolation and state variable realizations. Linear Algebra Appl., **137-138**, 479–509 (1990)
2. Anderson, B., Vongpanitlerd, S.: Network Analysis and Synthesis. Prentice Hall, Englewood Cliffs (1973)
3. Antoulas, A.: A new result on passivity preserving model reduction. Systems Control Lett., **54**, 361–374 (2005)
4. Antoulas, A.: Approximation of Large-Scale Dynamical Systems. SIAM, Philadelphia (2005)
5. Bai, Z.: Krylov subspace techniques for reduced-order modeling of large-scale dynamical systems. Appl. Numer. Math., **43**, 9–44 (2002)
6. Bai, Z., Slone, R.D., Smith, W.T., Ye, Q.: Error bound for reduced system model by Padé approximation via the Lanczos process. IEEE Trans. Comput. Aided Design, **18**, 133–141 (1999)
7. Bai, Z., Su, Y.: Dimension reduction of large-scale second-order dynamical systems via a second-order Arnoldi method. SIAM J. Sci. Comp., **26**, 1692–1709 (2005)
8. Baker Jr., G., Graves-Morris, P. Padé Approximants. Second edition. Encyclopedia of Mathematics and its Applications, 59. Cambridge University Press, Cambridge (1996)
9. Baur, U., Benner, P.: Factorized solution of Lyapunov equations based on hierarchical matrix arithmetic. , – (2006)
10. Bechtold, T.: Model order reduction of electro-thermal MEMS. Ph.D. Thesis, Albert-Ludwigs Universität Freiburg, Freiburg (2005)
11. Benner, P., Mehrmann, V., Sorensen, D. (eds): Dimension Reduction of Large-Scale Systems. Lecture Notes in Computational Science and Engineering, 45. Springer, Berlin Heidelberg (2005)
12. Benner, P., Sokolov, V.I.: Partial realization of descriptor systems. Systems Control Lett., **55**, 929–938 (2006)
13. Benner, P., Quintana-Ortí, E.: Solving stable generalized Lyapunov equations with the matrix sign function. Numerical Algorithms, **20**, 75–100 (1999)
14. Bindel, D.S., Bai, Z., Demmel, J.W.: Model reduction for RF MEMS simulation. In: Dongarra, J., Madsen, K., Wasniewski, J. (eds) Applied Parallel Computing: 7th International Conference (PARA 2004, Lyngby, Denmark, June 20-23, 2004), Lecture Notes in Computer Science, 3732, pages 286–295. Springer, Berlin Heidelberg (2006)
15. Bunse-Gerstner, A., Byers, R., Mehrmann, V., Nichols, N.: Feedback design for regularizing descriptor systems, Linear Algebra Appl., **299**, 119–151 (1999)
16. Craig Jr., R., Bampton, M.: Coupling of substructures for dynamic analysis. AIAA J., **6**, 1313–1319 (1968)
17. Curtain, R., Zwart, H.: An Introduction to Infinite Dimensional Linear Systems Theory. Springer, Berlin Heidelberg New York (1995)
18. Davison, E.: A method for simplifying linear dynamical systems. IEEE Trans. Automat. Control, **11**, 93–101 (1966)

19. Enns, D.: Model reduction with balanced realization: an error bound and a frequency weighted generalization. In: Proceedings of the 23rd IEEE Conference on Decision and Control (Las Vegas, 1984), pages 127–132. IEEE, New York (1984)
20. Feldmann, P., Freund, R.: Efficient linear circuit analysis by Padé approximation via the Lanczos process. IEEE Trans. Computer-Aided Design, **14**, 639–649 (1995)
21. Felippa, C., Park, C., Farhat, C.: Partitioned analysis of coupled mechanical systems. Comput. Methods Appl. Mech. Engrg., **190**, 3247–3270 (2001)
22. Freund, R.: Model reduction methods based on Krylov subspaces. Acta Numerica, **12**, 267–319 (2003)
23. Freund, R.: SPRIM: structure-preserving reduced-order interconnect macromodeling. In: Technical Digest of the 2004 IEEE/ACM International Conference on Computer-Aided Design, Los Alamos, CA, pages 80–87 (2004)
24. Freund, R., Feldmann, P.: The SyMPVL algorithm and its applications in interconnect simulation. In: Proceedings of the 1997 International Conference on Simulation of Semiconductor Processes and Devices, New York, pages 113–116 (1997)
25. Gallivan, K., Grimme, E., Van Dooren, P.: A rational Lanczos algorithm for model reduction. Numerical Algorithms, **12**, 33–63 (1996)
26. Gallivan, K., Vandendorpe, A., Van Dooren, P.: Sylvester equations and projection-based model reduction. J. Comp. Appl. Math., **162**, 213–229 (2004)
27. Glover, K.: All optimal Hankel-norm approximations of linear multivariable systems and their L^∞-errors bounds. Internat. J. Control, **39**, 1115–1193 (1984)
28. Gragg W., Lindquist, A.: On the partial realization problem. Linear Alg. Appl., **50**, 277–319 (1983)
29. Grasedyck, L.: Existence of a low rank of H-matrix approximation to the solution of the Sylvester equation. Numer. Linear Algebra Appl., **11**, 371–389 (2004)
30. Grimme, E.: Krylov projection methods for model reduction. Ph.D. Thesis, University of Illinois, Urbana-Champaign (1997)
31. Gugercin, S.: Projection methods for model reduction of large-scale dynamical systems. Ph.D. Thesis, Rice University, Houston (2003)
32. Gugercin, S., Antoulas, A.: A survey of model reduction by balanced truncation and some new results. Internat. J. Control, **77**, 748–766 (2004)
33. Gugercin, S., Sorensen, D., Antoulas, A.: A modified low-rank Smith method for large-scale Lyapunov equations. Numerical Algorithms, **32**, 27–55 (2003)
34. Günther, M., Feldmann, U.: CAD-based electric-circuit modeling in industry. I. Mathematical structure and index of network equations. Surveys Math. Indust., **8**, 97–129 (1999)
35. Hurty, W.: Dynamic analysis of structural systems using component modes. AIAA J., **3**, 678–685 (1965)
36. Ionescu, V., Oară, C., Weiss, M.: Generalized Riccati Theory and Robust Control: A Popov Function Approach. John Wiley and Sons, Chichester, UK (1999)
37. Karow, M., Hinrichsen, D., Pritchard, A.: Interconnected systems with uncertain couplings: explicit formulae for μ-values, spectral value sets and stability radii. SIAM J. Control Optim., **45**, 856–884 (2006)
38. Knockaert, L., De Zutter, D.: Laguerre-SVD reduced-order modeling. IEEE Trans. Microwave Theory Tech., **48**, 1469–1475 (2000)
39. Laub, A., Heath, M., Paige, C., Ward, R.: Computation of system balancing transformations and other applications of simultaneous diagonalization algorithms. IEEE Trans. Automat. Control, **32**, 115–122 (1987)
40. Li, R.-C.: Structural preserving model reductions. Technical Report 2004-02, Department of Mathematics, University of Kentucky (2004)

41. Liao, B.-S., Bai, Z., Gao, W.: Important modes of subsystems: a moment matching approach. Int. J. Numer. Meth. Engng, to appear.
42. Li, R.-C., Bai, Z.: Structure-preserving model reduction using a Krylov subspace projection formulation. Comm. Math. Sci., **3**, 179–199 (2005)
43. Li, J.-R., White, J.: Low rank solution of Lyapunov equations. SIAM J. Matrix Anal. Appl., **24**, 260–280 (2002)
44. Marschall, S.: An approximate method for reducing the order of a linear system. Contr. Eng., **10**, 642–648 (1966)
45. Mehrmann V., Stykel, T.: Balanced truncation model reduction for large-scale systems in descriptor form. In: Benner, P., Mehrmann, V., Sorensen, D. (eds) Dimension Reduction of Large-Scale Systems. Lecture Notes in Computational Science and Engineering, 45, pages 83–115. Springer, Berlin Heidelberg (2005)
46. Moore, B.C.: Principal component analysis in linear systems: controllability, observability, and model reduction. IEEE Trans. Automat. Control, **26**, 17–32 (1981)
47. Obinata, G., Anderson, B.: Model Reduction for Control System Design. Springer, London (2001)
48. Odabasioglu, A., Celik, M., Pileggi, L.: PRIMA: Passive reduced-order interconnect macromodeling algorithm. IEEE Trans. Circuits Systems, **17**, 645–654 (1998)
49. Penzl, T.: A cyclic low-rank Smith method for large sparse Lyapunov equations. SIAM J. Sci. Comput., **21**, 1401–1418 (1999/2000)
50. Perev K., Shafai, B.: Balanced realization and model reduction of singular systems. Internat. J. Systems Sci., **25**, 1039–1052 (1994)
51. Pernebo, L., Silverman, L.: Model reduction via balanced state space representation. IEEE Trans. Automat. Control, **27**, 382–387 (1982)
52. Reis, T.: Systems theoretic aspects of PDAEs and applications to electrical circuits. Ph.D. Thesis, Technische Universität Kaiserslautern, Kaiserslautern (2006)
53. Reis T., Stykel, T.: Stability analysis and model order reduction of coupled systems. Math. Comput. Model. Dyn. Syst., to appear.
54. Safonov, M., Chiang, R.: A Schur method for balanced-truncation model reduction. IEEE Trans. Automat. Control, **34**, 729–733 (1989)
55. Sorensen, D.: Passivity preserving model reduction via interpolation of spectral zeros. Systems Control Lett., **54**, 347–360 (2005)
56. Stykel, T.: Gramian-based model reduction for descriptor systems. Math. Control Signals Systems, **16**, 297–319 (2004)
57. Sun, T.-J., Graig Jr., R.: Model reduction and control of flexible structures using Krylov vectors. J. Guidance, Dynamics and Control, **14**, 260–267 (1991)
58. Tischendorf, C.: Coupled systems of differential algebraic and partial differential equations in circuit and device simulation: modeling and numerical analysis. Habilitation Thesis, Humboldt-Universität zu Berlin, Berlin (2003)
59. Tombs, M., Postlethweite, I.: Truncated balanced realization of a stable non-minimal state-space system. Internat. J. Control, **46**, 1319–1330 (1987)
60. Vandendorpe, A., Van Dooren, P.: On model reduction of interconnected systems. In: De Moore, B., Motmans, B., Willems, J., Van Dooren, P., Blondel, V. (eds) Proceedings of the 16th International Symposium of Mathematical Theory of Networks and Systems (Leuven, Belgium, July 5-9, 2004), Katholieke Universiteit Leuven (2004)
61. Varga, A.: Computation of coprime factorizations of rational matrices. Linear Algebra Appl., **271**, 83–115 (1998)
62. Vlach, J., Singhal, K.: Computer Methods for Circuit Analysis and Design. Van Nostrand Reinhold, New York (1994)

Space Mapping and Defect Correction

David Echeverría, Domenico Lahaye, and Piet W. Hemker

CWI, Kruislaan 413, 1054 AL Amsterdam, The Netherlands**
{D.Echeverria, P.W.Hemker, Domenico.Lahaye}@cwi.nl

Summary. In this chapter we present the principles of the space-mapping iteration techniques for the efficient solution of optimization problems. We also show how space-mapping optimization can be understood in the framework of defect correction.

We observe the difference between the solution of the optimization problem and the computed space-mapping solutions. We repair this discrepancy by exploiting the correspondence with defect correction iteration and we construct the manifold-mapping algorithm, which is as efficient as the space-mapping algorithm but converges to the true solution.

In the last section we show a simple example from practice, comparing space-mapping and manifold mapping and illustrating the efficiency of the technique.

1 Introduction

Space mapping is a technique, using simple surrogate models, to reduce the computing time in optimization procedures where time-consuming computer-models are needed to obtain sufficiently accurate results. Thus, space mapping makes use of both accurate (and time-consuming) models and less accurate (but cheaper) ones.

In fact, the original space-mapping procedure corresponds with right-preconditioning the coarse (inaccurate) model in order to accelerate the iterative procedure for the optimization of the fine (accurate) one. The iterative procedure used in space mapping for optimization can be seen as a defect correction iteration and the convergence can be analyzed accordingly. In this paper we show the structure of space mapping iteration. We also show that right-preconditioning is generally insufficient and (also) left-preconditioning is needed to obtain the solution for the accurate model. This leads to the improved space-mapping or 'manifold-mapping' procedure. This manifold mapping is shown in some detail in Section 5 and in the last section a few examples of an application are given.

The space-mapping idea was introduced by Bandler [3] in the context of microwave filter design and it has developed significantly over the last decade. In the

** This research was supported by the Dutch Ministry of Economic Affairs through the project IOP-EMVT-02201 B

rather complete survey [4] we see that the original idea has gone through a large number of changes and improvements. The reader is referred to the original literature [1, 2, 5] for a review of earlier achievements and for a classical introduction for engineers.

2 Fine and Coarse Models in Optimization

The Optimization Problem.

Let the specifications for the data of an optimization problem be denoted by $(\mathbf{t}, \mathbf{y}) \equiv (\{t_i\}, \{y_i\})_{i=1,\dots,m}$. The independent variable $\mathbf{t} \in \mathbb{R}^m$ could be, e.g., time, frequency, space, etc. The dependent variable $\mathbf{y} \in Y \subset \mathbb{R}^m$ represents the quantities that describe the behavior of the phenomena under study or design. The set $Y \subset \mathbb{R}^m$ is called the *set of possible aims*.

The behavior of the variable $\mathbf{y}$ not only depends on the independent variable $\mathbf{t}$ but also on an additional set of control/design variables. With $\mathbf{x}$ the vector of relevant control variables, we may write the components of $\mathbf{y}$ as $y_i \approx y(t_i, \mathbf{x})$. The behavior of the phenomenon is described by the function $y(t, \mathbf{x})$ and the difference between the measured data y_i and the values $y(t_i, \mathbf{x})$ may be the result of, e.g., measurement errors or the imperfection of the mathematical description.

Models to describe reality appear in several degrees of sophistication. Space mapping exploits the combination of the simplicity of the less sophisticated methods with the accuracy of the more complex ones. Therefore, we distinguish two types of model: fine and coarse.

The Fine Model.

The *fine model* response is denoted by $\mathbf{f}(\mathbf{x}) \in \mathbb{R}^m$, where $\mathbf{x} \in X \subset \mathbb{R}^n$ is the *fine model control variable*. The set X of possible control variables is usually a closed and bounded subset of $\mathbb{R}^n$. The set $\mathbf{f}(X) \subset \mathbb{R}^m$ of all possible fine model responses is the set of *fine model reachable aims*. The fine model is assumed to be *accurate* but *expensive* to evaluate. We also assume that $\mathbf{f}(\mathbf{x})$ is continuous.

For the optimization problem a *fine model cost function*, $\| \mathbf{f}(\mathbf{x}) - \mathbf{y} \|$, is defined, which is a measure for the discrepancy between the data and a particular response of the mathematical model. This cost function should be minimized. So we look for

$$\mathbf{x}^* = \underset{\mathbf{x} \in X}{\operatorname{argmin}} \, \| \mathbf{f}(\mathbf{x}) - \mathbf{y} \| \, . \tag{1}$$

A design problem, characterized by the model $\mathbf{f}(\mathbf{x})$, the aim $\mathbf{y} \in Y$, and the space of possible controls $X \subset \mathbb{R}^n$, is called a *reachable design* if the equality $\mathbf{f}(\mathbf{x}^*) = \mathbf{y}$ can be achieved for some $\mathbf{x}^* \in X$. □

The Coarse Model.

The *coarse model* is denoted by $\mathbf{c}(\mathbf{z}) \in \mathbb{R}^m$, with $\mathbf{z} \in Z \subset \mathbb{R}^n$ the *coarse model control variable.* This model is assumed to be *cheap* to evaluate but *less accurate* than the fine model. The set $\mathbf{c}(Z) \subset \mathbb{R}^m$ is the set of *coarse model reachable aims.* For the coarse model we have the *coarse model cost function,* $|\!|\!| \mathbf{c}(\mathbf{z}) - \mathbf{y} |\!|\!|$. We denote its minimizer by $\mathbf{z}^*$,

$$\mathbf{z}^* = \underset{\mathbf{z} \in Z}{\operatorname{argmin}} \; |\!|\!| \mathbf{c}(\mathbf{z}) - \mathbf{y} |\!|\!| \; . \tag{2}$$

We assume that the fine and coarse optimization problems, characterized by $\mathbf{y}$, $\mathbf{f}(\mathbf{x})$ and X, respectively $\mathbf{y}$, $\mathbf{c}(\mathbf{z})$ and Z, are uniquely solvable and well defined. If X and Z are closed and bounded non-empty sets in $\mathbb{R}^n$ and $\mathbf{f}$ and $\mathbf{c}$ continuous functions, the existence of the solutions is guaranteed. Generally, uniqueness can be achieved by properly reducing the sets X or Z. If the models are non-injective (or extremely ill-conditioned) in a small neighborhood of a solution, essential difficulties may arise.

The Space-Mapping Function.

The similarity or discrepancy between the responses of two models used for the same phenomenon is an important property. It is expressed by the *misalignment function*

$$r(\mathbf{z}, \mathbf{x}) = |\!|\!| \mathbf{c}(\mathbf{z}) - \mathbf{f}(\mathbf{x}) |\!|\!| \; . \tag{3}$$

For a given $\mathbf{x} \in X$ it is useful to know which $\mathbf{z} \in Z$ yields the smallest discrepancy. This information can be used to improve the coarse model. Therefore, the space-mapping function is introduced. The *space-mapping function* $\mathbf{p} : X \subset \mathbb{R}^n \to Z \subset \mathbb{R}^n$ is defined[1] by

$$\mathbf{p}(\mathbf{x}) = \underset{\mathbf{z} \in Z}{\operatorname{argmin}} \; r(\mathbf{z}, \mathbf{x}) = \underset{\mathbf{z} \in Z}{\operatorname{argmin}} \; |\!|\!| \mathbf{c}(\mathbf{z}) - \mathbf{f}(\mathbf{x}) |\!|\!| \; . \tag{4}$$

It should be noted that this evaluation of the space-mapping function $\mathbf{p}(\mathbf{x})$ requires both an evaluation of $\mathbf{f}(\mathbf{x})$ and a minimization process with respect to $\mathbf{z}$ in $|\!|\!| \mathbf{c}(\mathbf{z}) - \mathbf{f}(\mathbf{x}) |\!|\!|$. Hence, in algorithms we should make economic use of space-mapping function evaluations. In Figure 1 we see an example of a misalignment function and of a few space mapping functions.

Perfect Mapping.

In order to identify the cases where the accurate solution $\mathbf{x}^*$ is related with the less accurate solution $\mathbf{z}^*$ by the space mapping function, the following definition is introduced. A space-mapping function $\mathbf{p}$ is called a *perfect mapping* iff

---

[1] The process of finding $\mathbf{p}(\mathbf{x})$ for a given $\mathbf{x}$ is called *parameter extraction* or *single point extraction* because it finds the best coarse-model parameter that corresponds with a given fine-model control $\mathbf{x}$.

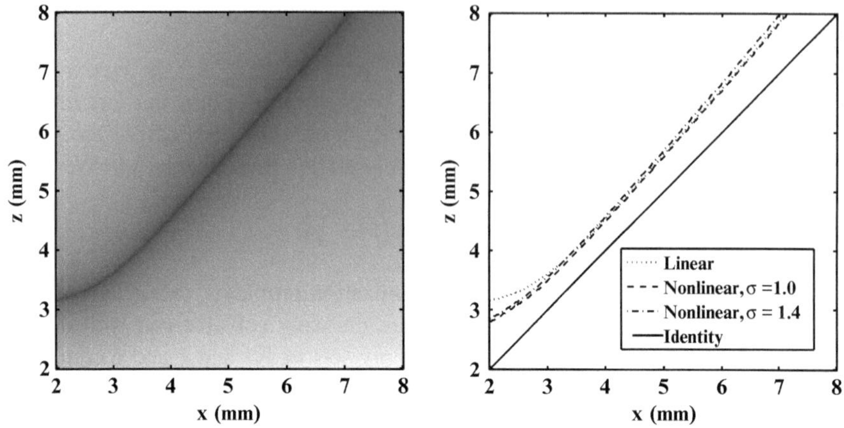

Fig. 1. Misalignment and space-mapping function

The left figure shows the misalignment function for a fine and a coarse model. Darker shading shows a smaller misalignment. The right figure shows the identity function and a few space-mapping functions for different coarse models (example taken from [9]).

$$\mathbf{z}^* = \mathbf{p}(\mathbf{x}^*) \ . \tag{5}$$

Using the definition of space mapping we see that (5) can be written as

$$\underset{\mathbf{z}\in Z}{\operatorname{argmin}} \ \| \mathbf{c}(\mathbf{z}) - \mathbf{y} \| = \underset{\mathbf{z}\in Z}{\operatorname{argmin}} \ \| \mathbf{c}(\mathbf{z}) - \mathbf{f}(\mathbf{x}^*) \| \ , \tag{6}$$

i.e., a perfect space-mapping function maps $\mathbf{x}^*$, the solution of the fine model optimization, exactly onto $\mathbf{z}^*$, the minimizer of the coarse model design.

Remark. We notice that *perfection* is not only a property of the space-mapping function, but it also depends on the data $\mathbf{y}$ considered. A space-mapping function can be perfect for one set of data but imperfect for a different data set. In this sense 'perfect mapping' can be a confusing notion.

3 Space-Mapping Optimization

In literature many space mapping based algorithms can be found [1, 4], but they all have the same basis. We first describe the original space-mapping idea and the resulting two principal approaches (primal and dual).

3.1 Primal and Dual Space-Mapping Solutions

The idea behind space-mapping optimization is the following: if either the fine model allows for an almost reachable design (i.e., $\mathbf{f}(\mathbf{x}^*) \approx \mathbf{y}$) or if both models are similar near their respective optima (i.e., $\mathbf{f}(\mathbf{x}^*) \approx \mathbf{c}(\mathbf{z}^*)$), we expect

$$\mathbf{p}(\mathbf{x}^*) = \underset{\mathbf{z}\in Z}{\operatorname{argmin}} \, \| \, \mathbf{c}(\mathbf{z}) - \mathbf{f}(\mathbf{x}^*) \| \, \approx \underset{\mathbf{z}\in Z}{\operatorname{argmin}} \, \| \, \mathbf{c}(\mathbf{z}) - \mathbf{y} \| \, = \mathbf{z}^* \, . \tag{7}$$

Based on this relation, the space-mapping approach assumes $\mathbf{p}(\mathbf{x}^*) \approx \mathbf{z}^*$. However, in general $\mathbf{p}(\mathbf{x}^*) \neq \mathbf{z}^*$ and even $\mathbf{z}^* \in \mathbf{p}(X)$ is not guaranteed. Therefore the *primal* space-mapping approach seeks for a solution of the minimization problem

$$\mathbf{x}_p^* = \underset{\mathbf{x}\in X}{\operatorname{argmin}} \, \|\mathbf{p}(\mathbf{x}) - \mathbf{z}^*\| \, . \tag{8}$$

An alternative approach can be chosen. The idea behind space-mapping optimization is the replacement of the expensive fine model optimization by a surrogate model. For the surrogate model we can take the coarse model $\mathbf{c}(\mathbf{z})$, and improve its accuracy by the space mapping function $\mathbf{p}$. Now the improved or *mapped coarse model* $\mathbf{c}(\mathbf{p}(\mathbf{x}))$ may serve as the better *surrogate model*. Because of (4) we expect $\mathbf{c}(\mathbf{p}(\mathbf{x})) \approx \mathbf{f}(\mathbf{x})$ and hence $\| \, \mathbf{f}(\mathbf{x}) - \mathbf{y} \| \, \approx \| \, \mathbf{c}(\mathbf{p}(\mathbf{x})) - \mathbf{y} \|$. Then the minimization of $\| \, \mathbf{c}(\mathbf{p}(\mathbf{x})) - \mathbf{y} \|$ will usually give us a value, $\mathbf{x}_d^*$, close to the desired optimum $\mathbf{x}^*$:

$$\mathbf{x}_d^* = \underset{\mathbf{x}\in X}{\operatorname{argmin}} \, \| \, \mathbf{c}(\mathbf{p}(\mathbf{x})) - \mathbf{y} \| \, . \tag{9}$$

This is the *dual* space-mapping approach.

We will see in Section 3.3 that both approaches coincide when $\mathbf{z}^* \in \mathbf{p}(X)$ and $\mathbf{p}$ is injective, and if the mapping is perfect both $\mathbf{x}_p^*$ and $\mathbf{x}_d^*$ are equal to $\mathbf{x}^*$. However, in general the space-mapping function $\mathbf{p}$ will not be perfect, and hence, a space mapping based algorithm will *not* yield the solution of the fine model optimization. The principle of the approach is summarized in Figure 2.

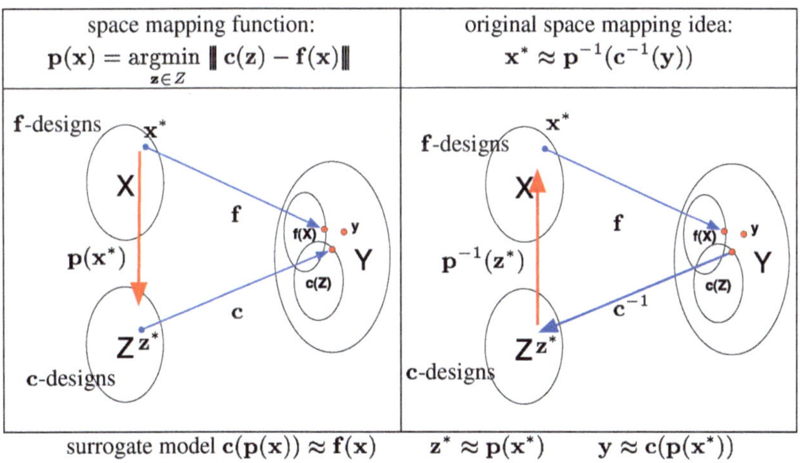

Fig. 2. Diagram showing the main idea of space mapping

$$
\begin{array}{l}
\mathbf{x}_0 = \mathbf{z}^* = \mathrm{argmin}_{\mathbf{z}\in Z} \parallel \mathbf{c}(\mathbf{z}) - \mathbf{y} \parallel \\
\mathbf{B}_0 = I \\
\textbf{for } k = 0, 1, \ldots \\
\textbf{while } \parallel \mathbf{p}(\mathbf{x}_k) - \mathbf{z}^* \parallel > \text{tolerance} \\
\textbf{do} \mathbf{h}_k = -\mathbf{B}_k^{-1}(\mathbf{p}(\mathbf{x}_k) - \mathbf{z}^*) \\
\quad \mathbf{x}_{k+1} = \mathbf{x}_k + \mathbf{h}_k \\
\quad \mathbf{B}_{k+1} = \mathbf{B}_k + \frac{(\mathbf{p}(\mathbf{x}_{k+1}) - \mathbf{z}^*)\mathbf{h}^T}{\mathbf{h}^T\mathbf{h}} \\
\textbf{enddo}
\end{array}
$$

Fig. 3. The ASM algorithm

3.2 Space-Mapping Algorithms

Because the evaluation of the space-mapping function is expensive, algorithms to compute $\mathbf{x}_p^*$ or $\mathbf{x}_d^*$ are based on iterative approximation of $\mathbf{p}(\mathbf{x})$. By the similarity of $\mathbf{f}(\mathbf{x})$ and $\mathbf{c}(\mathbf{z})$, a first approximation is the identity, $\mathbf{p}_0 = I$.

Linear approximations form the basis for the more popular space-mapping optimization algorithms. An extensive survey of available algorithms can be found in [4]. The most representative example is ASM (the 'Aggressive Space Mapping' shown in Figure 3), where the space-mapping function is approximated by linearisation to obtain

$$\mathbf{p}_k(\mathbf{x}) = \mathbf{p}(\mathbf{x}_k) + \mathbf{B}_k\,(\mathbf{x} - \mathbf{x}_k)\,. \tag{10}$$

In each space-mapping iteration step the matrix $\mathbf{B}_k$ is adapted by a rank-one update. For that purpose a Broyden-type approximation for the Jacobian of the space-mapping function $\mathbf{p}(\mathbf{x})$ is used,

$$\mathbf{B}_{k+1} = \mathbf{B}_k + \frac{\mathbf{p}(\mathbf{x}_{k+1}) - \mathbf{p}(\mathbf{x}_k) - \mathbf{B}_k\mathbf{h}}{\mathbf{h}^T\mathbf{h}}\,\mathbf{h}^T\,, \tag{11}$$

where $\mathbf{h} = \mathbf{x}_{k+1} - \mathbf{x}_k$. This is combined with original space mapping, so that $\mathbf{x}_{k+1} = \mathbf{x}_k - \mathbf{B}_k^{-1}(\mathbf{p}(\mathbf{x}_k) - \mathbf{z}^*)$.

3.3 Perfect Mapping, Flexibility and Reachability

By its definition, perfect mapping relates the similarity of the models and the specifications. If the fine model allows for a *reachable design*, then it is immediate that, independent of the coarse model used, the mapping is always perfect. Also if the coarse and the fine model optimal responses are identical, the space-mapping function is perfect. These two facts are summarized in the following lemma.

Lemma 1. *(i) If* $\mathbf{f}(\mathbf{x}^*) = \mathbf{y}$ *then* $\mathbf{p}(\mathbf{x}^*) = \mathbf{z}^*$*;*
(ii) If $\mathbf{f}(\mathbf{x}^*) = \mathbf{c}(\mathbf{z}^*)$ *then* $\mathbf{p}(\mathbf{x}^*) = \mathbf{z}^*$.

The following lemma [9] follows from the definitions (8) and (9).

Lemma 2. *(i) If* $\mathbf{z}^* \in \mathbf{p}(X)$, *then* $\mathbf{p}(\mathbf{x}_p^*) = \mathbf{p}(\mathbf{x}_d^*) = \mathbf{z}^*$*;*
(ii) If, in addition, $\mathbf{p}$ *is an injective perfect mapping then* $\mathbf{x}^* = \mathbf{x}_p^* = \mathbf{x}_d^*$.

In some cases we can expect that the sets of fine and coarse reachable aims overlap in a region of $\mathbb{R}^m$ close to their respective optima. The concept of model flexibility is introduced and from that some results concerning properties of the space-mapping functions can be derived.

Definition 1. *A model is called more* flexible *than another if the set of its reachable aims contains the set of reachable aims of the other. Two models are* equally flexible *if their sets of reachable aims coincide.* □

Thus, a coarse model $\mathbf{c}$ is more flexible than the fine one $\mathbf{f}$ if $\mathbf{c}(Z) \supset \mathbf{f}(X)$, i.e., if the coarse model response can reproduce all the fine model reachable aims. Similarly the fine model is more flexible if $\mathbf{f}(X) \supset \mathbf{c}(Z)$. Model flexibility is closely related to properties of the space-mapping function. This is shown in the following lemmas, where $\mathbf{p}$ denotes the space-mapping function. Proofs are found in [9].

Lemma 3. *If* $\mathbf{c}$ *is more flexible than* $\mathbf{f}$ *then*
(i) $\mathbf{c}(\mathbf{p}(\mathbf{x})) = \mathbf{f}(\mathbf{x}) \quad \forall \mathbf{x} \in X$;
(ii) $\mathbf{p} : X \to Z$ *is a perfect mapping* $\Leftrightarrow \mathbf{c}(\mathbf{z}^*) = \mathbf{f}(\mathbf{x}^*)$;
(iii) *if* $\mathbf{f} : X \to Y$ *is injective then* $\mathbf{p} : X \to Z$ *is injective;*
(iv) *if* $\mathbf{c}(Z) \setminus \mathbf{f}(X) \neq \emptyset$, *then* $\mathbf{p} : X \to Z$ *cannot be surjective.*

Remark. Because of (ii) generally we cannot expect space-mapping functions to be perfect for flexible coarse models unless the two models are equally flexible near the optimum. However, we remind that if the design is reachable, the perfect mapping property holds, even if $\mathbf{c}(Z) \setminus \mathbf{f}(X) \neq \emptyset$.

Lemma 4. *If* $\mathbf{f}$ *is more flexible than* $\mathbf{c}$ *then*
(i) $\mathbf{p} : X \to Z$ *is surjective;*
(ii) *if* $\mathbf{f}(X) \setminus \mathbf{c}(Z) \neq \emptyset$, *then* $\mathbf{p}$ *cannot be injective.*

We combine the previous two lemmas in the following.

Lemma 5. *If* $\mathbf{f}$ *and* $\mathbf{c}$ *are equally flexible and* $\mathbf{f} : X \to Y$ *is injective, then* (i) $\mathbf{p}$ *is a bijection, and* (ii) $\mathbf{p}$ *is a perfect mapping.*

The conclusions in Lemma 2 can now be derived from assumptions about model flexibility.

Lemma 6. (i) *If* $\mathbf{f}$ *is more flexible than* $\mathbf{c}$, *then* $\mathbf{p}(\mathbf{x}_p^*) = \mathbf{p}(\mathbf{x}_d^*) = \mathbf{z}^*$. (ii) *If* $\mathbf{f}$ *and* $\mathbf{c}$ *are equally flexible and* $\mathbf{f}$ *is injective, then* $\mathbf{x}^* = \mathbf{x}_p^* = \mathbf{x}_d^*$.

Remark. It is not really needed for the space-mapping function to be a bijection over the whole domain in which it is defined. In fact, perfect mapping is a property that concerns only a point, and it is enough if the function is injective in a (small) neighborhood. Thus the assumptions for the former lemmas can be relaxed and stated just locally.

4 Defect Correction and Space Mapping

The technique underlying space-mapping, i.e. the efficient solution of a complex problem by the iterative use of a simpler one, is known since long in computational mathematics. In numerical analysis it is known as defect correction iteration and studied in a number of papers [6, 7]. Below we first briefly summarize the defect correction principle for solving operator equations and then we apply the idea to optimization problems.

4.1 Defect Correction for Operator Equations

We first consider the problem of solving a nonlinear operator equation

$$\mathcal{F}\mathbf{x} = \mathbf{y}, \tag{12}$$

where $\mathcal{F} : D \subset E \to \widehat{D} \subset \widehat{E}$ is a continuous, generally nonlinear operator and E and $\widehat{E}$ are Banach spaces. In general, neither injectivity nor surjectivity of the mapping is assumed, but in many cases these properties can be achieved by a proper choice of the subsets D and $\widehat{D}$.

The classical *defect correction* iteration for the solution of equation (12) with $\mathbf{y} \in \mathcal{F}(D) \subset \widehat{D}$ is based on a sequence of operators $\widetilde{\mathcal{F}}_k : D \to \widehat{D}$ approximating $\mathcal{F}$. We assume that each $\widetilde{\mathcal{F}}_k$ has an easy-to-calculate inverse $\widetilde{\mathcal{G}}_k : \widehat{D} \to D$. Actually, it is the existence of the easy-to-evaluate operator $\widetilde{\mathcal{G}}_k$, rather than the existence of $\widetilde{\mathcal{F}}_k$, that is needed for defect correction and we do not need to assume $\widetilde{\mathcal{G}}_k$ to be invertible.

Defect correction comes in two brands [6], depending on the space, E or $\widehat{E}$, in which linear combinations for extrapolation are made. The two basic iterative defect correction procedures to generate a (hopefully convergent) sequence of approximations to the solution of (12) are

$$\begin{cases} \mathbf{x}_0 &= \widetilde{\mathcal{G}}_0\,\mathbf{y} \\ \mathbf{x}_{k+1} &= (I - \widetilde{\mathcal{G}}_{k+1}\,\mathcal{F})\,\mathbf{x}_k + \widetilde{\mathcal{G}}_{k+1}\,\mathbf{y} \end{cases} \quad \text{and} \quad \begin{cases} \mathbf{l}_0 &= \mathbf{y} \\ \mathbf{l}_{k+1} &= (I - \mathcal{F}\,\widetilde{\mathcal{G}}_k)\,\mathbf{l}_k + \mathbf{y}\,. \end{cases} \tag{13}$$

In the second, (13b), we identify the approximate solution as $\mathbf{x}_k \equiv \widetilde{\mathcal{G}}_k\mathbf{l}_k$. We see that the two iteration processes are dual in the sense that in the first, (13a), the extrapolation is in the space D, whereas the additions in (13b) are in $\widehat{D}$. If $\widetilde{\mathcal{G}}_k$ is injective, then an operator $\widetilde{\mathcal{F}}_k$ exists such that $\widetilde{\mathcal{F}}_k\widetilde{\mathcal{G}}_k = \mathcal{I}_{\widehat{D}}$, i.e., $\widetilde{\mathcal{F}}_k$ is the left-inverse of $\widetilde{\mathcal{G}}_k$. Then $\widetilde{\mathcal{F}}_k\mathbf{x}_k = \mathbf{l}_k$ and (13b) is equivalent with the iterative procedure

$$\begin{cases} \widetilde{\mathcal{F}}_0\mathbf{x}_0 &= \mathbf{y}\,, \\ \widetilde{\mathcal{F}}_{k+1}\mathbf{x}_{k+1} &= \widetilde{\mathcal{F}}_k\,\mathbf{x}_k - \mathcal{F}\widetilde{\mathcal{G}}_k\widetilde{\mathcal{F}}_k\,\mathbf{x}_k + \mathbf{y}\,. \end{cases} \tag{14}$$

In order to apply (14), the injectivity of $\widetilde{\mathcal{G}}_k$ is not really needed and it is immediately seen that neither (13b) nor (14) converges if $\mathbf{y} \notin \mathcal{F}(D)$. However, (14) can be modified so that it can be used for $\mathbf{y} \notin \mathcal{F}(D)$. Then we need injectivity for $\widetilde{\mathcal{F}}_k$ and we take $\widetilde{\mathcal{G}}_k$ its left-inverse, i.e., $\widetilde{\mathcal{G}}_k\widetilde{\mathcal{F}}_k = \mathcal{I}_D$. Then (14) leads to

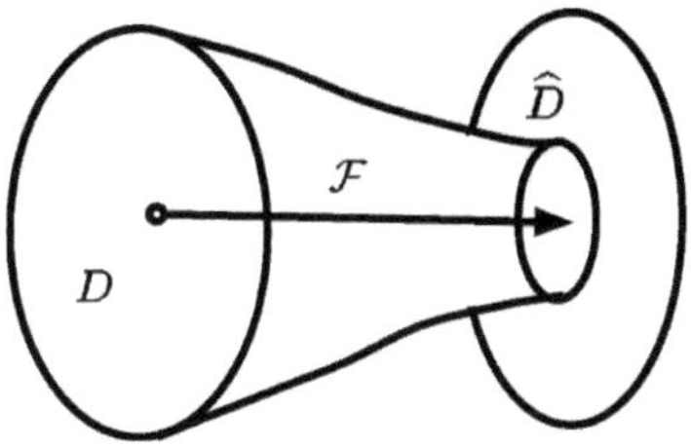

$\mathcal{F}$ not surjective $\rightarrow$ left-inverse $\mathcal{G}$: $\mathcal{G}\mathcal{F} = I_D$

Fig. 4. The non-surjective operator in the optimization problem

$$\begin{cases} \mathbf{x}_0 &= \widetilde{\mathcal{G}}_0\, \mathbf{y}\,, \\ \mathbf{x}_{k+1} &= \widetilde{\mathcal{G}}_{k+1} \left(\widetilde{\mathcal{F}}_k\, \mathbf{x}_k - \mathcal{F}\mathbf{x}_k + \mathbf{y} \right) . \end{cases} \tag{15}$$

Because (15) allows for a non-injective $\widetilde{\mathcal{G}}_k$, this iteration can be used for optimization purposes. In case of an invertible $\widetilde{\mathcal{G}}_{k+1}$ both (14) and (15) are equivalent with (13b).

For our optimization problems, where the design may be not reachable, $\mathbf{y} \in \widehat{D}$, but $\mathbf{y} \notin \mathcal{F}(D)$, i.e., $\mathcal{F}$ is no surjection so that no solution for (12) exists and (13b)-(14) cannot converge (Figure 4). Therefore, we drop the idea of finding an $\mathbf{x} \in D$ satisfying (12) and we replace the aim by looking for a solution $\mathbf{x}^* \in D$ so that the distance between $\mathcal{F}\mathbf{x}$ and $\mathbf{y}$ is minimal, i.e., we want to find

$$\mathbf{x}^* = \text{argmin}_{\mathbf{x}\in D} \left\| \mathcal{F}\mathbf{x} - \mathbf{y} \right\|_{\widehat{E}} . \tag{16}$$

For a compact non-empty D and a continuous $\mathcal{F}$, at least a solution exists and if the operators $\widetilde{\mathcal{G}}_k$ are such that (13a) or (15) converges, the stationary point $\overline{\mathbf{x}}$ satisfies $\widetilde{\mathcal{G}}\mathcal{F}\,\overline{\mathbf{x}} = \widetilde{\mathcal{G}}\mathbf{y}$ or $\overline{\mathbf{x}} = \widetilde{\mathcal{G}}(\widetilde{\mathcal{F}}\,\overline{\mathbf{x}} - \mathcal{F}\,\overline{\mathbf{x}} + \mathbf{y})$ respectively. (We assume that $\widetilde{\mathcal{G}}_k = \widetilde{\mathcal{G}}$ and $\widetilde{\mathcal{F}}_k = \widetilde{\mathcal{F}}$ for k large enough.)

Now we can associate with each defect correction iteration a process for iterative optimization by taking $E = \mathbb{R}^n$, $\widehat{E} = \mathbb{R}^m$, $D = X$, $\widehat{D} = Y$ and $\overline{\mathbf{p}} : X \rightarrow Z$, and by substitution of the corresponding operators:

$$\begin{aligned} \mathcal{F}\mathbf{x} = \mathbf{y} \quad &\Leftrightarrow \mathbf{f}(\mathbf{x}) = \mathbf{y}\,, \\ \mathbf{x} = \mathcal{G}\mathbf{y} \quad &\Leftrightarrow \mathbf{x} = \underset{\xi}{\text{argmin}} \left\| \mathbf{f}(\xi) - \mathbf{y} \right\| , \\ \widetilde{\mathcal{F}}\mathbf{x} = \mathbf{y} \quad &\Leftrightarrow \mathbf{c}(\overline{\mathbf{p}}(\mathbf{x})) = \mathbf{y}\,, \\ \mathbf{x} = \widetilde{\mathcal{G}}\mathbf{y} \quad &\Leftrightarrow \mathbf{x} = \underset{\xi}{\text{argmin}} \left\| \mathbf{c}(\overline{\mathbf{p}}(\xi)) - \mathbf{y} \right\| . \end{aligned} \tag{17}$$

Remark. Notice that $\overline{\mathbf{p}}$ is *not* the space mapping function but an arbitrary (easy to compute) bijection, e.g., the identity.

4.2 Defect Correction for Optimization

With (17) we derive from (13a) and (15) two defect-correction iteration schemes for optimization. Substitution of (17) yields the initial estimate and two iteration processes for $k = 0, 1, 2, \cdots$, with $\mathbf{p}_{k+1}$ for $\overline{\mathbf{p}}$ in every step,

$$\mathbf{x}_0 = \operatorname*{argmin}_{\mathbf{x}\in X} \|\, \mathbf{c}(\mathbf{p}_0(\mathbf{x})) - \mathbf{y} \| \;, \tag{18}$$

$$\mathbf{x}_{k+1} = \mathbf{x}_k - \operatorname*{argmin}_{\mathbf{x}\in X} \|\, \mathbf{c}(\mathbf{p}_{k+1}(\mathbf{x})) - \mathbf{f}(\mathbf{x}_k) \| + \operatorname*{argmin}_{\mathbf{x}\in X} \|\, \mathbf{c}(\mathbf{p}_{k+1}(\mathbf{x})) - \mathbf{y} \| \;, \tag{19}$$

$$\mathbf{x}_{k+1} = \operatorname*{argmin}_{\mathbf{x}\in X} \|\, \mathbf{c}(\mathbf{p}_{k+1}(\mathbf{x})) - \mathbf{c}(\mathbf{p}_k(\mathbf{x}_k)) + \mathbf{f}(\mathbf{x}_k) - \mathbf{y} \| \;. \tag{20}$$

The two processes (19) and (20) are still dual in the sense that extrapolation is applied in the space X for process (19) and in Y for process (20). The operators $\mathbf{p}_k$ are right-preconditioners for the coarse model, which may be adapted during the initial steps of the iteration. We take $\mathbf{p}_k$ non-singular and for the initial estimate (18), and if $X = Z$ we usually take $\mathbf{p}_0 = I$, the identity.

In the above iterations every minimization involves the surrogate model, $\mathbf{c} \circ \mathbf{p}_k$. However, it is the coarse model that was assumed to be cheaply optimized. Therefore, it is more convenient to write the procedures such that optimization over the coarse model becomes obvious. By taking in (13a) and (15) $\mathcal{F}\,\mathbf{z} = \mathbf{f}(\mathbf{q}(\mathbf{z}))$, $\widetilde{\mathcal{F}}_k\,\mathbf{z} = \mathbf{c}(\mathbf{z})$ and $\widetilde{\mathcal{G}}_k\,\mathbf{y} = \operatorname{argmin}_{\mathbf{z}\in Z} \|\, \mathbf{c}(\mathbf{z}) - \mathbf{y} \|$, with $\mathbf{q}$ and $\mathbf{q}_k$ bijections from Z to X fulfilling in every iteration $\mathbf{q}\,\mathbf{z}_k = \mathbf{q}_k\,\mathbf{z}_k$, we obtain, for $k = 0, 1, 2, \cdots$,

$$\mathbf{z}_0 = \mathbf{z}^* = \operatorname*{argmin}_{\mathbf{z}\in Z} \|\, \mathbf{c}(\mathbf{z}) - \mathbf{y} \| \;, \tag{21}$$

$$\mathbf{z}_{k+1} = \mathbf{z}_k - \operatorname*{argmin}_{\mathbf{z}\in Z} \|\, \mathbf{c}(\mathbf{z}) - \mathbf{f}(\mathbf{q}_k(\mathbf{z}_k)) \| + \mathbf{z}^* \;, \tag{22}$$

$$\mathbf{z}_{k+1} = \operatorname*{argmin}_{\mathbf{z}\in Z} \|\, \mathbf{c}(\mathbf{z}) - \mathbf{c}(\mathbf{z}_k) + \mathbf{f}(\mathbf{q}_k(\mathbf{z}_k)) - \mathbf{y} \| \;. \tag{23}$$

As the solution is wanted in terms of fine-model control variables, the procedures are complemented with $\mathbf{x}_k = \mathbf{q}_k(\mathbf{z}_k)$. The bijections can be interpreted as $\mathbf{q}_k = \mathbf{p}_k^{-1}$. For $k > k_0$, we assume the iteration process to be stationary: $\mathbf{p}_k = \overline{\mathbf{p}}$ and $\mathbf{q}_k = \overline{\mathbf{q}}$. It is a little exercise to see by proper simplifications of (19) and (20) that space-mapping iteration can be recovered from defect correction [9, Section 4.3.2].

Orthogonality and the Need for Left-preconditioning.

For the stationary points of the above processes, we can derive the following lemma [9].

Lemma 7. *In the case of convergence of (23), with fixed point* $\lim_{k\to\infty} \mathbf{x}_k = \overline{\mathbf{x}}$ *we obtain*

$$\mathbf{f}(\overline{\mathbf{x}}) - \mathbf{y} \in \mathbf{c}(Z)^{\perp}(\overline{\mathbf{p}}(\overline{\mathbf{x}}))\,. \tag{24}$$

In case of convergence of (22) with a fixed point $\overline{\mathbf{x}}$ *we obtain*

$$\mathbf{f}(\overline{\mathbf{x}}) - \mathbf{y} \in \mathbf{c}(Z)^{\perp}(\mathbf{z}^*)\;. \tag{25}$$

Like the space-mapping methods, the above iterations have the disadvantage that, in general, the fixed point of the iteration does not coincide with the solution of the fine model minimization problem. This is due to the fact that the

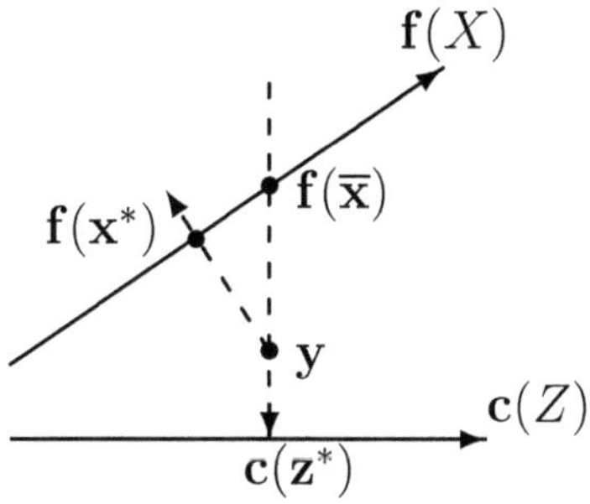

Fig. 5. The relative location of $\mathbf{c}(\mathbf{z}^*)$, $\mathbf{f}(\mathbf{x}^*)$ and $\mathbf{f}(\overline{\mathbf{x}})$

approximate solution $\overline{\mathbf{x}}$ satisfies either (24) or (25). whereas a (local) minimum $\mathbf{x}^* = \text{argmin}_{x \in X} \|\mathbf{f}(\mathbf{x}) - \mathbf{y}\|$ satisfies (see Figure 5)

$$\mathbf{f}(\mathbf{x}^*) - \mathbf{y} \in \mathbf{f}(X)^{\perp}(\mathbf{x}^*) \,. \tag{26}$$

Hence, differences between $\overline{\mathbf{x}}$ and $\mathbf{x}^*$ will be larger for larger distances between $\mathbf{y}$ and the sets $\mathbf{f}(X)$ and $\mathbf{c}(Z)$ and for larger angles between the linear manifolds tangential at $\mathbf{c}(Z)$ and $\mathbf{f}(X)$ near the optima.

By the orthogonality relations above, we see that it is advantageous, both for the conditioning of the problem and for the minimization of the residual, if the manifolds $\mathbf{f}(X)$ and $\mathbf{c}(Z)$ are found parallel in the neighborhood of the solution. However, by space mapping or by right-preconditioning the relation between the manifolds $\mathbf{f}(X)$ and $\mathbf{c}(Z)$ remains unchanged. This causes that the fixed point of traditional space mapping does generally not correspond with $\mathbf{x}^*$. This relation, however, can be improved by the introduction of an additional left-preconditioner. Therefore we introduce such a preconditioner $\mathbf{S}$ so that near $\mathbf{f}(\mathbf{x}^*) \in Y$ the manifold $\mathbf{c}(Z) \subset Y$ is mapped onto $\mathbf{f}(X) \subset Y$:

$$\mathbf{f}(\mathbf{x}) \approx \mathbf{S}(\mathbf{c}(\overline{\mathbf{p}}(\mathbf{x}))) \,. \tag{27}$$

In the next section we propose a new algorithm where an affine operator maps $\mathbf{c}(Z)$ onto $\mathbf{f}(X)$ in the neighborhood of the solution. (More precisely: it approximately maps one tangential linear manifold onto the other.) This restores the orthogonality relation $\mathbf{f}(\overline{\mathbf{x}}) - \mathbf{y} \perp \mathbf{f}(X)(\mathbf{x}^*)$. Thus it improves significantly the traditional approach and makes the solution $\mathbf{x}^*$ a stationary point of the iteration. Details on the convergence of the processes can be found in [10].

5 Manifold Mapping, the Improved Space Mapping Algorithm

We introduce the affine mapping $\mathbf{S} : Y \to Y$ such that $\mathbf{S}\,\mathbf{c}(\overline{\mathbf{z}}) = \mathbf{f}(\mathbf{x}^*)$ for a proper $\overline{\mathbf{z}} \in Z$, and the linear manifold tangential to $\mathbf{c}(Z)$ in $\mathbf{c}(\overline{\mathbf{z}})$ maps onto the one tangential to $\mathbf{f}(X)$ in $\mathbf{f}(\mathbf{x}^*)$. Because, in the non-degenerate case when $m \geq n$, both $\mathbf{f}(X)$ and $\mathbf{c}(Z)$ are n-dimensional sets in $\mathbb{R}^m$, the mapping $\mathbf{S}$ can be described by

$$\mathbf{S}\,\mathbf{v} = \mathbf{f}(\mathbf{x}^*) + S\ \ (\mathbf{v} - \mathbf{c}(\overline{\mathbf{z}}))\ , \tag{28}$$

where S is an $m \times m$-matrix S of rank n. This mapping $\mathbf{S}$ is not a priori available, but an approximation to it can be computed iteratively during the optimization. A full rank $m \times m$-matrix S can be constructed, which has a well-determined part of rank n, while a remaining part of rank $m - n$ is free to choose. Because of the supposed similarity between the models $\mathbf{f}$ and $\mathbf{c}$ we keep the latter part close to the identity. The meaning of the mapping $\mathbf{S}$ is illustrated in the Figures 6 and 7

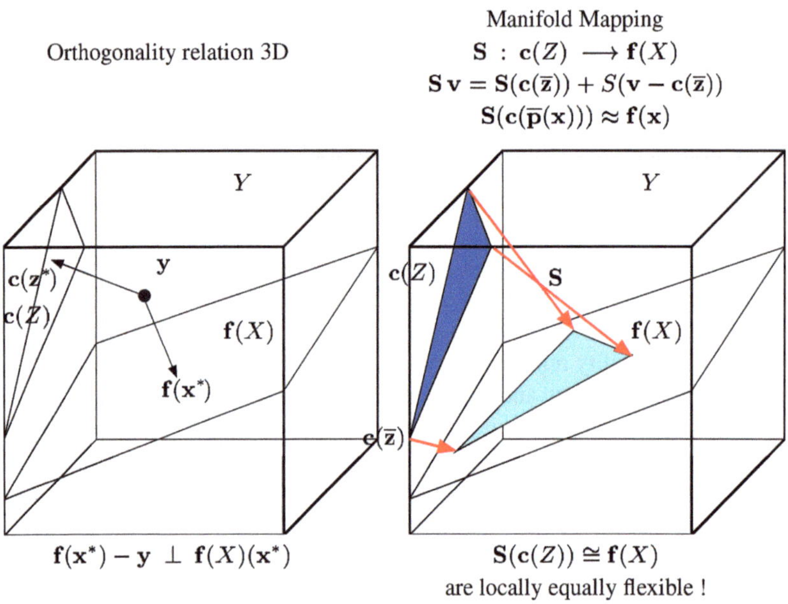

Fig. 6. Restoring the orthogonality relation by manifold mapping

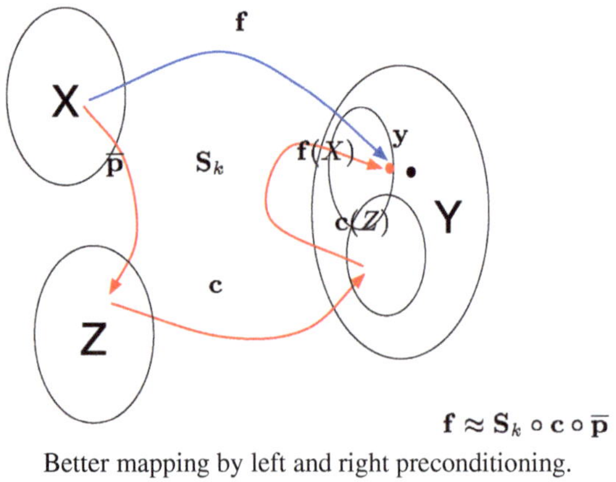

Fig. 7. Manifold Mapping

So we propose the following algorithm (where the optional right-preconditioner $\overline{\mathbf{p}} : X \rightarrow Z$ is still an arbitrary non-singular operator. It can be adapted to the problem. Often we will simply take the identity.)

1. Set $k = 0$, set $S_0 = I$ the $m \times m$ identity matrix, and compute
$$\mathbf{x}_0 = \text{argmin}_{\mathbf{x} \in X} \, \| \, \mathbf{c}(\overline{\mathbf{p}}(\mathbf{x})) - \mathbf{y} \| \, . \tag{29}$$
2. Compute $\mathbf{f}(\mathbf{x}_k)$ and $\mathbf{c}(\overline{\mathbf{p}}(\mathbf{x}_k))$.
3. If $k > 0$, with $\Delta\mathbf{c}_i = \mathbf{c}(\overline{\mathbf{p}}(\mathbf{x}_{k-i})) - \mathbf{c}(\overline{\mathbf{p}}(\mathbf{x}_k))$ and $\Delta\mathbf{f}_i = \mathbf{f}(\mathbf{x}_{k-i}) - \mathbf{f}(\mathbf{x}_k)$, $i = 1, \cdots, \min(n,k)$, we define ΔC and ΔF to be the rectangular $m \times \min(n,k)$-matrices with respectively $\Delta\mathbf{c}_i$ and $\Delta\mathbf{f}_i$ as columns. Their singular value decompositions are respectively $\Delta C = U_c \Sigma_c V_c^T$ and $\Delta F = U_f \Sigma_f V_f^T$.
4. The next iterand is computed as
$$\mathbf{x}_{k+1} = \underset{\mathbf{x} \in X}{\text{argmin}} \, \| \mathbf{c}(\overline{\mathbf{p}}(\mathbf{x})) - \mathbf{c}(\overline{\mathbf{p}}(\mathbf{x}_k)) + \left[\Delta C \Delta F^\dagger + I - U_c U_c^T \right] (\mathbf{f}(\mathbf{x}_k) - \mathbf{y}) \| . \tag{30}$$
5. Set $k := k + 1$ and goto 2.

Here, † denotes the pseudo-inverse: $\Delta F^\dagger = V_f \Sigma_f^{-1} U_f^T$. It can be shown that (30) is asymtotically equivalent to

$$\mathbf{x}_{k+1} = \text{argmin}_{\mathbf{x} \in X} \, \| \mathbf{S}_k(\mathbf{c}(\overline{\mathbf{p}}(\mathbf{x}))) - \mathbf{y} \| \, . \tag{31}$$

Above, the matrix $S_k = \Delta F \, \Delta C^\dagger + (I - U_\mathbf{f} \, U_\mathbf{f}^T)\,(I - U_\mathbf{c} \, U_\mathbf{c}^T)$ and the approximate affine mapping is

$$\mathbf{S}_k \, \mathbf{v} = \mathbf{f}(\mathbf{x}_k) + S_k(\mathbf{v} - \mathbf{c}(\overline{\mathbf{p}}(\mathbf{x}_k)) \, , \quad \forall \mathbf{v} \in Y,$$

which, for $l > 0$ and $l = k - 1, \cdots, \max(0, k - n)$, satisfies

$$S_k \left(\mathbf{c}(\overline{\mathbf{p}}(\mathbf{x}_l)) - \mathbf{c}(\overline{\mathbf{p}}(\mathbf{x}_k)) \right) = \mathbf{f}(\mathbf{x}_l) - \mathbf{f}(\mathbf{x}_k) \, .$$

In (30), the freedom in making S_k full-rank is used, replacing $\Delta C \;\; \Delta F^\dagger + (I - U_c U_c^T)(I - U_f U_f^T)$ by $\Delta C \;\; \Delta F^\dagger + I - U_c U_c^T$, in order to stabilize the algorithm. This does not change the solution.

If the above iteration converges with fixed point $\overline{\mathbf{x}}$ and mappings $\overline{\mathbf{S}}$ and $\overline{\mathbf{p}}$, we have

$$\mathbf{f}(\overline{\mathbf{x}}) - \mathbf{y} \in \overline{\mathbf{S}}(\mathbf{c}(\overline{\mathbf{p}}(X)))^\perp(\overline{\mathbf{x}}) = \mathbf{f}(X)^\perp(\overline{\mathbf{x}}) \, . \tag{32}$$

From this relation and the fact that $\mathbf{S}_k(\mathbf{c}(\overline{\mathbf{p}}(\mathbf{x}_k))) = \mathbf{f}(\mathbf{x}_k)$, it can be concluded that, under convergence to $\overline{\mathbf{x}}$, the fixed point is a (local) optimum of the fine model minimization.

The improved space-mapping scheme

$$\mathbf{x}_{k+1} = \underset{\mathbf{x}}{\text{argmin}} \, \| \, \mathbf{S}_k(\mathbf{c}(\overline{\mathbf{p}}_k(\mathbf{x})))) - \mathbf{y} \| \tag{33}$$

can also be recognized as defect correction iteration with either $\widetilde{\mathcal{F}}_k = \mathbf{S}_k \circ \mathbf{c} \circ \overline{\mathbf{p}}$ and $\mathcal{F} = \mathbf{f}$ in (19) or (20), or with $\widetilde{\mathcal{F}}_k = \mathbf{S}_k \circ \mathbf{c}$ and $\mathcal{F} = \mathbf{f} \circ \overline{\mathbf{p}}^{-1}$ in (22) or (23).

6 Examples

We illustrate the application of space-mapping and manifold-mapping by a design problem for a linear actuator. We compare the performance of these algorithms with that of two classical optimization methods: Nelder-Mead Simplex (NMS) and Sequential Quadratic Programming (SQP).

Linear actuators are electromechanical devices that convert electromechanical power into linear motion. An axi-symmetrical variant, called a *voice-coil* actuator, consisting of a permanent magnet, a current-carrying coil and a ferromagnetic core is shown in Figure 8. The permanent magnet is magnetized in the vertical direction. The coil, steered by the magnetic force, moves along the z-axis in the gap of the core, as illustrated in Figure 9. The position of the coil relative to the top of the

Fig. 8. A cylindrical voice-coil actuator consisting of a ferromagnetic core, permanent magnet and coil

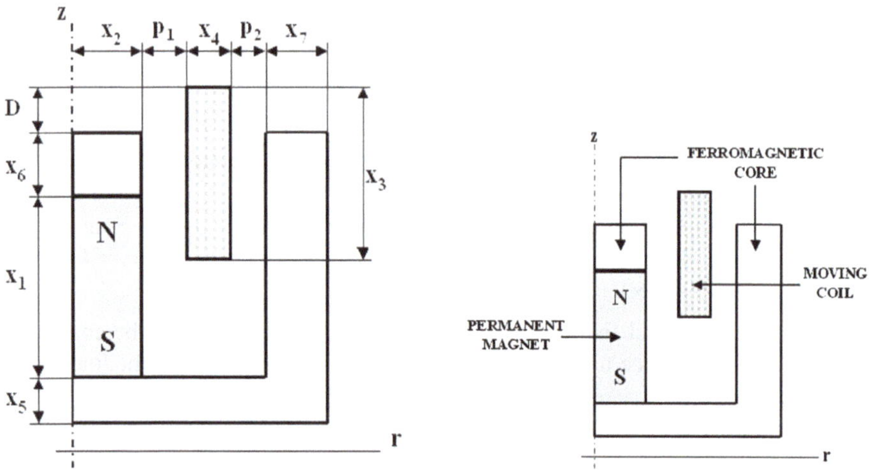

Design Variables. Geometry.

Fig. 9. Geometry and design variables of the cylindrical voice-coil actuator

core is denoted by D. Due to the axisymmetrical geometry, the force has an axial component only. It will be denoted by $F_z(D)$.

The design variables [13] are shown in Figure 9: x_1 and x_2 denote the height and radius of the magnet, x_3 and x_4 the height and thickness of the coil and x_5, x_6 and x_7 the sizes of the core. Two additional linear inequality constraints define feasible coil positions. The air-gap sizes p_1 and p_2 to the left and right of the coil are kept fixed. Remaining details are found in [11].

We allow the coil to move over a 4 mm range, i.e., $0 \leq \mathrm{D} \leq 4\,\mathrm{mm}$. The force on the coil is computed at nine equidistant points D_i in this interval. Values for the design variables have to be found such that the force response is flat and as close to $\mathbf{y} = 24\,\mathrm{N}$ as possible. The cost function is

$$\left(\sum_{i=1}^{9} [F_z(D_i) - \mathbf{y}(D_i)]^2 / \sum_{i=1}^{9} \mathbf{y}(D_i)^2 \right)^{1/2} . \tag{34}$$

The fine model is a second order Lagrangian finite element (FE) model in which the non-linear BH-curve of the ferromagnetic core is taken into account. The force is computed by means of the Lorentz Force Law [8]. The number of degrees of freedom in the FE model is between 8000 and 11000, yielding three digits of accuracy in the computed force.

The first of two coarse models is a FE model in which the BH-curve of the actuator core is linearized. Depending on the number of Newton iterations required in the non-linear case, this model is a factor between 30 and 50 cheaper than the fine one. The second coarse model is a lumped parameter model. This so-called magnetic equivalent circuit (MEC) [8] model has a negligible computational cost compared to the fine one. In both the FE and the MEC coarse models, the relative magnetic permeability in the core was overestimated and set equal to 1000. This was done for illustration purposes.

Below we will consider three variants of modelling approaches for this type of problem. The use of manifold mapping with the linearized finite element, respectively the MEC as coarse model, will be denoted by FE/MM and MEC/MM. Similar notations FE/SM and MEC/SM are used for space mapping.

6.1 A Variant with One Design Variable

We initially consider a design problem with a single design variable, only varying the radius of the permanent magnet. We denote the design variable x_2 simply by $\mathbf{x}$. As a starting guess we use the coarse model optimum, i.e., $x_0 = z^*$, as in Section 4.2, where the choice $\mathbf{p}_0 = I$ was made. For this one-parameter problem both space mapping (SM) and manifold mapping (MM), with either the linear FE or the MEC as coarse model, converge to the unique $\mathbf{x}^*$ in four iterations and both methods deliver a speed-up with a factor between four and five compared with the NMS or the SQP algorithm [11].

The cost function associated with the surrogate model that MM builds in the final iteration step approximates the fine model cost function in a neighbourhood

$\mathbf{x}^*$ much better than its SM counterpart. We illustrate the convergence of SM and MM by looking at the cost function of the surrogate models during successive iterations. Figure 10 (top) shows the cost functions of the surrogate model during the first

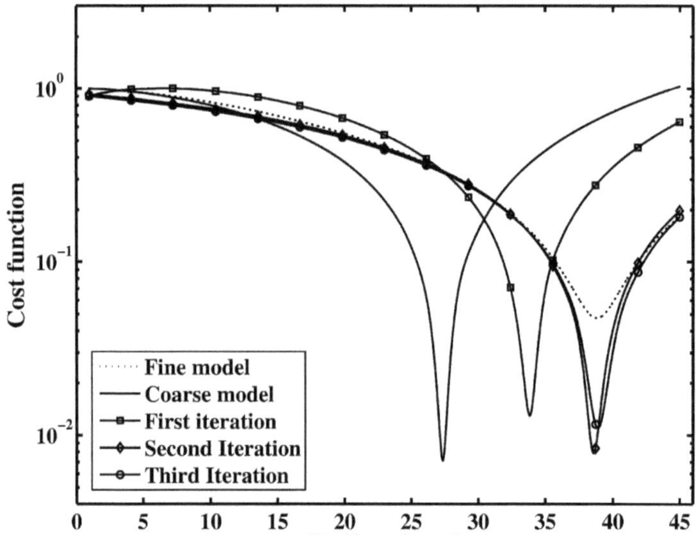

Space-Mapping (SM).

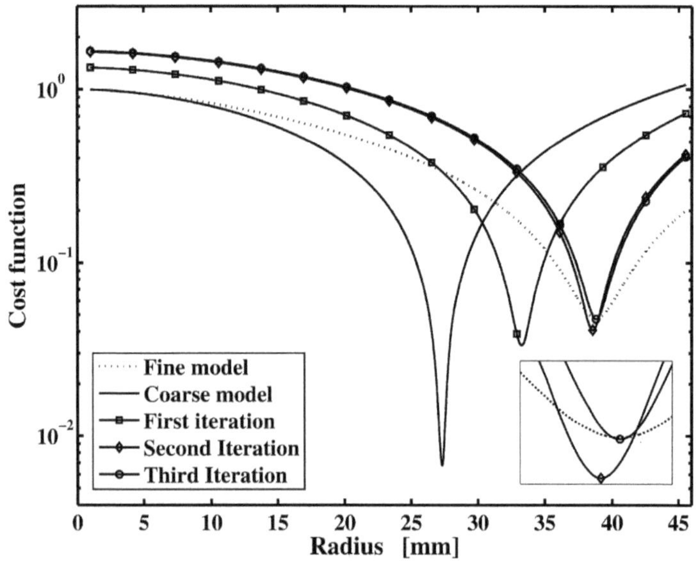

Manifold-Mapping (MM).

Fig. 10. Convergence history of SM and MM using the MEC as coarse model

Table 1. Computational efficiency of SM and MM for an example with a single design variable, compared with the NMS and the SQP method

	# iters	# $\mathbf{f}$ evals.	# $\mathbf{c}$ evals
NMS	10	20	20
SQP	5	18	20
MEC/SM	4	4	80
MEC/MM	4	4	80

MEC/SM iterations, i.e., $\|\mathbf{c}(\mathbf{p}_k(\mathbf{x})) - \mathbf{y}\|_2/\|\mathbf{y}\|_2$, for $k = 1, \ldots, 3$ as function of $\mathbf{x}$. The coarse ($k = 0$) and fine model cost functions are also shown. Figure 10 (bottom) shows the same for MEC/MM with $\|\mathbf{S}_k(\mathbf{c}(\mathbf{x})) - \mathbf{y}\|_2/\|\mathbf{y}\|_2$ for successive k. The overestimation of the magnetic permeability of the core in the coarse models is such that for these models a smaller radius is required to reach the design objective, i.e, $\mathbf{z}^* < \mathbf{x}^*$. The figures also illustrate the convergence of the iterands $\mathbf{x}_k$ to $\mathbf{x}^*$. They furthermore show that the mapping of the tangent manifold in MM provides a better approximation of the fine model cost function in a neighbourhood of $\mathbf{x}_f^*$.

To show the speed-ups that SM and MM-algorithms may yield, in Table 1 we show the number of fine and coarse model evaluations of MEC/SM and MEC/MM as well as the number required by NMS and SQP. For the latter two, the coarse model was used to generate an appropriate initial guess. In the other two algorithms each iteration requires one fine and twenty coarse model evaluations. From the table the computational speed-up is obvious. Even though the coarse model was chosen to be quite inaccurate, the SM based algorithms deliver a significant speed-up.

To quantify the difference between the two coarse models, in Figure 11 we show the decrease in cost function during SM and MM iteration with both coarse models. From this figure we conclude that the linear FE coarse model does not accelerate the converge of SM or MM better than the (much cheaper) MEC model. A linear FE coarse model can however be advantageous in more complex design problems.

6.2 A Variant with Two Design Variables

We now consider a design problem with two design variables, allowing changes in height (x_1) and radius (x_2) of the permanent magnet. Numerical results comparing the performance of SM and MM with NMS and SQP for this problem are given in Table 2. The first row in this table gives the total amount of work expressed in number of equivalent fine model evaluations. These figures are approximately proportional to the total computing time. As starting guess for the optimization procedures we used the values obtained by optimizing the MEC model. This design problem is extremely ill-conditioned and has a manifold of equivalent solutions. To stabilize the convergence of MM, the Levenberg-Marquardt method is used. The best results in terms of computational efficiency (speed-up by a factor of six) are obtained using MM with the MEC as coarse model. Full details about this problem and its solution by SM or MM are found in [11].

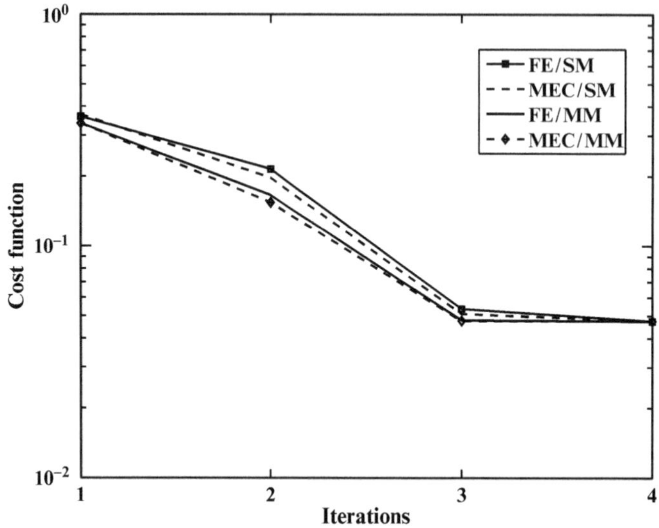

Fig. 11. Reduction in cost function value in successive iterations of SM and MM

Table 2. Computational efficiency of SM and MM for the example with two design variables. The total amount of computational work is approximately equal to the cost of the fine model function evaluations (# **f** evals.)

	NMS	SQP	FE/SM	MEC/SM	FE/MM	MEC/MM
# **f** evals.	24	31	9	6	9	4
cost function	0.046	0.046	0.046	0.065	0.046	0.046

6.3 A Variant with Seven Design Variables

In the last example we show the potential of MM and SM in the problem with all seven design variables and non-linear equality and inequality constraints. This design problem was introduced in [12] and details can be found in [11]. The total mass of the actuator has to be minimized, while the mass of the coil is constrained to 10 g. Thus, the cost function is the total mass of the device. The force at coil position $D = 4.25$ mm should be kept at 5 N and the magnetic flux density in three regions of the core should not exceed 1 T. In the fine model the constraints are evaluated by the same FE model as used in the two previous design problems. In the coarse model the constraints are based on a MEC model. Each coarse model related optimization is solved by SQP. Either MM or SM is applied for the constraints evaluation.

Numerical results for this problem are shown in Table 3. SM and MM show a similar behaviour: convergence is reached in seven or six fine constraint evaluations respectively. Having the coarse model optimum $\mathbf{z}^*$ as the initial guess, SQP converges within 56 fine constraint evaluations. MM offers an additional advantage over SM: the computation of the SM function $\mathbf{p}(\mathbf{x})$ is a very delicate issue [4], but MM replaces it simply by the identity.

Table 3. Computational efficiency of SM and MM for an example with seven design variables

	# evals.	total mass	final design (mm)
SQP	56	81.86 g	[8.543, 9.793, 11.489, 1.876, 3.876, 3.197, 2.524]
SM	7	81.11 g	[8.500, 9.786, 11.450, 1.883, 3.838, 3.200, 2.497]
MM	6	81.45 g	[8.500, 9.784, 11.452, 1.883, 3.860, 3.202, 2.515]

The initial guess for SQP is the coarse model optimum $\mathbf{z}^*$.
The total amount of work is approximately equal to the cost of the fine model constraint evaluations (# evals.).

7 Conclusions

The space-mapping technique aims at accelerating expensive optimization procedures by combining problem descriptions with different degrees of accuracy. In numerical analysis, for the solution of operator equations, the same principle is known as defect correction iteration.

When analyzing the behaviour of space-mapping iteration, it is important to know the notions of reachability of a design and flexibility of the underlying models. One can show that if neither the design is reachable nor the models are equally flexible, space mapping iteration does generally not converge to the (accurate) solution of the optimization problem.

Using the principle of defect correction iteration, we can repair this deficiency and construct the manifold-mapping iteration, which is as efficient as space mapping, but converges to the right solution.

Our findings are illustrated by an example from electromagnetics. Here parameters for the design of a voice coil actuator are determined, using a finite element discretization for the fine model and an equivalent magnetic circuit description for the coarse one.

References

1. M.H. Bakr, J.W. Bandler, K. Madsen, and J. Søndergaard. Review of the space mapping approach to engineering optimization and modeling. *Optimization and Engineering*, 1(3):241–276, 2000.
2. M.H. Bakr, J.W. Bandler, K. Madsen, and J. Søndergaard. An introduction to the space mapping technique. *Optimization and Engineering*, 2(4):369–384, 2001.
3. J.W. Bandler, R.M. Biernacki, S.H. Chen, P.A. Grobelny, and R.H. Hemmers. Space mapping technique for electromagnetic optimization. *IEEE Trans. Microwave Theory Tech*, 42:2536–2544, 1994.
4. J.W. Bandler, Q.S. Cheng, A.S. Dakroury, A.S. Mohamed, M.H. Bakr, K. Madsen, and J. Søndergaard. Space mapping: The state of the art. *IEEE Transactions on Microwave Theory and Techniques*, 52:337 360, 2004.
5. J.W. Bandler, Q.S. Cheng, D.M. Hailu, and N.K. Nikolova. A space-mapping design framework. *IEEE Trans. Microwave Theory Tech.*, 52(11):2601–2610, 2004.

6. K. Böhmer, P. W. Hemker, and H. J. Stetter. The defect correction approach. In K. Böhmer and H. J. Stetter, editors, *Defect Correction Methods: Theory and Applications*, Computing Suppl. 5, pages 1–32. Springer-Verlag, Berlin, Heidelberg, New York, Tokyo, 1984.
7. K. Böhmer and H.J. Stetter. *Defect Correction Methods: Theory and Applications*. Springer, Berlin, 1984.
8. D.K. Cheng. *Field and Wave Electromagnetics*. Cambridge University Press, 1989.
9. D. Echeverría and P.W. Hemker. Space mapping and defect correction. *Comp. Methods in Appl. Math.*, 5(2):107–136, 2005.
10. D. Echeverría and P.W. Hemker. On the manifold mapping optimization technique. Technical Report MAS-E0612, CWI, 2006. Submitted for publication.
11. D. Echeverría, D. Lahaye, L. Encica, E.A. Lomonova, P.W. Hemker, and A.J.A. Vandenput. Manifold mapping optimization applied to linear actuator design. *IEEE. Trans. on Magn.*, 42(4):1183–1186, 2006. (Also: Technical Report MAS-E0612, CWI).
12. L. Encica, D. Echeverría, E. A. Lomonova, A. J. A. Vandenput, P. W. Hemker, and D. Lahaye. Efficient optimal design of electromagnetic actuators using space-mapping. *Struct. Multidisc. Optim.*, 2007. DOI 10.1007/s00158-006-0054-6. To appear. (Also in: J. Herkovits, Mazorche S, and A. Canelas, editors, Sixth World Congress on Structural and Multidisciplinary Optimization (WCSMO6), Brazil, 2005. paper 5631).
13. L. Encica, J. Makarovic, E.A. Lomonova, and A.J.A. Vandenput. Space mapping optimization of a cylindrical voice coil actuator. *IEEE IEMDC-2005*, 2005. Conference proceedings.

Modal Approximation and Computation of Dominant Poles

Joost Rommes

NXP Semiconductors/Corp. I&T/DTF
High Tech Campus 37
5656 AE Eindhoven
The Netherlands
joost.rommes@nxp.com

1 Introduction

A large scale dynamical system can have a large number of modes. Like a general square matrix can be approximated by its largest eigenvalues, i.e. by projecting it onto the space spanned by the eigenvalues corresponding to the largest eigenvalues, a dynamical system can be approximated by its dominant modes: a reduced order model, called the *modal equivalent*, can be obtained by projecting the state space on the subspace spanned by the dominant modes. This technique, *modal approximation* or *modal model reduction*, has been successfully applied to transfer functions of large-scale power systems, with applications such as stability analysis and controller design, see [16] and references therein.

The dominant modes, and the corresponding dominant poles of the system transfer function, are specific eigenvectors and eigenvalues of the state matrix. Because the systems are very large in practice, it is not feasible to compute all modes and to select the dominant ones. This chapter is concerned with the efficient computation of these dominant poles and modes specifically, and their use in reduced order modeling. In Sect. 2 the concept of dominant poles and modal approximation is explained in more detail. Dominant poles can be computed with specialized eigensolution methods, as is described in Sect. 3. Some generalizations of the presented algorithms are shown in Sect. 4. The theory is illustrated with numerical examples in Sect. 5 and 6 concludes.

Part of the contents of this chapter is based on [15, 16]. The pseudocode algorithms presented in this chapter are written using Matlab-like [21] notation.

2 Transfer Functions, Dominant Poles and Modal Equivalents

Throughout this section and the next, only single-input single-output (SISO) transfer functions are considered. In Sect. 4, the theory is generalized to multi-input multi-output (MIMO) transfer functions.

The transfer function of a SISO linear, time invariant system

$$\begin{cases} \dot{\mathbf{x}}(t) = A\mathbf{x}(t) + \mathbf{b}u(t) \\ y(t) = \mathbf{c}^*\mathbf{x}(t) + du(t), \end{cases}$$

where $A \in \mathbb{R}^{n\times n}$, $\mathbf{x}(t), \mathbf{b}, \mathbf{c} \in \mathbb{R}^n$ and $u(t), y(t), d \in \mathbb{R}$, is defined as

$$H(s) = \mathbf{c}^*(sI - A)^{-1}\mathbf{b} + d, \tag{1}$$

where $I \in \mathbb{R}^{n\times n}$ is the identity matrix and $s \in \mathbb{C}$.

The eigenvalues $\lambda_i \in \mathbb{C}$ of the matrix A are the poles of transfer function (1). An eigentriplet $(\lambda_i, \mathbf{x}_i, \mathbf{y}_i)$ is composed of an eigenvalue λ_i of A and the corresponding right and left eigenvectors $\mathbf{x}_i, \mathbf{y}_i \in \mathbb{C}^n$:

$$\begin{aligned} A\mathbf{x}_i &= \lambda_i \mathbf{x}_i, \qquad & \mathbf{x}_i &\neq 0, \\ \mathbf{y}_i^* A &= \lambda_i \mathbf{y}_i^*, \qquad & \mathbf{y}_i &\neq 0. \end{aligned}$$

Assuming that A is a nondefective matrix, the right and left eigenvectors can be scaled so that $\mathbf{y}_i^*\mathbf{x}_i = 1$. Furthermore, it can be shown that left and right eigenvectors corresponding to distinct eigenvalues are orthogonal: $\mathbf{y}_i^*\mathbf{x}_j = 0$ for $i \neq j$. The transfer function $H(s)$ can be expressed as a sum of residues R_i over first order poles [10]:

$$H(s) = \sum_{i=1}^{n} \frac{R_i}{s - \lambda_i} + d, \tag{2}$$

where the residues R_i are

$$R_i = (\mathbf{c}^*\mathbf{x}_i)(\mathbf{y}_i^*\mathbf{b}).$$

A possible definition of a dominant pole follows from inspection of (2):

Definition 1. *A pole λ_i of $H(s)$ with corresponding right and left eigenvectors $\mathbf{x}_i$ and $\mathbf{y}_i$ ($\mathbf{y}_i^*\mathbf{x}_i = 1$) is called dominant if $\widehat{R}_i = |R_i|/|Re(\lambda_i)| = |(\mathbf{c}^*\mathbf{x}_i)(\mathbf{y}_i^*\mathbf{b})|/|Re(\lambda_i)|$ is relatively large compared to $\widehat{R}_j$, $j \neq i$.*

The quantity $\widehat{R}_i$ will be referred to as the dominance index of pole λ_i. It follows from this definition that a dominant pole is well observable and controllable. This can also be observed from the Bode magnitude plot of $H(s)$, where peaks occur at frequencies close to the imaginary parts of the dominant poles of $H(s)$. If the poles are ordered to decreasing $\widehat{R}_i$, a so called transfer function modal equivalent can be defined as follows.

Definition 2. *A transfer function modal equivalent $H_k(s)$ is an approximation of a transfer function $H(s)$ that consists of $k < n$ terms:*

$$H_k(s) = \sum_{j=1}^{k} \frac{R_j}{s - \lambda_j} + d. \tag{3}$$

A modal equivalent that consists of the most dominant terms determines the effective transfer function behavior [20]. If $X \in \mathbb{C}^{n\times k}$ and $Y \in \mathbb{C}^{n\times k}$ are matrices having the left and right eigenvectors $\mathbf{y}_i$ and $\mathbf{x}_i$ of A as columns, such that $Y^*AX = \Lambda = \text{diag}(\lambda_1, \ldots, \lambda_k)$, with $Y^*X = I$, then the corresponding (complex) reduced system follows by setting $\mathbf{x} = X\widetilde{\mathbf{x}}$:

$$\begin{cases} \dot{\widetilde{\mathbf{x}}}(t) = \Lambda\widetilde{\mathbf{x}}(t) + (Y^*\mathbf{b})u(t) \\ \widetilde{\mathbf{y}}(t) = (\mathbf{c}^*X)\widetilde{\mathbf{x}}(t) + du(t). \end{cases}$$

For stable systems, the error in the modal equivalent can be quantified as [7]

$$\begin{aligned} \|H - H_k\|_\infty &= \|\sum_{j=k+1}^{n} \frac{R_j}{s-\lambda_j}\|_\infty \\ &\le \sum_{j=k+1}^{n} \frac{|R_j|}{|\text{Re}(\lambda_j)|}, \end{aligned}$$

where $\|H\|_\infty$ is the operator norm induced by the 2-norm in the frequency domain [2,7]. An advantage of modal approximation is that the poles of the modal equivalent are also poles of the original system.

It should be stressed that there are more definitions of a dominant pole, see [1,7, 22]. The definition often depends on the application: in stability analysis for instance, the poles with positive real parts are considered as the dominant poles. Throughout this chapter, Def. 1 is used.

3 Computing Dominant Poles

3.1 Introduction

The poles of transfer function (1) are the $\lambda \in \mathbb{C}$ for which $\lim_{s\to\lambda} |H(s)| = \infty$. Consider now the function

$$G : \mathbb{C} \longrightarrow \mathbb{C} : s \mapsto \frac{1}{H(s)}. \tag{4}$$

For a pole λ of $H(s)$, $\lim_{s\to\lambda} G(s) = 0$. In other words, the poles are the roots of $G(s)$ and a good candidate to find these roots is Newton's method. This idea is the basis of the Dominant Pole Algorithm (DPA) [11]. Because the direct transmission term d has no influence on the dominance of a pole, $d = 0$ unless stated otherwise.

3.2 Dominant Pole Algorithm (DPA)

The derivative of $G(s)$ (4) with respect to s is given by

$$G'(s) = -\frac{H'(s)}{H^2(s)}. \tag{5}$$

The derivative of $H(s)$ (1) to s is

$$H'(s) = -\mathbf{c}^*(sI - A)^{-2}\mathbf{b}, \tag{6}$$

where it is used that the derivative of the inverse of a square matrix $A(s)$ is given by $d[A^{-1}(s)]/ds = -A^{-1}(s)A'(s)A^{-1}(s)$. Equations (5) and (6) lead to the following Newton scheme:

$$\begin{aligned} s_{k+1} &= s_k - \frac{G(s_k)}{G'(s_k)} \\ &= s_k + \frac{1}{H(s_k)} \frac{H^2(s_k)}{H'(s_k)} \\ &= s_k - \frac{\mathbf{c}^*(s_k I - A)^{-1}\mathbf{b}}{\mathbf{c}^*(s_k I - A)^{-2}\mathbf{b}}. \end{aligned} \tag{7}$$

The formula (7) was originally derived in [3]. Using $\mathbf{x} = (s_k I - A)^{-1}\mathbf{b}$ and $\mathbf{y} = (s_k I - A)^{-*}\mathbf{c}$, an implementation of this Newton scheme is Alg. 1, also known as the Dominant Pole Algorithm (DPA) [11]. The two linear systems that need to be solved in step 3 and 4 of Alg. 1 can be efficiently solved using one LU-factorization $LU = s_k I - A$, by noting that $U^*L^* = (s_k I - A)^*$. It will be assumed in this chapter that an exact LU-factorization is available, although this may not always be the case for real-life examples. If an exact LU-factorization is not available, one has to use inexact Newton schemes, such as Jacobi-Davidson style methods [9, 19].

For nondefective A, Alg. 1 converges asymptotically quadratically, but the solution depends on the initial estimate and hence is not guaranteed to be the most dominant pole. For defective A, the algorithm may fail, because the left and right eigenvector of a defective eigenvalue are orthogonal and hence $\mathbf{y}^*\mathbf{x} \to 0$ in step 5 of Alg. 1. The update in step 5 can be written as the two-sided Rayleigh quotient [14]

$$s_{k+1} = \frac{\mathbf{y}^* A\mathbf{x}}{\mathbf{y}^*\mathbf{x}}. \tag{8}$$

3.3 Subspace Accelerated Dominant Pole Algorithm

While DPA computes a single dominant pole, in practice usually more dominant poles are wanted. The Subspace Accelerated Dominant Pole Algorithm (SADPA) [16] is a generalization of DPA to compute more than one dominant pole. SADPA has three major improvements compared to DPA. Firstly, it uses subspace acceleration, a well-known technique for iterative methods. Secondly, a new selection strategy is used to select the most dominant pole approximation and corresponding right and left eigenvector approximation every iteration. Thirdly, deflation is used to avoid convergence to eigentriplets that are already found. The ideas, leading to SADPA (Alg. 2), are described in the following subsections.

Algorithm 1: The Dominant Pole Algorithm (DPA)

INPUT: System $(A, \mathbf{b}, \mathbf{c})$, initial pole estimate $s_1 \in \mathbb{C}$, tolerance $\varepsilon \ll 1$

OUTPUT: Approximate dominant pole λ and corresponding right and left eigenvectors $\mathbf{x}$ and $\mathbf{y}$

Set $k = 1$

while not converged **do**

Solve $\mathbf{x} \in \mathbb{C}^n$ from

$$(s_k I - A)\mathbf{x} = \mathbf{b}$$

Solve $\mathbf{y} \in \mathbb{C}^n$ from

$$(s_k I - A)^* \mathbf{y} = \mathbf{c}$$

Compute the new pole estimate

$$s_{k+1} = s_k - \frac{\mathbf{c}^* \mathbf{x}}{\mathbf{y}^* \mathbf{x}}$$

The pole $\lambda = s_{k+1}$ has converged if

$$\|A\mathbf{x} - s_{k+1}\mathbf{x}\|_2 < \varepsilon$$

Set $k = k + 1$

end while

Subspace Acceleration

A drawback of DPA is that information obtained in the current iteration is discarded at the end of the iteration. The only information that is preserved is contained in the new pole estimate s_{k+1}. The vectors $\mathbf{x}$ and $\mathbf{y}$, however, also contain information about other dominant eigentriplets (i.e., components in the direction of the corresponding eigenvectors) and the idea is to use this information as well. Reasoning this way leads to a generalization of DPA.

A global overview of SADPA is shown in Alg. 2. Starting with an estimate s_1, the first iteration is equivalent to the first iteration of DPA, but instead of discarding the corresponding right and left eigenvector approximations $\mathbf{x}_1$ and $\mathbf{y}_1$, they are kept in spaces X and Y. In the next iteration, these spaces are expanded orthogonally (step 5-6), by modified Gram-Schmidt (MGS) [6], with the approximations $\mathbf{x}_2$ and $\mathbf{y}_2$ corresponding to the new estimate s_2. In Sect. 3.3 it is explained how this new pole estimate is computed. The subspaces grow in dimension and may contain better approximations. This idea is known as subspace acceleration.

In the k-th iteration, k approximations $\widehat{\lambda}_i$ of the dominant poles are found by computing the eigentriplets of the projected matrix pencil (Y^*AX, Y^*X) (step 7-8). The question now is to determine which of these k approximations to use as estimate s_{k+1} in the next iteration.

Algorithm 2: Subspace Accelerated DPA

INPUT: System $(A, \mathbf{b}, \mathbf{c})$, initial pole estimate s_1 and the number of wanted poles p_{max}, tolerance $\varepsilon \ll 1$

OUTPUT: Approximate dominant pole triplets $(\lambda_i, \mathbf{r}_i, \mathbf{l}_i)$, $i = 1, \dots, p_{max}$

1: $k = 1$, $p_{found} = 0$, $\Lambda = [\,]^{1\times 0}$, $R = L = X = Y[\,]^{n\times 0}$ ($[\,]^{n\times 0}$ denotes an empty matrix of size $n \times 0$)

2: **while** $p_{found} < p_{max}$ **do**

3: Solve $\mathbf{x} \in \mathbb{C}^n$ from

$$(s_k I - A)\mathbf{x} = \mathbf{b}$$

4: Solve $\mathbf{y} \in \mathbb{C}^n$ from

$$(s_k I - A)^* \mathbf{y} = \mathbf{c}$$

5: $X = \text{Expand}(X, R, L, \mathbf{x})$ {Alg. 4}

6: $Y = \text{Expand}(Y, L, R, \mathbf{y})$ {Alg. 4}

7: Compute $T = Y^* A X$ and $G = Y^* X$

8: $(\widehat{\Lambda}, \widehat{X}, \widehat{Y}) = \text{Sort}(T, G, X, Y, \mathbf{b}, \mathbf{c})$ {Alg. 3}

9: **if** $\|A\widehat{\mathbf{x}}_1 - \widehat{\lambda}_1 \widehat{\mathbf{x}}_1\|_2 < \varepsilon$ **then**

10: $(\Lambda, R, L, X, Y) =$ $\text{Deflate}(\widehat{\lambda}_1, \widehat{\mathbf{x}}_1, \widehat{\mathbf{y}}_1, \Lambda, R, L, \widehat{X}_{2:k}, \widehat{Y}_{2:k})$ {Alg. 5}

11: $p_{found} = p_{found} + 1$

12: Set $\widehat{\lambda}_1 = \widehat{\lambda}_2$, $k = k - 1$

13: **end if**

14: Set $k = k + 1$

15: Set the new pole estimate $s_{k+1} = \widehat{\lambda}_1$

16: **end while**

Selection Strategy

In step 8 of Alg. 2, the new pole estimate s_{k+1} has to be determined. A possible choice is to use the two-sided Rayleigh quotient (8) as it is used in DPA, but this choice does not take full advantage of subspace acceleration. Here, however, also another choice is possible, that is closer to the goal of computing the dominant poles.

Because in iteration k the interaction matrices $T \in \mathbb{C}^{k\times k}$ and $G \in \mathbb{C}^{k\times k}$ are of low order $k \ll n$ (see step 7 in Alg. 2), it is relatively cheap to compute the full eigendecomposition of the pencil (T, G). This provides k approximate eigentriplets $(\widehat{\lambda}_i, \widehat{\mathbf{x}}_i, \widehat{\mathbf{y}}_i)$. A natural thing to do is to choose the triplet $(\widehat{\lambda}_j, \widehat{\mathbf{x}}_j, \widehat{\mathbf{y}}_j)$ with the most dominant pole approximation: compute the corresponding residues $\widehat{R}_i = (\mathbf{c}^*\widehat{\mathbf{x}}_i)(\widehat{\mathbf{y}}_i^*\mathbf{b})$ of the k pairs and use the pole with the largest $|\widehat{R}_j|/|\text{Re}(\widehat{\lambda}_j)|$ as new estimate. Numerically, it is more robust to normalize $\widehat{\mathbf{x}}_i$ and $\widehat{\mathbf{y}}_i$ such that $\|\widehat{\mathbf{x}}_i\|_2 = \|\widehat{\mathbf{y}}_i\|_2 = 1$. Algorithm 3 orders the k approximate eigentriplets in decreasing dominance. The SADPA then continues with the new estimate $s_{k+1} = \widehat{\lambda}_1$.

Algorithm 3: $(\widehat{\Lambda}, \widehat{X}, \widehat{Y}) = \text{Sort}(T, G, X, Y, \mathbf{b}, \mathbf{c})$

INPUT: $T, G \in \mathbb{C}^{k\times k}$, $X, Y \in \mathbb{C}^{n\times k}$, $\mathbf{b}, \mathbf{c} \in \mathbb{C}^n$
OUTPUT: $\widehat{\Lambda} \in \mathbb{C}^n$, $\widehat{X}, \widehat{Y} \in \mathbb{C}^{n\times k}$ with $\widehat{\lambda}_1$ the pole approximation with largest scaled residue magnitude and $\widehat{\mathbf{x}}_1$ and $\widehat{\mathbf{y}}_1$ the corresponding approximate right and left eigenvectors

1: Compute eigentriplets of the pair (T, G):

$$(\widetilde{\lambda}_i, \widetilde{\mathbf{x}}_i, \widetilde{\mathbf{y}}_i), \quad i = 1, \ldots, k$$

2: Compute approximate eigentriplets of A as (with $\|\widehat{\mathbf{x}}_i\|_2 = \|\widehat{\mathbf{y}}_i\|_2 = 1$)

$$(\widehat{\lambda}_i = \widetilde{\lambda}_i, \widehat{\mathbf{x}}_i = X\widetilde{\mathbf{x}}_i, \widehat{\mathbf{y}}_i = Y\widetilde{\mathbf{y}}_i), \quad i = 1, \ldots, k$$

3: $\widehat{\Lambda} = [\widehat{\lambda}_1, \ldots, \widehat{\lambda}_k]$
4: $\widehat{X} = [\widehat{\mathbf{x}}_1, \ldots, \widehat{\mathbf{x}}_k]$
5: $\widehat{Y} = [\widehat{\mathbf{y}}_1, \ldots, \widehat{\mathbf{y}}_k]$
6: Compute residues $\widehat{R}_i = (\mathbf{c}^*\widehat{\mathbf{x}}_i)(\widehat{\mathbf{y}}_i^*\mathbf{b})$
7: Sort $\widehat{\Lambda}$, $\widehat{X}$, $\widehat{Y}$ in decreasing $|\widehat{R}_i|/|\text{Re}(\widehat{\lambda}_i)|$ order

Deflation

At the end of every iteration, in step 9, a convergence test is done as in DPA: if for the selected eigentriplet $(\widehat{\lambda}_j, \widehat{\mathbf{x}}_j, \widehat{\mathbf{y}}_j)$ the norm of the residual $\|A\widehat{\mathbf{x}}_j - \widehat{\lambda}_j\widehat{\mathbf{x}}_j\|_2$ is smaller than some tolerance ε, it is considered to be converged. In general more dominant eigentriplets are wanted and during the computation of the next eigentriplets, components in the direction of already found eigenvectors may enter the search spaces X and Y again. This may lead to repeated computation of the same eigentriplet. A well known technique to avoid repeated computation is deflation [18].

If already the right and left eigenvectors $\mathbf{x}_j$ and $\mathbf{y}_j$ are found, then it can be verified that, if the eigenvectors are exact, the matrix

$$\widetilde{A} = \prod_j (I - \frac{\mathbf{x}_j\mathbf{v}_j^*}{\mathbf{v}_j^*\mathbf{x}_j}) \cdot A \cdot \prod_j (I - \frac{\mathbf{x}_j\mathbf{v}_j^*}{\mathbf{v}_j^*\mathbf{x}_j})$$

has the same eigentriplets as A, but with the found eigenvalues transformed to zero (see also [5, 9]): let $\widehat{\mathbf{x}}$ be one of the k found exact right eigenvectors, i.e. $\widehat{\mathbf{x}} \in \{\mathbf{x}_1, \ldots, \mathbf{x}_k\}$. Then it follows from the orthogonality relations (see Sect. 2) that

$$\prod_j (I - \frac{\mathbf{x}_j\mathbf{y}_j^*}{\mathbf{y}_j^*\mathbf{x}_j}) \cdot \widehat{\mathbf{x}} = \widehat{\mathbf{x}} - \widehat{\mathbf{x}} = 0,$$

and hence $\widetilde{A}\widehat{\mathbf{x}} = 0$. On the other hand, let $\widehat{\mathbf{x}} \notin \{\mathbf{x}_1, \ldots, \mathbf{x}_k\}$ be a right eigenvector of A with eigenvalue $\widehat{\lambda}$. Then

$$\prod_j (I - \frac{\mathbf{x}_j\mathbf{y}_j^*}{\mathbf{y}_j^*\mathbf{x}_j}) \cdot \widehat{\mathbf{x}} = \widehat{\mathbf{x}},$$

and hence $\widetilde{A}\widehat{\mathbf{x}} = \widehat{\lambda}\widehat{\mathbf{x}}$. The result for left eigenvectors follows in a similar way. In finite arithmetic only *approximations* to exact eigentriplets are available and hence the computed eigenvalues are transformed to $\eta \approx 0$ (see also Sect. 3.3).

Using this, the space X needs to be orthogonally expanded with

$$\prod_j (I - \frac{\mathbf{x}_j \mathbf{y}_j^*}{\mathbf{y}_j^* \mathbf{x}_j}) \cdot \mathbf{x},$$

and similarly, the space V needs to be orthogonally expanded with

$$\prod_j (I - \frac{\mathbf{y}_j \mathbf{x}_j^*}{\mathbf{x}_j^* \mathbf{y}_j}) \cdot \mathbf{y}.$$

These projections are implemented with modified Gram-Schmidt (MGS) (see Alg. 4).

Algorithm 4: $X = \text{Expand}(X, R, L, \mathbf{x})$

INPUT: $X \in \mathbb{C}^{n \times k}$ with $X^*X = I$, $R, L \in \mathbb{C}^{n \times p}$, $\mathbf{x} \in \mathbb{C}^n$
OUTPUT: $X \in \mathbb{C}^{n \times (k+1)}$ with $X^*X = I$ and
$\mathbf{x}_{k+1} = \prod_{j=1}^{p} (I - \frac{\mathbf{r}_j \mathbf{l}_j^*}{\mathbf{l}_j^* \mathbf{r}_j}) \cdot \mathbf{x}$

1: $\mathbf{x} = \prod_{j=1}^{p} (I - \frac{\mathbf{r}_j \mathbf{l}_j^*}{\mathbf{l}_j^* \mathbf{r}_j}) \cdot \mathbf{x}$
2: $\mathbf{x} = \text{MGS}(X, \mathbf{x})$
3: $X = [X, \mathbf{x}/\|\mathbf{x}\|_2]$

If a complex eigenvalue has converged, its complex conjugate is also a pole and the corresponding complex conjugate right and left eigenvectors can also be deflated. The complete deflation procedure is shown in Alg. 5.

Further Improvements and Remarks

It may happen that the subspaces X and Y become high-dimensional, especially when a large number of dominant poles is wanted. A common way to deal with this is to do an implicit restart [18]: if the subspaces X and Y reach a certain maximum dimension $k_{\text{max}} \ll n$, they are reduced to a dimension $k_{\text{min}} < k_{\text{max}}$ by keeping the k_{min} most dominant approximate eigentriplets; the process is restarted with the reduced X and Y (already converged eigentriplets are not part of the active subspaces X and Y). This procedure is repeated until all poles are found.

The approximate residues $\widehat{R}_i$ can be computed without computing the approximate eigenvectors explicitly (step 2 and step 6 of Alg. 3): if the $\widetilde{\mathbf{x}}_i$ and $\widetilde{\mathbf{y}}_i$ are scaled so that $\|\widetilde{\mathbf{y}}_i\|_2 = \|\widetilde{\mathbf{x}}_i\| = 1$, then it follows that the $\widehat{R}_i$ can be computed as $\widehat{R}_i = ((\mathbf{c}^*X)\widetilde{\mathbf{x}}_i)(\widetilde{\mathbf{y}}_i^*(Y^*\mathbf{b}))$ $(= (\mathbf{c}^*\widehat{\mathbf{x}}_i)(\widehat{\mathbf{y}}_i^*\mathbf{b}))$.

Algorithm 5:
$(\Lambda, R, L, \widetilde{X}, \widetilde{Y}) = \text{Deflate}(\lambda, \mathbf{x}, \mathbf{v}, \Lambda, R, L, X, Y)$

INPUT: $\lambda \in \mathbb{C}, \mathbf{x}, \mathbf{y} \in \mathbb{C}^n, \Lambda \in \mathbb{C}^p, R, L \in \mathbb{C}^{n \times p}, X, Y \in \mathbb{C}^{n \times k}$

OUTPUT: $\Lambda \in \mathbb{C}^q, R, L \in \mathbb{C}^{n \times q}, \widetilde{X}, \widetilde{Y} \in \mathbb{C}^{n \times k}$, where $q = p + 1$ if λ has zero imaginary part and $q = p + 2$ if λ has nonzero imaginary part

$\Lambda = [\Lambda, \lambda]$
$R = [R, \mathbf{x}]$
$L = [L, \mathbf{y}]$
if $\text{imag}(\lambda) \neq 0$ **then**
  {Also deflate complex conjugate}
  $\Lambda = [\Lambda, \bar{\lambda}]$
  $R = [R, \bar{\mathbf{x}}]$
  $L = [L, \bar{\mathbf{y}}]$
end if
$\widetilde{X} = \widetilde{Y} = [\,]^{n \times 0}$
for $j = 1, \ldots, k$ **do**
  $\widetilde{X} = \text{Expand}(\widetilde{X}, R, L, X_j)$
  $\widetilde{Y} = \text{Expand}(\widetilde{Y}, L, R, Y_j)$
end for

Furthermore, as more eigentriplets have converged, approximations of new eigentriplets may become poorer or convergence may be hampered, due to rounding errors in the orthogonalization phase and the already converged eigentriplets. It is therefore advised to take a small tolerance $\varepsilon < 10^{-10}$. Besides that, as the estimate converges to a dominant pole, the right and left eigenvectors computed in step 3 and 4 of algorithm 2 are usually more accurate than the approximations computed in the selection procedure (although in exact arithmetic they are equal). In the deflation phase, it is therefore advised to take the most accurate of both.

Deflation can be implemented more efficiently in the following way: let $\mathbf{x}$ and $\mathbf{y}$ be right and left eigenvectors for eigenvalue λ and scaled such that $\mathbf{y}^*\mathbf{x} = 1$, with residue $R = (\mathbf{c}^*\mathbf{x})(\mathbf{y}^*\mathbf{b})$. With $\mathbf{b}_d = (I - \mathbf{x}\mathbf{y}^*)\mathbf{b}$ and $\mathbf{c}_d = (I - \mathbf{y}\mathbf{x}^*)\mathbf{c}$, it follows that the residue of λ in $H_d(s) = \mathbf{c}_d(sI - A)^{-1}\mathbf{b}_d$ is transformed to $R_d = 0$, while the residues of the remaining poles are left unchanged. Since $(sI - A)^{-1}\mathbf{b}_d \perp \mathbf{y}$ and $(sI - A)^{-*}\mathbf{c}_d \perp \mathbf{x}$, the orthogonalizations against found eigenvectors in step 5 and 6 of Alg. 2 are not needed any more (provided $\mathbf{b}$ and $\mathbf{c}$ are replaced by $\mathbf{b}_d$ and $\mathbf{c}_d$, respectively).

SADPA requires only one initial estimate. If rather accurate initial estimates are available, one can take advantage of this in SADPA by setting the next estimate after deflation to a new initial estimate (step 15 of Alg. 2).

Every iteration, two linear systems are to be solved (step 3 and 4). As was also mentioned in Sect. 3.2, this can be efficiently done by computing one LU-factorization and solving the systems by using L and U, and U^* and L^*, respectively. Because in practice the system matrix A is often very sparse, computation of the LU-factorization can be relatively inexpensive.

The selection criterion can easily be changed to another of the several existing indices of modal dominance [1, 7, 22]. Furthermore, the strategy can be restricted to considering only poles in a certain frequency range. Also, instead of providing the number of wanted poles, the procedure can be automated even further by providing the desired maximum error $|H(s) - H_k(s)|$ for a certain frequency range: the procedure continues computing new poles until the error bound is reached. Note that such an error bound requires that the transfer function of the complete model can be computed efficiently.

4 Generalizations

In this section, three variants of the dominant pole algorithm presented in the previous section are briefly discussed. Section 4.1 generalizes the theory to descriptor systems. In Sect. 4.2, the theory is extended to multi-input multi-ouput systems. A variant of DPA that computes the dominant zeros of a transfer function is described in Sect. 4.3.

4.1 Descriptor Systems

A more general representation of a dynamical system is

$$\begin{cases} E\dot{\mathbf{x}}(t) = A\mathbf{x}(t) + \mathbf{b}u(t) \\ y(t) \quad = \mathbf{c}^*\mathbf{x}(t) + du(t), \end{cases} \tag{9}$$

where $A, E \in \mathbb{R}^{n\times n}$, $\mathbf{x}(t), \mathbf{b}, \mathbf{c} \in \mathbb{R}^n$ and $u(t), y(t), d \in \mathbb{R}$. The corresponding transfer function is

$$H(s) = \mathbf{c}^*(sE - A)^{-1}\mathbf{b} + d.$$

The case $E = I$ has been discussed already in Sect. 2. The descriptor system (9) arises for instance in electrical circuit simulation ($A = G, E = C$) and the sparse descriptor formulation of power systems (see for instance [15, 16]). The pencil (A, E) is assumed to be regular, that is, $A - \lambda E$ is singular only for a finite number of $\lambda \in \mathbb{C}$. If E is singular, system (9) is a system of differential-algebraic equations (DAE). If E is nonsingular, it is a system of ordinary differential equations (ODE).

The algorithms presented in this chapter can easily be adapted to handle (sparse) descriptor systems of the form (9). The changes essentially boil down to replacing I by E on most places and noting that for eigentriplets $(\lambda_j, \mathbf{x}_j, \mathbf{y}_j)$ with distinct finite λ_j, the relation $\mathbf{y}_i^* E\mathbf{x}_j = 0, i \neq j$ holds, and that for nondefective finite eigenvalues, the eigenvectors can be scaled so that $\mathbf{y}_i^* E\mathbf{x}_i = 1$. The modes corresponding to eigenvalues at infinity do not contribute to the effective transfer function behavior. For completeness, the changes are given for each algorithm:

- Algorithm 1:
  - Replace I by E in step 3 and 4.

- Step 5 becomes

$$s_{k+1} = s_k - \frac{\mathbf{c}^*\mathbf{x} + d}{\mathbf{y}^* E\mathbf{x}}.$$

- The criterion in step 6 becomes

$$\|A\mathbf{x} - s_{k+1}E\mathbf{x}\|_2 < \varepsilon.$$

- Algorithm 2:
  - Replace I by E in step 3 and 4.
  - Replace step 5 and 6 by

$$X = \text{Expand}(X, R, E^* \cdot L, \mathbf{x}),$$
$$Y = \text{Expand}(Y, L, E \cdot R, \mathbf{y}).$$

  - In step 7, use $G = Y^*EX$.
  - The criterion in step 9 becomes

$$\|A\widehat{\mathbf{x}}_1 - \widehat{\lambda}_1 E\widehat{\mathbf{x}}_1\|_2 < \varepsilon.$$

- Algorithm 5:
  - Replace step 12 and 13 by

$$\widetilde{X} = \text{Expand}(\widetilde{X}, R, E^* \cdot L, X_j),$$
$$\widetilde{Y} = \text{Expand}(\widetilde{Y}, L, E \cdot R, Y_j).$$

4.2 MIMO Systems

For a multi-input multi-output (MIMO) system

$$\begin{cases} E\dot{\mathbf{x}}(t) = A\mathbf{x}(t) + B\mathbf{u}(t) \\ \mathbf{y}(t) \quad = C^*\mathbf{x}(t) + D\mathbf{u}(t), \end{cases}$$

where $A, E \in \mathbb{R}^{n\times n}$, $B \in \mathbb{R}^{n\times m}$, $C \in \mathbb{R}^{n\times p}$, $\mathbf{x}(t) \in \mathbb{R}^n$, $\mathbf{u}(t) \in \mathbb{R}^m$, $\mathbf{y}(t) \in \mathbb{R}^p$ and $D \in \mathbb{R}^{p\times m}$, the transfer function $H(s) : \mathbb{C} \longrightarrow \mathbb{C}^{p\times m}$ is defined as

$$H(s) = C^*(sE - A)^{-1}B + D. \tag{10}$$

The dominant poles of (10) are those $s \in \mathbb{C}$ for which $\sigma_{\max}(H(s)) \to \infty$. For square transfer functions ($m = p$), there is an equivalent criterion: the dominant poles are those $s \in \mathbb{C}$ for which $\lambda_{\min}(H^{-1}(s)) \to 0$. This leads, for square transfer functions, to the following Newton scheme:

$$s_{k+1} = s_k - \frac{1}{\mu_{\min}} \frac{1}{\mathbf{v}^* C^*(s_k E - A)^{-2} B\mathbf{u}},$$

where $(\mu_{\min}, \mathbf{u}, \mathbf{v})$ is the eigentriplet of $H^{-1}(s_k)$ corresponding to $\lambda_{\min}(H^{-1}(s_k))$. An algorithm for computing the dominant poles of a MIMO transfer function can be readily derived from Alg. 1. The reader is referred to [13] for the initial MIMO DPA algorithm and to [15] for an algorithm similar to SADPA, generalizations to non-square MIMO systems and more details.

4.3 Computing Zeros of a Transfer Function

The zeros of a transfer function $H(s) = \mathbf{c}^*(sE - A)^{-1}\mathbf{b} + d$ are those $s \in \mathbb{C}$ for which $H(s) = 0$. An algorithm, very similar to Alg. 1, can be derived by noting that a Newton scheme for computing the zeros of a transfer function is given by

$$s_{k+1} = s_k + \frac{\mathbf{c}^*(s_k E - A)^{-1}\mathbf{b} + d}{\mathbf{c}^*(s_k E - A)^{-2}\mathbf{b}}.$$

A slightly different formulation can be found in [12].

5 Numerical Examples

5.1 A Small Test System

For illustrational purposes, SADPA was applied to a transfer function of the New England test system, a model of a power system. This small benchmark system has 66 state variables (for more information, see [11]). The tolerance used was $\varepsilon = 10^{-10}$ and no restarts were used. Every iteration, the pole approximation $\widehat{\lambda}_j$ with largest $|\widehat{R}_j|/|\mathrm{Re}(\widehat{\lambda}_j)|$ was selected. Table 1 shows the found dominant poles and the iteration number in which the pole converged. Bodeplots of two modal equivalents are shown in Fig. 1 and Fig. 2. The quality of the modal equivalent increases with the number of found poles, as can be observed from the better match of the exact and reduced transfer function.

5.2 A Large-Scale Descriptor System

The Brazilian Interconnected Power System (BIPS) is a year 1999 planning model that has been used in practice (see [16] for more technical details). The size of the sparse matrices A and E is $n = 13,251$ (the number of states in the dense state space realization is $1,664$). The corresponding transfer function has a non-zero direct transmission term d. Figure 3 shows the frequency response of the complete model and the reduced model (41 states) together with the error. Both the magnitude and the phase plot show good matches of the exact and the reduced transfer functions (a relative error of approximately $\|H(s) - H_k(s)\|/\|H_k(s)\| = 0.1$, also for the

Table 1. Results for SADPA applied to the New England test system ($s_1 = 1i$).

#poles	#states	new pole	iteration	Bodeplot
1	2	$-0.4672 \pm 8.9644i$	13	-
2	4	$-0.2968 \pm 6.9562i$	18	-
3	5	-0.0649	21	Fig. 1
4	7	$-0.2491 \pm 3.6862i$	25	-
5	9	$-0.1118 \pm 7.0950i$	26	-
6	11	$-0.3704 \pm 8.6111i$	27	Fig. 2

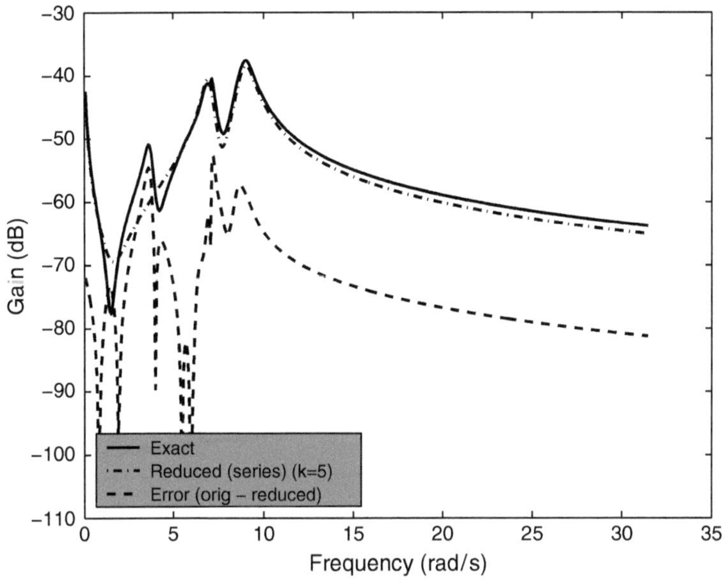

Fig. 1. Bode plot of modal equivalent, complete model and error for the transfer function of the New England test system (5 states in the modal equivalent, 66 in the complete model).

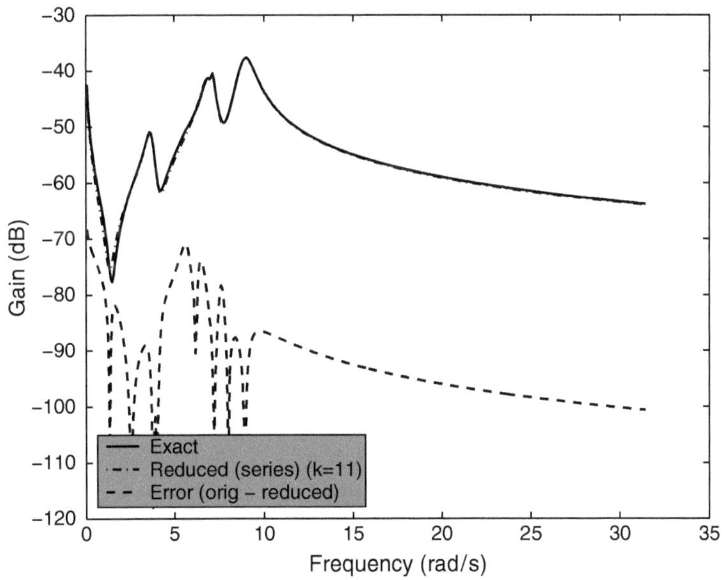

Fig. 2. Bode plot of modal equivalent, complete model and error for the transfer function of the New England test system (11 states in the modal equivalent, 66 in the complete model).

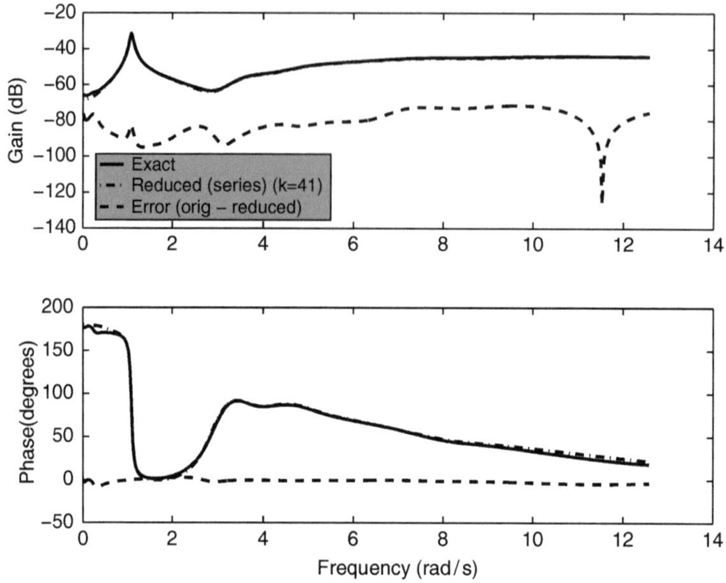

Fig. 3. Bode plot of modal equivalent, complete model and error for transfer function $P_{sc}(s)/B_{sc}(s)$ of BIPS (41 in the modal equivalent, 1664 in the complete model).

DC-gain $H(0)$). Figure 4 shows the corresponding step response (step $u = 0.01$)[1]. The reduced model nicely captures the system oscillations. The reduced model (30 poles, 56 states) was computed by SADPA in 341 LU-factorizations ($k_{\min} = 1$, $k_{\max} = 10$). This reduced model could be reduced further to 41 states (22 poles) by removing less dominant contributions, without decreasing the quality of the reduced model much.

5.3 A PEEC Example

This descriptor model arises from a partial element equivalent circuit (PEEC) of a patch antenna structure and the dimension of the matrices A and E is $n = 480$ (see [4, 17, 23] for more details and the model data). The system is known as a difficult problem, because it has many poles close to each other [8]. Figure 5 shows the Bodeplot of the complete model and the reduced model (45 poles, 89 states) computed by SADPA (initial estimate $s_1 = 100i$, $k_{\min} = 5$, $k_{\max} = 15$). The approximation is almost exact for the frequency range $[0.1, \ldots, 10^2]$ rad/sec. For frequencies higher than 10^2 rad/sec, the quality is less good.

[1] If $h_k(t)$ is the inverse Laplace transform of $H_k(s)$ (3), the step response for step $u(t) = c$ of the reduced model is given by $y(t) = \int_0^t h(t)u(t) = c(\sum_{i=1}^{k}(\frac{R_i}{\lambda_i}(\exp(\lambda_i t) - 1)) + d)$.

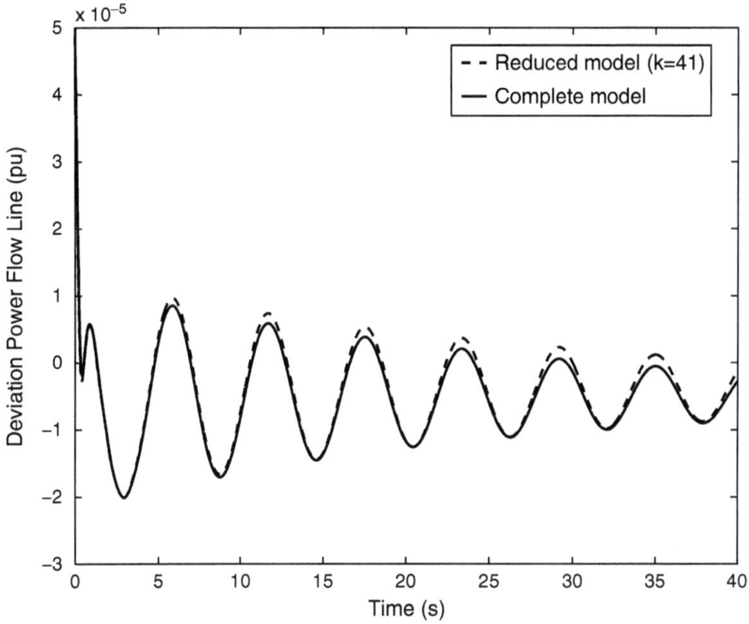

Fig. 4. Step responses for transfer function $P_{sc}(s)/B_{sc}(s)$ of BIPS, complete model and modal equivalent (41 states in the modal equivalent, 1664 in the complete model, step disturbance of 0.01 pu).

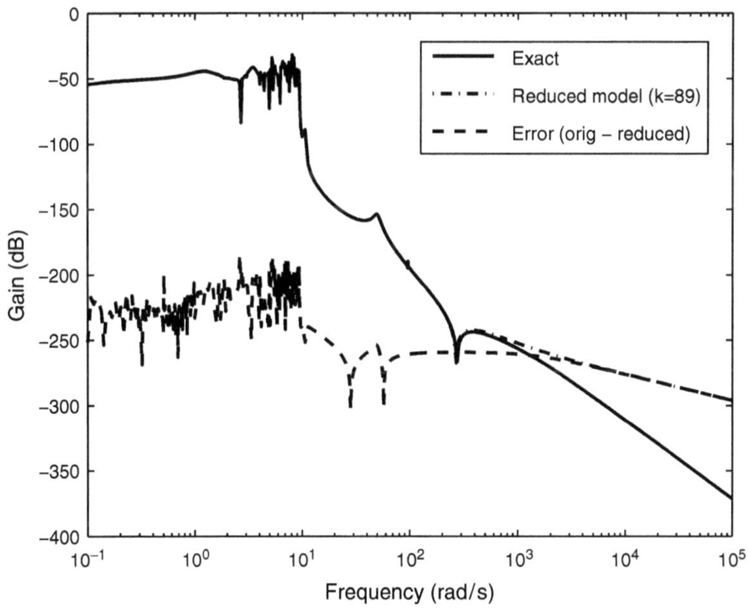

Fig. 5. Bode plot of modal equivalent, complete model and error for transfer function of the PEEC model (89 states in the modal equivalent, 480 in the complete model).

6 Conclusions

The algorithms presented in this chapter are efficient, automatic methods to compute dominant poles of large-scale transfer functions. It has been shown how the corresponding left and right eigenvectors can be used to construct a reduced order model, also known as the modal equivalent, of the original system. Although the methods may not be successful for every system, the numerical results for real-life systems and benchmarks indicate that the methods are applicable to a large class of systems.

Acknowledgement

Henk van der Vorst, Gerard Sleijpen and Nelson Martins are thanked for their valuable comments to earlier versions of this chapter.

References

1. AGUIRRE, L. A. Quantitative Measure of Modal Dominance for Continuous Systems. In *Proc. of the 32nd Conference on Decision and Control* (December 1993), pp. 2405–2410.
2. ANTOULAS, A. C. *Approximation of Large-Scale Dynamical Systems.* SIAM, 2005.
3. BEZERRA, L. H. Written discussion to [11]. *IEEE Trans. Power Syst. 11*, 1 (Feb. 1996), 168.
4. CHAHLAOUI, Y., AND VAN DOOREN, P. A collection of Benchmark examples for model reduction of linear time invariant dynamical systems. SLICOT Working Note 2002-2, 2002.
5. FOKKEMA, D. R., SLEIJPEN, G. L. G., AND VAN DER VORST, H. A. Jacobi-Davidson style QR and QZ algorithms for the reduction of matrix pencils. *SIAM J. Sc. Comp. 20*, 1 (1998), 94–125.
6. GOLUB, G. H., AND VAN LOAN, C. F. *Matrix Computations*, third ed. John Hopkins University Press, 1996.
7. GREEN, M., AND LIMEBEER, D. J. N. *Linear Robust Control.* Prentice-Hall, 1995.
8. HERES, P. J. *Robust and efficient Krylov subspace methods for model order reduction.* PhD thesis, Eindhoven University of Technology, 2005.
9. HOCHSTENBACH, M. E., AND SLEIJPEN, G. L. G. Two-sided and alternating Jacobi-Davidson. *Lin. Alg. Appl. 358*, 1-3 (2003), 145–172.
10. KAILATH, T. *Linear Systems.* Prentice-Hall, 1980.
11. MARTINS, N., LIMA, L. T. G., AND PINTO, H. J. C. P. Computing dominant poles of power system transfer functions. *IEEE Trans. Power Syst. 11*, 1 (Feb. 1996), 162–170.
12. MARTINS, N., PINTO, H. J. C. P., AND LIMA, L. T. G. Efficient methods for finding transfer function zeros of power systems. *IEEE Trans. Power Syst. 7*, 3 (Aug. 1992), 1350–1361.
13. MARTINS, N., AND QUINTÃO, P. E. M. Computing dominant poles of power system multivariable transfer functions. *IEEE Trans. Power Syst. 18*, 1 (Feb. 2003), 152–159.
14. PARLETT, B. N. The Rayleigh quotient iteration and some generalizations for nonnormal matrices. *Math. Comp. 28*, 127 (July 1974), 679–693.

15. ROMMES, J., AND MARTINS, N. Efficient computation of multivariable transfer function dominant poles using subspace acceleration. *IEEE Trans. Power Syst. 21*, 4 (Nov. 2006), 1471–1483.
16. ROMMES, J., AND MARTINS, N. Efficient computation of transfer function dominant poles using subspace acceleration. *IEEE Trans. Power Syst. 21*, 3 (Aug. 2006), 1218–1226.
17. RUEHLI, A. E. Equivalent circuit models for three dimensional multi-conductor systems. *IEEE Trans. Microwave Theory Tech. 22* (Mar. 1974), 216–221.
18. SAAD, Y. *Numerical Methods for Large Eigenvalue Problems: Theory and Algorithms.* Manchester University Press, 1992.
19. SLEIJPEN, G. L. G., AND VAN DER VORST, H. A. A Jacobi-Davidson iteration method for linear eigenvalue problems. *SIAM J. Matrix Anal. Appl. 17*, 2 (1996), 401–425.
20. SMITH, J. R., HAUER, J. F., TRUDNOWSKI, D. J., FATEHI, F., AND WOODS, C. S. Transfer function identification in power system application. *IEEE Trans. Power Syst. 8*, 3 (Aug. 1993), 1282–1290.
21. THE MATHWORKS, INC. Matlab.
22. VARGA, A. Enhanced modal approach for model reduction. *Math. Mod. Syst.*, 1 (1995), 91–105.
23. VERBEEK, M. E. Partial element equivalent circuit (PEEC) models for on-chip passives and interconnects. *Int. J. Num. Mod.: Electronic Networks, Devices and Fields 17*, 1 (Jan. 2004), 61–84.

Some Preconditioning Techniques for Saddle Point Problems

Michele Benzi[1**] and Andrew J. Wathen[2]

[1] Department of Mathematics and Computer Science, Emory University, Atlanta, Georgia 30322, USA
benzi@mathcs.emory.edu
[2] Oxford University Computing Laboratory, Wolfson Building, Parks Road, Oxford OX1 3QD, UK
andy.wathen@comlab.ox.ac.uk

1 Introduction

Saddle point problems arise frequently in many applications in science and engineering, including constrained optimization, mixed finite element formulations of partial differential equations, circuit analysis, and so forth. Indeed the formulation of most problems with constraints gives rise to saddle point systems. This paper provides a concise overview of iterative approaches for the solution of such systems which are of particular importance in the context of large scale computation. In particular we describe some of the most useful preconditioning techniques for Krylov subspace solvers applied to saddle point problems, including block and constraint preconditioners.

Many applied problems can be stated in the form of constrained minimization problems. Frequently, such problems are infinite-dimensional and highly nonlinear. Discretization results in finite-dimensional problems of large size. These problems are usually replaced by a sequence of quadratic minimization problems subject to linear equality constraints:

$$\min \; J(u) = \tfrac{1}{2}u^T A u - f^T u \tag{1}$$

$$\text{subject to} \;\; Bu = g\,. \tag{2}$$

Here $A \in \mathbb{R}^{n\times n}$ is symmetric positive semidefinite, and $B \in \mathbb{R}^{m\times n}$, with $m < n$; $f \in \mathbb{R}^n$ and $g \in \mathbb{R}^m$ are given vectors. The first-order optimality conditions are given by the linear system

$$\begin{bmatrix} A & B^T \\ B & O \end{bmatrix} \begin{bmatrix} u \\ p \end{bmatrix} = \begin{bmatrix} f \\ g \end{bmatrix}. \tag{3}$$

** The work of the first author was supported in part by the National Science Foundation grant DMS-0511336.

In (3), $p \in \mathbb{R}^m$ is a vector of Lagrange multipliers . Linear systems of the form (3) are known as *saddle point problems*, since any solution (u, p) of (3) is a saddle point of the Lagrangian function

$$\mathcal{L}(u,p) = \tfrac{1}{2}u^T Au - f^T u + (Bu - g)^T p\,.$$

Large linear systems in saddle point form also arise from inherently discrete physical models, such as mechanical structures [41] and RCL circuits [17].

More generally, we consider linear systems of the form

$$\mathcal{A}x = \begin{bmatrix} A & B^T \\ B & -C \end{bmatrix} \begin{bmatrix} u \\ p \end{bmatrix} = \begin{bmatrix} f \\ g \end{bmatrix} = b\,, \tag{4}$$

with A and B as before and $C \in \mathbb{R}^{m\times m}$ symmetric and positive semidefinite. Systems of the form (4) with a nonzero (2,2) block arise, for instance, in the context of interior point methods for constrained optimization [32]. Other examples are provided by mixed finite elements for incompressible flow problems, when some form of pressure stabilization is included in the discretization [13], and by the modeling of slightly compressible materials in linear elasticity theory [7].

Typically, $\mathcal{A}$ is large and sparse and (4) must be solved iteratively, usually by means of Krylov subspace algorithms [42]. Unfortunately, Krylov methods tend to converge very slowly when applied to saddle point systems, and good preconditioners are needed to achieve rapid convergence. In the last few years, much work has been devoted to developing effective preconditioners for saddle point systems. The goal of this paper is to provide a concise overview of such techniques. Due to space limitations, we focus mainly on three widely applicable classes of preconditioning techniques: block diagonal (or triangular) preconditioners, constraint preconditioners, and HSS preconditioning. For a more extensive survey of these and other techniques, see [3]. See further [13] for a thorough discussion of saddle point problems arising in fluid dynamics.

2 Properties of Saddle Point Systems

If A is nonsingular, the saddle point matrix $\mathcal{A}$ admits the following block triangular factorization:

$$\mathcal{A} = \begin{bmatrix} A & B^T \\ B & -C \end{bmatrix} = \begin{bmatrix} I & O \\ BA^{-1} & I \end{bmatrix} \begin{bmatrix} A & O \\ O & S \end{bmatrix} \begin{bmatrix} I & A^{-1}B^T \\ O & I \end{bmatrix}, \tag{5}$$

where $S = -(C + BA^{-1}B^T)$ is the *Schur complement* of A in $\mathcal{A}$. Several important properties of the saddle point matrix $\mathcal{A}$ can be derived on the basis of (5). To begin with, it is clear that $\mathcal{A}$ is nonsingular if and only if S is. Furthermore, since (5) defines a congruence transformation, we see that $\mathcal{A}$ is indefinite with n positive and m negative eigenvalues if A is symmetric positive definite (SPD).

There are some important applications in which A is symmetric positive semidefinite and singular, in which case there is no block factorization of the form (5). If $C = O$ and B has full rank, then $\mathcal{A}$ is invertible if and only if the null spaces of A and B satisfy $\mathcal{N}(A) \cap \mathcal{N}(B) = \{0\}$. In this case $\mathcal{A}$ is, again, indefinite with n positive and m negative eigenvalues. In some important applications A is SPD and B is rank deficient and the linear system (4) is singular but consistent. Generally speaking, the singularity of $\mathcal{A}$ does not cause any serious problem for iterative solvers; see [13, Section 5.3] for a discussion.

It is interesting to note that the simple stratagem of changing the sign of the last m equations in (4) leads to a linear system with completely different spectral properties. Indeed, assuming that A is SPD and C is symmetric positive semidefinite, it is easy to see that the (nonsymmetric) coefficient matrix

$$\widehat{\mathcal{A}} = \begin{bmatrix} A & B^T \\ -B & C \end{bmatrix} \tag{6}$$

is positive definite, in the sense that its spectrum is contained in the right half-plane $\Re(z) > 0$. Hence, $-\widehat{\mathcal{A}}$ is a *stable* matrix, an important property in circuit modeling; see [17, Section 4.3]. Furthermore, when certain (reasonable) conditions on A, B and C are met, it can be shown that $\widehat{\mathcal{A}}$ is diagonalizable and has all the eigenvalues real and positive. In other words, there exists a nonstandard inner product on $\mathbb{R}^{n+m}$ relative to which $\widehat{\mathcal{A}}$ is SPD; see [5] for details.

Regardless of the formulation of the saddle point system (symmetric indefinite or nonsymmetric positive definite), the convergence of Krylov subspace methods is almost always extremely slow unless a good preconditioner is available.

3 Preconditioned Krylov Subspace Methods

The well-known Conjugate Gradient method [25] which is widely used for the iterative solution of symmetric *definite* matrix systems is not in general robust for indefinite matrix systems. The main iterative approaches for indefinite matrix systems are the MINRES and SYMMLQ algorithms [31] which are based on the Lanczos procedure [28]. These algorithms (see [14] for a comprehensive and accessible description) require any preconditioner to be symmetric and positive definite. An alternative, which allows the use of symmetric and indefinite preconditioning (but has less clear theoretical convergence properties) is the Symmetric QMR (SQMR) method [19]. Even for indefinite problems, however, Conjugate Gradient methods can be employed with specific types of preconditioner: see the section on Constraint Preconditioning below.

The important feature of all of these methods is that at each iteration only one matrix times vector multiplication and a small number of vector operations (dot products and vector updates) are required. For sparse or structured matrices, the matrix times vector product may be efficiently computed and so the main issue concerning the overall computational work in the iterative solution of a linear system with such methods is the number of iterations it takes for convergence to an acceptable accuracy. Preconditioning is usually vital to ensure that this number is kept acceptably small. Methods which guarantee some monotonic reduction in a relevant quantity at each iteration are favoured in a number of situations: the MINRES method has such a property and so is sometimes regarded as the method of choice, however the SYMMLQ method has a related 'Petrov-Galerkin' property and is favoured for reasons of numerical stability when many iterations are required (see [40]).

For a generic linear system

$$\mathcal{A}x = b \tag{7}$$

where $\mathcal{A}$ is symmetric (and either indefinite or definite), the MINRES method computes a sequence of iterates $\{x_k\}$ for which the residual $r_k = b - \mathcal{A}x_k$ minimizes $\|r_k\|$ over the (shifted or affine) subspace

$$r_0 + \text{span}(\mathcal{A}r_0, \ldots, \mathcal{A}^k r_0). \tag{8}$$

The iterates themselves belong to the Krylov subspace

$$x_0 + \mathcal{K}_k(\mathcal{A}, r_0) = x_0 + \text{span}(r_0, \mathcal{A}r_0, \ldots, \mathcal{A}^{k-1}r_0) \tag{9}$$

where x_0 is the initial iterate (the initial 'guess') and r_0 the corresponding residual. This minimization property leads immediately to a description of the convergence properties of the MINRES method: since any vector, s say, in the space (8) can be written as $s = q(\mathcal{A})r_0$ where q is a polynomial of degree k with constant term equal to one (ie. $q(z) = 1 + \alpha_1 z + \ldots + \alpha_k z^k$ for some coefficients α_i), we have that

$$\|r_k\| \leq \|q(\mathcal{A})r_0\| \leq \|q(\mathcal{A})\|\|r_0\|. \tag{10}$$

Now the diagonalization of the symmetric matrix $\mathcal{A}$ as $\mathcal{A} = X\Lambda X^T$ where Λ is the diagonal matrix of eigenvalues and the matrix X is the orthogonal matrix of eigenvectors ensures that

$$\|q(\mathcal{A})\| = \|Xq(\Lambda)X^T\| = \|q(\Lambda)\| \tag{11}$$

because the Euclidean norm is invariant under orthogonal transformations. Further, since $q(\Lambda)$ is a diagonal matrix we have that

$$\|r_k\| \leq \min_{q \in \Pi_k, q(0)=1} \max_{z \in \sigma(\mathcal{A})} \|q(z)\|\|r_0\|. \tag{12}$$

Here, Π_k is the set of (real) polynomials of degree k and $\sigma(\mathcal{A})$ is the set of eigenvalues of $\mathcal{A}$. Thus for a real symmetric matrix, convergence depends only on its eigenvalues: if there are only a few distinct eigenvalues or they are sufficiently clustered away from the origin then there are polynomials of low degree which will be small at the eigenvalues. At each additional iteration the degree increases by one and so reasonable accuracy is quickly achieved in such cases. Various constructions based on the Chebyshev polynomials can give more explicit convergence bounds, but these are somewhat less straightforward to write down for indefinite rather than definite symmetric matrices (see for example [23] or [13]).

Preconditioning correponds to the application of a matrix (or linear operator), $\mathcal{P}$ to the original linear system to yield a different linear system for which convergence of the iterative method will be significantly faster. In most situations $\mathcal{P}$ must be constructed so that it is easy/fast to solve linear systems of the form $\mathcal{P}z = r$ for z when r is given. Conceptually one can think of preconditioned iteration as applying the original iteration to

$$\mathcal{P}^{-1}\mathcal{A}x = \mathcal{P}^{-1}b \tag{13}$$

however it would in almost all cases be a really bad move to create such a non-symmetric linear system when $\mathcal{A}$ is originally symmetric: the iterative solution of nonsymmetric linear systems is much less reliable and/or more expensive in general and most practitioners would believe that preserving symmetry is really valuable. For MINRES, a symmetric and positive definite preconditioner $\mathcal{P}$ must be employed so that we can write $\mathcal{P} = \mathcal{L}\mathcal{L}^T$ for some matrix $\mathcal{L}$ (eg. either the Cholesky factor or the matrix square root). We emphasize that this is only a mathematical artifact used to derive the method: no such factorization is required in practice—though of course such a factorization could be used if it were available. Where the preconditioner is not provided in factored form, the preconditioned MINRES method as given for example in [13, page 289] is used. In this way the MINRES iteration is effectively applied to the symmetric system

$$\mathcal{L}^{-1}\mathcal{A}\mathcal{L}^{-T}y = \mathcal{L}^{-1}b, \quad \mathcal{L}^T x = y \tag{14}$$

and convergence will depend on the eigenvalues of the symmetric and indefinite matrix $\mathcal{L}^{-T}\mathcal{A}\mathcal{L}^{-1}$. Via the obvious similarity tranformation

$$\mathcal{L}^{-T}\mathcal{L}^{-1}\mathcal{A}\mathcal{L}^{-T}\mathcal{L}^T = \mathcal{P}^{-1}\mathcal{A} \tag{15}$$

it is clear that the important eigenvalues are those of the matrix $\mathcal{P}^{-1}\mathcal{A}$, hence the convergence of the preconditioned MINRES iteration is described via (12) with the eigenvalue spectrum $\sigma(\mathcal{A})$ replaced in the preconditioned case by $\sigma(\mathcal{P}^{-1}\mathcal{A})$.

For SYMMLQ, there are similar considerations and good preconditioners should satisfy similar criteria. SQMR would generally only be used with a symmetric and indefinite preconditioner and there are no estimates of convergence in this case, though practical experience in a number of application areas indicates that SQMR convergence can be very good with a suitable indefinite preconditioner (see [18]).

In the next sections we discuss a number of possible approaches to preconditioning indefinite symmetric matices of saddle point type.

4 Block Preconditioners

Block preconditioners are based more or less explicitly on the block factorization (5). The performance of such preconditioners depends on whether fast, approximate solvers for linear systems involving A and the Schur complement S are available [34].

Assuming that A and $-S = C + BA^{-1}B^T$ are both SPD, the essentially ideal block diagonal preconditioner (as we shall see below) is

$$\mathcal{P}_d = \begin{bmatrix} A & O \\ O & -S \end{bmatrix}. \tag{16}$$

Preconditioning of $\mathcal{A}$ with $\mathcal{P}_d$ results in the matrix

$$\mathcal{M} = \mathcal{P}_d^{-1}\mathcal{A} = \begin{bmatrix} I & A^{-1}B^T \\ -S^{-1}B & O \end{bmatrix}. \tag{17}$$

The matrix $\mathcal{M}$ is nonsingular by assumption, is symmetrizable as described above and, as pointed out for example in [30], it satisfies

$$(\mathcal{M} - I)\left(\mathcal{M} - \frac{1}{2}(1+\sqrt{5})I\right)\left(\mathcal{M} - \frac{1}{2}(1-\sqrt{5})I\right) = O.$$

It follows that $\mathcal{M}$ is diagonalizable and has only three distinct eigenvalues, namely $1, \frac{1}{2}(1+\sqrt{5})$, and $\frac{1}{2}(1-\sqrt{5})$. Hence for each initial residual r_0, $\dim \mathcal{K}_{n+m}(\mathcal{M}, r_0) \leq 3$, which means that MINRES applied to the preconditioned system with preconditioner $\mathcal{P}_d$ will terminate after at most three steps.

Similarly, the essentially ideal block triangular preconditioner is

$$\mathcal{P}_t = \begin{bmatrix} A & B^T \\ O & \pm S \end{bmatrix}. \tag{18}$$

Choosing the minus sign in (18) results in a diagonalizable preconditioned matrix with only two distinct eigenvalues equal to ± 1. Choosing the plus sign yields a preconditioned matrix with all the eigenvalues equal to 1; this matrix is non-diagonalizable, but has minimum polynomial of degree two. For either choice of the sign in (18), the non-symmetric iterative solver GMRES [37] is guaranteed to converge in at most two steps in exact arithmetic.

Obviously, the ideal preconditioners $\mathcal{P}_d$ and $\mathcal{P}_t$ are not practical, since the exact Schur complement S is generally a dense matrix and is not available. In practice, A and S are replaced by some approximations, $\widehat{A} \approx A$ and $\widehat{S} \approx S$. If these approximations are chosen appropriately, the preconditioned matrices have most of their eigenvalues clustered around the eigenvalues of the ideally preconditioned matrices $\mathcal{P}_d^{-1}\mathcal{A}$ and $\mathcal{P}_t^{-1}\mathcal{A}$. Clearly, the choice of the approximations $\widehat{A}$ and $\widehat{S}$ is highly problem-dependent. Frequently $\widehat{A}$ and $\widehat{S}$ are not explicitly available matrices; rather, a prescription for computing the action of $\widehat{A}^{-1}$ and $\widehat{S}^{-1}$ on given vectors is given. For example, in mixed finite element formulations for incompressible flow problems the block A represents a discretization of a second-order elliptic operator, and the action of $\widehat{A}^{-1}$ on a vector can be computed by performing a small fixed number of iterations of some multigrid scheme. A varying number of iterations here would give a varying preconditioner for which a flexible outer iterative methods such as FGMRES [35] would be needed. The construction of good approximations $\widehat{S}$ to the Schur complement S is generally less straightforward and is highly problem-dependent; see [13, 38] for a detailed treatment in the case of incompressible flow problems.

Application of these techniques to more general saddle point problems arising in constrained optimization is more problematic. In particular, in the absence of well-understood elliptic operators it is unclear how to construct suitable approximations $\widehat{A} \approx A$ and $\widehat{S} \approx S$. One possibility is to use incomplete factorizations of A to build $\widehat{A}$, but it is unclear how to construct good approximations to the (typically dense) Schur complement S. Section 6 describes an alternative approach that has been applied succesfully in optimization.

We conclude this section with a brief discussion of a possible connection between block preconditioners based on approximate Schur complements and model order reduction of time-invariant linear dynamical systems. Following [17, Section 4.3], the (symmetric) transfer function of certain RCL subcircuits is the $m \times m$ matrix-valued rational function

$$H(s) = B\,(sE - A)^{-1}B\,, \quad \text{where} \quad A = A^T\,, \quad E = E^T \quad \text{and} \quad s \in \mathbb{C}\,. \tag{19}$$

In practice, n can be in the millions while m is of the order of a few hundreds or smaller. The goal of model order reduction is to find $m \times m$ approximations to the transfer function (19) of the form

$$\widehat{H}(s) = \widehat{B}\,(s\widehat{E} - \widehat{A})^{-1}\widehat{B}, \quad \text{where} \quad \widehat{A} = \widehat{A}^T, \quad \widehat{E} = \widehat{E}^T\,, \tag{20}$$

where the order $\widehat{n}$ of $\widehat{A}$ and $\widehat{E}$ is now small, typically of the same order as m. Furthermore, the approximate transfer function $\widehat{H}(s)$ must preserve certain properties of the original function $H(s)$ for the reduced-order model to be useful. A number of techniques have been developed to efficiently construct such approximations, including matrix Padé and Padé-type approximants. The approximants can be computed by means of (block) Lanczos methods; we refer the reader to [17] for a survey. These techniques have proved very effective in practice, and it would be interesting to investigate their use in constructing approximate Schur complements $\widehat{S} = \widehat{B}\widehat{A}^{-1}\widehat{B}^T \approx BA^{-1}B^T$. The approximate Schur complement could be used in turn to construct a block diagonal or block triangular preconditioner.

5 Augmented Lagrangian Formulations

The assumption that A is nonsingular may be too restrictive, and indeed A is singular in many applications. However, it is often possible to use augmented Lagrangian techniques [6, 15, 16] to replace the original saddle point system with an equivalent one having the same solution but in which the $(1,1)$ block A is now nonsingular. Thus, block diagonal and block triangular preconditioners based on appproximate Schur complement techniques may still be applicable. The augmented Lagrangian idea can also be useful in cases where the $(1,1)$ block is highly ill-conditioned and in order to trasform the original saddle point system into one that is easier to precondition.

The idea is to replace the original saddle point system (3) with the equivalent one

$$\begin{bmatrix} A+B^TWB & B^T \\ B & O \end{bmatrix} \begin{bmatrix} u \\ p \end{bmatrix} = \begin{bmatrix} f+B^TWg \\ g \end{bmatrix}. \tag{21}$$

The $m \times m$ matrix W, to be suitably determined, is symmetric positive semidefinite. The simplest choice is to take $W = \gamma I_m$ ($\gamma > 0$). In this case the $(1,1)$ block in (21) is nonsingular, and indeed positive definite, provided that A is positive definite on the null space of B. The goal is to choose W so that system (21) is easier to solve than the original one, particularly when using iterative methods. The choice of W is highly problem-dependent; see, e.g., [4, 20] for discussions of this issue in different settings.

It is important to keep in mind that there may be a trade-off between properties of the $(1,1)$ block and properties of the augmented system (21). Consider for instance the case where $W = \gamma I_m$. Then a possible preconditioner for (21) is given by

$$\mathcal{P}_\gamma = \begin{bmatrix} A+\gamma B^TB & B^T \\ O & -\gamma^{-1}I_m \end{bmatrix}.$$

It can be shown that the quality of this preconditioner, as measured in terms of number of iterations only, increases as γ tends to infinity; for large values of γ, however, the $(1,1)$ block becomes increasingly ill-conditioned. This is clear when one observes that for large γ the dominating term in the $(1,1)$ block becomes γB^TB, a singular matrix with a null space of dimension $n-m$. In practice, linear systems involving $A+\gamma B^TB$ will be solved inexactly, typically by some inner iteration, and finding efficient approximate solvers may become very difficult for large values of γ. The issue of variable preconditioning with an inner iteration would also arise. It is therefore important to strike a balance between the rate of convergence of the outer (preconditioned) iteration and the need for efficient approximate solution of linear systems involving $A+\gamma B^TB$.

Augmented Lagrangian techniques have been in use for many years in constrained optimization problems. Recent work indicates that the augmented Lagrangian approach may lead to powerful preconditioners for challenging problems in computational fluid mechanics and computational electromagnetics; see in particular [4] and [24].

6 Constraint Preconditioning

The second main type of preconditioner for saddle point problems are of the general form

$$\mathcal{P} = \begin{bmatrix} H & B^T \\ B & O \end{bmatrix} \tag{22}$$

where $H \in \mathbb{R}^{n\times n}$ ([27,29]). Since such an indefinite preconditioning matrix is itself a saddle point matrix which corresponds to a different quadratic energy but the same constraints as the original problem, it is called a 'constraint preconditioner'.

It is not evident that it is any easier to solve systems with this form of preconditioner than with the original matrix $\mathcal{A}$ in (3); since one such solution is required at each iteration this is a real issue. We will come back to this below, but firstly indicate what is known about the effect on iterative convergence of the use of preconditioners of the form (22).

The first point to notice is that the use of an indefinite preconditioner precludes the simple use of MINRES which requires a definite preconditioner. However a key observation is that by using the same constraint blocks in the preconditioner, the Hestenes–Stiefel Conjugate Gradient algorithm can be used: this is because solution of (3) with a preconditioner of the form (22) is equivalent to the solution of the positive definite symmetric system which would be derived by explicit elimination of the constraints with a positive definite symmetric preconditioner derived by direct elimination of these same constraints ([21]). This is a very attractive property since the Conjugate Gradient method is well known to be a very effective method with appropriate preconditioning for symmetric and positive definite systems. We emphasize that a constraint preconditioner *is required* here—for example it is clear that if no preconditioning were employed then Conjugate Gradients would not be a robust method for the indefinite saddle point system. Another consequence is that iterates for the primal variable u only are computed, so that the stopping criteria must reflect this. The Lagrange multipliers can be recovered if desired.

Thus the use of a constraint preconditioner with CG ensures (in exact arithmetic) that all of the iterates satisfy the constraints—only by employing a constraint preconditioner is this guarenteed. This appears to be a very desirable property in the context of Optimization when linear system solves are usually an inner part of an outer iterative optimization algorithm.

Given the equivalence to a symmetric positive definite problem, one might anticipate some special structure in the eigenvalues of the preconditioned matrix $\mathcal{P}^{-1}\mathcal{A}$; what is perhaps not expected is that this matrix should generically be non-diagonalizable! As shown in [27] this is always the case, but this is only due to a high multiplicity eigenvalue at 1: this eigenvalue has algebraic multiplicity $2m$ but only m independent eigenvectors. In the language of canonical forms, the Jordan form of this matrix has m 2×2 diagonal blocks. This means that $\mathcal{P}^{-1}\mathcal{A} - I$ has only an $m-$dimensional kernel, but $(\mathcal{P}^{-1}\mathcal{A} - I)^2$ has the full $2m-$dimensional kernel corresponding the the eigenvalue at 1. This is highly attractive from the standpoint of Krylov subspace iteration since only two iterations will eliminate the error in a $2m-$dimensional subspace.

The outcome is that iterative convergence depends on how well H approximates A in an $n - m$-dimensional subspace with only an additional two iterations required for the eigenvalue at 1.

Returning to the solution of systems with a constraint preconditioner, there are special situations where specific orthogonality properties enable easy solution: see for example [33]. A general approach, however, involves *not* preselecting the block H, but rather choosing it in an implicit fashion. One key approach is that based on Schilders' Factorization (see [9, 10, 12]); the idea is as follows. The factorization

$$\mathcal{P} = \begin{bmatrix} {B_1}^T & O & L_1 \\ {B_2}^T & L_2 & E \\ O & O & I \end{bmatrix} \begin{bmatrix} D_1 & O & I \\ O & D_2 & O \\ I & O & O \end{bmatrix} \begin{bmatrix} B_1 & B_2 & O \\ O & {L_2}^T & O \\ {L_1}^T & E^T & I \end{bmatrix}, \tag{23}$$

is exact for

$$\mathcal{A} = \begin{bmatrix} A & B^T \\ B & O \end{bmatrix} = \begin{bmatrix} A_{1,1} & A_{1,2} & B_1^T \\ A_{2,1} & A_{2,2} & B_2^T \\ B_1 & B_2 & O \end{bmatrix}$$

with $A_{1,1}, B_1 \in \mathbb{R}^{m\times m}$ (and other blocks correspondingly) when

$$\begin{aligned} D_1 &= {B_1}^{-T} A_{1,1} {B_1}^{-1} - {L_1}^T {B_1}^{-1} - {B_1}^{-T} L_1, \\ D_2 &= {L_2}^{-1}(A_{2,2} - {B_2}^T D_1 B_2 - E B_2 - {B_2}^T E^T){L_2}^{-T}, \\ E &= A_{2,1} {B_1}^{-1} - {B_2}^T D_1 - {B_2}^T {L_1}^T {B_1}^{-1}, \end{aligned}$$

but more importantly in our context, *any* choice of D_1, L_1 and E and any nonsingular choice of D_2, L_2 gives rise to a matrix of the form (22), i.e., gives rise to a constraint preconditioner in a reordered block triangular factored form. In this way by making choices for the blocks D_i, L_i and E in the factors in (23) a constraint preconditioner with an *implicitly defined* $(1,1)$ block H is obtained in a form in which solutions to preconditioner systems can easily be computed. The simplest choice would be

$$\begin{bmatrix} O & O & B_1^T \\ O & I & B_2^T \\ B_1 & B_2 & O \end{bmatrix} = \begin{bmatrix} {B_1}^T & O & O \\ {B_2}^T & I & O \\ O & O & I \end{bmatrix} \begin{bmatrix} O & O & I \\ O & I & O \\ I & O & O \end{bmatrix} \begin{bmatrix} B_1 & B_2 & O \\ O & I & O \\ O & O & I \end{bmatrix}. \tag{24}$$

It can be seen that it is always necessary to be able to compute the action of B_1^{-1}, thus it is important to be able to find a non-singular $m \times m$ leading block of the constraint matrix $B \in \mathbb{R}^{m\times n}$ possibly by reordering. A direct method (even for a dense system) will require $\mathcal{O}(m^3)$ computer (floating point) operations to achieve this, but sparsity will reduce this estimate considerably—and then the exact choice of which columns of B to reorder into B_1 also is likely to have an effect. There have been particular choices suggested for the special but important case of saddle point systems arising from interior point Optimization algorithms where large penalty parameters arise at least as convergence is approached (see [8]).

We comment that constraint preconditioners and Schilders-like factorisations for regularized saddle point systems of the form

$$\begin{bmatrix} A & B^T \\ B & -C \end{bmatrix} \tag{25}$$

where C is symmetric and positive semi-definite have also been described (see [10, 11]).

7 Other Techniques

Most preconditioning techniques that have been proposed in the literature on saddle point problems can be reduced to one of the main classes of methods described in the three sections above. For instance, the classical Uzawa method can be shown to be a special type of block triangular preconditioner. Similarly, preconditioning methods based on null-space (or *dual variable*) formulations, see for example [1], are closely related to constraint preconditioning. An exception is represented by the *HSS preconditioner* described in [2] and further analyzed in [39]. This preconditioner is based on the nonsymmetric formulation

$$\begin{bmatrix} A & B^T \\ -B & C \end{bmatrix} \begin{bmatrix} u \\ p \end{bmatrix} = \begin{bmatrix} f \\ -g \end{bmatrix}, \quad \text{or} \quad \widehat{\mathcal{A}} x = \widehat{b}. \tag{26}$$

Here we assume that A and C are symmetric positive semidefinite. We have the following splitting of $\widehat{\mathcal{A}}$ into its symmetric and skew-symmetric parts:

$$\widehat{\mathcal{A}} = \begin{bmatrix} A & B^T \\ -B & C \end{bmatrix} = \begin{bmatrix} A & O \\ O & C \end{bmatrix} + \begin{bmatrix} O & B^T \\ -B & O \end{bmatrix} = \mathcal{H} + \mathcal{K}. \tag{27}$$

Note that $\mathcal{H}$, the symmetric part of $\widehat{\mathcal{A}}$, is symmetric positive semidefinite since both A and C are. Let $\alpha > 0$ be a parameter. Similar in spirit to the classical ADI (Alternating-Direction Implicit) method, we consider the following two splittings of $\widehat{\mathcal{A}}$:

$$\widehat{\mathcal{A}} = (\mathcal{H} + \alpha\mathcal{I}) - (\alpha\mathcal{I} - \mathcal{K}) \qquad \text{and} \qquad \widehat{\mathcal{A}} = (\mathcal{K} + \alpha\mathcal{I}) - (\alpha\mathcal{I} - \mathcal{H}).$$

Here $\mathcal{I}$ denotes the identity matrix of order $n + m$. The stationary HSS iteration is then

$$x_{k+1} = x_k + \mathcal{P}_\alpha^{-1} r_k, \quad r_k = \widehat{b} - \widehat{\mathcal{A}} x_k,$$

where the matrix $\mathcal{P}$ is given by

$$\mathcal{P} \equiv \mathcal{P}_\alpha = \tfrac{1}{2\alpha}(\mathcal{H} + \alpha\mathcal{I})(\mathcal{K} + \alpha\mathcal{I}). \tag{28}$$

Assuming that A is SPD and B has full rank, it has been shown in [2] that the iterative process (28) is convergent to the unique solution of (26) for all $\alpha > 0$. However, the rate of convergence of the HSS iteration is rather slow, even with the "optimal" choice of α. For these reasons it was proposed in [2] that GMRES or other Krylov subspace methods should be used to accelerate the convergence of the HSS method. In other words, the HSS method is best used as a preconditioner for (say) GMRES rather than as a stationary iterative method. Note that as a preconditioner we can use $\mathcal{P}_\alpha = (\mathcal{H} + \alpha\mathcal{I})(\mathcal{K} + \alpha\mathcal{I})$ instead of the expression given in (28), since the factor $\frac{1}{2\alpha}$ has no effect on the preconditioned system. The spectral analysis of HSS preconditioning for general saddle point problems can be found in [39] and [5]. The analysis shows that the eigenvalues of the preconditioned matrix are all real and positive for all $\alpha > 0$, and furthermore as $\alpha \to 0$ they all fall within two small intervals $(0, \varepsilon_1)$ and $(2 - \varepsilon_2, 2)$, with $\varepsilon_1, \varepsilon_2 > 0$ and $\varepsilon_1, \varepsilon_2 \to 0$ as $\alpha \to 0$. This suggests that α should be taken to be small, but not too small; experience suggests that for a problem scaled so that A and C have unit nonzero diagonal entries, a value of α between 0.1 and 0.5 is often a good choice. In practice, solves with the shifted matrices $\mathcal{H} + \alpha\mathcal{I}$ and $\mathcal{K} + \alpha\mathcal{I}$ are performed inexactly for efficiency reasons. Approximately solving linear systems involving $\mathcal{H} + \alpha\mathcal{I}$ is usually straightforward, whereas solving linear systems involving the shifted skew-symmetric part $\mathcal{K} + \alpha\mathcal{I}$ is slighly more complicated. This step requires the solution of a linear system of the form

$$\begin{cases} \alpha\, u_{k+1} + B^T\, p_{k+1} = f_k\,, \\ -B\, u_{k+1} + \alpha\, p_{k+1} = g_k\,. \end{cases} \tag{29}$$

This can be accomplished by first eliminating u_{k+1} from the second equation using the first one (Schur complement reduction), leading to a smaller (order m) linear system of the form

$$(BB^T + \alpha^2 I)\, p_{k+1} = B\, f_k + \alpha\, g_k\,. \tag{30}$$

This is a linear system with an SPD coefficient matrix which can be approximately solved by, e.g., a preconditioned Conjugate Gradient method. In this case, it is necessary to use a flexible Krylov subspace method, such as FGMRES, for the outer iteration; see [36].

8 Numerical Examples

We firstly present an example of block diagonal preconditioning for a problem in incompressible fluid mechanics.

The underlying problem is the Stokes problem which is the particular case $\sigma = 0$ of the *generalized Stokes problem*:

$$\sigma\mathbf{u} - \nu\nabla^2\mathbf{u} + \nabla p = \mathbf{f} \quad \text{in} \quad \Omega \tag{31}$$

$$\operatorname{div}\mathbf{u} = 0 \quad \text{in} \quad \Omega \tag{32}$$

$$\mathbf{u} = \mathbf{g} \quad \text{on} \quad \partial\Omega\,. \tag{33}$$

Here $\mathbf{u}$ is the velocity and p the pressure (the Lagrange multiplier in this application). $\Omega \subset \mathbb{R}^d$ $(d = 2, 3)$ is the domain of the partial differential equation with boundary $\partial\Omega$ on which we have assumed simple Dirichlet conditions. The parameter ν is the kinematic viscosity which is taken to have the value one for the classical Stokes problem. See [13] for details.

This first example is computed with a common mixed finite element formulation: the block preconditioner combines a single simple multigrid V-cycle approximation of A and a diagonal matrix to approximate S and is run using the freely available IFISS software ([26]). We include iteration counts (which are seen to be essentially

Table 1. Block dimensions and number of MINRES iterations needed for 10^{-6} reduction in residual for locally stabilized $Q1-P0$ mixed finite elements for Stokes flow in a cavity. Block diagonal preconditioner: $\widehat{A}$ is one multigrid V-cycle with 1,1 relaxed Jacobi smoothing and $\widehat{S}$ is the diagonal pressure mass matrix. The cpu time (in seconds) is that required on the same computer (a Sun sparcv9 502 MHz processor with 1024 Mb of memory). The cpu time is also given for a sparse direct solve (UMFPACK in MATLAB).

grid	n	m	iterations	cpu time	sparse direct cpu
64×64	8450	4096	38	14.3	6.8
128×128	33282	16384	37	37.7	48.0
256×256	132098	65536	36	194.6	897
512×512	526339	263169	35	6903	out of memory

Table 2. Block dimensions and number of Conjugate Gradient iterations needed for 10^{-6} reduction in the preconditioned residual for the simplest Schilders' factorization preconditioner (24).

test problem	n	m	iterations
CVXQP1_S	100	50	44
CVXQP1_M	1000	500	28
CVXQP1_L	10000	5000	10

constant—indeed to reduce slightly—over a range of increasing problem dimension) and cpu times on the same workstation. Timings for a direct solution are given for comparison.

We can notice from Table 1 that for the largest-dimensional problem memory becomes an issue: the sparse direct method runs out of memory completely and fails for this problem and the timing for the iterative method is much greater than expected presumably because of slower memory access times for the more remote levels of cache which are needed for this problem.

To give an example of constraint preconditioning, we turn to problems from Optimization, specifically to a family of test problems from the CUTEr test set ([22]). We present results only for the simplest Schilders' factorization (24) for three of the family of CVXQP1 test problems. As indicated in the section above, Conjugate Gradient iteration is applicable with constraint preconditioning and this is applied here. The number of Conjugate Gradient iterations to achieve a 10^{-6} reduction in the preconditioned residual (defined only on the $n-$dimensional space of the primal variable u as described above) are given in Table 2. Note that the three problems are different: the comparison here is for the same relative reduction which gives the decreasing iteration counts indicated. For these problems, the iteration counts would be more similar for an absolute residual tolerance.

Our final numerical example demonstrates the performance of the HSS preconditioner on the generalized Stokes problem.

In Table 3 we report the numerical results for Flexible GMRES with inexact HSS preconditioning applied to a set of generalized Stokes problems. The discrete

Table 3. Iteration count for 3D generalized Stokes problem, inexact HSS preconditioning.

σ	$\nu = 0.1$	$\nu = 0.01$	$\nu = 0.001$	$\nu = 10^{-6}$
1	45	27	16	13
10	32	19	15	12
20	30	18	14	11
50	28	15	13	11
100	25	14	12	10

Table 4. Results for 3D unsteady Stokes problem, $\nu = 0.001$.

grid	n	m	iterations	cpu time
$10 \times 10 \times 10$	2700	1000	12	0.42
$20 \times 20 \times 20$	22800	8000	12	4.66
$30 \times 30 \times 30$	78300	27000	12	20.97
$40 \times 40 \times 40$	187200	64000	13	66.02

saddle point problems were generated in this case by the Marker-and-Cell (MAC) finite difference discretization on a $40 \times 40 \times 40$ grid for different values of σ ($= 1/\Delta t$ in the context of implicit solution of time-dependent problems) and ν. Homogeneous Dirichlet boundary conditions were imposed on the velocities. Here $\Omega = [0,1] \times [0,1] \times [0,1]$; the discrete problem has over 250,000 unknowns. The parameter α was set to 0.5, and a zero initial guess was used. The outer iteration was stopped when a reduction of the initial residual by six orders of magnitude was reached. For the inexact inner solves we used Conjugate Gradients with incomplete Cholesky preconditioning; the inner iterations were stopped as soon as a reduction of the initial residual by one order of magnitude was attained. This only required 1-2 PCG iterations per inner linear solve. The iteration counts, which can be shown to be largely independent of the grid size, improve for increasing σ and decreasing ν.

In Table 4 we show timings (in seconds) for an unsteady Stokes problem with $\nu = 0.001$ for different grids. Denoted by h the grid size, we let $\sigma = h^{-1}$. We use HSS preconditioning with $\alpha = 0.5$. We also report the dimensions n and m and the total number of FGMRES iterations. The test runs were done on one processor of a SunFire V880 workstation with 8 CPUs and 16 GB of memory.

9 Conclusions

Saddle point problems arise naturally in many large scale computations, particularly in the solution of PDEs by mixed finite elements, interior point methods for constrained optimization, weighted least squares, and so forth. The last decade has seen considerable progress in the development of iterative solvers and preconditioners for this class of problems. In this paper we have given a concise overview of some of the most promising preconditioning techniques for linear systems in saddle point form, in particular block and constraint preconditioning. We have also pointed out a possible connection between preconditioners based on approximate Schur complements and the approximation of matrix-valued transfer functions, an essential component of model order reduction for time-invariant linear dynamical systems.

Acknowledgements. This paper originated in the stimulating atmosphere fostered by Wil Schilders and Henk van der Vorst at the Workshop on Model Reduction, Coupled Problems and Optimization held in Leiden, The Netherlands, on September 19–23, 2005. We would like to thank Wil and Henk for inviting us to the Workshop, for suggesting that we write the present overview, and for their careful reading of the original manuscript.

References

1. M. Arioli, J. Mariška, M. Rozložník and M. Tůma, *Dual variable methods for mixed-hybrid finite element approximation of the potential fluid flow problem in porous media*, Electr. Trans. Numer. Anal., 22 (2006), pp. 17–40.
2. M. Benzi and G. H. Golub, *A preconditioner for generalized saddle point problems*, SIAM J. Matrix Anal. Appl., 26 (2004), pp. 20–41.
3. M. Benzi, G. H. Golub, and J. Liesen, *Numerical solution of saddle point problems*, Acta Numerica, 14 (2005), pp. 1–137.
4. M. Benzi and M. A. Olshanskii, *An augmented Lagrangian approach to the Oseen problem*, SIAM J. Sci. Comput., 28 (2006), pp. 2095–2113.
5. M. Benzi and V. Simoncini, *On the eigenvalues of a class of saddle point matrices*, Numer. Math., 103 (2006), pp. 173–196.
6. P. Bochev and R. B. Lehoucq, *Regularization and stabilization of discrete saddle-point variational problems*, Electr. Trans. Numer. Anal., 22 (2006), pp. 97–113.
7. D. Braess, *Finite Elements. Theory, Fast Solvers, and Applications in Solid Mechanics*. Second Edition, Cambridge University Press, 2001.
8. H. S. Dollar, *Iterative Linear Algebra for Constrained Optimization*, DPhil (PhD) thesis, Oxford University Computing Laboratory, 2005.
9. H. S. Dollar, N. I. M. Gould and A. J. Wathen, *On implicit-factorization constraint preconditioners*, in 'Large Scale Nonlinear Optimization', Eds. G. Di Pillo, Gianni and M. Roma, Springer Verlag, Heidelberg, Berlin, New York (2006), pp. 61-82.
10. H. S. Dollar, N. I. M. Gould, W. H. A. Schilders and A. J. Wathen, *Implicit-factorization preconditioning and iterative solvers for regularized saddle-point systems*, SIAM J. Matrix Anal. Appl., 28 (2006), pp. 170–189.
11. H. S. Dollar, N. I. M. Gould, W. H. A. Schilders and A. J. Wathen, *Using constraint preconditioners with regularized saddle-point problems*, Comput. Optim. Appl., to appear.
12. H. S. Dollar and A. J. Wathen, *Approximate factorization constraint preconditioners for saddle-point matrices*, SIAM J. Sci. Comput., 27 (2006), pp. 1555–1572.
13. H. C. Elman, D. J. Silvester and A. J. Wathen, *Finite Elements and Fast Iterative Solvers with Applications in Incompressible Fluid Dynamics*, Numerical Mathematics and Scientific Computation, Oxford University Press, Oxford, UK, 2005.
14. B. Fischer, *Polynomial Based Iteration Methods for Symmetric Linear Systems*, Wiley-Teubner, Chichester and Stuttgart, 1996.
15. R. Fletcher, *Practical Methods of Optimization (Second Edition)*, J. Wiley & Sons, Chichester, 1987.

16. M. FORTIN AND R. GLOWINSKI, *Augmented Lagrangian Methods: Application to the Solution of Boundary-Value Problems*, Stud. Math. Appl., Vol. 15, North-Holland, Amsterdam, 1983.
17. R. W. FREUND, *Model reduction based on Krylov subspaces*, Acta Numerica, 12 (2003), pp. 267–319.
18. R. W. FREUND AND N. M. NACHTIGAL, *A new Krylov-subspace method for symmetric indefinite linear systems*, in proceedings of 14th IMACS World Congress on Computational and Applied Mathematics, W. F. Ames, ed., IMACS, 1994, pp. 1253–1256.
19. R. W. FREUND AND N. M. NACHTIGAL, *Software for simplified Lanczos and QMR algorithms*, Appl. Numer. Math., 19 (1995), pp. 319–341.
20. G. H. GOLUB AND C. GREIF, *On solving block-structured indefinite linear systems*, SIAM J. Sci. Comput., 24 (2003), pp. 2076–2092.
21. N. I. M. GOULD, M. E. HRIBAR, AND J. NOCEDAL, *On the solution of equality constrained quadratic programming problems arising in optimization*, SIAM J. Sci. Comput., 23 (2001), pp. 1376–1395.
22. N. I. M. GOULD, D. ORBAN, AND PH. L. TOINT, *CUTEr (and SifDec), a Constrained and Unconstrained Testing Environment, Revisited*, Tech. Report TR/PA/01/04, CERFACS, Toulouse, France, 2001.
23. A. GREENBAUM, *Iterative Methods for Solving Linear Systems*, SIAM, Philadelphia, 1997.
24. C. GREIF AND D. SCHÖTZAU, *Preconditioners for the discretized time-harmonic Maxwell equations in mixed form*, Numer. Linear Algebra Appl., to appear.
25. M. R. HESTENES AND E. STIEFEL, *Methods of conjugate gradients for solving linear systems*, J. Res. Nat. Bur. Stand., 49 (1952), pp. 409–436.
26. H. C. ELMAN, A. RAMAGE, AND D. J. SILVESTER, *IFISS: a Matlab toolbox for modelling incompressible flow*, Manchester University Numerical Analysis Report No. 474 (2005), ACM Transactions on Mathematical Software, to appear.
27. C. KELLER, N. I. M. GOULD AND A. J. WATHEN, *Constraint preconditioning for indefinite linear systems*, SIAM J. Matrix Anal. Appl., 21 (2000), pp. 1300-1317.
28. C. LANCZOS, *An iteration method for the solution of the eigenvalue problem of linear differential and integral operators*, J. Res. Nat. Bur. Stand., 45 (1950), pp. 255-282.
29. L. LUKŠAN AND J. VLČEK, *Indefinitely preconditioned inexact Newton method for large sparse equality constrained non-linear programming problems*, Numer. Linear Algebra Appl., 5 (1998), pp. 219–247.
30. M. F. MURPHY, G. H. GOLUB, AND A. J. WATHEN, *A note on preconditioning for indefinite linear systems*, SIAM J. Sci. Comput., 21 (2000), pp. 1969–1972.
31. C. C. PAIGE AND M. A. SAUNDERS, *Solution of sparse indefinite systems of linear equations*, SIAM J. Numer. Anal., 12 (1975), pp. 617–629.
32. J. NOCEDAL AND S. J. WRIGHT, *Numerical Optimization*, Springer, New York, 1999.
33. I. PERUGIA AND V. SIMONCINI, *Block-diagonal and indefinite symmetric preconditioners for mixed finite element formulations*, Numer. Linear Algebra Appl., 7 (2000), pp. 585–616.
34. T. RUSTEN AND R. WINTHER, *A preconditioned iterative method for saddlepoint problems*, SIAM J. Matrix Anal. Appl., 13 (1992), pp. 887–904.
35. Y. SAAD, *A flexible inner-outer preconditioned GMRES*, SIAM J. Sci. Comput., 14 (1993), pp. 461–469.
36. Y. SAAD, *Iterative Methods for Sparse Linear Systems, Second Edition*, SIAM, Philadelphia, PA, 2003.
37. Y. SAAD AND M. H. SCHULTZ, *GMRES: A generalized minimal residual algorithm for solving nonsymmetric linear systems*, SIAM J. Sci. Stat. Comput., 7 (1986), pp. 856–869.

38. D. J. SILVESTER AND A. J. WATHEN, *Fast iterative solution of stabilised Stokes systems. II: Using general block preconditioners*, SIAM J. Numer. Anal., 31 (1994), pp. 1352–1367.
39. V. SIMONCINI AND M. BENZI, *Spectral properties of the Hermitian and skew-Hermitian splitting preconditioner for saddle point problems*, SIAM J. Matrix Anal. Appl., 26 (2004), pp. 377–389.
40. G. L. G. SLEIJPEN, H. A. VAN DER VORST, AND J. MODERSITZKI, *Effects of rounding errors in determining approximate solutions in Krylov solvers for symmetric indefinite linear systems*, SIAM J. Matrix Anal. Appl., 22 (2000), pp. 726-751.
41. G. STRANG, *Introduction to Applied Mathematics*, Wellesley, Cambridge, MA, 1986.
42. H. A. VAN DER VORST, *Iterative Krylov Methods for Large Linear Systems*, Cambridge Monographs on Applied and Computational Mathematics, Cambridge University Press, Cambridge, UK, 2003.

Time Variant Balancing and Nonlinear Balanced Realizations

E.I. Verriest

School of Electrical and Computer Engineering, Georgia Institute of Technology, Atlanta, GA 30332-0250, USA
erik.verriest@ee.gatech.edu

Summary. Balancing for linear time varying systems and its application to model reduction via projection of dynamics (POD) are briefly reviewed. We argue that a generalization for balancing nonlinear systems may be expected to be based upon three sound principles: 1) Balancing should be defined with respect to a nominal flow; 2) Only Gramians defined over small time intervals should be used in order to preserve the accuracy of the linear perturbation model and; 3) Linearization should commute with balancing, in the sense that the linearization of a globally balanced model should correspond to the balanced linearized model in the original coordinates.

The first two principles lead to local balancing, which provides useful information about the dynamics of the system and the topology of the state space. It is shown that an integrability condition generically provides an obstruction towards a notion of a globally balanced realization in the strict sense. By relaxing the conditions of "strict balancing" in various ways useful system approximations may be obtained.

1 Introduction

Reduction techniques are routinely used to replace the relevant discretized PDE's to ODE models of much smaller dimension. Most existing methods pertain to linear models and dynamics and fail to correctly model the nonlinear couplings and dynamics. Balancing for linear time invariant systems has been applied to problems of model reduction (via the Projection of Dynamics (POD), also called "balanced truncation"), in parameterization, sensitivity analysis and system identification. With these initial successes, extensions a balanced realization to other classes of systems soon followed. Balancing for linear time-varying systems [SSV83, VK83, SR02, VH98], and an alternative form of balancing, suitable for controller reduction, called LQG-balancing [Ve81a, Ve81b, JS83] were the first generalizations. This chapter presents a possible approach towards generalizing balanced realizations to nonlinear systems. In a series of papers [VG00, VG01b, Ve04, VG04] we extended the linear balancing method to a class of nonlinear systems. While sharing many similarities with the method first proposed by Scherpen [Sch94], there are some fundamental differences. Whereas Scherpen took the notions of observability and reachability[1] functions as a starting point, our method is rooted in earlier work for linear time-varying systems. In particular, the Sliding-Interval-Balancing (SIB), proposed in [Ve80, VK83], is here used as a stepping stone to the nonlinear case.

In Section 2, the principles behind balancing are motivated. Section 3 reviews SIB, and some new results are presented. The proposed approach to (local) nonlinear balancing is given in Section 4, which forms the main part of the chapter. The extension to global balancing is given in Section 5, where also an 'obstruction' is encountered. In Section 6 we introduce Mayer-Lie interpolation as a way around this obstruction. Application to nonlinear model reduction is briefly discussed in Section 7. Final comments regarding the proposed solution are formulated in Section 8.

Some of this work was performed and evolved over many years in collaboration with Professor W. Steven Gray from the Old Dominion University.

2 Time Varying Linear Systems

2.1 Finite Time Gramians

For general time varying linear systems, we define the reachability and observability map, and use adjoint operator techniques to solve various problems related to energy, ambiguity and uncertainty. This sheds light on the role played by the Gramian matrices (reachability and observability Gramian), and their subsequent importance in model reduction. It is assumed that there are n states, m inputs and p outputs.

$$\dot{x}(t) = A(t)x(t) + B(t)u(t) \tag{1}$$

$$y(t) = C(t)x(t) \tag{2}$$

Let $\Phi(t,\tau)$ be the transition matrix, satisfying for all t and τ, $\frac{\partial}{\partial t}\Phi(t,\tau) = A(t)\Phi(t,\tau)$ with $\Phi(\tau,\tau) = I$. The finite time reachability and observability Gramians (for an interval of length $\delta > 0$) are defined by

$$\mathcal{R}(t,\delta) = \int_{t-\delta}^{t} \Phi(t,\tau)B(\tau)B(\tau)^T\Phi(t,\tau)^T \,\mathrm{d}\tau \tag{3}$$

$$\mathcal{O}(t,\delta) = \int_{t}^{t+\delta} \Phi(\tau,t)^T C(\tau)^T C(\tau)\Phi(\tau,t)\,\mathrm{d}\tau. \tag{4}$$

In general, these $n \times n$ matrices depend explicitly on time t. It is well known that the system is completely reachable and observable if these Gramians have full rank for all t. In fact, the following results are standard, and are stated here without proof:

The Gramians are naturally obtained via the definition of adjoint operators. Let K be an arbitrary operator taking vectors in a Hilbert space $\mathcal{H}_1$ to a Hilbert space $\mathcal{H}_2$. Then the *adjoint* of K, denoted K^*, is a map from $\mathcal{H}_2$ to $\mathcal{H}_1$ defined via $\langle y, Kx\rangle_2 = \langle K^*y, x\rangle_1$. In what follows we make use of the finite dimensional vector space $\mathbb{R}^n$

[1] This author has a preference for 'reachability' over the term 'controllability', which appears in the original work.

with its standard inner product $\langle x, y\rangle_{\mathbb{R}^n} = x^T y$. We also consider the input space $\mathrm{L}_2([t-\delta, t], \mathbb{R}^m)$ and the output space $\mathrm{L}_2([t, t+\delta], \mathbb{R}^p)$ of square integrable vector functions.

2.2 Reachability Map

Definition 1. *Associated with the system (2) we have the time-indexed family of* reachability maps

$$L_t : \mathrm{L}_2([t-\delta, t], \mathbb{R}^m) \to \mathbb{R}^n; \qquad L_t(u(\cdot)) = \int_{t-\delta}^{t} \Phi(t,\tau) B(\tau) u(\tau)\, \mathrm{d}\tau. \tag{5}$$

Let the metric in the input (function) space be

$$\forall\, u, v \in \mathrm{L}_2^m : \quad \langle u(\cdot), v(\cdot)\rangle_{\mathrm{L}_2^m} = \int_{t-\delta}^{t} u(\tau)^T v(\tau)\, \mathrm{d}\tau. \tag{6}$$

It follows that the *adjoint maps* are given by

$$L_t^* : \mathbb{R}^n \to \mathrm{L}_2([t-\delta, t], \mathbb{R}^m); \qquad L_t^* x = B(\cdot)^T \Phi(t, \cdot)^T x. \tag{7}$$

This gives at once some interpretations for the above defined Gramians. Two 'energy' related problems and one 'uncertainty' related problem will shed light on the role of the reachability Gramian. We define an *event* as a state-time pair: (x, t) means that at time t the system is in the state x.

Energy Interpretation 1

Reachability: Here we consider the reachability of the event (x, t) from the event $(0, t-\delta)$. The minimum norm solution to $L_t(u(\cdot)) = x$ is given by $u(\cdot) = L_t^*(z)$, where z is any solution to $L_t L_t^*(z) = x$. Note that $L_t L_t^* : \mathbb{R}^n \to \mathbb{R}^n$ is precisely the reachability Gramian $\mathcal{R}(t, \delta)$, and hence this problem is solvable in general if the system is reachable (invertibility of the Gramian). In this case, it is easily shown that this solution is in fact the one minimizing the squared norm (energy) $\|u\|^2$:

$$\|u\|^2 = \|x\|^2_{(L_t L_t^*)^{-1}} = x^T (L_t L_t^*)^{-1} x \stackrel{\text{def}}{=} \mathcal{E}_i(x). \tag{8}$$

Interpretation: The cost associated with transfer of event $(0, t-\delta) \in \mathbb{R}^n \times \mathbb{R}$ to the event (x, t) is the above quadratic form in x, with weight $\mathcal{R}^{-1}(t, \delta)$. It relates to the *effort* needed to reach x at time t from the origin at $t - \delta$.

Energy Interpretation 2

Let S_{n-1} be the unit sphere in n dimensions. Given a direction $\nu \in S_{n-1}$, what is the maximal distance that can be reached from the past event $(0, t-\delta)$ at time t in the direction ν with an input of unit energy? This means that we are now looking for the solution to

$$\max_{\|u\|=1} |\langle \nu, L_t u\rangle| \stackrel{\text{def}}{=} \mathcal{P}_i(\nu) \tag{9}$$

It is easily shown that the solution is given by $u = kL_t^*\nu$, with $k = \langle L_t^*\nu, L_t^*\nu\rangle^{-1/2}$, and the corresponding excursion in the direction ν has magnitude

$$\mathcal{P}_i(\nu) = (\nu^T \mathcal{R}(t,\delta)\nu)^{1/2}. \tag{10}$$

Interpretation: The Gramian is the weight matrix for the worst case excursion in a given direction, ν, due to an input disturbance of unit energy. If this were small, the input has little influence in the direction ν. Hence $\mathcal{P}_i(\nu)$ measures the *influencability* of the dynamics in that direction in the state space.

Uncertainty Interpretation

We now use a result from stochastic system theory. Let a white noise input have zero mean and unit variance (=standard white noise). If the system starts with the event $(0, t-\delta)$, then the resulting state at time t is a random vector $L_t u$ in the Hilbert space $\mathrm{L}_2(\Omega, \mathbb{R}^n)$. The inner product in this space is $\langle x, y\rangle = \mathbf{E}\, x^T y$, where $\mathbf{E}$ is the expectation operator. This random vector defines an additive measure on the subspaces of $\mathbb{R}^n$ by

$$\mu_i(A) \stackrel{\text{def}}{=} \|P^A L_t u\|^2 = \mathrm{Tr}\,(L_t L_t^* P^A) = \mathrm{Tr}\,(\mathcal{R}(t,\delta)P^A), \tag{11}$$

where P^A is the projector onto the subspace A.

Interpretation: The uncertainty of the state reached from the origin by the white noise process from $t-\delta$ to t in the direction ν is given by $\nu^T\mathcal{R}(t,\delta)\nu$. We conclude that this uncertainty relates again to the *influencability* of the subspace spanned by ν.

We now turn to the *dual* property of observability.

2.3 Observability Map

Definition 2. *Associate the* observability maps, *M_t, defined for all t and fixed $\delta > 0$, with the system (2):*

$$M_t : \mathbb{R}^n \to \mathrm{L}_2^p([t, t+\delta], \mathbb{R}); \qquad M_t(x) = C(\cdot)\Phi(\cdot, t)\, x. \tag{12}$$

With the inner product in the output space L_2^p defined by

$$\forall\, u, v \in \mathrm{L}_2^p : \quad \langle u(\cdot), v(\cdot)\rangle_{\mathrm{L}_2^p} = \int_t^{t+\delta} u^T(\tau)v(\tau)\,\mathrm{d}\tau, \tag{13}$$

the adjoint maps are,

$$M_t^* : L_2([t, t+\delta], \mathbb{R}) \to \mathbb{R}^n; \qquad M_t^* u(\cdot) = \int_t^{t+\delta} \Phi(\tau, t)^T C(\tau)^T u(\tau)\, \mathrm{d}\tau. \quad (14)$$

We now clarify the natural role of the observability Gramian (4) by illustrating its appearance in some standard problems. Again we have two energy type interpretations and an uncertainty interpretation.

Energy Interpretation-1

We start with a standard result. Given $x \in \mathbb{R}^n$, the energy in the output $x \to y(\cdot) = M_t x$ is

$$\mathcal{E}_o(x) \stackrel{\text{def}}{=} \langle M_t x, M_t x \rangle = \langle M_t^* M_t x, x \rangle, \tag{15}$$

where $M_t^* M_t = \mathcal{O}(t, \delta) : \mathbb{R}^n \to \mathbb{R}^n$, is the observability Gramian

$$\mathcal{O}(t, \delta) = \int_t^{t+\delta} \Phi(\tau, t)^T C(\tau)^T C(\tau) \Phi(\tau, t)\, \mathrm{d}\tau. \tag{16}$$

Interpretation: The cost associated with transfer of event $(x, t) \in \mathbb{R}^n \times \mathbb{R}$ to the event $(x_f, t+\delta)$, where $x_f = \Phi(t+\delta, t)x$, is the above quadratic form in x with $\mathcal{O}(t, \delta)$ as weight matrix. It relates to the signal energy available to detect the initial condition x, or the Signal-to-Noise Ratio (SNR) if embedded in unit variance white noise.

Energy Interpretation-2

Given a direction $\nu \in S_{n-1}$, let an output signal, $y = M_t x_0$, be embedded in a unit energy disturbance, w, thus giving an observed signal $z = y + w$. What is the maximal ambiguity of the initial state component in the direction ν? This problem may be reformulated as

$$\max_{\|w\|=1} |\langle \nu, (x_0 - \widehat{x}) \rangle| \stackrel{\text{def}}{=} \mathcal{P}_o(\nu), \;\; \text{where } \widehat{x} = \operatorname{argmin} \|M\widehat{x} - z\|, \;\; z = Mx_0 + w. \tag{17}$$

This problem has not been discussed often and we shall therefore present all details of its solution. First obtain $\min_{\widehat{x} \in \mathbb{R}^n} \|M_t \widehat{x} - z\|$. We use a variational method: assume that $\widehat{x}$ is the optimal solution. Let $\widetilde{x} \in \mathbb{R}^n$ be an arbitrary perturbation, and consider the vector $x = \widehat{x} + \epsilon \widetilde{x}$. By the optimality assumption: $\|M_t x - z\| \geq \|M_t \widehat{x} - z\|$. This inequality is squared, giving

$$\langle M_t \widehat{x} + \epsilon M_t \widetilde{x} - z, M_t \widehat{x} + \epsilon M_t \widetilde{x} - z \rangle \geq \langle M_t \widehat{x} - z, M_t \widehat{x} - z \rangle. \tag{18}$$

Using the linearity and symmetry of the inner product, it follows that $2\epsilon \langle M_t \widetilde{x}, M_t \widehat{x} - z \rangle + \epsilon^2 \langle M_t \widetilde{x}, M_t \widetilde{x} \rangle \geq 0$. For $|\epsilon|$ sufficiently small, this implies (since $\widetilde{x}$ is arbitrary) that $\langle M_t \widetilde{x}, M_t \widehat{x} - z \rangle = 0$. The adjoint relation implies $\langle \widetilde{x}, M_t^* (M_t \widehat{x} - z) \rangle = 0$. This can only happen if $M_t^* M_t \widehat{x} = M_t^* z$. Hence, the best estimate in the deterministic least squares sense of the initial condition is $\widehat{x} = (M_t^* M_t)^{-1} M_t^* z$. The ambiguity $x_0 - \widehat{x}$ in the direction ν is, since $z = M_t x + w$, equal to

$$|\nu^T(x_0 - (M_t^* M_t)^{-1} M_t^* z)| = |\nu^T \mathcal{O}(t,\delta)^{-1} M_t^* w| = |\langle \nu, \mathcal{O}(t,\delta)^{-1} M_t^* w\rangle| \quad (19)$$

The worst case is obtained for the specific w of unit norm that maximizes (19) or, equivalently

$$\max_{\|w\|=1} |\langle M_t \mathcal{O}(t,\delta)^{-1}\nu, w\rangle|. \quad (20)$$

By the Cauchy-Schwarz inequality, this is maximal if w is proportional to $M_t\nu$, say, $w = \lambda M_t \mathcal{O}(t,\delta)^{-1}\nu$, where λ follows from the normalization condition

$$\|w\|^2 = \lambda^2 \langle M_t \mathcal{O}(t,\delta)^{-1}\nu, M_t \mathcal{O}(t,\delta)^{-1}\nu\rangle = \lambda^2 \langle \mathcal{O}(t,\delta)^{-1}\nu, \nu\rangle. \quad (21)$$

Thus $\lambda = 1/\sqrt{\nu^T \mathcal{O}(t,\delta)^{-1}\nu}$, and finally

$$\mathcal{P}_\wr(\nu) = \left| \left\langle M_t \mathcal{O}(t,\delta)^{-1}\nu, \frac{M_t \mathcal{O}(t,\delta)^{-1}\nu}{\sqrt{\nu^T \mathcal{O}(t,\delta)^{-1}\nu}} \right\rangle \right| = \{\nu^T \mathcal{O}(t,\delta)^{-1}\nu\}^{1/2}. \quad (22)$$

Interpretation: The induced ambiguity of the initial state (x_0, t) in direction ν is the quadratic form $\mathcal{P}_\wr(\nu)$ in ν with weight matrix $\mathcal{O}(t,\delta)^{-1}$. It relates to the *ambiguity* in determining the initial condition x_0.

Uncertainty Interpretation

Given $y(\cdot) \in \mathrm{L}_2^p([t, t+\delta], \mathbb{R})$, the error norm $\|y - M_t\widehat{x}\|_{\mathrm{L}_2^p}$ is minimal for

$$\widehat{x} = (M_t^* M_t)^{-1} M_t^* y(\cdot). \quad (23)$$

We now present a stochastic interpretation. The proof is standard in an estimation course. If $y = M_t x + u$ (signal + standard white noise), the estimate $\widehat{x}$ is a random vector with covariance $\mathcal{O}(t,\delta) = M_t^* M_t$. Hence the uncertainty in the subspace A is then given by the additive measure

$$\mu_o(A) \stackrel{\text{def}}{=} \|P^A (M_t^* M_t)^{-1} M_t^* u\|^2 = \mathrm{Tr}\,((M_t^* M_t)^{-1} P^A) = \mathrm{Tr}\,(\mathcal{O}^{-1}(t,\delta) P^A), \quad (24)$$

where P^A is the projector onto A.

Interpretation: The residual uncertainty of the state at t in the direction ν after observation of the output from t to $t+\delta$ is $\nu^T \mathcal{O}^{-1}(t,\delta)\nu$. It relates to the *difficulty* of observing the ν-component of the state.

2.4 Summary

As the above interpretations are of quintessential importance in the motivation for balanced realizations, we reproduce them here (assuming δ is fixed, we drop it now from the notation):

- Energy maps - 1
$$\mathcal{E}_i : \mathbb{R}^n \to \mathbb{R}_+ : x \to x'\mathcal{R}_t^{-1}x$$
$$\mathcal{E}_o : \mathbb{R}^n \to \mathbb{R}_+ : x \to x'\mathcal{O}_t x$$

- Energy maps - 2
$$\mathcal{P}_i : S_{n-1} \to \mathbb{R}_+ : \nu \to (\nu'\mathcal{R}_t\nu)^{1/2}$$
$$\mathcal{P}_o : S_{n-1} \to \mathbb{R}_+ : \nu \to (\nu'\mathcal{O}_t^{-1}\nu)^{1/2}$$

- Uncertainty maps
$$\mu_i : \mathrm{Proj}\,\mathbb{R}^n \to \mathbb{R}_+ : A \to \mathrm{Tr}\,(\mathcal{R}_t P^A)$$
$$\mu_o : \mathrm{Proj}\,\mathbb{R}^n \to \mathbb{R}_+ : A \to \mathrm{Tr}\,(\mathcal{O}_t^{-1} P^A)$$

  or, for $\nu \in S_{n-1}$, (one dimensional subspaces) respectively
$$\mathcal{U}_i : S_{n-1} \to \mathbb{R}_+ : \nu \to \nu'\mathcal{R}_t\nu$$
$$\mathcal{U}_o : S_{n-1} \to \mathbb{R}_+ : \nu \to \nu'\mathcal{O}_t^{-1}\nu$$

3 Sliding Interval Balancing

The interpretations in Sect. 2 now lead to the following disparate measures. The eigenspace of $\mathcal{O}_t$ corresponding to the smallest eigenvalue gives a rather *unimportant* direction in the state space. If this eigenvalue were zero, it would be an exact *unobservable* direction, and consequently decouples the state (hence also the input) from the output. We may simply discard this component from the model. Hence the idea is to discard the *hard to observe* state components as a rationale for model reduction. However, this is not quite right as we may now look from the input to state perspective, and similarly discard states components that are *hard to reach.* When looking at the same system with two metrics, conflicts necessarily arise: What to do with states that are hard to reach but easy to observe, or vice versa? The way around this is to realize that an invertible state space transformation, *similarity transformation*, transforms the Gramians by a *congruence*:

$$\mathcal{R}_t \to T(t)\mathcal{R}_t T(t)^T \tag{25}$$
$$\mathcal{O}_t \to T(t)^{-T}\mathcal{O}_t T(t)^{-1}. \tag{26}$$

Based on this fact, an invertible $T(t)$ can always be found, point wise, such that the transformed Gramians are equal and diagonal. In addition, it was shown in [Ve80, VK80, VK83] that for an *analytic* system (i.e., one where the entries of the matrices $A(\cdot), B(\cdot)$ and $C(\cdot)$ are analytic functions of time), a similarity $T(t)$ exists as a *differentiable* function of t such that the new realization $(TAT^{-1} + \dot{T}T^{-1}, TB, CT^{-1})$ satisfies the property that the Gramians of the transformed system are *equal* and *diagonal.* Such a realization is called a sliding interval balanced (SIB) realization, and the corresponding (reachability and/or observability) Gramian is defined as the *canonical Gramian*, denoted by $\Lambda(t)$. Unlike the time invariant case, the elements of the canonical Gramian may cross at certain times. If one insists on *ordering* the canonical Gramian elements such that

$\lambda_1(t) \geq \lambda_2 \geq \cdots \geq \lambda_n(t)$ then continuity of the resulting balanced realization may be lost at the crossing points giving *umbilic* points for the parameterization. (At an umbilic point we have that $\lambda_i(t) = \lambda_{i+1}(t)$ for some i). In the remainder of this chapter we will restrict SIB to intervals where such crossover does not happen. Now one has a uniform measure (input and output) for the relative importance of subspaces of the state space, and model reduction may proceed by discarding these less important states.

Rationale for Model Reduction: In balanced coordinates, for each coordinate direction (more generally, for each subspace) the degrees of reachability and observability (as determined by the canonical Gramian) are equal. Assume that the elements of $\Lambda(t) = \text{diag}\,[\lambda_1(t), \cdots, \lambda_n(t)]$ are ordered in magnitude, i.e., for all $t \in [t_i, t_f]$: $\lambda_1(t) \geq \lambda_2(t) \geq \cdots \geq \lambda_n(t)$, then by *projection of dynamics* (POD) we shall understand the model reduction as

$$\begin{pmatrix} \begin{bmatrix} A_{11} & A_{12} \\ A_{21} & A_{22} \end{bmatrix} & \begin{bmatrix} B_1 \\ B_2 \end{bmatrix} \\ \\ \begin{bmatrix} C_1 & C_2 \end{bmatrix} & \end{pmatrix} \overset{\text{POD}}{\longrightarrow} \begin{pmatrix} \begin{bmatrix} A_{11} & 0 \\ 0 & 0 \end{bmatrix} & \begin{bmatrix} B_1 \\ 0 \end{bmatrix} \\ \\ \begin{bmatrix} C_1 & 0 \end{bmatrix} & \end{pmatrix} \tag{27}$$

according to

$$\begin{bmatrix} \Lambda_1 & \\ & \Lambda_2 \end{bmatrix} \overset{\text{POD}}{\longrightarrow} \begin{bmatrix} \Lambda_1 & 0 \\ 0 & 0 \end{bmatrix} \tag{28}$$

The resulting reduced order model is then (A_{11}, B_1, C_1) with canonical Gramian Λ_1. Note that it is essential that POD is combined with balancing, otherwise any arbitrary reduced model may result.

Suppose that the original state variables x have a well defined physical meaning (e.g., potentials and currents in a VLSI circuit, stresses and strains in mechanical structures, etc). Then one may want to know what the effect is of the balanced model reduction in terms of these original coordinates. Once we have the balancing transformation in partitioned form (x^b is the state in balanced coordinates)

$$\begin{bmatrix} T_{11} & T_{12} \\ T_{21} & T_{22} \end{bmatrix} \begin{bmatrix} x_1 \\ x_2 \end{bmatrix} = \begin{bmatrix} x_1^b \\ x_2^b \end{bmatrix} \tag{29}$$

consistent with $\Lambda = \text{diag}\,[\Lambda_1, \Lambda_2]$, with $\Lambda_2 < \Lambda_1$, the reduced model via POD sets all components $\widetilde{x}_2^b$ equal to zero. This means that in terms of the original coordinate system the relation

$$T_{21}x_1 + T_{22}x_2 = 0 \tag{30}$$

is induced, and results in a parameterization of these original state variables in terms of the reduced order state

$$\begin{bmatrix} x_1 \\ x_2 \end{bmatrix} = \begin{bmatrix} (T^{-1})_{11} \\ (T^{-1})_{21} \end{bmatrix} x_1^b. \tag{31}$$

These states can thus be appended to the reduced order model in balanced form as *output* equations. Alternatively, the reduced model may be re-formulated as an equation in the states x_1 (assuming for simplicity the invertibility of the submatrices T_{11} and T_{22}),

$$\dot{x}_1 = [(\dot{T}^{-1})_{11} + (T^{-1})_{11}A^b_{11}][T_{11} - T_{12}T_{22}^{-1}T_{21}]x_1 + (T^{-1})_{11}B^b_1 u. \quad (32)$$

3.1 SIB: Balanced in the "Sense of Moore"

Originally, Moore [M81] defined a balanced realization for a linear time-invariant (LTI) system as a realization that is SIB for $\delta \to \infty$. Obviously, this only makes sense for *asymptotically stable* LTI systems.

The balancing transformation T_{bal} is obtained by simultaneous diagonalization of the Gramian matrices which satisfy in this case the *algebraic Lyapunov equations*. For continuous time systems these are:

$$A\mathcal{R} + \mathcal{R}A^T + BB^T = 0$$
$$A^T\mathcal{O} + \mathcal{O}A + C^TC = 0.$$

For discrete time systems, the Lyapunov equations read:

$$A\mathcal{R}A^T + BB^T = \mathcal{R}$$
$$A^T\mathcal{O}A + C^TC = \mathcal{O}.$$

It is readily shown that in continuous time, a reduced order model via POD satisfies the truncated Lyapunov equations. In discrete time this property breaks down, so that the POD reduced order model for a balanced realization itself is not necessarily balanced.

For a given (unbalanced) realization, the Gramians can be computed either by solving the Lyapunov equations (usually iteratively), or in some simple cases by explicitly computing the matrix exponential e^{At} (e.g. by Laplace transform techniques) and integration. Computation of the balancing transformation T hinges on the singular value decomposition (SVD). Since the singular value decomposition is unique modulo a signature matrix, the balanced realizations are not unique. However, it was shown that in the single-input single-output case all balanced realizations are sign symmetric. That is, there exists a *signature* matrix S such that $A^T = SAS$ and $Sb = c^T$.

3.2 Approximate SIB

In the time varying case, computation of the Gramians may be problematic. Indeed, the transition matrix is a nice *representation*, useful in establishing various results, but its direct computation by integration may be prohibitive. The alternative of computing the solution of the Lyapunov equation is not much better, since these equations are now also time varying differential equations (we consider δ as fixed and denote $\mathcal{R}(t,\delta)$ and $\mathcal{O}(t,\delta)$ simply by $\mathcal{R}_t$ and $\mathcal{O}_t$ respectively):

$$\dot{\mathcal{R}}_t = A(t)\mathcal{R}_t + \mathcal{R}_t A(t)^T + B(t)B(t)^T - \Phi(t,t-\delta)B(t-\delta)B(t-\delta)^T\Phi(t,t-\delta)^T \quad (33)$$
$$-\dot{\mathcal{O}}_t = A(t)^T\mathcal{O}_t + \mathcal{O}_t A(t) + C(t)^T C(t) - \Phi(t+\delta,t)^T C(t+\delta)^T C(t+\delta)\Phi(t+\delta,t). \quad (34)$$

However, when δ is sufficiently small, the solution is readily obtained by a series expansion. Indeed, as observed in Sect. 1, the transfer from $(0, t-\delta)$ to (x_f, t), requires an input, $u(\tau), t-\delta < \tau < t$, satisfying

$$x_f = \int_{t-\delta}^{t} \Phi(t,\tau)B(\tau)\, u(\tau)\, \mathrm{d}\tau. \quad (35)$$

Note that by definition of the transition matrix,

$$\frac{\partial}{\partial\tau}\Phi(t,\tau)B(\tau) = \Phi(t,\tau)\left(\frac{\mathrm{d}}{\mathrm{d}\tau} - A(\tau)\right)B(\tau) \quad (36)$$

and by iteration

$$\left(\frac{\partial}{\partial\tau}\right)^k \Phi(t,\tau)B(\tau) = \Phi(t,\tau)\left(\frac{\mathrm{d}}{\mathrm{d}\tau} - A(\tau)\right)^k B(\tau). \quad (37)$$

Hence, expanding the factor $\Phi(t,\tau)B(\tau)$ in the integrand of (35) gives

$$x_f = \sum_{k=0}^{\infty}\left[\left(\frac{\mathrm{d}}{\mathrm{d}t} - A(t)\right)^k B(t)\right]\left(\int_{t-\delta}^{t} \frac{(\tau-t)^k}{k!}\, u(\tau)\,\mathrm{d}\tau\right). \quad (38)$$

The term between the square brackets is the $(k+1)$-st block column of the infinite *instantaneous reachability matrix* of the realization defined as (with $\mathbf{D} = \frac{\mathrm{d}}{\mathrm{d}t}$)

$$\mathbf{R}_t^{(\infty)}(A,B) = [B; (A-\mathbf{D})B; (A-\mathbf{D})^2B; \cdots; (A-\mathbf{D})^nB; \cdots]_t. \quad (39)$$

Indeed, with a generalized (impulsive) input, $u(\tau) = \sum u_i\delta^{(i)}(\tau - t)$, the state jumps instantaneously by the amount $\Delta x|_t = \mathbf{R}_t^{(\infty)}(A,B)\ [u_0^T; u_1^T; \ldots; u_n^T; \cdots]^T$. It follows that the reachability Gramian can be expressed in terms of the reachability matrix via its expansion around t.

$$\mathcal{R}_t = \mathbf{R}_t^{(\infty)}(A,B)\,\Delta_{\delta,m}^{(\infty)}\,\mathbf{R}_t^{(\infty)}(A,B)^T, \quad (40)$$

where $\Delta_{\delta,m}^{(\infty)}$ is the symmetric infinite dimensional matrix with (i,j)-th block entry

$$\left[\Delta_{\delta,m}^{(\infty)}\right]_{ij} = \int_0^{\delta} \frac{\theta^{i+j-2}}{(i-1)!(j-1)!}\,\mathrm{d}\theta\, I_m = \frac{\delta^{i+j-1}}{(i+j-1)(i-1)!(j-1)!}\, I_m. \quad (41)$$

The first N terms in the series of $\mathcal{R}_t$ coincide with the first N terms in the series of

$$\mathbf{R}_t^{(N)}(A,B)\Delta_{\delta,m}^{(N)}\mathbf{R}_t^{(N)}(A,B)^T$$

where we define the *finite* dimensional reachability matrix as

$$\mathbf{R}_t^{(N)}(A,B) = [B; (A-\mathbf{D})B; (A-\mathbf{D})^2 B; \cdots; (A-\mathbf{D})^{N-1}B]_t. \tag{42}$$

and the $Nm \times Nm$ matrix $\Delta_{\delta,m}^{(N)}$ as

$$\left[\Delta_{\delta,m}^{(N)}\right]_{ij} = \frac{\delta^{i+j-1}}{(i+j-1)(i-1)!(j-1)!} I_m. \tag{43}$$

Likewise, for small δ, the observability Gramian is factored as

$$\mathcal{O}_t = \mathbf{O}_t^{(\infty)}(A,C)^T \Delta_{\delta,p}^{(\infty)} \mathbf{O}_t^{(\infty)}(A,C), \tag{44}$$

where

$$\mathbf{O}_t^{(\infty)}(A,C) = [C^T; (A^T+\mathbf{D})C^T; (A^T+\mathbf{D})^2 C^T; \cdots; (A^T+\mathbf{D})^N C^T; \cdots]_t^T, \tag{45}$$

and subsequently approximated to N-th order by

$$\mathbf{O}_t^{(N)}(A,C) = [C^T; (A^T+\mathbf{D})C^T; (A^T+\mathbf{D})^2 C^T; \cdots; (A^T+\mathbf{D})^{N-1} C^T]_t^T. \tag{46}$$

The significance of the observability matrix $\mathbf{O}_t^{(N)}(A,C)$ lies in the fact that the successive derivatives of the output at time t, given the state at time t, is indeed determined by $[y^T, (y')^T, \ldots, (y^{(N-1)})^T] = \left[\mathbf{O}_t^{(N)}(A,C)x(t)\right]^T$.

The above approximations of small time Gramians are accomplished by (symbolic) differentiation and algebraic operations only. This greatly simplifies their computation.

Our early research on balancing for time varying systems is presented in [VK80, VK83]. Other research on time varying balancing includes [SSV83, LB03, SR04], and for periodic linear systems [LO99, V00, VH98].

3.3 SIB Time-Weighted Gramians and M-Balancing

The results in Sect. 3.1 can be generalized. In the single input single output (siso) case it is known that such balanced time varying realizations have an interesting sign symmetry [VK83]. Moreover, they are also related to the so called time-weighted balanced realizations [GA89] and the general class of M-weighted balanced realizations [VG04]. For any positive definite matrix M, consider the M-reachability and M-observability Gramians defined by

$$\mathcal{R}_M(\delta) = \mathbf{R} M \mathbf{R}^T \tag{47}$$

$$\mathcal{O}_M(\delta) = \mathbf{O}^T M \mathbf{O}. \tag{48}$$

Obviously these matrices are positive definite if (A, b, c) is minimal. Define an M-*balanced realization* as the realization for which $\mathcal{R}_M(\delta)$ and $\mathcal{O}_M(\delta)$ are equal and diagonal.

It was shown in [VG04] that any time-invariant siso M-balanced realization is sign symmetric and that conversely, all minimal sign symmetric realizations of a Markov parameter sequence are M-balanced for some appropriate M. Summarizing:

SIB **Sign Symmetry**

$$\mathcal{R}(\delta) = \mathcal{O}(\delta) = \Lambda(\delta) \quad \Longleftrightarrow \quad \begin{aligned} A^T &= SAS \\ b &= Sc^T \end{aligned}$$

$$\searrow\!\nwarrow \qquad\qquad \nearrow\!\swarrow$$

M-balanced

$$\begin{aligned} \mathcal{R}_M &= \mathbf{R}M\mathbf{R}^T \\ \mathcal{O}_M &= \mathbf{O}^T M\mathbf{O} \end{aligned}$$

Since for time weighted Gramians the reachability Gramian associated with the problem of minimizing an energy

$$\mathcal{E}_i = \int_0^\infty \gamma^2(t)u^2(t)\,\mathrm{d}t, \tag{49}$$

satisfies $\mathcal{R}_\gamma = \mathbf{R}M_\gamma\mathbf{R}^T$, M-weighted balanced realizations can be interpreted as time weighted balanced realizations and vice versa. For stable systems, balancing in the sense of Moore is already well approximated by SIB if δ is about twice the characteristic time (largest time constant or oscillation period) of the system. This has interesting repercussions for the numerical methods applicable to obtain the balanced realization and its subsequent use in model reduction.

4 Nonlinear Balancing

Because of the successes in model reduction via POD of balanced realizations in the linear case, it is obvious to try to extend the method to nonlinear systems. Since nonlinearity refers more to the absence of a property rather than to a property, we shall restrict our analysis in what follows to systems of the form

$$\begin{aligned} \dot{x} &= F(x, u) \\ y &= H(x, u) \end{aligned}$$

where F and H are sufficiently *smooth*, in order to retain some structure. Potential time variance can be dealt with by augmenting the state with, x_{time}, for which the dynamics is $\dot{x}_{\text{time}} = 1$.

As a starting point we shall assume that some nominal input has been chosen, for instance by feedback over a nonlinear map $u = k(x)$. Now we consider small input perturbations about this nominal solution. In addition, we consider only output maps that do not depend explicitly on the input u. Thus, the system is in the affine form

$$\dot{x} = f(x) + g(x)u \tag{50}$$
$$y = h(x). \tag{51}$$

In this model, the maps f, g and h are considered smooth.

4.1 NLB: Tenets of Nonlinear Balancing

We propose three principles on which we shall base balancing:

1. **Nominal Flow**
   Balancing should be defined for a *perturbation* system with respect to some *nominal flow*, as opposed to a single equilibrium point (for a linear system the two coincide).
2. **Perturbation → short time**
   Only small perturbations are permitted for the linear variational equation (the perturbation system) along the nominal trajectory to remain sufficiently accurate. This motivates the use of short time SIB-like balancing.
3. **Commutation: balancing/linearization**
   It seems reasonable to strive for the following commutative diagram:

$$\begin{array}{ccc} (f, g, h) & \overset{\text{global balancing}}{\longrightarrow} & (\widehat{f}, \widehat{g}, \widehat{h}) \\ \downarrow \textit{linearization} & & \textit{linearization} \downarrow \\ (A(t), B(t), C(t)) & \overset{\text{local balancing}}{\longrightarrow} & (\widehat{A}(t), \widehat{B}(t), \widehat{C}(t)) \end{array} \tag{52}$$

4.2 Linear Variational Model

Since balancing for linear systems is well understood, the approach towards nonlinear balancing should proceed via the balancing of the linear variational system. For this reason, we shall first develop the equivalent for the small time reachability and observability Gramians in the nonlinear case.

Let the nonlinear system (51) have the nominal flow (for $u = 0$), denoted $x^o(t)$, which satisfies

$$\dot{x^o} = f(x^o) \tag{53}$$
$$y^o = h(x^o). \tag{54}$$

With $\widetilde{x} = x - x^o$, and a 'processed' output $\widetilde{y} = y - h(x^o)$, the linear variational equations, evaluated along the nominal trajectory, are

$$\dot{\widetilde{x}} = \left.\frac{\partial f}{\partial x}\right|_{x^O} \widetilde{x} + g(x^o)u \tag{55}$$
$$\widetilde{y} = \left.\frac{\partial h}{\partial x}\right|_{x^O} \widetilde{x}, \tag{56}$$

and the resulting model is thus the *linear time-varying* system.

$$\dot{\widetilde{x}} = A(t)\widetilde{x} + B(t)u \tag{57}$$

$$\widetilde{y} = C(t)\widetilde{x}, \tag{58}$$

where $(A(t), B(t), C(t))$ is the triplet $\left(\left.\frac{\partial f}{\partial x}\right|_{x^o(t)}, g(x^o(t)), \left.\frac{\partial h}{\partial x}\right|_{x^o(t)} \right)$. The machinery developed in Sect. 3 can now be used. This also explains our second principle. We cannot let the interval length δ for observation and reaching become too large, because the linearization might stray too far from the exact result.

What is the meaning of this linear variational system? For the smooth nonlinear system (51) denote the state space (where x lives) by the manifold $\mathbf{M}$. At each point x of $\mathbf{M}$, the linear variational state $\widetilde{x}$ lives in the tangent space $\mathbf{T}_x\mathbf{M}$ to $\mathbf{M}$ at x. The nominal flow can be envisioned by the flow lines on $\mathbf{M}$. Except at equilibria, these flow lines do not intersect. If $x \in \mathbf{M}$ is the nominal state at time t, we shall again speak of the *event* (x, t).

Consider a dynamical system evolving with a differentiable manifold as state space. The nominal (autonomous) system is characterized by the map $f : \mathbf{M} \to \mathbf{TM}$ where, as usual, $\mathbf{TM} = \{\mathbf{T}_x\mathbf{M} | x \in \mathbf{M}\}$ denotes the tangent bundle. We assume that the system is smooth, so that f is differentiable. Thus we get (for $x \in \mathbf{M}$)

$$\dot{x} = f(x). \tag{59}$$

Denote by $\Psi_t^f(x)$ the flow of the vector field f, i.e., the smooth function of t and x with the property that $x^o(t) = \Psi_t^f(x)$ solves the ordinary differential equation $\dot{x^o} = f(x^o)$ with initial condition $x^o(0) = x$. In other words, $\Psi_t^f(x)$ is a smooth function of t and x satisfying the evolution equation[2], i.e.,

$$\frac{\partial}{\partial t}\Psi_t^f(x) = f(\Psi_t^f(x)), \quad \Psi_0^f(x) = x. \tag{60}$$

We point out that since we assumed that the global nonlinear system is time invariant, all properties that hold for the event (x, t), may be shifted to the event $(x, 0)$.

The vector field f induces another map

$$\widehat{f} : \mathbf{M} \to C^1\left([-\delta, \delta]; \widetilde{L}(\mathbf{TM}, \mathbf{TM})\right), \tag{61}$$

defining for each state a map from $[-\delta, \delta] \subset \mathbb{R}$ to the set of maps from the tangent bundle to itself. In particular, with some abuse of notation, by reusing $\widetilde{L}$,

$$\widehat{f} : x \mapsto \widehat{f}(x) \in C^1\left([-\delta, \delta]; \widetilde{L}(\mathbf{T}_x\mathbf{M}, \mathbf{TM})\right). \tag{62}$$

[2] Usually this is denoted by $\Phi_t^f(x)$ [I89], but Φ is here reserved for the linear system transition matrix.

For each $t \in [-\delta, \delta]$, this defines a linear map

$$[\widehat{f}(x)](t) \in \widetilde{L}(\mathbf{T}_x\mathbf{M}, \mathbf{T}_{\Psi_t^f(x)}\mathbf{M}), \tag{63}$$

with its action on the tangent vector (a bound vector) $(x, \xi) \in \mathbf{T}_x\mathbf{M}$ defined by

$$\left[[\widehat{f}(x)](t)\right](x, \xi) = \left(\Psi_t^f(x), \Phi_x^f(t, 0)\xi\right). \tag{64}$$

In (64), $\Phi_x^f(t, \tau)$ is the transition matrix of the linear variational system associated with the nominal trajectory through x at time $t = 0$, and thus satisfies

$$\frac{\partial}{\partial t}\Phi_x^f(t, 0) = \left.\frac{\partial f}{\partial x}\right|_{\Psi_t^f(x)} \Phi_x^f(t, 0), \quad \Phi_x^f(0, 0) = I. \tag{65}$$

Observe that $\widetilde{L}$ is linear in its second argument. In fact, the maps $[\widehat{f}(x)](t)$ depict the natural evolution of the tangent space along the nominal flow with initial condition at x.

Note that since f is differentiable, its Jacobian exists and is continuous. Hence Φ_x^f is differentiable for all x. Also, $x^o(t) = \Psi_t^f(x)$ is twice differentiable with respect to t. Hence $\widehat{f}(x)$ in (62) is differentiable, justifying that the map is indeed in C^1.

Since the original nonlinear system is time invariant, with respect to state $x \in \mathbf{M}$, we may shift time and consider the nominal trajectory passing through x at time $t = 0$. We treat the nonlinear reachability problem as follows: For $\delta \in \mathbb{R}$ and $\widetilde{x}_f \in \mathbb{R}^n$ sufficiently small, associate with the event $\left(\psi_{-\delta}^f(x), -\delta\right) \in \mathbf{M} \times \mathbb{R}$, the event $(0, -\delta) \in \mathbf{T}_{\psi_{-\delta}^f(x)}\mathbf{M} \times \mathbb{R}$, and with $(x, 0) \in \mathbf{M} \times \mathbb{R}$ associate $(\widetilde{x}_f, 0) \in \mathbf{T}_x\mathbf{M} \times \mathbb{R}$.

We will further on use some standard tools from differential geometry, see [I89]. The first is the **Lie-product**: $[f, g] = \frac{\partial g}{\partial x} f - \frac{\partial f}{\partial x} g$. We set also $\mathrm{ad}_f g = [f, g]$ and define by iteration,

$$\mathrm{ad}_f^k g = [f, \mathrm{ad}_f^{k-1} g]. \tag{66}$$

The **Gradient** d of a vector function stands for $[\mathrm{d}f]_{ij} = \frac{\partial f_i}{\partial x_j}$. The gradient of a map $f : \mathbb{R}^n \to \mathbb{R}^n$ is referred to as the *Jacobian* $\mathrm{d}f = \frac{\partial f}{\partial x}$.

Finally, the **Lie-derivative** $L_f h$ is the *directional derivative* $\frac{\partial h}{\partial x} f$.

4.3 Nonlinear Local Reachability

Consider now the transition from the event $(0, -\delta) \in \mathbf{T}_{\Psi_{-\delta}^f(x)}\mathbf{M} \times \mathbb{R}$ to the event $(\widetilde{x}_f, 0) \in \mathbf{T}_x\mathbf{M} \times \mathbb{R}$ as an *approximation* of the nonlinear transition from $\left(\psi_{-\delta}^f(x), -\delta\right) \in \mathbf{M} \times \mathbb{R}$ to $(x + \widetilde{x}_f, 0) \in \mathbf{M} \times \mathbb{R}$ (the latter notation with some abuse of notation: addition may not be defined in $\mathbf{M}$). From *linear* system theory we know that, with $(A(t), B(t))$ the variational system associated with $(x, 0)$, an input, $u(\tau), -\delta < \tau < 0$ is required that satisfies

$$\widetilde{x}_f = \int_{-\delta}^{0} \Phi_x^f(0, \tau) B(\tau)\, u(\tau)\, \mathrm{d}\tau. \tag{67}$$

Note that

$$\frac{\partial}{\partial \tau}\Phi_x^f(0,\tau)B(\tau) = \Phi_x^f(0,\tau)\left(\frac{\partial}{\partial \tau} - A(\tau)\right)B(\tau) \tag{68}$$

and by iteration

$$\left(\frac{\partial}{\partial \tau}\right)^k \Phi_x^f(0,\tau)B(\tau) = \Phi_x^f(0,\tau)\left(\frac{\partial}{\partial \tau} - A(\tau)\right)^k B(\tau). \tag{69}$$

Hence, expanding the factor $\Phi_x^f(0,\tau)B(\tau)$ in the integrand about $\tau = 0$ gives

$$\widetilde{x}_f = \int_{-\delta}^{0} \sum_{k=0}^{\infty} \left[\left(\frac{\partial}{\partial t} - A(t)\right)^k B(t)\right]_{t=0} \frac{\tau^k}{k!}\, u(\tau)\, \mathrm{d}\tau. \tag{70}$$

In reference to the original nonlinear system we obtain

$$\begin{aligned}
\left(\frac{\partial}{\partial t} - A(t)\right) B(t) &= \frac{\partial}{\partial t} g(\Psi_t^f(x)) - \left.\frac{\partial f}{\partial x}\right|_{\Psi_t^f(x)} g(\Psi_t^f(x)) \\
&= \left.\frac{\partial g}{\partial x}\right|_{\Psi_t^f(x)} f(\Psi_t^f(x)) - \left.\frac{\partial f}{\partial x}\right|_{\Psi_t^f(x)} g(\Psi_t^f(x)) \\
&= \left.\mathrm{ad}_f g\right|_{\Psi_t^f(x)}.
\end{aligned}$$

Likewise, iteration and evaluation at $t = 0$ gives for all $k > 0$

$$\left[\left(\frac{\partial}{\partial t} - A(t)\right)^k B(t)\right]_{t=0} = \left.\mathrm{ad}_f^k g\right|_x. \tag{71}$$

Substituting (71), the variational equation (70) gives

$$\widetilde{x}_f = \int_{-\delta}^{0} \sum_{k=0}^{\infty} \frac{\tau^k}{k!} \left[\mathrm{ad}_f^k g\right]_x u(\tau)\, \mathrm{d}\tau. \tag{72}$$

Defining an ad-*exponential* formally by

$$\left.e^{t\,\mathrm{ad}_f} g\right|_x = \sum_{k=0}^{\infty} \frac{t^k}{k!} \left[\mathrm{ad}_f^k g\right]_x, \tag{73}$$

we obtain the simple representation

$$\widetilde{x}_f = \int_{-\delta}^{0} \left.e^{\tau\,\mathrm{ad}_f} g\right|_x u(\tau)\, \mathrm{d}\tau. \tag{74}$$

Linear system theory, discussed in Sect. 1, tells us that the optimal (in the sense of minimizing the L_2-norm of u) solution is

$$u^*(\tau) = \left(e^{\tau\,\mathrm{ad}_f} g\right)_x^T \left[\mathcal{R}_{f,g}^{(\delta)}(x)\right]^{-1} \widetilde{x}_f, \tag{75}$$

where $\mathcal{R}^{(\delta)}_{f,g}(x)$ is the δ-Gramian of the nonlinear system (f, g) at the point x in the state space, and is defined explicitly by

$$\mathcal{R}^{(\delta)}_{f,g}(x) = \int_{-\delta}^{0} \left(e^{\theta\, \mathrm{ad}_f} g\right)_x \left(e^{\theta\, \mathrm{ad}_f} g\right)_x^T \, d\theta. \tag{76}$$

Note that the right hand side is indeed independent of t. All that is required is that the nominal value of the state at the end of the control (when $\widetilde{x} = \widetilde{x}_f$) is x.

Recall that the response to an impulsive input for a linear system $u(t)=\sum_{i=0}^{\infty} u_i \delta^{(i-1)}(t-\tau)$ is a jump in the state at time τ, given by

$$\Delta x_\tau = \mathbf{R}^{(\infty)}_{f,g}(x)\, u, \qquad u = \begin{bmatrix} u_1 \\ u_2 \\ \vdots \end{bmatrix}, \tag{77}$$

where $\mathbf{R}^{(\infty)}_{f,g}(x) = [g, \mathrm{ad}_{-f} g, \ldots, \mathrm{ad}^{n-1}_{-f} g \ldots]_x$ is the local infinite dimensional *reachability matrix* evaluated at $x(\tau)$. This follows from (39) and the fact that

$$(A_x - \mathbf{D})B_x = \frac{\partial f}{\partial x} g - \frac{\partial g}{\partial x} f = [-f, g]. \tag{78}$$

Remark 1. What is solved is the reachability problem of the flow linearized equation, and not the reachability problem for the original nonlinear system. Of course, for small δ and $\widetilde{x}_f$ this solution may be a relatively good *approximation*. We refer to papers by Gilbert [G77], Lesiak and Krener [LK78], Monaco and Normand-Cyrot [MNC84], and Fliess [FLL83], for Volterra series expressions for the solution of the nonlinear system. Furthermore, the control input we considered is actually a perturbation of a nominal input (here considered to be fixed as open or closed loop). Hence, minimizing the energy of the deviation is also not the same as minimizing the total energy.

Example 1. Consider the nonlinear system $\dot{x} = x^2 + (x+1)u$. The nominal flow for $u \equiv 0$ is given by $\dot{x^o} = (x^o)^2$ with solution passing through x at $t = 0$,

$$x^o(t) = \frac{x}{1 - xt}. \tag{79}$$

If $x > 0$, the system has a finite escape time $t_{\mathrm{esc}} = \frac{1}{x} > 0$. The linear variational system associated with the event $(0, x)$ is

$$\begin{aligned} \dot{\widetilde{x}}(t) &= 2x^o(t)\,\widetilde{x}(t) + (x^o(t) + 1)u(t) \\ &= \frac{2x}{1 - xt}\widetilde{x}(t) + \frac{1 + x(1-t)}{1 - xt} u(t). \end{aligned}$$

Thus $A(t) = \frac{2x}{1-xt}$ and $b(t) = \frac{1+x(1-t)}{1-xt}$. The transition matrix is

$$\Phi_x(t, \tau) = \frac{(1 - x\tau)^2}{(1 - xt)^2}. \tag{80}$$

We find for this system $f(x) = x^2$ and $g(x) = x + 1$ and

$$\begin{aligned}
\mathrm{ad}_f g &= x^2 - 2x(x+1) = -x^2 - 2x \\
\mathrm{ad}_f^2 g &= -(2x+2)x^2 - 2x(-x^2 - 2x) = 2x^2 \\
\mathrm{ad}_f^3 g &= 4x^3 - 4x^3 = 0.
\end{aligned}$$

Hence, $\mathrm{e}^{t\,\mathrm{ad}_f} g = 1 + x - t(x+2)x + t^2x^2$, which coincides with $\Phi_x(t,\tau)b(\tau)$. The most reachable direction in the state space corresponds to the eigen direction associated with the largest eigenvalue of the reachability Gramian.

4.4 Nonlinear Local Observability

In this section we consider the observability problem for the system (51), with the nominal input $u = 0$. Denote the nominal state by $x^o(t)$. The solution of the nominal system may be expanded in the Lie-exponential [I89]. Hence, for a nominal trajectory passing through x at time $t = 0$, we get $x^o(t) = \Psi_t^f(x) = \sigma_x \mathrm{e}^{tL_f} I$, where we denote the identity function $I(x) = x$ by I, and σ_x is the *evaluator functional* at x. This was established in [FLL83, MNC85].

Example 2. The nominal system $\dot{x}^o = (x^o)^\alpha$ is Lipshitz for $\alpha \geq 1$. For $\alpha = 1$, it simply has linear dynamics. For $\alpha > 1$, the equation is readily integrated from the initial condition $x^o(0) = x$, and gives

$$x^o(t) = \left[x^{1-\alpha} + (1-\alpha)t\right]^{1/1-\alpha}. \tag{81}$$

Constructing the Lie series, we find $L_f I = x^\alpha$, $L_f^2 I = \alpha x^{2\alpha-1}$, $L_f^3 I = \alpha(2\alpha - 1)x^{3\alpha-2}$. At this point we may conjecture

$$\sigma_x L_f^k I = \alpha(2\alpha-1)(3\alpha-2)\cdots((k-1)\alpha - (k-2))x^{k\alpha-(k-1)}. \tag{82}$$

This is easily proven with induction. It follows that the Lie-exponential is

$$\mathrm{e}^{tL_f} I = x\left[1 + \sum_{k=1}^{\infty} x^{(\alpha-1)k}(k-1)! \prod_{i=0}^{k-2}\left(\alpha - \frac{i}{i+1}\right)\right]. \tag{83}$$

The product can be expressed in closed form as

$$\sigma_x L_f^k I = x\left[1 + \sum_{k=1}^{\infty} (tx^{\alpha-1})^k(\alpha-1)^{k-1} \frac{\Gamma\left(k-1+\frac{\alpha}{\alpha-1}\right)}{\Gamma(k+1)\Gamma\left(\frac{\alpha}{\alpha-1}\right)}\right]. \tag{84}$$

The latter expression simplifies to the expression for $x(t)$ in (81).

The nominal output is given by $h(x^o(t))$, but this expression is not so useful because it is not a series expansion in t. We establish the following concise representation:

Theorem 1. *The output of the autonomous analytic system* (f, h) *with initial condition* $x^o(0) = x$ *is given for all t less than some* $T > 0$ *(an escape time may exists) by*

$$h(x^o(t)) = h\left(\Psi_t^f(x)\right) = \sigma_x \, \mathrm{e}^{tL_f} h. \tag{85}$$

Proof: follows from the Fliess series expansion [FLL83]. ◇

In [KT93], the series representation (85) was applied to sampled data systems. Since $x^o(t) = \sigma_x \mathrm{e}^{tL_f} I$, the left hand side of (85) is $\sigma_{(\sigma_x \mathrm{e}^{tL_f} I)} h$. By theorem 1 this telescoping series equals $\sigma_x \mathrm{e}^{tL_f} h$. The energy in the output over an interval of duration δ for the nominal nonlinear system is

$$\mathcal{E}_y(\delta, x) = \int_0^\delta \left(\sigma_x \mathrm{e}^{tL_f} h\right) \left(\sigma_x \mathrm{e}^{tL_f} h\right)^T \, \mathrm{d}t. \tag{86}$$

Unlike for a linear system, this is not a quadratic form in x, despite the similarity of (86) to a Gramian.

Consider now the linear variational system (65) for $u \equiv 0$ along the arbitrary nominal trajectory associated with the event $(x, 0) \in \mathbf{M} \times \mathbb{R}$. For sufficiently small $\delta \in \mathbb{R}$ and $\widetilde{x}_0 \in \mathbb{R}^n$ consider the initial event $(\widetilde{x}_0, 0) \in \mathbf{T}_x\mathbf{M} \times \mathbb{R}$. Linear system theory gives that this initial perturbation at time 0 induces the events $(\widetilde{x}(t), t) \in \mathbf{T}_{\psi_x^f(t)}\mathbf{M} \times \mathbb{R}$, where

$$\widetilde{x}(t) = \Phi_x^f(t, 0)\widetilde{x}_0, \tag{87}$$

and hence

$$\widetilde{y}(t) = \left.\frac{\partial h}{\partial x}\right|_{x^o(t)} \Phi_x^f(t, 0)\, \widetilde{x}_0. \tag{88}$$

As discussed in Sect. 1, a measure for the observability is given by a quadratic form with the (time-varying) observability δ- Gramian as weight. Hence,

$$\mathcal{O}_{f,h}^{(\delta)}(x) = \int_0^\delta \left(\left.\frac{\partial h}{\partial x}\right|_{x^o(\tau)} \Phi_x^f(\tau, 0) \right)^T \left(\left.\frac{\partial h}{\partial x}\right|_{x^o(\tau)} \Phi_x^f(\tau, 0) \right) \mathrm{d}\tau. \tag{89}$$

From the properties of the transition matrix,

$$\left(\frac{\partial}{\partial \tau}\right) \Phi_x^f(\tau, t)^T C^T(\tau) = \Phi_x^f(\tau, 0)^T \left(\frac{\mathrm{d}}{\mathrm{d}\tau} + A^T(\tau)\right) C^T(\tau). \tag{90}$$

and by iteration, it follows that

$$\left(\frac{\partial}{\partial \tau}\right)^k \Phi_x^f(\tau, t)^T C^T(\tau) = \Phi_x^f(\tau, t)^T \left(\frac{\mathrm{d}}{\mathrm{d}\tau} + A^T(\tau)\right)^k C^T(\tau). \tag{91}$$

Hence, a Taylor expansion gives

$$\Phi_x^f(\tau, t)^T C^T(\tau) = \sum_{k=0}^{\infty} \frac{(\tau - t)^k}{k!} \left.\left(\frac{\mathrm{d}}{\mathrm{d}\tau} + A^T(\tau)\right)^k C^T(\tau)\right|_{\tau = t}. \tag{92}$$

Note that since $C(t) = \frac{\partial h}{\partial x}|_x$ is the gradient of h at x (also denoted as $\mathrm{d}h$) it follows that,

$$\left(\frac{\mathrm{d}}{\mathrm{d}t} + A^T(t)\right) C^T(t) = \left(\frac{\mathrm{d}}{\mathrm{d}t} + \left(\frac{\partial f}{\partial x}\right)^T\right)\left(\frac{\partial h}{\partial x}\right)^T$$
$$= \mathrm{d}\, L_f h.$$

One easily shows by induction that

$$\left(\frac{\mathrm{d}}{\mathrm{d}\tau} + A^T\right)^k C^T\Big|_{t=\tau} = \mathrm{d}\, L_f^k h, \tag{93}$$

and defining the *Lie-exponential* (operator) in an obvious way as

$$\mathrm{e}^{tL_f} = \sum_{k=0}^{\infty} \frac{t^k}{k!} L_f^k, \tag{94}$$

we get finally

$$\Phi_x^f(\tau,0)^T C^T(\tau) = \sum_{k=0}^{\infty} \frac{\tau^k}{k!}\ \sigma_x\, \mathrm{d}L_f^k h = \sigma_x\, \mathrm{d}\,\mathrm{e}^{L_f} h. \tag{95}$$

This yields the following expression for the observability δ-Gramian, $\mathcal{O}_{f,h}^{(\delta)}(x)$, for the nonlinear system (f, h) at the point x in the state space,

$$\mathcal{O}_{f,h}^{(\delta)}(x) = \int_0^\delta \left(\mathrm{d}\,\mathrm{e}^{\theta\, L_f} h\right)_x^T \left(\mathrm{d}\,\mathrm{e}^{\theta\, L_f} h\right)_x \mathrm{d}\theta. \tag{96}$$

Note that the right hand side is indeed independent of t. All that is required is that the nominal value of the state at the onset of the observation (when $\widetilde{x} = \widetilde{x}_0$) is x.

Note also that an initial condition x at τ gives the output and its derivatives

$$\mathcal{Y}(\tau) = \begin{bmatrix} y(\tau) \\ \dot{y}(\tau) \\ \vdots \end{bmatrix} = \mathbf{O}^{(\infty)}(\tau)x, \tag{97}$$

by virtue of (45). Hence, $\mathbf{O}_{f,h}^{(\infty)}(x) = \begin{bmatrix} \mathrm{d}h \\ \mathrm{d}\, L_f h \\ \vdots \\ \mathrm{d}L_f^{n-1} h \\ \vdots \end{bmatrix}_x$ is the local infinite dimensional *observability matrix* for the nonlinear system.

4.5 Nonlinear Local Lyapunov Equations

The nonlinear Gramians (76) and (96) derived in Sects. 4.3 and 4.4 are Gramians of a linear time-varying system. Lyapunov differential equations for them were established in [VK83].

For the observability Gramian, we obtain

$$-L_f \mathcal{O}^{(\delta)}_{f,h}(x) = \left(\frac{\partial f}{\partial x}\right)^T_x \mathcal{O}^{(\delta)}_{f,h}(x) + \mathcal{O}^{(\delta)}_{f,h}(x) \left(\frac{\partial f}{\partial x}\right)_x + \\ + \mathrm{d}h(x)^T \,\mathrm{d}h(x) \;-\; \left(\mathrm{d}\,\mathrm{e}^{\delta\, L_f} h\right)^T_x \left(\mathrm{d}\,\mathrm{e}^{\delta\, L_f} h\right)_x, \qquad (98)$$

by substituting the time varying perturbational system (87) and (88) in the Lyapunov equation (34) for a general linear time varying system.

Note that the left hand side is the Lie-derivative along the flow. If x is an equilibrium point for the flow, then $f(x_{eq}) = 0$, and consequently the Lie derivative, $L_f \mathcal{O}^{(\delta)}_{f,h}(x_{eq})$, vanishes. The nonlinear observability Gramian at the equilibrium is determined by the *algebraic* Lyapunov equation:

$$\left(\frac{\partial f}{\partial x}\right)^T_{x_{eq}} \mathcal{O}^{(\delta)}_{f,h}(x_{eq}) + \mathcal{O}^{(\delta)}_{f,h}(x_{eq}) \left(\frac{\partial f}{\partial x}\right)_{x_{eq}} + \mathrm{d}h(x_{eq})^T \,\mathrm{d}h(x_{eq}) + \\ - \left(\mathrm{d}\,\mathrm{e}^{\delta\, L_f} h\right)^T_{x_{eq}} \left(\mathrm{d}\,\mathrm{e}^{\delta\, L_f} h\right)_{x_{eq}} = 0. \qquad (99)$$

Alternatively, the Lyapunov equations may be set up directly from their definition. We illustrate this route for the reachability Gramian.

Defining the Lie derivative of a matrix as the matrix of its Lie derivatives, we get (e_i is the i-th column of the identity matrix):

$$\begin{aligned}
[L_f \mathcal{R}^{(\delta)}_{f,g}(x)]_{ij} &= \int_{-\delta}^{0} e_i' \left[L_f \left(\mathrm{e}^{\theta\, \mathrm{ad}_f} g\right)_x\right] e_j' \left(\mathrm{e}^{\theta\, \mathrm{ad}_f} g\right)_x \,\mathrm{d}\theta + \\
&\quad + \int_{-\delta}^{0} e_i' \left(\mathrm{e}^{\theta\, \mathrm{ad}_f} g\right)_x e_j' \left[L_f \left(\mathrm{e}^{\theta\, \mathrm{ad}_f} g\right)_x\right] \mathrm{d}\theta \\
&= \int_{-\delta}^{0} e_i' \left[\left(\mathrm{ad}_f + \frac{\partial f}{\partial x}\right) \left(\mathrm{e}^{\theta\, \mathrm{ad}_f} g\right)_x\right] e_j' \left(\mathrm{e}^{\theta\, \mathrm{ad}_f} g\right)_x \,\mathrm{d}\theta + \\
&\quad + \int_{-\delta}^{0} e_i' \left(\mathrm{e}^{\theta\, \mathrm{ad}_f} g\right)_x e_j' \left[\left(\mathrm{ad}_f + \frac{\partial f}{\partial x}\right) \left(\mathrm{e}^{\theta\, \mathrm{ad}_f} g\right)_x\right] \mathrm{d}\theta \\
&= e_i' \left\{\int_{-\delta}^{0} \frac{d}{d\theta} \left[\left(\mathrm{e}^{\theta\, \mathrm{ad}_f} g\right)_x \left(\mathrm{e}^{\theta\, \mathrm{ad}_f} g\right)_x^T\right] \mathrm{d}\theta\right\} e_j \; + \\
&\quad + e_i' \left\{\left(\frac{\partial f}{\partial x}\right) \mathcal{R}^{(\delta)}_{f,g}(x) + \mathcal{R}^{(\delta)}_{f,g}(x) \left(\frac{\partial f}{\partial x}\right)^T\right\} e_j,
\end{aligned}$$

and hence:

$$L_f\mathcal{R}^{(\delta)}_{f,g}(x) = \left(\frac{\partial f}{\partial x}\right)\mathcal{R}^{(\delta)}_{f,g}(x) + \mathcal{R}^{(\delta)}_{f,g}(x)\left(\frac{\partial f}{\partial x}\right)^T + \\ + g(x)g(x)^T - \left(\mathrm{e}^{-\delta\,\mathrm{ad}_f}g\right)_x\left(\mathrm{e}^{-\delta\,\mathrm{ad}_f}g\right)_x^T. \quad (100)$$

We used the properties that for any scalar functions g and h,

$$L_f(gh) = (L_f g)h + g(L_f h) \quad (101)$$

and for any vector fields f and g

$$\mathrm{ad_f}\left(\mathrm{e}^{\theta\,\mathrm{ad}_f}g\right)_x = \frac{\mathrm{d}}{\mathrm{d}\theta}\left(\mathrm{e}^{\theta\,\mathrm{ad}_f}g\right)_x \quad (102)$$

and for any vector fields f and k

$$L_f\,k_x = \left(\mathrm{ad}_f + \frac{\partial f}{\partial x}\right)k_x. \quad (103)$$

At an an equilibrium point, x_{eq}, for the nominal flow, the Lie derivative, $L_f\mathcal{R}^{(\delta)}_{f,g}(x_{eq})$, vanishes and the nonlinear reachability Gramian is determined by the solution of an *algebraic* Lyapunov equation

$$\left(\frac{\partial f}{\partial x}\right)_{x_{eq}}\mathcal{R}^{(\delta)}_{f,g}(x_{eq}) + \mathcal{R}^{(\delta)}_{f,g}(x_{eq})\left(\frac{\partial f}{\partial x}\right)^T_{x_{eq}} + g(x_{eq})g(x_{eq})^T + \\ - \left(\mathrm{e}^{-\delta\,\mathrm{ad}_f}g\right)_{x_{eq}}\left(\mathrm{e}^{-\delta\,\mathrm{ad}_f}g\right)^T_{x_{eq}} = 0. \quad (104)$$

4.6 Local Balancing

Let x be a point in the state space where the local reachability and observability matrices have full rank. The nonlinear Gramians, $\mathcal{R}^{(\delta)}_{f,g}(x)=\int_0^\delta \left(e^{\theta\,\mathrm{ad}_{-f}}g\right)_x\left(e^{\theta\,\mathrm{ad}_{-f}}g\right)_x^T d\theta$, and $\mathcal{O}^{(\delta)}_{f,h}(x)=\int_0^\delta\left(\mathrm{d}\,\mathrm{e}^{\theta\,L_f}h\right)_x^T\left(\mathrm{d}\,\mathrm{e}^{\theta\,L_f}h\right)_x\mathrm{d}\theta$, transform under similarity in the state space by a pointwise congruence. Hence a nonsingular $T(x)$ exists at x such that the transformed Gramians, at x,

$$T(x)\mathcal{R}^{(\delta)}_{f,g}(x)T^T(x),\text{ and } T(x)^{-T}\mathcal{O}^{(\delta)}_{f,h}(x)T^{-1}(x)$$

are equal and diagonal. In general this T will depend on x. Note that we expressed the reachability Gramian in a slightly different form, for reasons that will become clear in Sect. 4.7.

Thus, we arrive at a map $T:\mathbf{M}\to GL_n(\mathbb{R})$. At this point, this prompts three questions:

1. Can $T(x)$ be computed in an efficient way?
2. Can the pointwise defined transformation $T(x)$, which is a similarity transformation on $\mathbf{T}_x\mathbf{M}$ (hence local), be extended to a global transformation on $\mathbf{M}$?
3. What can be inferred from *local* balancing?

We note that in answer to the first question, balancing proceeds via computation of the Gramians. This involves either explicit computation through integration of the Lie-series and the ad-series. If these series are computed symbolically, then only in exceptional cases the series terminate or have a nice explicit form. In a general symbolic computation, the number of terms soon becomes excessive. Alternatively, the Lyapunov equations may be used to derive the solution. This however also requires the solution of a PDE. In the next section we propose a simple *approximation* to the problem, which allows for simple computations as only differentiations and matrix algebra are involved.

We defer the second problem to Sect. 5.

As to the third question, we note that the information comprised in the local balancing transformation $T(x)$, is already very useful. Indeed from the nature of the problem, the local balancing tells us which directions *in the tangent space*, $\mathbf{T}_x\mathbf{M}$, are important from a natural point of view (the balanced point of view, symmetrizing local reachability and observability). In the locally balanced coordinates, the canonical Gramian $\Lambda(x)$ is therefore an *importance measure* for the directions in the tangent space, as illustrated in Sect. 1. But by nature of the problem, this corresponds to the directions in the state space $\mathbf{M}$ at the nominal point $x \in \mathbf{M}$ under consideration. The input-output properties of the nonlinear system may therefore neglect this unimportant state component, at least locally in space (neighborhood of x, and time balanced reduction only justified for times less than δ). The canonical Gramian $\Lambda(x)$ shows this importance for each *principal direction* at x. Furthermore, points and directions on the tangent space are mapped back to the controlled trajectories on the manifold.

In order to illustrate the method, let us consider the two-dimensional torus as example. Assume that the coordinate lines on the torus are the ones following the lines of curvature. Toroidal surfaces are the only ones for which a single global coordinate patch exists [M67]. Assume that the nominal flow is such that the trajectories wind around the torus. Assume also, for simplicity, that the canonical Gramian at each point coincides with the coordinate direction in the meridian plane, but shows negligible dynamics in the plane spanned by the symmetry axis and the radial direction (imagine the spoke of a bicycle wheel). Obviously, in this case the local balancing is trivially extendable in a consistent way to the global state space. Furthermore, the 1-D reduced model would then essentially model the motion in the plane of the spokes but neglect dynamics winding around the tube, even though he nominal trajectory winds around in the full order model. A problem requiring several coordinate charts is given in [GV06a].

4.7 Approximate Local Reachability and Observability

We defined the infinite *reachability matrix* by

$$\mathbf{R}_{f,g}^{(\infty)}(x) = \Big[g, \mathrm{ad}_{-f}\, g, \mathrm{ad}_{-f}^2\, g, \ldots, \mathrm{ad}_{-f}^k\, g, \ldots\Big]_x , \tag{105}$$

and the reachability Gramian for the linear variational system may be expressed via the series expansion of the integrand. Thus,

$$\mathcal{R}_{f,g}^{(\delta)}(x) = \int_{-\delta}^{0} \left[\sum_{k=0}^{\infty} \frac{\theta^k}{k!} \left(\mathrm{ad}_f^k\, g\right)_x\right] \left[\sum_{k=0}^{\infty} \frac{\theta^k}{k!} \left(\mathrm{ad}_f^k\, g\right)_x\right]^T \mathrm{d}\theta \tag{106}$$

$$= \sum_{k=0}^{\infty}\sum_{\ell=0}^{\infty} \left(\mathrm{ad}_{-f}^k\, g\right)_x \left[\int_0^{\delta} \frac{\theta^k}{k!}\frac{\theta^\ell}{\ell!}\,\mathrm{d}\theta\right] \left(\mathrm{ad}_{-f}^\ell\, g\right)_x^T. \tag{107}$$

Defining the symmetric infinite-dimensional block matrix with (i, j)-th block entry

$$\left[\Delta_{\delta,m}^{(\infty)}\right]_{ij} = \int_0^{\delta} \frac{\theta^{i+j-2}}{(i-1)!(j-1)!}\,\mathrm{d}\theta\, I_m = \frac{\delta^{i+j-1}}{(i+j-1)(i-1)!(j-1)!} I_m, \tag{108}$$

we obtain a symmetric factorization of the reachability Gramian in terms of the reachability matrix

$$\mathcal{R}_{f,g}^{(\delta)}(x) = \mathbf{R}_{f,g}^{(\infty)}(x)\Delta_{\delta,m}^{(\infty)}\mathbf{R}_{f,g}^{(\infty)}(x)^T. \tag{109}$$

Note that for a linear time invariant system with $f(x) = Ax$ and $g(x) = b$ we get $\mathrm{ad}_{-f}\, g = Ab$ and by iteration $\mathrm{ad}_{-f}^k\, g = A^k b$ so that the reachability matrix reduces to the known (but extended) reachability matrix for the pair (A, b)

$$\mathbf{R}_{f,g}^{(\delta)}(x) = [b, Ab, A^2 b, \cdots, A^k b, \cdots] = \mathbf{R}^{(\infty)}(A, b). \tag{110}$$

It follows from

$$\mathcal{R}_{f,g}^{(\delta)}(x) = \sum_{k=0}^{\infty}\sum_{\ell=0}^{\infty} \left(\mathrm{ad}_{-f}^k\, g\right)_x \left(\mathrm{ad}_{-f}^\ell\, g\right)_x^T \frac{\delta^{k+\ell+1}}{(k+\ell+1)k!\ell!} \tag{111}$$

$$= \sum_{n=0}^{\infty} \left[\sum_{k=0}^{\infty} \binom{n}{k} \left(\mathrm{ad}_{-f}^k\, g\right)_x \left(\mathrm{ad}_{-f}^{n-k}\, g\right)_x^T\right] \frac{\delta^{n+1}}{(n+1)!} \tag{112}$$

that the first N terms in the series of $\mathcal{R}_{f,g}^{(\delta)}(x)$ coincide with the first N terms in the series of $\mathbf{R}_{f,g}^{(N)}(x)\Delta_{\delta,m}^{(N)}\mathbf{R}_{f,g}^{(N)}(x)^T$, where we introduced the *finite* dimensional reachability matrix

$$\mathbf{R}_{f,g}^{(N)}(x) = \left[g, \mathrm{ad}_{-f}\, g, \mathrm{ad}_{-f}^2\, g, \ldots, \mathrm{ad}_{-f}^{N-1}\, g\right]_x \tag{113}$$

and the $N \times N$ block matrix $\Delta^{(N)}(\delta)$

$$\left[\Delta_{\delta,m}^{(N)}\right]_{ij} = \frac{\delta^{i+j-1}}{(i+j-1)(i-1)!(j-1)!} I_m. \tag{114}$$

Hence,

$$\mathcal{R}_{f,g}^{(\delta)}(x) = \mathbf{R}_{f,g}^{(N)}(x)\Delta_{\delta,m}^{(N)}(\mathbf{R}_{f,g}^{(N)})^T(x) + o(\delta^N), \tag{115}$$

which defines the N-th degree approximation of the reachability Gramian, $\mathcal{R}_{f,g}^{(\delta)}(x)$, as

$$\mathcal{R}_{f,g}^{(\delta,N)}(x) \stackrel{\text{def}}{=} \mathbf{R}_{f,g}^{(N)}(x)\Delta_{\delta,m}^{(N)}(\mathbf{R}_{f,g}^{(N)})^T(x). \tag{116}$$

It is precisely this property that makes the approximation interesting. Whereas the infinite Gramians must indeed be computed either by doing the explicit integration (requiring first computation of the ad_f-exponential), or by solving a partial differential equation (the Lyapunov equation), the approximation only requires a computation of a finite reachability matrix. This is much simpler as it only involves differential operations. We may get an idea of the quality of the approximation by investigating the LTI case (e_i denotes the i-th column of the identity matrix).

Theorem 2. *The exact reachability Gramian for the LTI system $\dot{x} = Ax + bu$ is given by*

$$\int_0^\delta \mathrm{e}^{A\theta} bb' \mathrm{e}^{A'\theta}\,\mathrm{d}\theta = \mathbf{R}(A,b)\mathcal{R}_\delta \mathbf{R}^T(A,b), \tag{117}$$

where

$$\mathcal{R}_\delta = \int_0^\delta \mathrm{e}^{A_{re}\theta} e_1 e_1^T \mathrm{e}^{A_{re}^T\theta}\,\mathrm{d}\theta \tag{118}$$

is the reachability Gramian of the similar reachability canonical form, and only depends on the coefficients of the characteristic polynomial of A.

Proof: First consider $\mathrm{e}^{At}b$. This may be expressed as $\mathbf{R}(A,b)\psi(t)$. Taking the time derivative of both expressions, we find $A\mathbf{R}(A,b)\psi(t) = \mathbf{R}(A,b)\dot{\psi}(t)$. Hence $\dot{\psi}(t) = \mathbf{R}^{-1}(A,b)A\mathbf{R}(A,b)\psi(t)$. But $\mathbf{R}^{-1}(A,b)$ is the similarity transformation that maps the pair (A,b) to the reachability canonical form (A_{re}, b_{re}). Since also $\psi(0) = b_{re} = e_1$, we find by uniqueness of the solution of a linear ODE that $\psi(t) = \mathrm{e}^{A_{re}t}e_1$ and the theorem follows. ◇

Hence it suffices to compare $\mathcal{R}_\delta$ to the approximation $\Delta_\delta^{(n)}$.

Theorem 3. *The integrand of the reachability Gramian for a single input LTI system in the reachability canonical form is given by the the outer product of the inverse Laplace transform of its Leverrier polynomials.*

Before we prove this theorem, we first establish a lemma.

Lemma 1. *The adjunct matrix* Adj $(sI - A_{re})$*, with A_{re} a companion matrix in the* reachability *canonical form, is*

$$\left[l_1(s), l_2(s), \ldots, l_n(s)\right]^T \left[1, s, s^2, \ldots, s^{n-1}\right] - l_0(s)T_{\text{upper}}(s). \tag{119}$$

The terms $l_i(s)$ denote the Leverrier polynomials, associated with $a(s) = \det(sI - A_{re})$

$$\begin{aligned} l_1(s) &= s^{n-1} + a_1 s^{n-2} + a_2 s^{n-3} + \cdots + a_{n-1} \\ l_2(s) &= s^{n-2} + a_1 s^{n-3} + a_2 s^{n-4} + \cdots + a_{n-2} \\ &\vdots \\ l_{n-1}(s) &= s + a_1 \\ l_n(s) &= 1 \end{aligned}$$

Proof: Multiplication of (119) with $sI - A_{re}$ gives the right hand side $a(s)I$. ◇

Note that $l_0(s) = a(s)$, the characteristic polynomial of A_{re}, and $T_{\text{upper}}(s)$ is an upper triangular Toeplitz matrix with first row $[0, 1, s, s^2, \ldots, s^{n-1}]$. The vector $\mathrm{l}(s) = [l_1(s), \ldots, l_n(s)]^T$ is expressible as $\mathrm{l}(s) = T_{\text{upper}}(a)\left[s^{n-1}, \ldots, 1\right]^T$, with $T_{\text{upper}}(a)$ upper triangular Toeplitz matrix with first row $[1, a_1, \ldots, a_{n-1}]$.

Proof of Theorem 3: It follows from the lemma 1 that the first column of $(sI - A_{re})^{-1}$ is $\frac{1}{a(s)}[l_1(s), l_2(s), \ldots, l_n(s)]^T$. Let its inverse Laplace transform be $L(t)$. The result now follows from $\mathrm{e}^{A_{re}t} e_1 e_1^T \mathrm{e}^{A_{re}^T t} = L(t)L(t)^T$. ◇

The comparison is thus between $\mathcal{R}_\delta = \int_0^\delta L(t)L^T(t)\,\mathrm{d}t$ and $\Delta_\delta^{(n)}$. Therefore it is conceivable to improve the approximation by determining a matrix $\Omega(\delta)$ perhaps itself approximated as $\int_0^\delta L(t)L^T(t)\,\mathrm{d}t$ from the frozen LTI system by replacing $\mathcal{R}_{f,g}^{(\delta,n)}(x)$ in (116) by $\mathbf{R}_{f,g}^{(n)}(x)\Omega_\delta^{(n)}\mathbf{R}_{f,g}^{(n)}(x)^T$.

4.8 Approximate Local Nonlinear Balancing

The approximation for the local nonlinear balanced realization is based upon simultaneous diagonalization of the approximate Gramians. By construction, the realization is balanced and is much simpler to compute (only a finite number of differentiations are required). This offers a tremendous advantage over other nonlinear balancing methods.

Define the *pseudo-Hankel matrix* $\mathcal{H} = \mathcal{O}^{1/2}\mathcal{R}^{1/2}$, where $\mathcal{O}^{1/2}$ and $\mathcal{R}^{1/2}$ are the symmetric positive definite square root matrices of the local Gramians or their approximations. Note that the spectrum $\mathrm{Spec}\,(\mathcal{H}\mathcal{H}^T) = \mathrm{Spec}\,(\mathcal{O}\mathcal{R}) = \mathrm{Spec}\,(\Lambda^4)$. The following was shown in [VG04]:

Theorem 4. *The dominant direction of (f, g, h) at x maximizes the form*

$$\frac{\xi'(\mathcal{H}\mathcal{H}^T)^{1/2}\xi}{\xi'(\mathcal{H}\mathcal{H}^T)^{-1/2}\xi}. \tag{120}$$

4.9 Local Balancing - Discrete Case

The above ideas have also been worked out for discrete time affine systems of the form

$$x_{k+1} = f(x_k) + g(x_k)u_k \tag{121}$$
$$y_k = h(x_k). \tag{122}$$

If we consider again the nominal system to be the one for $u \equiv 0$, then the finite nonlinear local reachability and observability matrices are

$$\mathbf{R}_n(f,g) = [\, g|_{-(n-1)}, \mathrm{it}_f^1 g|_{-(n-2)}, \ldots, \mathrm{it}_f^{n-1} g|_0 \,], \tag{123}$$

$$\mathbf{O}_n(f,g) = \begin{bmatrix} \mathrm{d}h|_0 \\ \mathrm{d}h|_1\, \mathrm{d}f|_0 \\ \vdots \\ \mathrm{d}h|_{n-2}\, \mathrm{d}f|_{n-1} \cdots \mathrm{d}f|_0 \end{bmatrix}, \tag{124}$$

where

$$\mathrm{it}_f^k g|_\ell = \mathrm{d}f|_\ell \cdots \mathrm{d}f|_{\ell-k+1} g|_{\ell-k}. \tag{125}$$

The nonlinear Gramians are then computed as in the continuous case, but with $\Delta = I$. This was presented in [VG01a] and applied to nonlinear discrete periodic systems in [Ve01].

5 Global Balancing

In Sect. 4 we derived at each point of the state space a local balanced realization obtained by a local transformation of the linear variational system. Now we want to find a global coordinate transformation so that the corresponding linear variational systems coincide with the local balanced form. Note that by local we mean *pointwise*, whereas *global* may still refer to some proper subset $\mathcal{D}$ of the state manifold. Thus we want to find a diffeomorphism ξ such that

$$\frac{\partial \xi}{\partial x} = T(x), \quad x \in \mathcal{D}. \tag{126}$$

This step corresponds exactly with our third requirement that global balancing followed by linearization matches linearization followed by (linear) local balancing, i.e., the postulated commutation of balancing with linearization. This is the *Jacobian Problem*. The set (126) is a *Mayer-Lie* system of PDE's, and it is known *not* to be generically solvable. In fact, necessary and sufficient conditions for its solvability are

$$\frac{\partial T_{ij}(x)}{\partial x_k} - \frac{\partial T_{ik}(x)}{\partial x_j} = 0, \tag{127}$$

for all $i, j, k = 1, \ldots, n$ and $x \in \mathcal{D}$. These conditions are also known as the *Frobenius conditions*.

We proceed in two ways. If the Frobenius conditions are satisfied, we show in Sect. 5.1 how to solve the Jacobian problem, and hence obtain a solution to the global nonlinear balancing problem. We refer to this as Mayer-Lie *integration*. In Sect. 6

we relax the problem. Suppose we only have a finite set of points x_i; $i = 1, \ldots, N$, in the state space where the Jacobian matrix $T(x_i)$ is defined. Then we derive an *approximate* global balancing transformation by starting with a sufficiently large parameterized class of diffeomorphisms and identify the parameters by matching these N Jacobians. We refer to this problem as the Mayer-Lie *interpolation* problem.

5.1 Mayer-Lie Integration

When the Frobenius conditions hold, a diffeomorphism can be found on some domain $\mathcal{D}_1 \subset \mathcal{D}$ by the *method of characteristics*: Transform the Mayer-Lie system to a system of nonlinear equations. It is somewhat easier to work with the inverse of the Jacobian, since if $G(x) = T^{-1}(x)$, where $G(x) : \mathbb{R}^n \to \mathbb{R}^n$ is an element of $L(\mathbb{R}^n, \mathbb{R}^n)$, then $G : \mathbb{R}^n \to L(\mathbb{R}^n, \mathbb{R}^n)$, and the Frobenius conditions imply the existence of a vector field $F_x : \mathbb{R}^n \to \mathbb{R}^n$, such that

$$\frac{\partial F_x(\xi)}{\partial \xi} = G(F_x(\xi)) \tag{128}$$

$$F_x(0) = x_0. \tag{129}$$

If F_x is an invertible map, define $h_x : \mathbb{R}^n \to \mathbb{R}^n$ by $h_x(z) = F_x^{-1}(z)$, i.e., $h_x(z) = \xi$ iff $F_x(\xi) = z$, then

$$\frac{\partial h_x(z)}{\partial z} = T(z), \tag{130}$$

so that $h_x(z)$ is a solution for ξ with $h_x(x_0) = 0$.

The Frobenius condition can be expressed in an alternative form: The derivative $\mathbf{D}G$ of G is a map $\mathbf{D}G(x) : \mathbb{R}^n \to L(\mathbb{R}^n, L(\mathbb{R}^n, \mathbb{R}^n))$. i.e., this is 'unravelled' as follows: $(\mathbf{D}G(x))[u]$ is a linear map in $L(\mathbb{R}^n, \mathbb{R}^n)$, so that $((\mathbf{D}G(x))[u])[v] \in \mathbb{R}^n$. Choose $u = G(x)w$.

The Frobenius condition is now a *symmetry* condition. For all $v, w \in \mathbb{R}^n$:

$$((\mathbf{D}G(x))[G(x)w])[v] = ((\mathbf{D}G(x))[G(x)v])[w]. \tag{131}$$

Example 3. Frobenius Condition: Let

$$G(x) = \begin{bmatrix} 1 & 0 \\ x_1^2 & 1 \end{bmatrix} \tag{132}$$

then, $G(x)w = [w_1, x_1^2 w_1 + w_2]^T$, and

$$(\mathbf{D}G(x))\,[G(x)w] = \begin{bmatrix} 0 & 0 \\ 2x_1 w_1 & 0 \end{bmatrix}, \tag{133}$$

giving the form $((\mathbf{D}G(x))\,[G(x)w])[v] = [0, 2x_1 w_1 v_1]^T$, which is symmetric upon permutation of v and w, so that G is integrable.

Example 4. Mayer-Lie Integration: In order to illustrate the Mayer-Lie integration, we give a simple planar example. Assume the local balancing transformation on $\mathbb{R}^2$ is:

$$T(x,y) = \begin{bmatrix} \cos y & -x\sin y \\ \sin y & x\cos y \end{bmatrix}. \tag{134}$$

Note that this transformation is singular along the y-axis ($x = 0$). Inversion yields

$$G(x,y) = \begin{bmatrix} \cos y & \sin y \\ -\frac{1}{x}\sin y & \frac{1}{x}\cos y \end{bmatrix}. \tag{135}$$

Solve by the method of characteristics (taking two steps: first along the path $(0,0) \to (\xi,0)$, then along $(\xi,0) \to (\xi,\eta)$. The integration along the first segment proceeds as follows:

$$\begin{cases} \frac{dF_1}{d\xi} = \cos F_2 \\ \frac{dF_2}{d\xi} = -\frac{1}{F_1}\sin F_2 \end{cases} \quad \begin{cases} F_1(0,0) = x_0 \\ F_2(0,0) = y_0 \end{cases}. \tag{136}$$

Noting that $F_1 F_1'' + (F_1')^2 = 1$, The solution is readily found to be

$$\begin{cases} F_1(\xi,0) = \sqrt{(\xi + x_0\cos y_0)^2 + x_0^2\sin^2 y_0} \\ F_2(\xi,0) = \arctan\frac{x_0\sin y_0}{\xi + x_0\cos y_0} \end{cases}. \tag{137}$$

Next we solve along the second path $\begin{cases} \frac{dF_1}{d\eta} = \sin F_2 \\ \frac{dF_2}{d\eta} = \frac{1}{F_1}\cos F_2 \end{cases}$,with the initial conditions $F_1(\xi,0)$ and $F_2(\xi,0)$ obtained at the end of the first path. Thus,

$$\begin{cases} F_1(\xi,0) = \sqrt{(\xi + x_0\cos y_0)^2 + (\eta + x_0\sin y_0)^2} \\ F_2(\xi,0) = \arctan\frac{\eta + x_0\sin y_0}{\xi + x_0\cos y_0} \end{cases}. \tag{138}$$

Finally, inversion gives, upon resetting the original variables $F_1 = x$ and $F_2 = y$,

$$\begin{cases} \xi = x\cos y - x_0\cos y_0 \\ \eta = x\sin y - x_0\sin y_0 \end{cases}, \tag{139}$$

satisfying $(\xi,\eta) = (0,0)$ if $(x,y) = (x_0,y_0)$.

We conclude with the following observations:

For scalar systems, there is no obstruction to global balancing.

For second order (planar) systems, generically the Mayer-Lie conditions do not hold. However, integrating factors $S(x) = \operatorname{diag}\{S_1(x), S_2(x)\}$ can be defined such that $S(x)T(x)$ is integrable. Whereas equality of the Gramians is destroyed in this form, note that such a *non-uniform* scaling retains the *diagonality* of the Gramians. Moreover, the product of the Gramians specifies the *canonical Gramian* and therefore the relative importance of the balanced local state components. Since this is the required information for model reduction such a scaled global balanced realization is still useful. We referred to such a realization, which is still useful for model reduction via POD, as an *uncorrelated* realization [VG00].

$$\text{Balanced Realizations} \subset \text{Uncorrelated Realizations}$$

For order three and higher also the set of uncorrelated realizations is not sufficiently rich. There is a real topological obstruction. This however, is not the end of our story.

6 Mayer-Lie Interpolation

Generically, the Mayer-Lie conditions form an obstruction to the application of the third principle, the commutation of balancing and linearization. To get around this, we propose an approximate solution via a special interpolation.

Thus motivated, suppose one is given a distinct set of points $\{x_1, \ldots, x_N\} \subset \mathcal{D}$, with the corresponding matrices $T(x_i)$. We want to find a diffeomorphism, η defined on $\mathcal{D}$ such that

$$\left.\frac{\partial \eta}{\partial x}\right|_{x_i} = T(x_i), \quad \text{for } i = 1, \ldots, N. \tag{140}$$

For the special case of a discrete periodic or chain recurrent orbit, the proposed interpolation approach is easily justified. Only the neighborhoods of the successive points in the nominal orbit are important, hence the true form in between is rather immaterial [Ve01].

One potential solution is to start from a parameterized set of diffeomorphisms, and identify the necessary parameters in order to satisfy the N point constraints (140). We follow [Ve02].

For instance, for a planar system consider a class of transformations of the form

$$\xi(x, y) = \sum_{i=0}^{2N-1} c_i^{(\xi)} x^{2N-1-i} y^i \tag{141}$$

$$\eta(x, y) = \sum_{i=0}^{2N-1} c_i^{(\eta)} x^{2N-1-i} y^i. \tag{142}$$

These are homogeneous polynomials in x and y of degree $2N-1$, parameterized by the coefficients $c_j^{\{\xi\}}$ and $c_j^{\{\eta\}}$. The interpolation constraints imply

$$\begin{bmatrix} T^T(x_0, y_0) \\ \vdots \\ T^T(x_{N-1}, y_{N-1}) \end{bmatrix} = \begin{bmatrix} \mathcal{Z}(x_0, y_0) \\ \vdots \\ \mathcal{Z}(x_{N-1}, y_{N-1}) \end{bmatrix} C. \tag{143}$$

where $C = [c^{(\xi)}, c^{(\eta)}] \in \mathbb{R}^{(2N-1)\times 2}$. and $\mathcal{Z}(x, y)$ is

$$\begin{bmatrix} (2N-1)x^{2N-2} & (2N-2)x^{2N-3}y & \cdots & y^{2N-2} & 0 \\ 0 & x^{2N-2} & \cdots & (2N-2)xy^{2N-3} & (2N-1)y^{2N-2} \end{bmatrix}.$$

In a more compact form we rewrite this as $\mathcal{T}^T = \mathbf{Z}(x_0, y_0, \dots x_{N-1}, y_{N-1})\, C$. If the matrix $\mathbf{Z}(x_0, y_0, \dots x_{N-1}, y_{N-1})$ is invertible, then $C = \mathbf{Z}(x_0, y_0, \dots x_{N-1}, y_{N-1})^{-1}\mathcal{T}^T$, and a candidate diffeomorphism is

$$\begin{bmatrix} \xi(x,y) \\ \eta(x,y) \end{bmatrix} = \mathcal{T}\mathbf{Z}^{-T}(x_0, y_0, \dots, x_{N-1}, y_{N-1}) \begin{bmatrix} x^{2N-1} \\ \vdots \\ y^{2N-1} \end{bmatrix}. \tag{144}$$

Since $\mathbf{Z}$ is homogeneous of degree $2N-2$, the Jacobian determinant has degree $4(N-1)$. Consequently, there are at most $4(N-1)$ lines through the origin where the full rank property will fail. These lines define wedges in the original state space coordinates. Hence, if all interpolation states $\overline{x}_0, \dots, \overline{x}_{N-1}$ fall inside a wedge, $\mathcal{D}$, a *globally* defined balanced realization can be defined in $\mathcal{D}$. The following theorem is proven in [Ve04].

Theorem 5. *If no two states are collinear with the origin, then the matrix* $\mathbf{Z}(x_0, y_0, \dots, x_{N-1}, y_{N-1})$ *is an invertible matrix.*

Example 5. Delayed Logistic Equation [Ve02]*:* Consider the discrete time delayed system with observation of x.

Let the control be the deviation of μ from a nominal value 2.1.

$$\begin{aligned} x_{k+1} &= \mu x_k(1 - y_k) \\ y_{k+1} &= x_k. \end{aligned}$$

For $\mu = 2.1$, the system exhibits an attracting limit cycle enclosed in $[0,1]^2$. Starting at a point on the limit cycle, the seventh iterate overtakes it (Fig. 1).

The approximate (two-step) local reachability Gramian and observability Gramian for this discrete system are

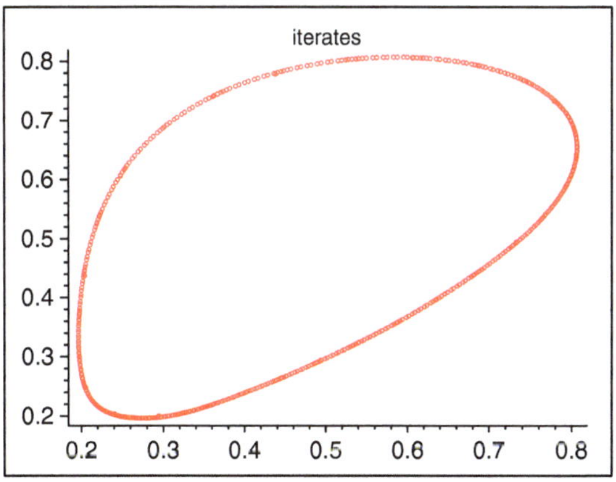

Fig. 1. Limit cycle for delayed logistic equation

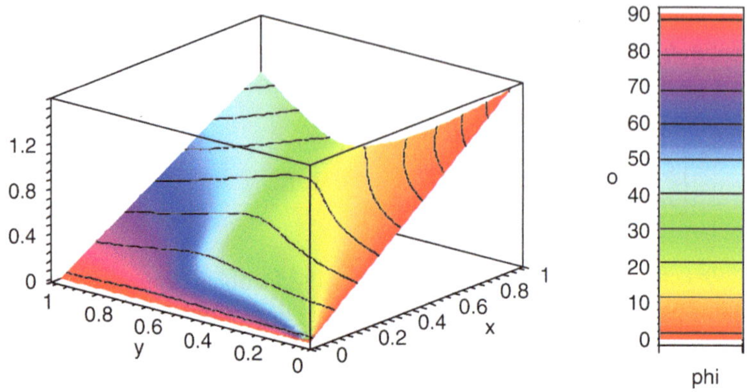

Fig. 2. Dominant λ and corresponding direction.

$$\mathcal{R}(x,y) = \begin{bmatrix} 2\frac{x^2}{\mu^2} & \frac{xy}{\mu^2} \\ \frac{xy}{\mu^2} & \frac{y^2}{u^2} \end{bmatrix}$$

$$\mathcal{O}(x,y) = \begin{bmatrix} 1+\mu^2(1-y)^2 & -\mu^2(1-y)x \\ -\mu^2(1-y)x & \mu^2x^2 \end{bmatrix},$$

respectively. With interpolation points $\{(0.2, 0.2), (0.5, 0.2)(0.2, 0.5)\}$, which are close to but not on the limit cycle, the transformation is found to be

$$\xi(x,y) = 2115x^5 - 13823x^4y + 31269x^3y^2 - 29100x^2y^3 + 11596xy^4 - 1649y^5$$
$$\eta(x,y) = 1048x^5 - 7187x^4y + 17472x^3y^2 - 18101x^2y^3 + 7821xy^4 - 1180y^5.$$

As this is not the balancing transformation in the strict sense of Sect. 5, we refer to this as a *pseudo-balancing* transformation. The resulting system is then likewise *pseudo-balanced*. The dominant value $\lambda_1(x,y)$ of the canonical Gramian is displayed in Fig. 2. The height of the plot indicates the value, the coloring is modulated by the angle of the dominant direction (mapped back to the original (x,y) coordinates). Note that on the x-axis, this angle is zero, while on the y-axis it reaches 90 degrees. If T_{loc} is the local balancing transformation, this is the direction of the first column of T_{loc}^{-1}, i.e., the 'jointly most observable and reachable' direction. As expected, the dynamics is almost one dimensional along the limit cycle.

7 Nonlinear Model Reduction

What is a reduced model for a nonlinear system? As shown earlier, a local balancing transformation exists at each point of $\mathbf{M}$. The *canonical Gramian* contains important information suitable for approximating the topological structure of the system. It is the information one needs for model reduction.

Once we have the local balancing transformation in partitioned form

$$\begin{bmatrix} T_{11}(x) & T_{12}(x) \\ T_{21}(x) & T_{22}(x) \end{bmatrix} \begin{bmatrix} \widetilde{x}_1 \\ \widetilde{x}_2 \end{bmatrix} = \begin{bmatrix} \widetilde{x}_1^b \\ \widetilde{x}_2^b \end{bmatrix}, \tag{145}$$

consistent with $\Lambda(x) = \text{diag}\,[\Lambda_1(x), \Lambda_2(x)]$, with $\Lambda_2 < \Lambda_1$, the reduced model via POD sets all components $\widetilde{x}_2^b$ equal to zero. This means that in terms of the original coordinate system, the equality

$$T_{21}\widetilde{x}_1 + T_{22}\widetilde{x}_2 = 0 \tag{146}$$

is acquired. The local reduced dynamical equations (equations for the perturbation system) are

$$\begin{aligned} \dot{\widetilde{x}}_1 &= \frac{\partial f_1}{\partial x_1}\widetilde{x}_1 + g_1\widetilde{x}_1 \\ \widetilde{y} &= \frac{\partial h_1}{\partial x_1}\widetilde{x}_1. \end{aligned}$$

Reconstituting the equations in the original coordinates gives

$$\dot{x}_1 = f_1(x_1, \overline{x}_2) + g_1(x_1, \overline{x}_2)u \tag{147}$$

$$\dot{\overline{x}}_2 = f_2(\overline{x}_1, \overline{x}_2) \tag{148}$$

$$y = h(x_1, \overline{x}_2). \tag{149}$$

The nonlinear dynamics of x_2 is only specified by its behavior under the nominal condition, hence it can be precomputed (off line), giving the vector function $\overline{x}_2(t)$, which is fixed once and for all. Hence the reduced order nonlinear model is in fact a nonautonomous system of the form

$$\dot{x}_1 = F_1(x_1, t) + g_1(x_1, t)u \tag{150}$$

$$y = H(x_1, t), \tag{151}$$

where $F(z,t) = f_1(z, \overline{x}_2(t))$ and $H(z,t) = h(z, \overline{x}_2(t))$.

Should one be interested in the model near an equilibrium point, then $\overline{x}_2$ is constant, and an autonomous reduced model results. It should be pointed out that different subdomains in **M** may suggest the use of different dimensions for the lower order approximations. This is explored in more detail in [VG06]. In this case the domain boundaries are characterized by umbilic points.

8 How Far Can You Go?

One of the crucial problems in the above is the management of the two approximations. First there is the interval length δ, then there is the truncation of the Lie(f)-and the ad(f)-series to N terms. Here simulation and analysis in the LTI case may suggest some heuristics. For instance, it was found in the LTI case that for asymptotically stable systems the δ Gramians where δ is at least twice the largest characteristic time (time-constant or oscillation period) of the system, pretty much captures the same behavior as $\delta \to \infty$, which was the original set up by Moore.

We suggest the following heuristic for the nonlinear system. The nominal system $\dot{x} = f(x)$ has *exact* solution, given by the Lie-series, $x(t) = \sigma_x \, \mathrm{e}^{tL_f} I$. If this series is truncated after N terms (degree $N-1$ in t), then denote this time function by $x^{(N)}$(t). Note that the superscript N does not refer to the N-th derivative of $x(t)$ here. The error is then $x(t) - x^{(N)}(t)$. The norm of this error should be small compared to either norm of the difference, $x(t) - x$, between initial and final (nominal) state, or the length of the path traveled,

$$\int_0^t \|\dot{x}(\theta)\| \, \mathrm{d}\theta = \int_0^t \|f(x(\theta))\| \, \mathrm{d}\theta = \int_0^t \|\sigma_x \mathrm{e}^{\theta L_f} f\| \, \mathrm{d}\theta. \tag{152}$$

But having an error bound criterion in terms of the exact solution is not practical, as we would then need the exact instead of the approximation in the first place. The following theorem shows how this can be overcome.

Theorem 6. *Given* $0 < \epsilon < 1$, *if for all* $\theta \in (0, t)$ *it holds that*

$$\frac{t^{N+1}}{(N+1)!} \left\| \sigma_{x(\theta)} \left(L_f^{N+1} I \right) \right\| < \frac{\epsilon}{1+\epsilon} \|x^{(N)}(t) - x\|, \tag{153}$$

then

$$\|x^{(N)}(t) - x(t)\| < \epsilon \|x(t) - x\|. \tag{154}$$

Proof: By Taylor's remainder theorem, for some θ_0 in $(0, t)$

$$\|x(t) - x^{N)}(t)\| = \left\| \frac{t^{N+1}}{(N+1)!} \left(\frac{\mathrm{d}}{\mathrm{d}t} \right)^{N+1} x(\theta_0) \right\| = \frac{t^{N+1}}{(N+1)!} \left\| \sigma_{x(\theta_0)} \left(L_f^{N+1} I \right) \right\| \tag{155}$$

By the assumption, this yields

$$\|x(t) - x^{N)}(t)\| < \frac{\epsilon}{1+\epsilon} \|x^{(N)}(t) - x(t)\| + \frac{\epsilon}{1+\epsilon} \|x(t) - x\|,$$

from which the assertion follows. ◇

If N is fixed a priori, then the criterion in Theorem 8.1 provides a reasonable heuristic to choose δ.

Does one really need global balancing? From a point of view of sensitivity analysis and model reduction, all information (e.g., the approximate dimension for the dynamics in the neighborhood of that point, and the dominant subspace) that one would like to have from a globally balanced form is already contained in the local (or pointwise) balanced form. So why bother with this difficult and unnecessary step? In fact, it is shown that global model reduction may not be possible at all on the grounds that different patches of the state space may suggest different reduced order dimensions, a phenomenon that is nonexistent in linear balancing (which is necessarily global over the entire state space). The argument is simple: we always consider two metrics associated with a realization: the reachability metric and the observability metric. Each one separately gives the state space the structure of a Riemann manifold. Balancing bends and stretches these two Riemann spaces locally without tearing. In fact, balancing consists in transforming both Riemann spaces until they coalesce. But there may never exist a transformation that makes the two spaces coincide *everywhere*: a sphere cannot be bend into a torus! This is the essence of the topological obstruction to global balancing.

9 Conclusions

In this chapter we presented some old and new results on balancing for time varying systems. For nonlinear systems, a reasonable set of principles for general NL balancing was suggested, and partially implemented based on the approach for the linear time-varying case. The results is partial because of the obstruction in the form of the Mayer-Lie conditions. When the conditions hold, integration of the Jacobian is possible and a global balanced or uncorrelated balanced realization exists. Alternatively, we showed how an approximation via Mayer-Lie interpolation may be obtained, leading to a pseudo-balanced realization. The advantage of the given method for balancing lies in its computational feasibility: only differentiation is involved in obtaining a local balanced form (as opposed to the solution of a PDE). It is further consistent with the linear theory, and it is also not restricted to stable systems. Other approaches to NL balancing exist: The proper orthogonal decomposition based approach [Lal02] is restricted to asymptotically stable nonlinear systems. The approach by Scherpen et al. [Sch94] requires the solution of nonlinear PDE's, and is also restricted to stable nonlinear systems. The RKHS approach in [Ve84] seems quite conservative in its approximation. See also [GV06b].

We have not documented past and ongoing research on extensions and applications such as LQG-balancing [Ve81b, Ve81a, Ve86a, Ve86b, JS83, OC04] and balanced completion with applications in robust control and uncertainty equivalence [Ve85, VP92, Ve94, SR02, Ve06]. The latter topic seeks to quantify the uncertainty associated with the reduction of a given system via projection of dynamics of the balanced form. Indeed, if the full order system provides the complete information about the system, then necessarily a reduced model overlooks some of this information in one form or other. Discarding information means that uncertainty is introduced in the model. Traditionally, reduced order models have not been cast with associated uncertainty bounds. Maximum likelihood techniques and stochastic system theory are used to specify uncertainty bounds for the balanced reduced order models.

Acknowledgement

The valuable comments and suggestions by Henk A. van der Vorst to the preliminary version of this chapter are gratefully acknowledged.

References

[FLL83] Fliess, M., Lamnahbi M. and Lamnabhi-Lagarrigue, F.: An algebraic approach to nonlinear functional expansions. IEEE Trans. Circ. Syst., **30**,554–570 (1983)

[G77] Gilbert, E.G.: Functional expansions for the response of nonlinear differential systems. IEEE Trans. Automatic Control, **22**, 6 909–921 (1977)

[GV04] Gray, W.S. and Verriest, E.I.: Balanced realizations near stable invariant manifolds. Proc. MTNS-04, Leuven, Belgium, (2004).

[GV06a] Gray, W.S. and Verriest, E.I.: Balanced realizations near stable invariant manifolds. Automatica **42** 653-659 (2006)

[GV06b] Gray, W.S. and Verriest, E.I.: Algebraically defined Gramians for nonlinear Systems. Proceedings of the 45-th IEEE Conference on Decision and Control, San Diego, CA 3730-3735 (2006).

[GA89] Gucercin, S. and Antoulas, A.: A survey of model reduction by balanced truncation and some new results. Int. J. Control, **77** 8 748–766 (2004)

[I89] Isidori, A., Nonlinear Control Systems, 2-nd edition. Springer-Verlag (1989)

[JS83] Jonckheere, E. and Silverman, L.: A new set of invariants for linear systems - Application to reduced order compensator design. IEEE Transactions on Automatic Control, **28** 10 953–964 (1983)

[KT93] Khayatian, A. and Taylor, D.G.: Sampled-data modeling and control design for nonlinear pulse-width modulated systems. Proc. American Control Conference, 660–664 (1993)

[LB03] Lall, S. and Beck, C.: Error-bounds for balanced reduction of linear time-varying systems. IEEE Trans. Automatic Control, **48** 946–956 (2003)

[Lal02] Lall, L., Marsden, J.E. and Glavaški, S.: A subspace approach to balanced truncation for model reduction of nonlinear control systems. Int. J. Robust and Nonlinear Control, **12** 6 519–535 (2002)

[LK78] Lesiak, C. and Krener, A.J.: The existence and uniqueness of Volterra series for nonlinear systems. IEEE Trans. Auto. Ctr. **23** 6 1090–1095 (1978)

[LO99] Longhi, S. and Orlando, G.: Balanced reduction of linear periodic systems. Kybernetika, **35** 737–751 (1999)

[MNC84] Monaco, S. and Normand-Cyrot, D.: Input-output approximation of a nonlinear discrete time system from an equilibrium point. Proc. 23-th Conference on Decision and Control, Las Vegas, NV 90–96 (1984)

[MNC85] Monaco, S. and Normand-Cyrot, D.: On the sampling of a linear analytic control system. Proc. 24-th Conference on Decision and Control, Ft. Lauderdale, FL 1457–1462 (1985)

[M81] Moore, B.: Principal component analysis in linear systems: Controllability, observability and model reduction. IEEE Transactions on Automatic Contro;, **26** 17–32 (1981)

[M67] Morse, M.: What is analysis in the large? In Chern, S.S. (ed) Studies in Global Geometry and Analysis. Mathematical Association of America (1967)

[NK98] Newman A. and Krishnaprasad, P.S.: Computation for nonlinear balancing. Proceedings of the 37th Conference on Decision and Control, Tampa, FL 4103–4104 (1998)

[OC04] Opmeer. M.R. and Curtain, R.F.: Linear quadratic Gaussian balancing for discrete-time infinite-dimensional linear systems. SIAM Journal on Control and Optimization, **43** 4 119–122 (2004)

[SR02] Sandberg H. and Rantzer, A.: Error bounds for balanced truncation of linear time-varying systems. Proc. 39-th Conf. Decision and Control, Sydney, Australia 2892–2897 (2002)

[SR04] Sandberg H. and Rantzer, A.: Balanced truncation of linear time-varying systems. IEEE Transactions on Automatic Control, **49** 2 217–229 (2004)

[Sch94] Scherpen, J.M.A.: Balancing for nonlinear systems, Ph.D. Dissertation, University of Twente (1994)

[SSV83] Shokoohi, S., Silverman, L. and van Dooren, P.: Linear time-variable systems: Balancing and model reduction. IEEE Transactions on Automatic Control, **28**, 833–844 (1983)

[V99] Varga, A.: Balancing related methods for minimal realization of periodic systems. Systems & Control Lett., **36** 339–349 (1999)

[V00] Varga, A.: Balanced truncation model reduction for periodic systems. Proc. 39-th Conf. Dec. and Contr., Sydney, Australia 2379–2484 (2000)

[Ve80] Verriest, E.I.: On generalized balanced realizations. PhD Dissertation, Stanford University (1980)

[Ve81a] Verriest, E.I.: Low sensitivity design and optimal order reduction for the LQG-Problem. Proceedings of the 1981 Midwest Symposium on Stochastic Systems, Albuquerque, NM 365–369 (1981)

[Ve81b] Verriest, E.I.: Suboptimal LQG-design via balanced realizations. Proc. 20th IEEE Conf. on Dec. and Contr., San Diego, CA 686–687 (1981)

[Ve84] Verriest, E.I.: Approximation and order reduction in nonlinear models using an RKHS approach. Proc. 18th Conf. on Information Sciences and Systems, Princeton, NJ 197–201 (1984)

[Ve85] Verriest, E.I.: Stochastic reduced order modeling of deterministic systems. Proc. 1985 American Control Conf., Boston, MA 1003–1004 (1985)

[Ve86a] Verriest, E.I.: Model reduction via balancing and connections with other methods. In Desai, U.B. (ed) Modeling and Application of Stochastic Processes. Kluwer Academic Publishers (1986)

[Ve86b] Verriest, E.I.: Reduced-order LQG design: Conditions for feasibility. Proceedings of the 1986 IEEE Conference on Decision and Control, Athens, Greece 1765–1769 (1986)

[Ve94] Verriest, E.I.: Balancing for robust modeling and uncertainty equivalence. In Helmke, U., Mennicken, R. and Saurer, J. (eds) Systems and Networks: Mathematical Theory and Applications, Uni. Regensburg, 543–546 (1994)

[Ve01] Verriest, E.I.: Balancing for discrete periodic nonlinear systems. IFAC Workshop Prepr.: Periodic Control Systems, Como, Italy 253–258 (2001)

[Ve02] Verriest, E.I.: Pseudo balancing for discrete nonlinear systems. Proceedings of the MTNS 2002, Notre Dame (2002)

[Ve04] Verriest, E.I.: Nonlinear balancing and Mayer-Lie interpolation. Proc. SSST-04, Atlanta, GA 180–184 (2004)

[Ve06] Verriest, E.I.: Uncertainty equivalent model reduction. Proceedings of the MTNS 2006, Kyoto, Japan 1987–1992 (2006)

[VG00] Verriest, E.I., and Gray, W.S.: Flow balancing nonlinear systems. Proceedings of the MTNS 2000, Perpignan, France (2000)

[VG01a] Verriest, E.I. and Gray, W.S.: Discrete time nonlinear balancing. Proceedings IFAC NOLCOS 2001, St Petersburg, Russia (2001)

[VG01b] Verriest, E.I. and Gray, W.S.: Nonlinear balanced realizations. Proc. 40-th IEEE Conference on Decision and Control, Orlando, 3250–3251 (2001)

[VG04] Verriest, E.I. and Gray, W.S.: Nonlinear balanced realizations. Proc. 43d IEEE Conf. Dec. and Contr., Paradise Island, Bahamas, 1164-1169 (2004)

[VG06] Verriest, E.I. and Gray, W.S.: Geometry and topology of the state space via balancing. Proceedings of the MTNS 2006, Kyoto, Japan 840–848 (2006)

[VH98] Verriest, E.I. and Helmke, U.: Periodic balanced realizations. Proceedings of the IFAC Conference on System Structure and Control, Nantes, France 519–524 (1998)

[VK80] Verriest, E.I. and Kailath, T.: On generalized balanced realizations. Proceedings of the 19th Conference on Decision and Control, Albuquerque, NM, 504–505 (1980)

[VK83] Verriest, E.I. and Kailath, T.: On generalized balanced realizations. IEEE Transactions on Automatic Control, **28** 8 833–844 (1983)

[VP92] Verriest, E.I. and Pajunen, G.A.: Robust control of systems with input and state constraints via LQG balancing. Proceedings of the 1992 IEEE Conference on Decision and Control, Tucson, AZ 2607–2610 (1992)

Singular Value Analysis and Balanced Realizations for Nonlinear Systems

Kenji Fujimoto[1] and Jacquelien M.A. Scherpen[2]

[1] Department of Mechanical Science and Engineering, Nagoya University
`fujimoto@nagoya-u.jp`
[2] Faculty of Mathematics and Natural Sciences, University of Groningen
`j.m.a.scherpen@rug.nl`

1 Introduction

For linear control systems minimal realization theory and the related model reduction methods play a crucial role in understanding and handling the system. These methods are well established and have proved to be very successful, e.g., [Antoulas05, OA01, ZDG96]. In particular the method called balanced truncation gives a good reduced order model with respect to the input-output behavior, [Moore81, Glover84]. This method relies on the relation with the system Hankel operator, which plays a central role in minimal realization theory. Specifically, the Hankel operator supplies a set of similarity invariants, the so called Hankel singular values, which can be used to quantify the importance of each state in the corresponding input-output system [JS82]. The Hankel operator can also be factored into a composition of observability and controllability operators, from which Gramian matrices can be defined and the notion of balanced realization follows, first introduced in [Moore81] and further studied by many authors, e.g. [JS82, ZDG96]. This linear theory is rather complete and the relations between and interpretations in the state-space and input-output settings are fully understood.

A nonlinear extension of the state-space concept of balanced realizations has been introduced in [Scherpen93], mainly based on studying the past input nenergy and the future output energy. Since then, many results on state-space balancing, modifications, computational issues for model reduction and related minimality considerations for nonlinear systems have appeared in the literature, e.g., [GS01, HE02, LMG02, NK00, NK98, SG00, VG00]. In particular, singular value functions which are a nonlinear state-space extension of the Hankel singular values for linear systems play an important role for nonlinear balanced realizations. However, the original characterization in [Scherpen93] was incomplete in a sense that the defined singular value functions are not unique, the relation with the nonlinear Hankel operator was not clarified, and the resulting model reduction procedure gives different reduced order models depending on the choice of different set of singular value functions, e.g. [GS01].

Balanced realization and the related model order reduction technique rely on singular value analysis. This analysis investigates the singular values and the corresponding singular vectors for a given operator. The analysis is important since it extracts the gain structure of the operator, that is, it characterizes the largest input-output ratio and the corresponding input [Stewart93]. Since linear singular values are defined as eigenvalues of the composition of the given operator and its adjoint, it is natural to introduce a nonlinear version of adjoint operators to obtain a nonlinear counterpart of a singular value. There has been done quite some research on the nonlinear extension of adjoint operators, e.g. [Batt70, SG02, FSG02] and the references therein. Here we do not explicitly use these definitions of nonlinear adjoint operators. We rely on a characterization of singular values for nonlinear operators based on the gain structure as studied in [Fujimoto04]. The balanced realization based on this analysis yields a realization that is based on the singular values of the corresponding Hankel operator, and results in a method which can be viewed as a complete extension of the linear methods, both from an input-output and a state-space point of view, [FS05].

The related model order reduction technique, nonlinear balanced truncation, preserves several important properties of the original system and corresponding input-output operator, such as stability, controllability, observability and the gain structure [FS03].

This paper gives an overview of the series of research on balanced realization and the related model order reduction method based on nonlinear singular value analysis. Section 2 explains the taken point of view on singular value analysis for nonlinear operators. Section 3 briefly reviews the linear balancing method and balanced truncation in order to show the way of thinking for the nonlinear case. Section 4 treats the state-space balancing method stemming from [Scherpen93]. Then, in Section 5 we continue with balanced realizations based on the singular value analysis of the nonlinear Hankel operator. Furthermore, in Section 6 balanced truncation based on the method of Section 5 is presented. Finally, in Section 7 a numerical simulation illustrates how the proposed model order reduction method works for real-world systems.

2 Singular Value Analysis of Nonlinear Operators

Singular value analysis plays an important role in the characterizations of the principal behavior of linear operators. Here we formulate a nonlinear counterpart of singular value analysis. It is a basic ingredient for considering balanced realizations for nonlinear systems explained further on in this paper.

Let us consider a linear operator $A : U \to Y$ with Hilbert spaces U and Y. Then

$$A^* A\, v = \sigma^2 v \tag{1}$$

holds with $\sigma(\geq 0) \in \mathbb{R}$ and $v \in U$ where σ and v are called a *singular value* and a (right) *singular vector* of the operator A. Here A^* is the adjoint of A satisfying

$$\langle y, A\,u\rangle_Y = \langle A^* y, u\rangle_U \tag{2}$$

for all $u \in U$ and $y \in Y$ where $\langle\cdot,\cdot\rangle_X$ denotes the the inner product of the space X. For a finite dimensional signal space U, the operator A can be described by

$$A = \sum_{i=1}^{n} \sigma_i\, w_i\, v_i^*$$

with the singular values $\sigma_1 \geq \sigma_2 \geq \ldots \geq \sigma_n$, the corresponding right singular vectors v_i's. and the left singular vectors w_i's. Then we can obtain an approximation of A with rank $m < n$ by

$$A^a := \sum_{i=1}^{m} \sigma_i\, w_i\, v_i^*.$$

We can easily observe that this approximation preserves the gain of the original operator A

$$\|A^a\| = \sigma_1 = \|A\|.$$

Furthermore, the error bound is obtained by

$$\|A - A^a\| = \sigma_{m+1}.$$

For the generalization to nonlinear systems, we consider the following interpretation of singular values for linear operators. The largest singular value of the operator A characterizes the gain of the operator and the corresponding singular vector $v_{\max}$ represents the input maximizing the input-output ratio. Namely, the following equations hold.

$$\sigma_{\max} = \sup_{u\neq 0} \frac{\|A\,u\|}{\|u\|}, \quad v_{\max} = \arg\sup_{u\neq 0} \frac{\|A\,u\|}{\|u\|} \tag{3}$$

Now, let us consider a smooth nonlinear operator $f : U \to Y$ with Hilbert spaces U and Y. How to define singular values of the nonlinear operator $f(u)$ is not immediately clear because there does not exist an operator $f^*(y)$ such that Equation (2) holds with $A = f$. Several papers define a nonlinear counterpart of an adjoint operator, e.g., [Batt70, SG02, FSG02]. For our nonlinear balancing purpose we generalize the linear way of thinking given by Equation (3). More precisely we consider the following definitions

$$\sigma^c_{\max} = \sup_{\|u\|=c} \frac{\|f(u)\|}{\|u\|}, \quad v^c_{\max} = \arg\sup_{\|u\|=c} \frac{\|f(u)\|}{\|u\|} \tag{4}$$

where the gain of the operator f is characterized for each input magnitude c. The property that the gain of a nonlinear operator depends on the magnitude of its input is quite natural in the nonlinear setting and, for instance, this idea can be found in the input-to-state stability literature, e.g., [JTP94, SW96]. If $\sigma^c_{\max}$ is obtained, then we can calculate the largest singular value $\sigma_{\max}$ and the corresponding singular vector $v_{\max}$ of f by

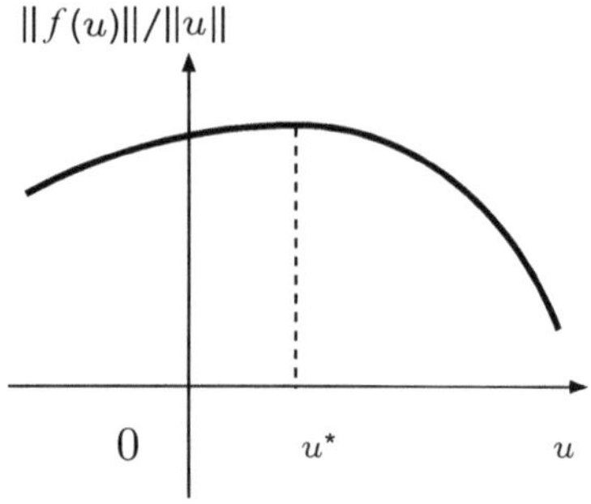

Fig. 1. Maximizing input $u = u^\star$ of $f(u)$

$$\sigma_{\max} = \sup_{c>0} \sigma^c_{\max}, \quad v_{\max} = v^c_{max}\big|_{c=\arg\sup_{c\neq 0} \sigma^c_{\max}}. \tag{5}$$

In the linear case, the largest singular value $\sigma_{\max}$ coincides with $\sigma^c_{\max}$ for all $c > 0$.

Now we are ready to define the singular value σ and the corresponding singular vector v for the operator f fulfilling the relationship (4). This is obtained by simply differentiating the condition in Equation (4). Figure 1 depicts the (locally) largest singular vector $u^\star$ when f is a mapping of $\mathbb{R} \to \mathbb{R}$. At the point $u = u^\star$ where the input-output takes its maximum value, the derivative of the input-output ratio has to be 0. Therefore the following equation has to hold for all u satisfying $\|u\| = c$.

$$\mathrm{d}\left(\frac{\|f(u)\|}{\|u\|}\right)(\mathrm{d}u) = 0 \tag{6}$$

Here the Fréchet derivative[1] is adopted to describe the problem. This equation is equivalent to

$$\langle (\mathrm{d}f(u))^* f(u) - \frac{\|f(u)\|^2}{\|u\|^2} u, \ \mathrm{d}u\rangle = 0. \tag{7}$$

On the other hand, the derivative of $\|u\| = c$ yields

$$\langle u, \ \mathrm{d}u\rangle = 0. \tag{8}$$

Combining Equations (7) and (8), we obtain the condition for the singular vector v.

Theorem 1. *[Fujimoto04] Consider a nonlinear operator $f : U \to Y$ with Hilbert spaces U and Y. Then the input-output ratio of $\|f(u)\|/\|u\|$ has a critical value for an arbitrary input magnitude $\|u\| = c$ if and only if*

$$(\mathrm{d}f(v))^* f(v) = \lambda\, v \tag{9}$$

with a scalar $\lambda \in \mathbb{R}$ and $v \in U$.

[1] The Fréchet derivative of an operator $T : U \to Y$ is an operator $T : U \times U \to Y$ satisfying $f(u+v) = f(u) + \mathrm{d}f(u)(v) + o(\|v\|)$ such that $f(u)(v)$ depends linearly on v.

We now define v as a *singular vector* for a nonlinear operator f if it fulfills Equation (9). Immediate extension of the linear case by defining the singular value of f by $\sigma := \sqrt{\lambda}$ is not appropriate. This can be seen from the fact that, e.g., λ can be negative. A better extension is given by using the singular vector v, and defining the corresponding singular value by

$$\sigma = \frac{\|f(v)\|}{\|v\|}. \tag{10}$$

In the remainder of this paper, investigating the solutions of the pair of Equations (9) and (10) is called *singular value analysis* of the nonlinear operator f. Here σ is called a *singular value* of f, and v is called the corresponding *singular vector*. It can be readily observed that

$$\lambda = \sigma^2$$

holds in the linear case. However, this equation does not hold in the nonlinear case. Although the scalar λ is always real, it can be negative in the nonlinear case [Fujimoto04].

A more detailed discussion on nonlinear singular value analysis is given in [Fujimoto04].

3 Balanced Realization for Linear Systems

This section briefly reviews balanced realizations in the linear systems case in order to show the way of thinking in the nonlinear case. See standard textbooks for the detail, e.g., [OA01, ZDG96]. Consider the following controllable, observable, and asymptotically stable linear system

$$\Sigma : \begin{cases} \dot{x} = Ax + Bu \; x(0) = 0 \\ y = Cx \end{cases} \tag{11}$$

where $x(t) \in \mathbb{R}^n$, $u(t) \in \mathbb{R}^m$ and $y(t) \in \mathbb{R}^p$. The controllability Gramian P and the observability Gramian Q of the system Σ in Equation (11) are obtained by the solutions to the Lyapunov equations

$$AP + PA^{\mathrm{T}} + BB^{\mathrm{T}} = 0 \tag{12}$$
$$A^{\mathrm{T}}Q + QA + C^{\mathrm{T}}C = 0. \tag{13}$$

It is known that the positive definiteness of the Gramians P and Q is equivalent to controllability and observability of the system Σ in Equation (11), respectively. Furthermore, the matrices P and Q themselves are quantitative indicators of the controllability and observability, that is, P and Q describe the behavior of input-to-state and that of state-to-output, respectively.

A *balanced realization* of Σ is a state-space realization which has the following Gramians

$$P = Q = \mathrm{diag}(\sigma_1, \sigma_2, \ldots, \sigma_n) \tag{14}$$

where $\sigma_1 \geq \sigma_2 \geq \ldots \geq \sigma_n > 0$ are called *Hankel singular values*. Here the system is balanced because $P = Q$ implies that relation between input-to-state and state-to-output is balanced and diagonalized P and Q implies that the importance of each coordinate axis is balanced. There is another realization called an *input-normal form* which has the following Gramians

$$P = I, \;\; Q = \mathrm{diag}(\sigma_1^2, \sigma_2^2, \ldots, \sigma_n^2) \tag{15}$$

where only the balancing between the coordinate axes is achieved.

If $\sigma_i > \sigma_j$ then the coordinate axis x_i is more important to the input-output behavior, i.e., better controllable and observable, than the axis x_j. Therefore if $\sigma_k \gg \sigma_{k+1}$ holds for a certain k $(1 \leq k < n)$, then we can obtain a k-dimensional reduced order model by neglecting the dynamics of $x_{k+1}, \ldots, x_n$. This model reduction procedure is called *balanced truncation*. More precisely, balanced truncation is executed as follows. Suppose that the system is in a balanced realization and divide the coordinate as follows

$$\begin{aligned} x &= (x^a, x^b) \\ x^a &= (x_1, \ldots, x_k) \\ x^b &= (x_{k+1}, \ldots, x_n). \end{aligned}$$

Further divide the state-space system

$$\begin{pmatrix} \dot{x}^a \\ \dot{x}^b \end{pmatrix} = \begin{pmatrix} A^a & A^{ab} \\ A^{ba} & A^b \end{pmatrix} \begin{pmatrix} x^a \\ x^b \end{pmatrix} + \begin{pmatrix} B^a \\ B^b \end{pmatrix} u$$

$$y = \begin{pmatrix} C^a, & C^b \end{pmatrix} \begin{pmatrix} x^a \\ x^b \end{pmatrix}.$$

Then the reduced order model is obtained by

$$\Sigma^a : \begin{cases} \dot{x}^a &= A^a x^a + B^a u \\ y &= C^a x^a \end{cases}.$$

By balanced truncation it is readily obtained that several properties are preserved. This can be seen by studying the Lyapunov equations (12) and (13), and their truncated versions, e.g.,

Theorem 2. *[Moore81] The controllability Gramian P^a and the observability Gramian Q^a of the reduced order model Σ^a are given by*

$$P^a = Q^a = \mathrm{diag}(\sigma_1, \ldots, \sigma_k).$$

The controllability operator $\mathcal{C}$ and the observability operator $\mathcal{O}$ of the system Σ as in (11) are given by

$$\begin{aligned}\mathcal{C} : u \mapsto x^0 &:= \int_0^\infty \mathrm{e}^{\tau A} B u(\tau) \mathrm{d}\tau \\ \mathcal{O} : x^0 &\mapsto y := C\mathrm{e}^{tA} x^0 .\end{aligned}$$

Furthermore, their composition is defined as the *Hankel operator* $\mathcal{H}$ of the original system Σ.

$$\mathcal{H} = \mathcal{O}\,\mathcal{C} \tag{16}$$

These operators are closely related to the Gramians, i.e.,

$$\begin{aligned}P &= \mathcal{C}\,\mathcal{C}^* \\ Q &= \mathcal{O}^* \mathcal{O}\end{aligned}$$

Now consider a linear system given by Equation (11), which is not necessarily observable and/or controllable. The relation between the Gramians and the observability and controllability operator allows one to prove the following theorem.

Theorem 3. *[ZDG96] The operator $\mathcal{H}^*\mathcal{H}$ and the matrix PQ have the same nonzero eigenvalues.*

Proof: The proof of this theorem is easily obtained and instructive for the nonlinear extension case. We first prove the '⇒' part. Due to (16), the eigenvalue problem of $\mathcal{H}^*\mathcal{H}$ reduces to

$$\mathcal{C}^*\mathcal{O}^*\mathcal{O}\,\mathcal{C}\,v = \lambda\, v, \quad v \in U,\ \lambda \in \mathbb{R}$$

with $\lambda = \sigma^2$. Defining $\xi := \mathcal{C}\, v \in \mathbb{R}^n$ and premultiplying $\mathcal{C}$ to the above equation, we obtain

$$\mathcal{C}\,\mathcal{C}^*\mathcal{O}^*\mathcal{O}\,\mathcal{C}\,v = \mathcal{C}\,\lambda\, v$$

which reduces to

$$PQ\,\xi = \lambda\,\xi \tag{17}$$

characterizing the eigenvalues of PQ. Furthermore, the '⇐' part can be proved in a similar way. Suppose that we have the above equation. Then premultiplying $\mathcal{C}^*\mathcal{O}\,\mathcal{O}^*$ and defining $\bar{v} := \mathcal{C}^*\mathcal{O}\,\mathcal{O}^*\xi$ we obtain

$$\mathcal{H}^*\mathcal{H}\,\bar{v} = \lambda\,\bar{v}$$

which coincides with the eigenvalue problem of $\mathcal{H}^*\mathcal{H}$. □

Thus the singular value problem of the operator $\mathcal{H}$ is closely related to the eigenvalue problem of the matrix PQ, and a singular vector v of $\mathcal{H}$ is characterized by an eigenvector ξ of PQ.

Due to this property, the constants σ_i's in Equation (14) are called Hankel singular values. Furthermore, the Hankel norm $\|\Sigma\|_H$ of the operator Σ is defined by the L_2 gain of the corresponding Hankel operator as

$$\|\Sigma\|_H := \sup_{\substack{u \in L_2(0,\infty) \\ u \neq 0}} \frac{\|\mathcal{H}(u)\|_{L_2}}{\|u\|_{L_2}} = \sigma_1 . \tag{18}$$

Theorem 2 implies that the balanced truncation procedure preserves the Hankel norm of the original system, that is,

$$\|\Sigma^a\|_H = \|\Sigma\|_H. \tag{19}$$

It is also known that the error bound of this model order reduction procedure is given by

$$\|\Sigma - \Sigma^a\|_\infty \leq 2 \sum_{i=k+1}^{n} \sigma_i. \tag{20}$$

The relation between the Gramians and the Hankel, controllability and observability operators gives rise to both input-output operator interpretations as well as state-space interpretations of Hankel singular values and balanced truncation. These interpretations are crucial for the extension to nonlinear systems.

4 Basics of Nonlinear Balanced Realizations

This section gives a nonlinear extension of balanced realization introduced in the previous section. Let us consider the following asymptotically stable input-affine nonlinear system

$$\Sigma : \begin{cases} \dot{x} = f(x) + g(x)u \; x(0) = x^0 \\ y = h(x) \end{cases} \tag{21}$$

where $x(t) \in \mathbb{R}^n$, $u(t) \in \mathbb{R}^m$ and $y(t) \in \mathbb{R}^p$. The controllability operator $\mathcal{C} : U \to X$ with $X = \mathbb{R}^n$ and $U = L_2^m[0, \infty)$, and the observability operator $\mathcal{O} : X \to Y$ with $Y = L_2^p[0, \infty)$ for this system are defined by

$$\mathcal{C} : u \mapsto x^0 : \begin{cases} \dot{x} = -f(x) - g(x)u \; x(\infty) = 0 \\ x^0 = x(0) \end{cases}$$

$$\mathcal{O} : x^0 \mapsto y : \begin{cases} \dot{x} = f(x) \; x(0) = x^0 \\ y = h(x) \end{cases}.$$

This definition implies that the observability operator $\mathcal{O}$ is a map from the initial condition $x(0) = x^0$ to the output L_2 signal when no input is applied. To interpret the meaning of $\mathcal{C}$, let us consider a time-reversal behavior of the $\mathcal{C}$ operator as

$$\mathcal{C} : u \mapsto x^0 : \begin{cases} \dot{x} = f(x) + g(x)u(-t) \; x(-\infty) = 0 \\ x^0 = x(0) \end{cases}. \tag{22}$$

Then the controllability operator $\mathcal{C}$ can be regarded as a mapping from the input L_2 signal to the terminal state $x(0) = x^0$ when the initial state is $x(-\infty) = 0$. Therefore, as in the linear case, $\mathcal{C}$ and $\mathcal{O}$ represent the input-to-state behavior and the state-to-output behavior, respectively. As in the linear case, the Hankel operator for the nonlinear operator Σ in (16) is given by the composition of $\mathcal{C}$ and $\mathcal{O}$

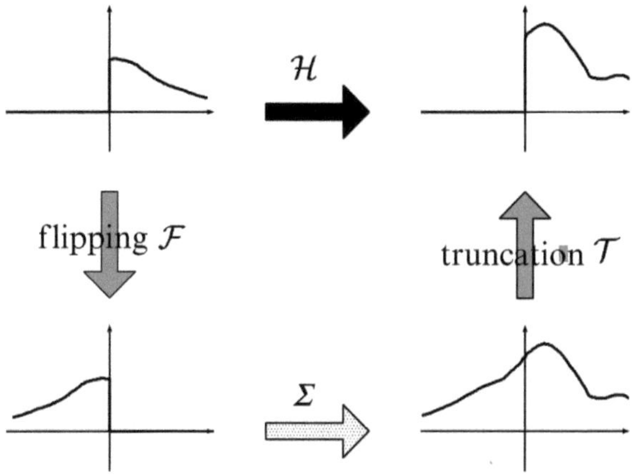

Fig. 2. Hankel operator $\mathcal{H}$ of Σ

$$\mathcal{H} := \mathcal{O} \circ \mathcal{C}. \tag{23}$$

The input-output mapping of a Hankel operator is explained in Figure 2. The lower side of the figure depicts the input-output behavior of the original operator Σ in Equation (21). The upper side depicts the input-output behavior of the Hankel operator of Σ, where the signal in the upper left side is the time-flipped signal of the lower left side signal. The flipping operator is defined by

$$\mathcal{F}(u(t)) := u(-t).$$

The upper right side signal is the truncated signal (to the space $L_2[0,\infty)$) of the lower left side signal. The corresponding truncation operator is given by

$$\mathcal{T}(y(t)) := \begin{cases} 0 & (t < 0) \\ y(t) & (t \geq 0) \end{cases}.$$

The definition of a Hankel operator implies that it describes the mapping from the input to the output generated by the state at $t = 0$. Hence we can analyze the relationship between the state and the input-output behavior of the original operator Σ by investigating its Hankel operator.

To this end, we need to define certain operators and functions related to Gramians in the linear case. First a norm-minimizing inverse $\mathcal{C}^\dagger : X \to U$ of $\mathcal{C}$ is introduced.

$$\mathcal{C}^\dagger : x^0 \mapsto u := \arg \min_{\mathcal{C}(u)=x^0} \|u\|$$

The operators $\mathcal{C}^\dagger$ and $\mathcal{O}$ yield the definitions of the controllability function $L_c(x)$ and the observability function $L_o(x)$ that are generalization of the controllability and observability Gramians, respectively.

$$L_c(x^0) := \frac{1}{2}\|\mathcal{C}^\dagger(x^0)\|^2 = \min_{\substack{u\in L_2(-\infty,0] \\ x(-\infty)=0,x(0)=x^0}} \frac{1}{2}\int_{-\infty}^{0} \| u(t) \|^2 \, dt \tag{24}$$

$$L_o(x^0) := \frac{1}{2}\|\mathcal{O}(x^0)\|^2 = \frac{1}{2}\int_{0}^{\infty} \| y(t) \|^2 \, dt, \; x(0)=x^0, \; u(t)\equiv 0, \quad 0\leq t<\infty. \tag{25}$$

These definitions imply that the controllability function $L_c(x^0)$ is the minimum input energy (in the L_2 norm sense) required to move from the initial state $x(-\infty) = 0$ to the terminal state $x(0) = x^0$, and that the observability function $L_o(x^0)$ is the output energy generated by the initial state $x(0) = x^0$ with zero input, respectively. If the system Σ is linear as in (11), then those functions are described by

$$L_c(x) = \frac{1}{2}x^{\mathrm{T}}P^{-1}x, \;\; L_o(x) = \frac{1}{2}x^{\mathrm{T}}Qx \tag{26}$$

with the controllability Gramian P and the observability Gramian Q the solutions of the Lyapunov equations (12) and (13). Here the inverse of P appears in the above equation because $\mathcal{C}^\dagger$ appears in the definition (24), whereas $\mathcal{C}$ can be used in the linear case. In order to obtain those functions $L_c(x)$ and $L_o(x)$, we need to solve a Hamilton-Jacobi equation and a Lyapunov equation.

Theorem 4. *[Scherpen93] Consider the system (21). Suppose that* 0 *is an asymptotically stable equilibrium point and that a smooth observability function $L_o(x)$ exists. Then $L_o(x)$ is the unique smooth solution of*

$$\frac{\partial L_o(x)}{\partial x} f(x) + \frac{1}{2}\, h(x)^{\mathrm{T}} h(x) = 0$$

with $L_o(0) = 0$. Furthermore, assume that a smooth controllability function $L_c(x)$ exists. Then $L_c(x)$ is the unique smooth solution of

$$\frac{\partial L_c(x)}{\partial x} f(x) + \frac{1}{2}\,\frac{\partial L_c(x)}{\partial x} g(x)g(x)^{\mathrm{T}}\frac{\partial L_c(x)}{\partial x}^{\mathrm{T}} = 0$$

with $L_c(0) = 0$ such that 0 *is an asymptotically stable equilibrium point of $\dot{x} = -f(x) - g(x)g(x)^{\mathrm{T}}(\partial L_c(x)/\partial x)^{\mathrm{T}}$.*

Similar to the linear case, the positive definiteness of the controllability and observability functions implies strong reachability and zero-state observability of the system Σ in (21), respectively. Combining these two properties, we can obtain the following result on the minimality of the system.

Theorem 5. *[SG00] Consider the system (21). Suppose that*

$$0 < L_c(x) < \infty$$
$$0 < L_o(x) < \infty$$

hold for all $x \neq 0$. Then the system is a minimal realization as defined in [Isidori95].

Similar to the linear case, $L_c(x)$ and $L_o(x)$ can be used to "measure the minimality" of a nonlinear dynamical system. Furthermore, a basis for nonlinear balanced realization is obtained as a nonlinear generalization of the relationship (15) in the linear case. For that, a factorization of $L_o(x)$ into a semi-quadratic form needs to be done, i.e., in a convex neighborhood of the equilibrium point 0 we can write

$$L_o(x) = \frac{1}{2}x^{\mathrm{T}}M(x)x, \quad \text{with} \quad M(0) = \frac{\partial^2 L_o}{\partial x^2}(0). \tag{27}$$

Now, an input-normal/output-diagonal form can be obtained.

Theorem 6. *[Scherpen93] Consider the system (21) on a neighborhood W of 0. Suppose that 0 is an asymptotically stable equilibrium point, that it is zero-state observable, that smooth controllability and observability functions $L_c(x)$ and $L_o(x)$ exist on W, and that $(\partial^2 L_c/\partial x^2)(0) > 0$ and $(\partial^2 L_o/\partial x^2)(0) > 0$ hold. Furthermore, assume that the number of distinct eigenvalues of $M(x)$ is constant on W. Then there exists coordinates such that the controllability and observability functions $L_c(x)$ and $L_o(x)$ satisfy*

$$L_c(x) = \frac{1}{2}\sum_{i=1}^{n} x_i^2 \tag{28}$$

$$L_o(x) = \frac{1}{2}\sum_{i=1}^{n} x_i^2 \tau_i(x) \tag{29}$$

where $\tau_1(x) \geq \tau_2(x) \geq \ldots \geq \tau_n(x)$.

A state-space realization satisfying the conditions (28) and (29) is called an *input-normal form*, and the functions $\tau_i(x)$, $i = 1, 2, \ldots, n$ are called singular value functions. We refer to [Scherpen93] for the construction of the coordinate transformation that brings the system in the form of Theorem 6. If a singular value function $\tau_i(x)$ is larger than $\tau_j(x)$, then the coordinate axis x_i plays more important role than the coordinate axis x_j does. Thus this realization is similar to the linear input-normal/output-diagonal realization (15), and it directly yields a tool for model order reduction of a nonlinear systems. However, a drawback of the above realization is that the the singular value functions $\tau_i(x)$'s and consequently, the corresponding realization are not unique, e.g. [GS01]. For example, if the observability function is given by

$$L_o(x) = \frac{1}{2}(x_1^2\tau_1(x) + x_2^2\tau_2(x)) = \frac{1}{2}(2x_1^2 + x_2^2 + x_1^2 x_2^2),$$

with the state-space $x = (x_1, x_2)$, then the corresponding singular value functions are

$$\tau_1(x) = 2 + kx_2^2$$
$$\tau_2(x) = 1 + (1-k)x_1^2$$

with an arbitrary scalar constant k. This example reveals that the singular value function are not uniquely determined by this characterization. To overcome these problems, balanced realization based on nonlinear singular value analysis introduced in Section 2 is investigated in the following section.

5 Balanced Realizations Based on Singular Value Analysis of Hankel Operators

In this section, application of singular value analysis to nonlinear Hankel operators determines a balanced realization with a direct input-output interpretation whereas the balanced realization of Theorem 6 is completely determined based on state-space considerations only. To this end, we consider the Hankel operator $\mathcal{H} : U \to Y$ as defined in (23) with $U = L_2^m[0,\infty)$ and $Y = L_2^p[0,\infty)$. Then Equation (9) yields

$$(\mathrm{d}\mathcal{H}(v))^* \, \mathcal{H}(v) = \lambda \, v, \quad \lambda \in \mathbb{R}, \quad v \in U. \tag{30}$$

Since we consider a singular value analysis problem on L_2 spaces, we need to find state trajectories of certain Hamiltonian dynamics, see e.g., [FS05]. In the linear case, Theorem 3 shows that we only need to solve an eigenvalue problem (17) on a finite dimensional space $X = \mathbb{R}^n$ to obtain the singular values and singular vectors of the Hankel operator $\mathcal{H}$. Here we provide its nonlinear counterpart as follows.

Theorem 7. *[FS05] Consider the Hankel operator defined by Equation (23). Suppose that the operators $\mathcal{C}^\dagger$ and $\mathcal{O}$ exist and are smooth. Suppose moreover that $\lambda \in \mathbb{R}$ and $\xi \in X$ satisfy the following equation*

$$\frac{\partial L_o(\xi)}{\partial \xi} = \lambda \, \frac{\partial L_c(\xi)}{\partial \xi}, \quad \lambda \in \mathbb{R}, \quad \xi \in X. \tag{31}$$

Then λ and

$$v := \mathcal{C}^\dagger(\xi) \tag{32}$$

satisfy Equation (9). That is, v defined above is a singular vector of $\mathcal{H}$.

Though the original singular value analysis problem (9) is a nonlinear problem on an infinite dimensional signal space $U = L_2^m[0,\infty)$, the problem to be solved in the above theorem is a nonlinear algebraic equation on a finite dimensional space $X = \mathbb{R}^n$ which is also related to a nonlinear eigenvalue problem on X, see [Fujimoto04].

In the linear case, where $L_c(x)$ and $L_o(x)$ are given by (26), Equation (31) reduces to

$$\xi^{\mathrm{T}} Q = \lambda \, \xi^{\mathrm{T}} P^{-1}$$

where P and Q are the controllability and observability Gramians. This equation is equivalent to (17), i.e., λ and ξ are an eigenvalue and an eigenvector of PQ. Furthermore, Equation (32) characterizes the relationship between a singular vector v of $\mathcal{H}$ and an eigenvector ξ of PQ as in the linear case result. Thus Theorem 7 can be regarded as a nonlinear counterpart of Theorem 3.

In the linear case, there always exist n independent pairs of eigenvalues and eigenvectors of PQ. What happens in the nonlinear case? The answer is provided in the following theorem.

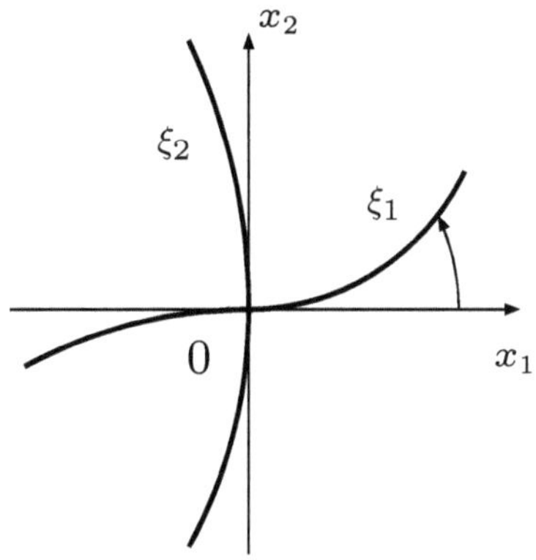

Fig. 3. Configuration of $\xi_1(s)$ and $\xi_2(s)$ in the case $n = 2$

Theorem 8. *[FS05] Consider the system Σ in (21) and the Hankel operator $\mathcal{H}$ in Equation (23) with $X = \mathbb{R}^n$. Suppose that the Jacobian linearization of the system has n distinct Hankel singular values. Then Equation (31) has n independent solution curves $\xi = \xi_i(s)$, $s \in \mathbb{R}$, $i = 1, 2, \ldots, n$ intersecting to each other at the origin and satisfying the condition*

$$\|\xi_i(s)\| = |s|.$$

In the linear case, the solutions of Equation (31) are the lines (orthogonally) intersecting to each other at the origin. The above theorem shows that instead of these lines, in the nonlinear case n independent curves $x = \xi_i(s)$, $i = 1, 2, \ldots, n$ exist. For instance, if the dimension of the state is $n = 2$, the solution of Equation (31) is illustrated in Figure 3.

We can relate the solutions $\xi_i(s)$ to the singular values of the Hankel operator $\mathcal{H}$. Let $v_i(s)$ and $\sigma_i(s)$ denote the singular vector and the singular value parameterized by s corresponding to $\xi_i(s)$. Then we have

$$\begin{aligned} v_i(s) &:= \mathcal{C}^\dagger(\xi_i(s)) \\ \sigma_i(s) &:= \frac{\|\mathcal{H}(v_i(s))\|_{L_2}}{\|v_i(s)\|_{L_2}} = \frac{\|\mathcal{O}(\xi_i(s))\|_{L_2}}{\|\mathcal{C}^\dagger(\xi_i(s))\|_{L_2}} \\ &= \sqrt{\frac{L_o(\xi_i(s))}{L_c(\xi_i(s))}}. \end{aligned}$$

By this equation, we can obtain an explicit expression of the singular values $\sigma_i(s)$'s of the Hankel operator $\mathcal{H}$. These functions $\sigma_i(s)$'s are called *Hankel singular values*. Without loss of generality we assume that the following equation holds for $i = 1, 2, \ldots, n$ in a neighborhood of the origin

$$\min\{\sigma_i(s), \sigma_i(-s)\} > \max\{\sigma_{i+1}(s), \sigma_{i+1}(-s)\}. \tag{33}$$

As in the linear case, the solution curves $\xi_i(s)$'s play the roles of the coordinate axes of balanced realization. By applying an isometric coordinate transformation which maps the solution curves $\xi_i(s)$'s into the coordinate axes, we obtain a realization whose (new) coordinate axes x_i are the solution of Equation (31), i.e.,

$$\left.\frac{\partial L_o(x)}{\partial x}\right|_{x=(0,\ldots,0,x_i,0,\ldots,0)} = \lambda \left.\frac{\partial L_c(x)}{\partial x}\right|_{x=(0,\ldots,0,x_i,0,\ldots,0)} \tag{34}$$

$$\sigma_i(x_i) = \sqrt{\frac{L_o(0,\ldots,0,x_i,0,\ldots,0)}{L_c(0,\ldots,0,x_i,0,\ldots,0)}}. \tag{35}$$

Equation (35) implies that the new coordinate axes x_i, $i = 1,\ldots,n$ are the solutions of Equation (31) for Hankel singular value analysis. Therefore the Hankel norm defined in (18) can be obtained by

$$\begin{aligned}
\|\Sigma\|_H &= \sup_{u\neq 0} \frac{\|\mathcal{H}(u)\|_{L_2}}{\|u\|_{L_2}} \\
&= \sup_{s\in\mathbb{R}} \max_i \sigma_i(s) \\
&= \sup_{x_1\in\mathbb{R}} \sqrt{\frac{L_o(x_1,0,\ldots,0)}{L_c(x_1,0,\ldots,0)}}
\end{aligned}$$

provided the ordering condition (33) holds for all $s \in \mathbb{R}$. Furthermore, apply this coordinate transformation recursively to all lower dimensional subspaces such as $(x_1, x_2, \ldots, x_k, 0, \ldots, 0)$, then we can obtain a state-space realization satisfying Equation (35) and

$$x_i = 0 \iff \frac{\partial L_o(x)}{\partial x_i} = 0 \iff \frac{\partial L_c(x)}{\partial x_i} = 0. \tag{36}$$

This property is crucial for balanced realization and model order reduction. Using tools from differential topology, e.g. [Milnor65], we can prove that this realization is diffeomorphic to the following *precise* input-normal/output-diagonal realization.

Theorem 9. *[FS03] Consider the system Σ in (21). Suppose that the assumptions in Theorem 8 hold. Then the there exists a coordinates in a neighborhood of the origin such that the system is in input-normal/output-diagonal form satisfying*

$$\begin{aligned}
L_c(x) &= \frac{1}{2}\sum_{i=1}^{n} x_i^2 \\
L_o(x) &= \frac{1}{2}\sum_{i=1}^{n} x_i^2\, \sigma_i(x_i)^2
\end{aligned}$$

This realization is much more precise than that in Theorem 6 in the following senses: (a) The solutions of Equation (31) coincide with the coordinate axes, that is,

Equation (34) holds. (b) The ratio of the observability function L_o to the controllability function L_c equals the singular values $\sigma_i(x_i)$'s on the coordinate axes, that is Equation (35) holds. (c) Furthermore, an exact balanced realization can be obtained by a coordinate transformation

$$z_i = \phi_i(x_i) := x_i \sqrt{\sigma_i(x_i)} \tag{37}$$

which is well-defined in a neighborhood of the origin.

Corollary 1. *[FS03] The coordinate change (37) transforms the input-normal realization in Theorem 9 into the following form*

$$L_c(z) = \frac{1}{2}\sum_{i=1}^{n} \frac{z_i^2}{\sigma_i(z_i)}$$
$$L_o(z) = \frac{1}{2}\sum_{i=1}^{n} z_i^2 \sigma_i(z_i).$$

Since we only use the coordinate transformation (37) preserving the coordinate axes, the realization obtained here also satisfies the properties (a) and (b) explained above. The controllability and observability functions can be written as

$$L_c(z) = \frac{1}{2}\, z^{\mathrm{T}} \underbrace{\mathrm{diag}(\sigma_1(z_1), \ldots, \sigma_n(z_n))^{-1}}_{P(z)} z$$
$$L_o(z) = \frac{1}{2}\, z^{\mathrm{T}} \underbrace{\mathrm{diag}(\sigma_1(z_1), \ldots, \sigma_n(z_n))}_{Q(z)} z$$

Here $P(z)$ and $Q(z)$ can be regarded as nonlinear counterparts of the controllability and observability Gramians as observed in Equation (14) with the relation (26) since

$$P(z) = Q(z) = \mathrm{diag}(\sigma_1(z_1), \sigma_2(z_2), \ldots, \sigma_n(z_n)). \tag{38}$$

The axes of this realization are uniquely determined. We call this state-space realization a *balanced realization* of the nonlinear system Σ in Equation (21). As in the linear case, both the relationship between the input-to-state and state-to-output behavior and that among the coordinate axes are balanced.

6 Model Order Reduction

An important application of balanced realizations is that it is a tool for model order reduction called *balanced truncation*. Here, a model order reduction method preserving the Hankel norm of the original system is proposed. Suppose that the system (21) is balanced in the sense that it satisfies Equations (35) and (36). Note that the realizations in Theorem 9 and Corollary 1 satisfy these conditions. Suppose moreover that

$$\min\{\sigma_k(s), \sigma_k(-s)\} \gg \max\{\sigma_{k+1}(s), \sigma_{k+1}(-s)\}$$

holds with a certain k ($1 \le k < n$). Divide the state into two vectors $x = (x^a, x^b)$

$$\begin{aligned} x^a &:= (x_1, \dots, x_k) \in \mathbb{R}^k \\ x^b &:= (x_{k+1}, \dots, x_n) \in \mathbb{R}^{n-k}, \end{aligned}$$

and the vector field into two vector fields accordingly

$$\begin{aligned} f(x) &= \begin{pmatrix} f^a(x) \\ f^b(x) \end{pmatrix} \\ g(x) &= \begin{pmatrix} g^a(x) \\ g^b(x) \end{pmatrix}, \end{aligned}$$

and truncate the state by substituting $x^b = 0$. Then we obtain a k-dimensional state-space model Σ^a with the state x^a (with a $(n-k)$-dimensional residual model Σ^b with the state x^b).

$$\Sigma^a : \begin{cases} \dot{x}^a = f^a(x^a, 0) + g^a(x^a, 0)u^a \\ y^a = h(x^a, 0) \end{cases} \tag{39}$$

$$\Sigma^b : \begin{cases} \dot{x}^b = f^b(0, x^b) + g^b(0, x^b)u^b \\ y^b = h(0, x^b) \end{cases} \tag{40}$$

This procedure is called *balanced truncation*. The obtained reduced order models have preserved the following properties.

Theorem 10. *[FS01, FS06] Suppose that the system Σ satisfies Equations (35) and (36) and apply the balanced truncation procedure explained above. Then the controllability and observability functions of the reduced order models Σ^a and Σ^b denoted by L_c^a, L_c^b, L_o^a and L_o^b, respectively, satisfy the following equations*

$$\begin{aligned} L_c^a(x^a) &= L_c(x^a, 0), \quad L_o^a(x^a) = L_o(x^a, 0) \\ L_c^b(x^b) &= L_c(0, x^b), \quad L_o^b(x^b) = L_o(0, x^b) \end{aligned}$$

which implies

$$\begin{aligned} \sigma_i^a(x_i^a) &= \sigma_i(x_i^a), \quad i = 1, 2, \dots, k \\ \sigma_i^b(x_i^b) &= \sigma_{i+k}(x_i^b), \quad i = 1, 2, \dots, n-k \end{aligned}$$

with the singular values σ^a's of the system Σ^a and the singular values σ^b of the system Σ^b. In particular, if σ_1 is defined globally, then

$$\|\Sigma^a\|_H = \|\Sigma\|_H. \tag{41}$$

Theorem 10 states that the important characteristics of the original system such as represented by the controllability and observability functions and Hankel singular values are preserved. Moreover, by Theorem 5, this implies that the controllability, observability, minimality and the gain property is preserved under the model reduction. These preservation properties hold for truncation of any realization satisfying the conditions (35) and (36), such as the realizations in Theorem 9 and Corollary 1 [FS01]. Furthermore, concerning the stability, (global) Lyapunov stability and local asymptotic stability are preserved with this procedure as well. Note that this theorem is a natural nonlinear counterpart of Theorem 2 and Equation (19). However, a nonlinear counterpart of the error bound of the reduced order model as in (20) has not been found yet.

7 Numerical Example

In this section, we apply the proposed model order reduction procedure to a double pendulum (an underactuated two degrees of freedom robot manipulator) as depicted in Figure 4.

Here m_i denotes the mass located at the end of the i-th link, l_i denotes the length of the i-th link, μ_i denotes the friction coefficient of the i-th link, and x_i denotes the angle of the i-th link. We select the physical parameters as $l_1 = l_2 = 1$, $m_1 = m_2 = 1$, $\mu_1 = \mu_2 = 1$, $g_0 = 9.8$ with g_0 the gravity coefficient. The dynamics of this system can be described by an input-affine nonlinear system model (21) with 4 dimensional state-space

$$x = (x_1, x_2, x_3, x_4) := (x_1, x_2, \dot{x}_1, \dot{x}_2). \tag{42}$$

The input u denotes the torque applied to the first link at the first joint and the output y denotes the horizontal and the vertical coordinates of the position of the mass m_2. The potential energy $V(x)$ and the kinetic energy $T(x)$ for this system are described by

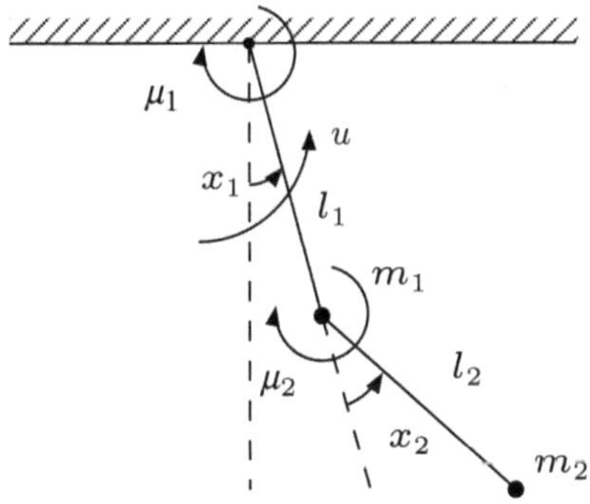

Fig. 4. The double pendulum

$$
\begin{aligned}
V(x) &= -m_1 g_0 l_1 \cos x_1 - m_2 g_0 l_1 \cos x_1 - m_2 g_0 l_2 \cos(x_1 + x_2) \\
T(x) &= \frac{1}{2} (\dot{x}_1, \dot{x}_2) M(x) \begin{pmatrix} \dot{x}_1 \\ \dot{x}_2 \end{pmatrix} \\
M(x) &= \\
&\begin{pmatrix} m_1 l_1^2 + m_2 l_1^2 + m_2 l_2^2 + 2 m_2 l_1 l_2 \cos x_2 & m_2 l_2^2 + m_2 l_1 l_2 \cos x_2 \\ m_2 l_2^2 + m_2 l_1 l_2 \cos x_2 & m_2 l_2^2 \end{pmatrix}
\end{aligned} \tag{43}
$$

where $M(x)$ denotes the inertia matrix. Then the dynamics of this system is obtained by the Lagrange's method as follows

$$
\frac{\mathrm{d}}{\mathrm{d}t} \frac{\partial L(x)}{\partial(\dot{x}_1, \dot{x}_2)}^{\mathrm{T}} - \frac{\partial L(x)}{\partial(x_1, x_2)}^{\mathrm{T}} = \begin{pmatrix} u - \mu_1 \dot{x}_1 \\ -\mu_2 \dot{x}_2 \end{pmatrix} \tag{44}
$$

with the Lagrangian $L(x) := T(x) - V(x)$. This equation reduces to the system (21) with

$$
\begin{aligned}
f(x) &= \begin{pmatrix} x_3 \\ x_4 \\ M^{-1} \left(\frac{\partial(T-V)}{\partial(x_1,x_2)}^{\mathrm{T}} - \dot{M} \begin{pmatrix} \dot{x}_1 \\ \dot{x}_2 \end{pmatrix} - \begin{pmatrix} \mu_1 \dot{x}_1 \\ \mu_2 \dot{x}_2 \end{pmatrix} \right) \end{pmatrix} \\
g(x) &= \begin{pmatrix} 0 \\ 0 \\ M^{-1} \begin{pmatrix} 1 \\ 0 \end{pmatrix} \end{pmatrix} \\
h(x) &= \begin{pmatrix} l_1 \sin x_1 + l_2 \sin(x_1 + x_2) \\ l_1(1 - \cos x_1) + l_2(1 - \cos(x_1 + x_2)) \end{pmatrix}.
\end{aligned}
$$

See [FT06] for the details of the model.

For computing L_o and L_c, we use the method based on Taylor series expansion proposed in [Lukes69]. Then we need to solve the nonlinear algebraic equation (31). Although it is much easier to be solved compared with the original singular value analysis problem in (30), it is still difficult to obtain a closed form solution. Again using Taylor series expansion we can prove that the computation of Equation (31) reduces to solving linear algebraic equations recursively. Applying this procedure and calculating the balancing coordinate transformation up to the 4-th order terms of the Taylor series expansion, results in the following Hankel singular value functions.

$$
\begin{aligned}
\sigma_1(x_1)^2 &= 1.98 \times 10^{-1} + 4.14 \times 10^{-4} x_1^2 + o(|x_1|^3) \\
\sigma_2(x_2)^2 &= 1.72 \times 10^{-1} + 3.28 \times 10^{-4} x_2^2 + o(|x_2|^3) \\
\sigma_3(x_3)^2 &= 5.83 \times 10^{-5} + 1.51 \times 10^{-4} x_3^2 + o(|x_3|^3) \\
\sigma_4(x_4)^2 &= 9.37 \times 10^{-6} + 9.22 \times 10^{-6} x_4^2 + o(|x_4|^3)
\end{aligned}
$$

These functions are depicted in Figure 5 where the solid line denotes σ_1, the dotted line denotes σ_2, the dashed line denotes σ_3 and the dashed and dotted line denotes σ_4.

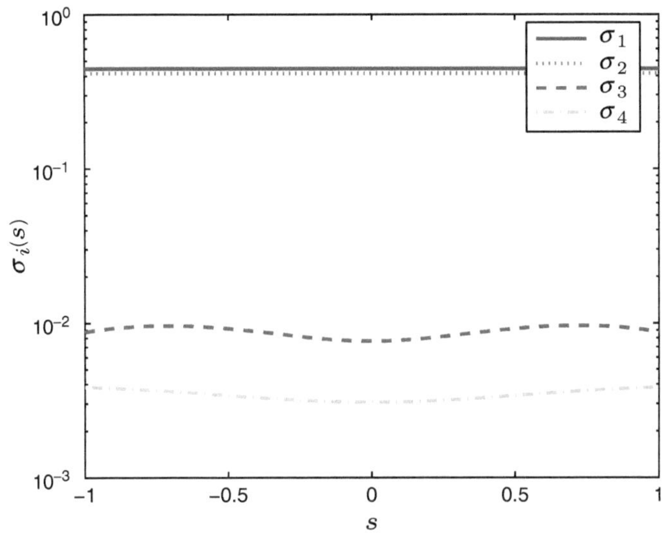

Fig. 5. Hankel singular value functions $\sigma_i(s)$, $i = 1, \ldots, 4$

From this figure, we conclude that in a neighborhood of 0 $\sigma_2(x_2) \gg \sigma_3(x_3)$, and thus that an appropriate dimension of the reduced order model is 2. We can now apply the balanced truncation procedure as presented in the previous section.

We have executed some simulations of the original and reduced models to evaluate the effectiveness of the proposed model order reduction method. Here the time responses for impulsive inputs are depicted in the figures, i.e., Figure 6 describes the response of the horizontal movement and Figure 7 describes the response of the vertical movement. In the figures, the solid line denotes the response of the original system, the dashed line denotes the response of the linearized reduced order model, and the dashed/dotted line denotes the response of the nonlinear reduced order model.

In Figure 6, all trajectories are identical which indicates that both linear and nonlinear reduced order models can approximate the behavior of the original model well. However, in Figure 7, one can observe that the trajectory of the linear reduced order model is quite different from the original whereas the trajectory of the nonlinear reduced order model is almost identical with that of the original system. This is due to the fact that the linearization of the vertical displacement of the mass m_2 is 0 since it consists of a cosine function of the state. These simulations demonstrate the effectiveness of our nonlinear balanced truncation method. It is noted that the proposed computation algorithm is currently only applicable to systems whose size is relatively small. A big progress on computation of nonlinear balanced realization is required to make it be applicable to real-world large scale systems.

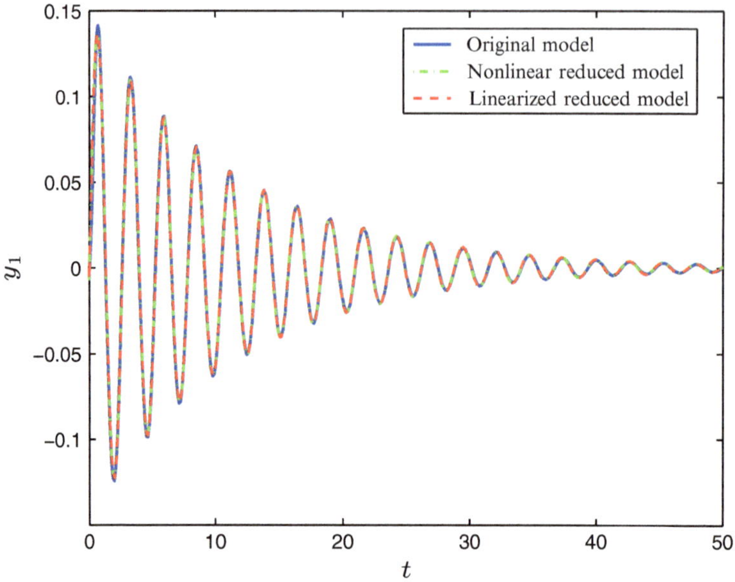

Fig. 6. The horizontal displacement

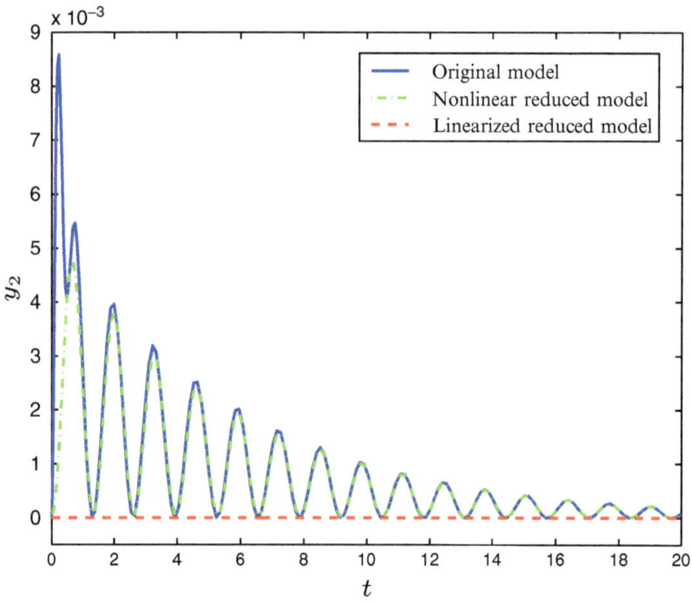

Fig. 7. The vertical displacement

8 Conclusion

In this paper, we have presented an overview of singular value analysis of nonlinear operators and its application to balanced realizations and model order reduction methods for nonlinear systems. Recent development in this area of research provides a precise and complete basis for model oder reduction of nonlinear dynamical systems. A reduced order model derived by this technique preserves many important properties of the original system such as controllability, observability, stability and the Hankel norm. Compared with the theoretical results, however, computational developments are still in their infancy, meaning that large scale nonlinear systems are still difficult to handle. Future research should thus include a strong focus on the computational algorithms for making nonlinear balanced truncation a useful tool in large scale applications.

References

[Antoulas05] Antoulas, A.C.: Approximation of large-scale dynamical systems. SIAM, Philadelphia (2005)

[Batt70] Batt, J.: Nonlinear compact mappings and their adjoints. Math. Ann., **189**, 5–25 (1970)

[Fujimoto04] Fujimoto, K.: What are singular values of nonlinear operators?. Proc. 43rd IEEE Conf. on Decision and Control, 1623–1628 (2004)

[FS01] Fujimoto, K., Scherpen, J.M.A.: Balancing and model reduction for nonlinear systems based on the differential eigenstructure of Hankel operators. Proc. 40th IEEE Conf. on Decision and Control, 3252–3257 (2001)

[FS03] Fujimoto, K., Scherpen, J.M.A.: Nonlinear balanced realization based on singular value analysis of Hankel operators. Proc. 42nd IEEE Conf. on Decision and Control, 6072–6077 (2003)

[FS05] Fujimoto, K., Scherpen, J.M.A.: Nonlinear input-normal realizations based on the differential eigenstructure of Hankel operators. IEEE Trans. Autom. Contr., **50**, 2–18 (2005)

[FS06] Fujimoto, K., Scherpen, J.M.A.: Balanced realization and model order reduction for nonlinear systems based on singular value analysis. Submitted (2006)

[FSG02] Fujimoto, K., Scherpen, J.M.A., Gray, W.S.: Hamiltonian realizations of nonlinear adjoint operators. Automatica, **38**, 1769–1775 (2002)

[FT06] Fujimoto, K., Tsubakino, D.: On computation of nonlinear balanced realization and model reduction. Proc. American Control Conference, 460–465 (2006)

[Glover84] Glover, K.: All optimal Hankel-norm approximations of linear multivariable systems and their l^{∞}-error bounds. Int. J. Control, **39**, 1115–1193 (1984)

[GS01] Gray, W.S., Scherpen, J.M.A.: State dependent matrices in quadratic forms. Systems & Control Letters, **44**, 219–232 (2001)

[HE02] Hahn, J., Edgar, T.F.: An improved method for nonlinear model reduction using balancing of empirical gramians. Comp. Chem. Eng., **26**, 1379–1397 (2002)

[Isidori95] Isidori, A.: Nonlinear Control Systems. Springer-Verlag, Berlin, third edition (1995)

[JTP94] Jiang, Z.P., Teel, A.R., Praly, L.: Small-gain theorem for ISS systems and applications. Mathematics of Control, Signals, and Systems, **7**, 95–120 (1994)

[JS82] Jonckheere, E.A., Silverman, L.M.: Singular value analysis of deformable systems. Circuits, Systems and Signal Processing, **1**, 447–470 (1982)

[LMG02] Lall, S., Marsden, J.E., Glavaski, S.: A subspace approach to balanced truncation for model reduction of nonlinear control systems. Int. J. Robust and Nonlinear Control, **12**, 519–535 (2002)

[Lukes69] Lukes, D.L.: Optimal regulation of nonlinear dynamical systems. SIAM J. Control, **7**, 75–100 (1969)

[Milnor65] Milnor, J.W.: Topology from the Differential Viewpoint. Princeton University Press, New Jersey (1965)

[Moore81] Moore, B.C.: Principal component analysis in linear systems: controllability, observability and model reduction. IEEE Trans. Autom. Contr., **26**, 17–32 (1981)

[NK00] Newman, A.J., Krishnaprasad, P.S.: Computing balanced realizations for nonlinear systems. Proc. Symp. Mathematical Theory of Networks and Systems (2000)

[NK98] Newman, A.J., Krishnaprasad, P.S.: Computation for nonlinear balancing. Proc. 37th IEEE Conf. on Decision and Control, 4103–4104 (1998)

[OA01] Obinata, G., Anderson, B.D.O.: Model Reduction for Control System Design. Springer-Verlag, London (2001)

[Scherpen93] Scherpen, J.M.A.: Balancing for nonlinear systems. Systems & Control Letters, **21**, 143–153 (1993)

[SG00] Scherpen, J.M.A., Gray, W.S.: Minimality and local state decompositions of a nonlinear state space realization using energy functions. IEEE Trans. Autom. Contr., **45**, 2079–2086 (2000)

[SG02] Scherpen, J.M.A., Gray, W.S.: Nonlinear Hilbert adjoints: Properties and applications to Hankel singular value analysis. Nonlinear Analysis: Theory, Methods and Applications, **51**, 883–901 (2002)

[SW96] Sontag, E.D., Wang, Y.: New characterization of the input to state stability property. IEEE Trans. Autom. Contr., **41**, 1283–1294 (1996)

[Stewart93] Stewart, G.W.: On the early history of the singular value decomposition. SIAM Review, **35**, 551–566 (1993)

[VG00] Verriest, E.I., Gray, W.S.: Flow balancing nonlinear systems. Proc. 2000 Int. Symp. Math. Th. Netw. Syst. (2000)

[ZDG96] Zhou, K., Doyle, J.C., Glover, K.: Robust and Optimal Control. Prentice-Hall, Inc., Upper Saddle River, N.J. (1996)

Part III

Research Aspects and Applications

Matrix Functions

Andreas Frommer[1] and Valeria Simoncini[2]

[1] Fachbereich Mathematik und Naturwissenschaften, Universität Wuppertal, D-42097 Wuppertal, Germany
frommer@math.uni-wuppertal.de
[2] Dipartimento di Matematica, Università di Bologna, Piazza di Porta S. Donato 5, I-40127 Bologna, and CIRSA, Ravenna, Italy
valeria.simoncini@unibo.it

1 Introduction

In this chapter, we give an overview on methods to compute functions of a (usually square) matrix A with particular emphasis on the matrix exponential and the matrix sign function. We will distinguish between methods which indeed compute the entire matrix function, i.e. they compute a matrix, and those which compute the action of the matrix function on a vector. The latter task is particularly important in the case where we have to deal with a very large (and possibly sparse) matrix A or in situations, where A is not available as a matrix but just as a function which returns Ax for any input vector x. Computing the action of a matrix function on a vector is a typical model reduction problem, since the resulting techniques usually rely on approximations from small-dimensional subspaces.

This chapter is organized as follows: In section 2 we introduce the concept of a matrix function $f(A)$ in detail, essentially following [38] and [27]. Section 3 gives an assessment of various general computational approaches for either obtaining the whole matrix $f(A)$ or its action $f(A)v$ on a vector v. Sections 4 and 5 then give much more details for two specific functions, the exponential and the sign functions, which, as we will show, are particularly important in many areas like control theory, simulation of physical systems and other application fields involving the solution of certain ordinary or partial differential equations. The applicability of matrix functions in general, and of the exponential and the sign functions in particular, is vast. However, we will limit our discussion to characterizations and to application problems that are mostly related to Model Order Reduction. For a comprehensive analysis of matrix functions and their computation we refer to the recent book by Nick Higham [34].

2 Matrix Functions

In this section we address the following general question: Given a function $f : \mathbb{C} \to \mathbb{C}$, is there a canonical way to extend this function to square matrices, i.e.

to extend f to a mapping from $\mathbb{C}^{n\times n}$ to $\mathbb{C}^{n\times n}$? If f is a polynomial p of degree d, $f(z) = p(z) = \sum_{k=0}^{d} a_k z^k$, the canonical extension is certainly given by

$$p : \mathbb{C}^{n\times n} \to \mathbb{C}^{n\times n}, \quad p(A) = \sum_{k=0}^{d} a_k A^k. \tag{1}$$

If $f(z)$ can be expressed by a power series, $f(z) = \sum_{k=0}^{\infty} a_k z^k$, a natural next step is to put

$$f(A) = \sum_{k=0}^{\infty} a_k A^k, \tag{2}$$

but for (2) to make sense we must now discuss convergence issues. The main result is given in the following theorem, the proof of which gives us valuable further information on matrix functions. Recall that the spectrum $\operatorname{spec}(A)$ is the set of all eigenvalues of A.

Theorem 1. *Assume that the power series $f(z) = \sum_{k=0}^{\infty} a_k z^k$ is convergent for $|z| < \rho$ with $\rho > 0$ and assume that* $\operatorname{spec}(A) \subset \{z \in \mathbb{C} : |z| < \rho\}$. *Then the series (2) converges.*

Proof. Let T be the transformation matrix occuring in the Jordan decomposition

$$A = TJT^{-1}, \tag{3}$$

with

$$J = \begin{bmatrix} J_{m_1}(\lambda_1) & & 0 \\ & \ddots & \\ 0 & & J_{m_\ell}(\lambda_\ell) \end{bmatrix} =: \operatorname{diag}(\, J_{m_1}(\lambda_1), \ldots, J_{m_\ell}(\lambda_\ell)\,). \tag{4}$$

Here, $\lambda_1, \ldots, \lambda_\ell$ are the (not necessarily distinct) eigenvalues of A and m_j is the size of the jth Jordan block associated with λ_j, i.e.

$$J_{m_j}(\lambda_j) = \begin{pmatrix} \lambda_j & 1 & 0 & \cdots & 0 \\ 0 & \lambda_j & 1 & \ddots & \vdots \\ \vdots & \ddots & \ddots & \ddots & 0 \\ \vdots & & \ddots & \ddots & 1 \\ 0 & \cdots & \cdots & 0 & \lambda_j \end{pmatrix} =: \lambda_j I + S_{m_j} \in \mathbb{C}^{m_j \times m_j}, \tag{5}$$

and $\sum_{j=1}^{\ell} m_j = n$. For each λ_j, the powers of $J_{m_j}(\lambda_j)$ are given by

$$J_m(\lambda_j)^k = \sum_{\nu=0}^{k} \binom{k}{\nu} \lambda_j^{k-\nu} \cdot S_{m_j}^{\nu}.$$

Note that $S_{m_j}^{\nu}$ has zero entries everywhere except for the ν-th upper diagonal, whose entries are equal to 1. In particular, $S_{m_j}^{\nu} = 0$ for $\nu \geq m_j$. Therefore,

$$f(J_{m_j}(\lambda_j)) = \sum_{k=0}^{\infty} a_k \sum_{\nu=0}^{k} \binom{k}{\nu} \lambda_j^{k-\nu} \cdot S_{m_j}^{\nu},$$

and for ν and j fixed we have

$$\sum_{k=0}^{\infty} a_k \binom{k}{\nu} \lambda_j^{k-\nu} = \sum_{k=0}^{\infty} \frac{1}{\nu!} \cdot a_k \cdot (k \cdot \ldots \cdot (k-\nu+1))\lambda_j^{k-\nu} = \frac{1}{\nu!} f^{(\nu)}(\lambda_j).$$

Note that the last cquality holds in the sense of absolute convergence because λ_j lies within the convergence disk of the series. This shows that the series $f(J_{m_j}(\lambda_j))$ converges. Plugging these expressions into the series from (2) we obtain the value of the original (now convergent) series,

$$\begin{aligned} f(A) &= T\mathrm{diag}\left(f(J_{m_1}(\lambda_1)), \ldots, f(J_{m_\ell}(\lambda_\ell))\right) T^{-1} \\ &= T\mathrm{diag}\left(\sum_{\nu=0}^{m_1-1} \frac{1}{\nu!} f^{(\nu)}(\lambda_1) \cdot S_{m_1}^{\nu}, \ldots, \sum_{\nu=0}^{m_\ell-1} \frac{1}{\nu!} f^{(\nu)}(\lambda_\ell) \cdot S_{m_\ell}^{\nu} \right) T^{-1}. \end{aligned} \tag{6}$$

□

It may happen that a function f cannot be expressed by a series converging in a large enough disk. If f is sufficiently often differentiable at the eigenvalues of A, then the right-hand side of (6) is still defined. We make it the basis of our final definition of a matrix function.

Definition 1. *Let $A \in \mathbb{C}^{n\times n}$ be a matrix with* $\mathrm{spec}(A) = \{\lambda_1, \ldots, \lambda_\ell\}$ *and Jordan normal form*

$$J = T^{-1}AT = \mathrm{diag}(\, J_{m_1}(\lambda_1), \ldots, J_{m_\ell}(\lambda_\ell)\,). \tag{7}$$

Assume that the function $f : \mathbb{C} \to \mathbb{C}$ is $m_j - 1$ times differentiable at λ_j for $j = 1, \ldots, \ell$. Then the matrix function $f(A)$ is defined as $f(A) = Tf(J)T^{-1}$ where

$$f(J) = \mathrm{diag}(f(J_{m_1}(\lambda_1)), \ldots, f(J_{m_\ell}(\lambda_\ell))),$$

with

$$f(J_{m_j}(\lambda_j)) = \sum_{\nu=0}^{m_j-1} \frac{1}{\nu!} f^{(\nu)}(\lambda_j) \cdot S_{m_j}^{\nu}.$$

This definition makes explicit use of the Jordan canonical form and of the associated transformation matrix T. Neither T nor J are unique, but it can be shown – as is already motivated by (2) – that $f(A)$ as introduced in Definition 1 does not depend on the particular choice of T or J.

As a first consequence of Definition 1 we note the following important property.

Proposition 1. *With the notation above, it holds $f(A) = p(A)$, where p is the polynomial of degree not greater than $n-1$ which interpolates the eigenvalues λ_j of A in the Hermite sense (i.e. $f^{(\nu)}(\lambda_j) = p^{(\nu)}(\lambda_j)$ for all relevant ν's and j's).*

The polynomial p in Proposition 1 will not only depend on f, but also on A or, more precisely, on the minimal polynomial of A (of which the multiplicity of an eigenvalue λ determines the maximal block size m_j for the Jordan blocks corresponding to this eigenvalue). When A is normal, T is an orthogonal matrix and all Jordan blocks have size one, i.e. we have

$$J = \text{diag}(\lambda_1, \dots, \lambda_n). \tag{8}$$

So, in this particular case, we do not need any differentiability assumption on f.

A further representation of $f(A)$ can be derived in the case when f is analytic in a simply connected region Ω containing $\text{spec}(A)$. Let γ be a curve in Ω with winding number +1 w.r.t. a point $z \in \Omega$. The Residue Theorem tells us

$$\frac{f^{(\nu)}(z)}{\nu!} = \frac{1}{2\pi i} \oint_\gamma \frac{f(t)}{(t-z)^{\nu+1}} dt. \tag{9}$$

Let $J_{m_j}(\lambda_j)$ be a Jordan block associated with λ_j and let $z \neq \lambda_j$. Then

$$(zI - J_{m_j})^{-1} = ((z - \lambda_j)I - S_{m_j})^{-1} = \frac{1}{z - \lambda_j} \cdot \sum_{\nu=0}^{m_j - 1} \left(\frac{1}{z - \lambda_j} \cdot S_{m_j} \right)^\nu, \tag{10}$$

from which we get

$$\begin{aligned} \frac{1}{2\pi i} \oint_\gamma f(z)(zI - J_{m_j})^{-1} dz &= \sum_{\nu=0}^{m_j-1} \frac{1}{2\pi i} \oint_\gamma \frac{f(z)}{(z - \lambda_j)^{\nu+1}} S_{m_j}^\nu dz \\ &= \sum_{\nu=0}^{m_j-1} \frac{f^{(\nu)}(\lambda_j)}{\nu!} \cdot S_{m_j}^\nu, \end{aligned}$$

the second line holding due to (9). Using this for each Jordan block in Definition 1 and recombining terms we obtain the following integral representation of $f(A)$,

$$f(A) = \frac{1}{2\pi i} \oint_\gamma f(t)(tI - A)^{-1} dt. \tag{11}$$

3 Computational Aspects

It is not necessarily a good idea to stick to one of the definitions of matrix function given in the previous section when it comes to numerically compute a matrix function $f(A)$. In this section we will discuss such computational issues, describing several numerical approaches having their advantages in different situations, basically depending on spectral properties of A, on the dimension and sparsity of A and on whether we really want to obtain the matrix $f(A)$ rather than "just" its action $f(A)v$ on a vector v.

3.1 Normal Matrices

A matrix $A \in \mathbb{C}^{n \times n}$ is said to be *normal* if it commutes with its adjoint, $AA^{\mathrm{H}} = A^{\mathrm{H}}A$. Normal matrices may also be characterized as being unitarily diagonalizable, i.e. we have the representation

$$A = Q\Lambda Q^{\mathrm{H}} \quad \text{with } Q^{-1} = Q^{\mathrm{H}},\ \Lambda = \mathrm{diag}(\lambda_1, \ldots, \lambda_n),\ \mathrm{spec}(A) = \{\lambda_1, \ldots, \lambda_n\}. \tag{12}$$

This representation is also the Jordan decomposition of A from (3), so that

$$f(A) = Qf(\Lambda)Q^{\mathrm{H}}, \qquad f(\Lambda) = \mathrm{diag}(f(\lambda_1), \ldots, f(\lambda_n)). \tag{13}$$

Normal matrices have the very attractive property that their eigenvalues λ_i and the corresponding invariant subspaces are well conditioned (see [16], for example), i.e. small changes in A yield only small changes in Λ and Q. Therefore, if we use a numerically (backward) stable algorithm to compute Λ and Q, like, for example, the standard Householder reduction to upper Hessenberg form followed by the QR-iteration, we may safely use the so computed Λ and Q to finally compute $f(A)$ via (13). The computational cost of this approach is $\mathcal{O}(n^3)$ due to the various matrix-matrix multiplications and to the cost for computing the eigendecomposition.

If A is not normal, its eigenvalues are not necessarily well conditioned, the condition number being related to $\|T\|_2 \cdot \|T^{-1}\|_2$ with T from the Jordan decomposition (3). It is also important to realize that the size of the Jordan blocks may widely vary under infinitesimal perturbations in A. Therefore, if A is not normal, Definition 1 does not provide a numerically stable means for computing $f(A)$.

3.2 Quadrature Rules

Assume that f is analytic in Ω and that γ and Ω are as in (11) so that we have

$$f(A) = \frac{1}{2\pi i} \oint_{\gamma} f(t)(tI - A)^{-1} dt. \tag{14}$$

We apply a quadrature rule with m nodes $t_j \in \gamma$ and weights ω_j to the right-hand side to get

$$\frac{1}{2\pi i} \oint_{\gamma} \frac{f(t)}{t - z} dt = \sum_{j=1}^{m} \omega_j \frac{f(t_j)}{t_j - z} + r. \tag{15}$$

This shows that we can approximate

$$f(A) \approx \sum_{j=1}^{m} \omega_j f(t_j) \cdot (t_j I - A)^{-1}. \tag{16}$$

For such quadrature rules, the approximation error r can be expressed or bounded using higher derivatives of f. Actually, since we integrate over a closed curve, taking the right nodes the quadrature error is usually much smaller than what one would expect from quadrature formulas over finite (real) intervals, and the accuracy often increases exponentially with the number of nodes, see [14, 15]. In principle, this can then be used to obtain bounds on the approximation error in (16), but to do so we usually need some knowledge about the norms of T and T^{-1} in (3), as well as on the size of the eigenvalues of A. See also section 3.6.

For specific functions, other integral representations may be used. For example, for $z \in \mathbb{C}$, z not on the non-positive real line, we have (see [14])

$$\log(z) = \int_0^1 (z-1)[t(z-1)+1]^{-1} dt, \tag{17}$$

so that using a quadrature rule for the interval $[0, 1]$, we can use the approximation

$$\log(A) \approx \sum_{j=1}^{m} \omega_j \cdot (A-I)[t_j(A-I)+I]^{-1}. \tag{18}$$

As another example, for $z > 0$ we can write

$$z^{-1/2} = \frac{2}{\pi} \cdot \int_0^\infty \frac{1}{t^2+z}\, dt, \tag{19}$$

and use a quadrature rule on $[0, \infty]$ to approximate $A^{-1/2}$ when $\mathrm{spec}(A) \subset (0, \infty]$.

Similar approaches have been proposed for various other functions like the p-th root or the sign function, see [6], [58], for example.

Within this quadrature framework, the major computational cost will usually be due to the inversion of several matrices. As is explained in [14], this cost can often be reduced if we first compute a unitary reduction to upper Hessenberg form (which can be done in a numerically stable manner using Householder transformations), i.e.

$$A = QHQ^{\mathrm{H}}, \quad Q \text{ unitary}, \quad H \text{ zero below the first subdiagonal}. \tag{20}$$

Then, for example,

$$(t_jI - A)^{-1} = Q \cdot (t_jI - H)^{-1} \cdot Q^{\mathrm{H}} \text{ for all } j, \tag{21}$$

with the inversion of the matrix $t_jI - H$ having cost $\mathcal{O}(n^2)$ rather than $\mathcal{O}(n^3)$.

3.3 Matrix Iterations

Sometimes, it is convenient to regard $f(z)$ as the solution of a fixed point equation $g_z(f) = f$ with g_z being contractive in a neighbourhood of the fixed point $f(z)$. The method of successive approximations

$$f_{k+1} = g_z(f_k) \tag{22}$$

can then be turned into a corresponding matrix iteration

$$F_{k+1} = g_A(F_k). \tag{23}$$

Approaches of this kind have, for example, been proposed for the matrix square root [31], [32], where Newton's method

$$f_{k+1} = \frac{1}{2} \cdot \left(f_k + \frac{z}{f_k} \right) \tag{24}$$

to compute $\sqrt{z}$ results in the iteration

$$F_{k+1} = \frac{1}{2} \cdot \left(F_k + A \cdot F_k^{-1} \right). \tag{25}$$

Similar other iterations, not always necessarily derived from Newton's method, have been proposed for the matrix p-th root [6] or for the matrix sign function [41]. A major catch with these approaches is that numerical stability of the matrix iteration (23) is not always guaranteed, even when the scalar iteration (22) is perfectly stable. Then, some quite subtle modifications, like e.g. the coupled two-term iteration for the square root analyzed in [31] must be used in order to achieve numerical stability. The iteration (23) is usually also quite costly. For example, (25) requires the inversion of F_k at every step, so that each step has complexity $\mathcal{O}(n^3)$. Therefore, for these methods to be efficient, convergence should be fast, at least superlinear.

3.4 Rational Approximations

Polynomial approximations for a function f often require a quite high degree of the approximating polynomial in order to achieve a reasonable quality of approximation. *Rational* approximations typically obtain the same quality with substantially fewer degrees of freedom.

Assume that we have the rational approximation

$$f(z) \approx \frac{\mathcal{N}_{\mu\nu}(z)}{\mathcal{D}_{\mu\nu}(z)}, \tag{26}$$

where $\mathcal{N}_{\mu\nu}, \mathcal{D}_{\mu\nu}$ are polynomials of degree μ and ν, respectively. (The use of the two indices μ and ν in both polynomials may appear abusive at this point, but it will be very convenient when discussing Padé approximations to the exponential in section 4.2). Then

$$f(A) \approx \mathcal{N}_{\mu\nu}(A) \cdot \left(\mathcal{D}_{\mu\nu}(A)\right)^{-1}. \tag{27}$$

Assume that A is diagonalizable. If we know

$$\left| f(z) - \frac{\mathcal{N}_{\mu\nu}(z)}{\mathcal{D}_{\mu\nu}(z)} \right| \leq \epsilon \text{ for } z \in \operatorname{spec}(A), \tag{28}$$

for some $\epsilon > 0$, we get

$$\| f(A) - \mathcal{N}_{\mu\nu}(A) \cdot \left(\mathcal{D}_{\mu\nu}(A)\right)^{-1} \|_2 \leq \epsilon \cdot \|T\|_2 \cdot \|T^{-1}\|_2 \tag{29}$$

which further simplifies when A is normal, since then T is unitary so that $\|T\|_2 \cdot \|T^{-1}\|_2 = 1$. Rational functions can be expressed as partial fraction expansions. Simplifying our discussion to the case of single poles, this means that we can expand

$$\frac{\mathcal{N}_{\mu\nu}(z)}{\mathcal{D}_{\mu\nu}(z)} = p(z) + \sum_{j=1}^{\nu} \frac{\omega_j}{z - \tau_j}, \tag{30}$$

with $p(z)$ being a polynomial of degree $\mu - \nu$ if $\mu \geq \nu$ and $p \equiv 0$ if $\mu < \nu$. This representation is particularly useful if we are interested only in $f(A)v$ for some vector v, as we will discuss later in section 3.6. Note also that the quadrature rules from (16) immediately give a partial fraction expansion, so that the two approaches are very closely related. For a recent investigation, see [66].

3.5 Krylov Subspace Approaches

When A has large dimension, the action of $f(A)$ on a vector v, namely $f(A)v$, may be effectively approximated by projecting the problem onto a subspace of possibly much smaller dimension. The Krylov subspace

$$K_k(A, v) = \text{span}\{v, Av, \ldots, A^{k-1}v\}$$

has been extensively used to this purpose, due to its favourable computational and approximation properties, see, e.g., van der Vorst [68], [69] for a discussion for general f. Let V_k be a full column rank $n \times k$ matrix whose columns span $K_k(A, v)$, and assume the following Arnoldi type recurrence holds for V_k,

$$AV_k = V_{k+1}H_{k+1,k} = V_k H_k + h_{k+1,k} v_{k+1} e_k^T. \tag{31}$$

An approximation to $x = f(A)v$ may be obtained as

$$x_k = V_k f(H_k) e_1 \|v\|. \tag{32}$$

The procedure amounts to projecting the matrix onto the much smaller subspace $K_k(A, v)$, by means of the representation matrix H_k and $v = V_k e_1 \|v\|$. If V_k has orthonormal columns then $H_k = V_k^{\mathrm{H}} A V_k$. If in addition A is Hermitian, the iteration (31) reduces to the Lanczos three-term recurrence, in which case H_k is tridiagonal and Hermitian.

The functional evaluation is carried out within this reduced space, and the obtained solution is expanded back to the original large space. Assume now that $k = n$ iterations can be carried out, so that the square matrix V_n is orthogonal. Then (31) gives $AV_n = V_n H_n$ and thus $A = V_n H_n V_n^{\mathrm{H}}$. Using this relation, for $k < n$, the approximation in $K_k(A, v)$ may be viewed as a problem order reduction to the first k columns of V_n and corresponding portion of H_n as

$$x = f(A)v = V_n f(H_n) V_n^{\mathrm{H}} v \approx V_k f(H_k) V_k^{\mathrm{H}} v.$$

For k small compared to n, the quality of the approximation strongly depends on the spectral properties of A and on the capability of $K_k(A, v)$ to capture them. A first characterization in this sense is given by the following result, which can be deduced from Proposition 1 applied to the matrix H_k and the fact that $p(A)v = V_k p(H_k)v$ for all polyomials of degree less than or equal to $k - 1$; see [61, Proposition 6.3]. This is a generalization of [60, Theorem 3.3].

Proposition 2. *Let the columns of V_k, with $V_k^{\mathrm{H}} V_k = I_k$ span $K_k(A, v)$ and let $H_k = V_k^{\mathrm{H}} A V_k$. Then, the approximation $V_k f(H_k) e_1 \|v\|$ represents a polynomial approximation $p(A)v$ to $f(A)v$, in which the polynomial p of degree $k - 1$ interpolates the function f in the Hermite sense on the set of eigenvalues of H_k.*

Other polynomial approximations have been explored, see, e.g., [18]; approaches that interpolate over different sets have been proposed for the exponential function [53]. Note that the projection nature of the approach allows one to derive estimates for $\|f(A)\|$ as $\|f(A)\| \approx \|f(H_k)\|$ which may be accurate even for small k when A is Hermitian.

All these results assume exact precision arithmetic. We refer to [17] for an analysis of finite precision computation of matrix functions with Krylov subspace methods when A is Hermitian.

It should be mentioned that the projection onto a Krylov subspace does not require A to be stored explicitly, but it only necessitates a function that given v, returns the action of A, namely $y = Av$. This operational feature is of paramount importance in applications where, for instance, A is the (dense) product or other combination of sparse matrices, so that the operation $y = Av$ may be carried out by a careful application of the given matrix combination.

Another practical aspect concerns the situation where k, the dimension of the Krylov subspace, becomes large. Computing $f(H_k)$ with one of the methods presented in the previous sections can then become non-negligible. Moreover, we may run into memory problems, since approximating $f(A)v$ via (32) requires the whole matrix V_k to be stored. This is needed even when, for istance, A is Hermitian, in which case (31) is the Lanczos recurrence and H_k is tridiagonal. In such a situation, however, we can resort to a "two–pass" procedure which crucially reduces the amount of memory needed: In the first pass, we run the short-term recurrence Lanczos process. Here, older columns from V_k can be discarded, yet the whole (tridiagonal) matrix H_k can be built column by column. Once $f(H_k)$ has been generated, we compute $y_k = f(H_k)e_1 \cdot \|v\|$. Then we run the short-term recurrence Lanczos process once again to recompute the columns of V_k and use them one at a time to sum up $V_k f(H_k) e_1 = V_k y_k$. Of course, this two-stage approach essentially doubles the computational work for generating the Lanczos basis.

For a general matrix A the Arnoldi process cannot be turned into a short-term recurrence, so one must search for alternatives in the case that k gets too large. Recently, Eiermann and Ernst [20] have developed an interesting scheme that allows one to *restart* Krylov subspace methods for computing $f(A)v$, in the same flavour as with linear system solvers; in fact, the two approaches are tightly related; see [47]. Having computed a not yet sufficiently good approximation x_k via (32), the idea is to start again a Krylov subspace approximation based on the error $x_k - f(A)v$ which is expressed as a *new* matrix function of A. The algorithmic formulation is non-trivial, particularly since special care has to be taken with regard to numerical stability, see [20].

Other alternatives include acceleration procedures, that aim at improving the convergence rate of the approximation as the Krylov subspace dimension increases. Promising approaches have been recently proposed in the Hermitian case by Druskin and Knizhnerman [19], by Moret and Novati [52] and by Hochbruck and van den Eshof [37].

3.6 Krylov Subspaces and Rational Approximations

As a last contribution to this section, let us turn back to rational approximations for f which we assume to be given in the form of a partial fraction expansion (no multiple poles for simplicity)

$$f(z) \approx p(z) + \sum_{j=1}^{\nu} \frac{\omega_j}{z - \tau_j}. \tag{33}$$

Then $f(A)v$ can be approximated as

$$f(A)v \approx p(A)v + \sum_{j=1}^{\nu} \omega_j (A - \tau_j I)^{-1} v. \tag{34}$$

Since evaluating $p(A)v$ is straightforward, let us assume $p \equiv 0$ in the sequel.

The computation of $(A-\tau_j I)^{-1}v$ means that we have to solve a linear system for each j, where all linear systems have the same right-hand side, while the coefficient matrix only differs for the shift τ_j. In general, shifts may be complex even for real and symmetric A, although they appear in conjugate pairs. Interestingly, the particular "shifted" structure of these systems can be exploited in practical computation. If we solve each system iteratively using a Krylov subspace method with initial zero guess for all j, the kth iterate for each system lies in $K_k(A - \tau_j I, v)$ which is identical to $K_k(A, v)$. The fact that Krylov subspaces are invariant with respect to shifts can now be exploited in various Krylov subspace solvers like CG, BiCG, FOM and QMR (and also with modifications in BiCGStab and restarted GMRES) to yield very efficient procedures which require only *one* matrix-vector multiplication with A, and possibly with A^{H}, in order to update the iterates for *all* m systems simultaneously; see [63] for a survey of these methods for shifted systems and also [21–24]. Denote by $x_k^{(j)}$ the iterate of the Krylov solver at step k for system j. Then the linear combination

$$x_k = \sum_{j=1}^{\nu} \omega_j x_k^{(j)} \in K_k(A, v) \tag{35}$$

is an approximation to $f(A)v$. In fact, it is an approximation to the action of the rational function approximating $f(A)$. Therefore, what we obtained in (35) is an approximation to $f(A)v$ in $K_k(A, v)$,which is different from (32) presented before.

A special case is when f is itself a rational function. In such a situation, the two approaches may coincide if, for instance, a Galerkin method is used to obtain the approximate solutions $x_k^{(j)}$. Indeed, for $f = \mathcal{R}_{\mu\nu} = \mathcal{N}_{\mu\nu}/\mathcal{D}_{\mu\nu}$,

$$f(A)v = \mathcal{N}_{\mu\nu}(A)(\mathcal{D}_{\mu\nu}(A))^{-1}v = \sum_{j=1}^{\nu} \omega_j (A - \tau_j I)^{-1} v \tag{36}$$

$$\approx \sum_{j=1}^{\nu} \omega_j V_k (H_k - \tau_j I)^{-1} e_1 \|v\| = V_k f(H_k) e_1 \|v\|.$$

The approach outlined above has several attractive features for a general function f. Firstly, if we have a bound for the error between $x_k^{(j)}$ and the solution $(A - \tau_j)^{-1}v$ for each j, we can combine these bounds with the approximation error of the rational approximation to get an overall a posteriori bound for $\|f(A)v - x^{(k)}\|$. Sometimes, such bounds might be obtained quite easily. For example, if A is Hermitian and positive definite and all shifts τ_j are real and negative, the norm of the inverse $(A - \tau_j I)^{-1}$ is bounded by $1/|\tau_j|$. Since the residuals $r_k^{(j)} = (A - \tau_j I)x_k^{(j)} - v$ are usually available in the Krylov solver in use, we can use the bound

$$\|x_k^{(j)} - (A - \tau_j I)^{-1} v\|_2 \leq \frac{1}{|\tau_j|} \|r_k^{(j)}\|_2. \tag{37}$$

Similar bounds that require estimates of the spectrum of A may be obtained also for complex poles τ_j, see [47].

Secondly, in the Hermitian case, the memory requirements of this approach only depend on m, the number of poles in the rational approximation, but not on k, the dimension of the Krylov subspace. Indeed, the symmetry of the problem can be exploited to devise a short-term recurrence which dynamically updates the solution x_k without storing the whole Krylov subspace basis. So even if k has to be sensibly large in order to get a good approximation, we will not run into memory problems. This is in contrast to the approach from section 3.5, although the two approaches are strictly related. Indeed, using x_k in (35), by the triangle inequality we have

$$\begin{aligned} \big| \|f(A)v - x_k\| - \|f(A)v - V_k f(H_k) e_1 \|v\| \| \big| &\leq \|V_k f(H_k) e_1 \|v\| - x_k\| \\ &= \| (f(H_k) - \mathcal{R}_{\mu\nu}(H_k)) e_1 \| \|v\| . \end{aligned}$$

Therefore, whenever the chosen rational function $\mathcal{R}_{\mu\nu}$ accurately approximates f, the two approaches evolve similarly as the Krylov subspace dimension increases.

4 The Exponential Function

We next focus our attention on methods specifically designed to approximate the matrix exponential, $\exp(A)$, and its action on a vector v. We start by briefly discussing the role of this function within Model Order Reduction applications. Depending on the setting, we shall use either of the two equivalent notations $\exp(A)$ and e^A. We explicitly observe that Definition 1 ensures that $\exp(A)$ is nonsingular for any matrix A.

4.1 The Exponential Matrix in Model Order Reduction Applications

In this section we briefly review some application problems whose numerical solution benefits from the approximate computation of the exponential.

Numerical solution of time-dependent differential equations. The numerical solution of ordinary and time-dependent partial differential equations (ODEs and PDEs, respectively) may involve methods that effectively employ the matrix exponential. Recent developments in the efficient approximation of $\exp(A)v$ have increased the use of numerical "exponential-based" (or just "exponential") techniques that allow one to take larger time steps. More precisely, consider the system of ODEs of the form

$$u'(t) = Au(t) + b(t), \qquad u(0) = u_0,$$

where A is a negative semidefinite matrix. The analytic solution is given by

$$u(t) = e^{tA}u_0 + \int_0^t e^{(\tau - t)A} b(\tau) d\tau.$$

Whenever a good approximation to the propagation operator e^{sA} is available, it is possible to approximate the analytic solution by simply approximating the integral above with convenient quadrature formulas, leading to stable solution approximations. The generalization of this approach to the numerical solution of partial differential equations can be obtained, for instance, by employing a semidiscretization (in space) of the given problem. Consider the following self-adjoint parabolic equation

$$\frac{\partial u(x,t)}{\partial t} = \mathrm{div}(a(x)\nabla u(x,t)) - b(x)u(x,t) + c(x),$$

with $x \in \Omega$, Dirichlet boundary conditions and $b(x) \geq 0$, $a(x) > 0$ in Ω, with a, b, c sufficiently regular functions. A continuous time – discrete space discretization leads to the ordinary differential equation

$$E\frac{d\mathbf{u}(t)}{dt} = -A\mathbf{u}(t) + \mathbf{c}, \quad t \geq 0,$$

where A, E are positive definite Hermitian matrices, so that the procedure discussed above can be applied; see, e.g., [11,25,51,65,70]. Further attempts to generalize this procedure to non-selfadjoint PDEs can be found in [25, section 6.2], although the theory behind the numerical behavior of the ODE solver in this case is not completely understood yet.

The use of exponential integrators is particularly effective in the case of certain stiff systems of nonlinear equations. Consider, e.g., the initial value problem

$$\frac{du(t)}{dt} = f(u), \qquad u(t_0) = u_0.$$

If the problem is stiff, standard integrators perform very poorly. A simple example of an exponential method for this system is the *exponentially fitted Euler* scheme, given by

$$u_1 = u_0 + h\phi(hA)f(u_0),$$

where h is the step size, $\phi(z) = \frac{e^z-1}{z}$, and $A = f'(u_0)$. The recurrence $\{u_k\}_{k=0,1,\ldots}$ requires the evaluation of $\phi(hA)v$ at each iteration, for some vector v; see, e.g., [36].

An application that has witnessed a dramatic increase in the use of the matrix exponential is Geometric Integration. This research area includes the derivation of numerical methods for differential equations whose solutions are constrained to belong to certain manifolds equipped with a group structure. One such example is given by linear Hamiltonian problems of the form

$$\begin{cases} \dot{Y}(t) = \mathcal{J}A(t)Y(t), \\ Y(t_0) = Y_0, \end{cases}$$

where $\mathcal{J}$ is the matrix $[0, I; -I, 0]$, A is a continuous, bounded, symmetric matrix function, and $Y_0 \in \mathbb{R}^{N\times p}$ is symplectic, that is it satisfies $Y_0^{\mathrm{H}}\mathcal{J}Y_0 = \mathcal{J}$. The solution $Y(t)$ is symplectic for any $t \geq t_0$. Using the fact that $\mathcal{J}A$ is Hamiltonian, it can be shown that $\exp(\mathcal{J}A(t))$ is symplectic as well. Numerical methods that aim at approximating $Y(t)$ should also preserve its symplecticity property. This is achieved for instance by the numerical scheme $Y_{k+1} = \exp(h\mathcal{J}A(t_k))Y_k$, $t_{k+1} = t_k + h$, $k = 0, 1, \ldots$. Structure preserving methods associated with small dimensional problems have received considerable attention, see, e.g., [10,29,39,71] and references therein. For large problems where order reduction is mandatory, approximations obtained by specific variants of Krylov subspace methods can be shown to maintain these geometric properties; see, e.g., [48].

Analysis of dynamical systems. The exponential operator has a significant role in the analysis of linear time-invariant systems of the form

$$\begin{cases} \dot{x}(t) = Ax(t) + Bu(t) \\ y(t) = Cx(t) \end{cases} \tag{38}$$

where A, B and C are real matrices of size $n \times n$, $n \times m$ and $p \times n$, respectively. In the following we assume that A is stable, that is its eigenvalues are in the left half plane $\mathbb{C}^-$, and that the system is controllable and observable; see, e.g., [1].

The matrix of the states of the system for impulsive inputs is $x(t) = e^{tA}B$, whereas in general, for an initial state x_0 at time t_0, the resulting state at time $t \geq t_0$ is given by

$$x(t) = e^{(t-t_0)A}x_0 + \int_{t_0}^{t} e^{(t-\tau)A}Bu(\tau)d\tau.$$

Therefore, an approximation to the state involves the approximation of the matrix exponential. Moreover, the state function is used to define the first of the following two matrices which are called the controllability and the observability Gramians, respectively,

$$P = \int_0^\infty e^{tA}BB^{\mathrm{H}}e^{tA^{\mathrm{H}}}dt, \quad Q = \int_0^\infty e^{tA^{\mathrm{H}}}C^{\mathrm{H}}Ce^{tA}dt. \tag{39}$$

The following result shows that these are solutions to Lyapunov equations.

Theorem 2. *Given the linear time-invariant system (38), let P, Q be as defined in (39). Then they satisfy*

$$AP + PA^{\mathrm{H}} + BB^{\mathrm{H}} = 0, \qquad A^{\mathrm{H}}Q + QA + C^{\mathrm{H}}C = 0.$$

Proof. The proof follows from substituting the definition of P and Q into the corresponding expressions $AP+PA^{\mathrm{H}}$, $A^{\mathrm{H}}Q+QA$. By using the fact that $e^{tA}A = \frac{d}{dt}(e^{tA})$ and integrating, we obtain, e.g., for Q,

$$\begin{aligned} QA + A^{\mathrm{H}}Q &= \int_0^\infty \left(e^{tA^{\mathrm{H}}}C^{\mathrm{H}}Ce^{tA}A + A^{\mathrm{H}}e^{tA^{\mathrm{H}}}C^{\mathrm{H}}Ce^{tA}\right) dt \\ &= \int_0^\infty \left(e^{tA^{\mathrm{H}}}C^{\mathrm{H}}C\frac{de^{tA}}{dt} + \frac{de^{tA^{\mathrm{H}}}}{dt}C^{\mathrm{H}}Ce^{tA}\right) dt \\ &= \int_0^\infty \frac{d(e^{tA^{\mathrm{H}}}C^{\mathrm{H}}Ce^{tA})}{dt} dt = \lim_{\tau\to\infty} (e^{tA^{\mathrm{H}}}C^{\mathrm{H}}Ce^{tA})\Big|_0^\tau = -C^{\mathrm{H}}C. \quad \square \end{aligned}$$

It can also be shown that the solution to each Lyapunov equation is unique. In a more general setting, the matrix $M := -(A^{\mathrm{H}}Q + QA)$ is not commonly given in factored form. In this case, if it can be shown that M is positive semidefinite and that the pair (A, M) is observable, then Q is positive definite (a corresponding result holds for P); see, e.g., [1,4,13].

The Lyapunov equation may be used to compute estimates for $\|e^{tA}\|$, which in turn provides information on the stability of the original system in the case of $C^{\mathrm{H}}C$ full rank; see, e.g., [13, Th. 3.2.2] for a proof.

Theorem 3. *Let A be stable and $C^{\mathrm{H}}C$ full rank. Then the unique solution Q to the Lyapunov equation $A^{\mathrm{H}}Q + QA + C^{\mathrm{H}}C = 0$ satisfies*

$$\|e^{tA}\| \le \left(\frac{\lambda_{\max}(Q)}{\lambda_{\min}(Q)}\right)^{\frac{1}{2}} e^{-\alpha t},$$

where $\alpha = \lambda_{\min}(Q^{-1}C^{\mathrm{H}}C)/2 > 0$.

For large problems, other devices can be used to directly approximate $\|e^{tA}\|$ without first resorting to the solution of a Lyapunov equation; cf. section 3.5. We also refer to [46] for a general discussion on the norm $\|e^{tA}\|$ and some of its bounds.

4.2 Computing the Exponential of a Matrix

Over the years, several methods have been devised and tested for the computation of the matrix exponential; we refer to [50] for a recent survey of several approaches and for a more complete bibliographic account. The algorithmic characteristics may be very different depending on whether the matrix has small or large dimension, or whether it is dense or sparse; the structural and symmetry properties also play a crucial role; see, e.g., the discussion in [62]. In this section we discuss the case of small matrices. When A is normal, the spectral decomposition discussed in section 3.1 can be employed, namely $A = TJT^{\mathrm{H}}$ with T unitary. This gives $\exp(A) = T\exp(J)T^{\mathrm{H}}$, once the decomposition of A is computed.

In the non-normal case, one method has emerged in the last decade, for its robustness and efficiency: Padé approximation with scaling and squaring. The basic method employs a rational function approximation to the exponential function as

$$\exp(\lambda) \approx \mathcal{R}_{\mu\nu}(\lambda) = \frac{\mathcal{N}_{\mu\nu}(\lambda)}{\mathcal{D}_{\mu\nu}(\lambda)},$$

where $\mathcal{N}_{\mu\nu}, \mathcal{D}_{\mu\nu}$ are polynomials of degree μ and ν, respectively. One attractive feature of the $[\mu/\nu]$ Padé approximation is that the coefficients of the two polynomials are explicitly known, that is

$$\mathcal{N}_{\mu\nu}(\lambda) = \sum_{j=0}^{\mu} \frac{(\mu+\nu-j)!\mu!}{(\mu+\nu)!(\mu-j)!j!}\lambda^j, \quad \mathcal{D}_{\mu\nu}(\lambda) = \sum_{j=0}^{\nu} \frac{(\mu+\nu-j)!\nu!}{(\mu+\nu)!(\nu-j)!j!}(-\lambda)^j.$$

These two polynomials have a rich structure. For example, one has the relation $\mathcal{N}_{\mu\nu}(\lambda) = \mathcal{D}_{\nu\mu}(-\lambda)$ as well as several other important properties which can be found, e.g., in [26, section 5.2].

Diagonal Padé approximation ($\mu = \nu$), is usually preferred because computing $\mathcal{R}_{\mu\nu}$ with say, $\mu > \nu$, is not cheaper than computing the more accurate $\mathcal{R}_{\nu_*\nu_*}$ where $\nu_* = \max\{\mu, \nu\}$. Nonetheless, because of their stability properties, Padé $[\nu+1/\nu]$ approximations are used, together with $[\nu/\nu]$ approximations, in the numerical solution of initial value problems with one-step methods. Another attractive property of the diagonal Padé approximation is that if A has eigenvalues with negative real part, then the spectral radius of $\mathcal{R}_{\nu\nu}(A)$ is less than one, for any ν. In the following, diagonal rational approximation will be denoted by $\mathcal{R}_{\nu\nu} = \mathcal{R}_\nu$. The accuracy of the approximation can be established by using the following result.

Theorem 4. [26, Theorem 5.5.1] *Let the previous notation hold. Then*

$$e^\lambda - \mathcal{R}_{\mu\nu}(\lambda) = (-1)^\nu \frac{\mu!\,\nu!}{(\mu+\nu)!\,(\mu+\nu+1)!}\lambda^{\mu+\nu+1} + O(\lambda^{\mu+\nu+2}).$$

This error estimate shows that the approximation degrades as λ gets away from the origin. This serious limitation motivated the introduction of the scaling and squaring procedure. By exploiting the property $e^A = (e^{A/k})^k$, for any square matrix A and scalar k, the idea is to determine k so that the scaled matrix A/k has norm close to one, and then employ the approximation

$$e^{A/k} \approx \mathcal{R}_\nu(A/k).$$

The approximation to the original matrix e^A is thus recovered as $e^A \approx \mathcal{R}_\nu(A/k)^k$. The use of powers of two in the scaling factor is particularly appealing. Indeed, by writing $k = 2^s$, the final approximation $\mathcal{R}_\nu(A/2^s)^{2^s}$ is obtained by repeated squaring. The scalar s is determined by requiring that $\|A\|_\infty/2^s$ is bounded by some small constant, say 1/2. In fact, this constant could be allowed to be significantly larger with no loss in stability and accuracy; see [33]. The approach oulined here is used in Matlab 7.1. [49].

A rational function that is commonly used in the case of symmetric negative semidefinite matrices, is given by the Chebychev rational function. The Chebychev approximation $\mathcal{R}^{\star}_{\mu\nu}$ determines the best rational function approximation in $[0, +\infty)$ to $e^{-\lambda}$ by solving the problem

$$\min_{\mathcal{R}_{\mu\nu}} \max_{\lambda \in [0,+\infty)} \left| e^{-\lambda} - \mathcal{R}_{\mu\nu}(\lambda) \right|,$$

where the minimum is taken over all rational functions. In particular, the cases $\mu = 0$ and $\mu = \nu$ have been investigated in greater detail, and the coefficients of the polynomials of $\mathcal{R}^{\star}_{\nu}$ have been tabulated first by Cody, Meinardus and Varga in [12] for $\nu \leq 14$ and then in [9] for degree up to 30. Setting $\mathcal{E}_{\nu} = \max_{\lambda \in [0,+\infty)} \left| e^{-\lambda} - \mathcal{R}^{\star}_{\nu}(\lambda) \right|$, great efforts in the approximation theory community have been devoted to show the following elegant result on the error asymptotic behavior,

$$\lim_{\nu \to \infty} \mathcal{E}_{\nu}^{1/\nu} = \frac{1}{9.28903...},$$

disproving the so-called “1/9” conjecture. From the result above it follows that $\sup_{\lambda \in [0,+\infty)} \left| e^{-\lambda} - \mathcal{R}_{\nu}(\lambda) \right| \approx 10^{-\nu}$.

Other rational function approximations that have recently received renewed interest are given by rational functions with real poles, such as $\mathcal{R}_{\mu\nu}(\lambda) = \mathcal{N}_{\mu}(\lambda)/(1+h\lambda)^{\nu}$; see, e.g., [7], [52], [55]. An advantage of these functions is that they avoid dealing with *complex* conjugate poles.

4.3 Reduction Methods for Large Matrices

In many application problems where A is large, the action of $\exp(A)v$ is required, rather than $\exp(A)$ itself, so that the methods of section 3.5 and of section 3.6 can be used. We first discuss some general convergence properties, and then show the role of the Krylov subspace approximation to $\exp(A)v$ in various circumstances. Note that time dependence can, in principle, be easily acommodated in the Krylov approximation as, for instance, $\exp(tA)v \approx V_k \exp(tH_k)e_1\|v\|$. In the following, we shall assume that A already incorporates time dependence. In particular, estimates involving spectral information on the matrix will be affected by possible large values of t.

An analysis of the Krylov subspace approximation $V_k \exp(H_k)e_1\|v\|$ to $\exp(A)v$ was given by Saad [60], where the easily computable quantity

$$h_{k+1,k} \cdot \left| e_k^T \exp(H_k)e_1\|v\| \right|$$

was proposed as stopping criterion for the iterative Arnoldi process; a higher order estimate was also introduced in [60]. Further study showed that the convergence rate of the approximation is often superlinear. In the Hermitian negative semidefinite case, a complete characterization of this superlinear behavior can be derived using the following bounds. We refer to [18, 65] for qualitatively similar, although asymptotic bounds.

Theorem 5 (see Hochbruck and Lubich [35]). *Let A be a Hermitian negative semidefinite matrix with eigenvalues in the interval $[-4\rho, 0]$, with $\rho > 0$. Then the error in the approximation (32) of* $\exp(A)v$ *is bounded as follows:*

$$\|\exp(A)v - V_k \exp(H_k)e_1\| \le 10e^{-k^2/(5\rho)}, \qquad \sqrt{4\rho} \le k \le 2\rho,$$
$$\|\exp(A)v - V_k \exp(H_k)e_1\| \le \frac{10}{\rho}e^{-\rho}\left(\frac{e\rho}{k}\right)^k, \qquad k \ge 2\rho.$$

Other bounds that emphasize the superlinear character of the approximation have also been proposed in [64], and earlier in [25]. Similar results also hold in the case when A is skew-symmetric, or when A is non-symmetric, under certain hypotheses on the location of its spectrum, see [18, 35].

A typical convergence curve of the error together with the bounds of Theorem 5 (called HL bound) are shown in Figure 1, for a diagonal 1001×1001 matrix A with entries uniformly distributed in $[-40, 0]$ and a random vector v with uniformly distributed values in $[0, 1]$ and unit norm; this example is taken from [35].

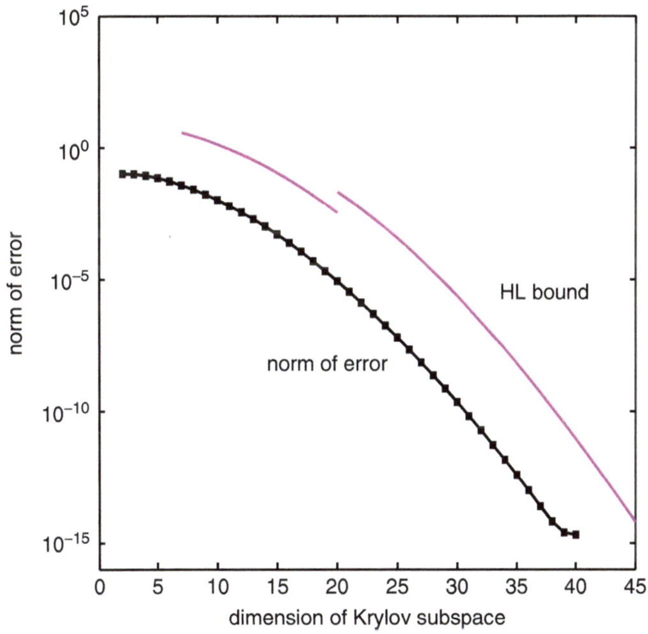

Fig. 1. Typical convergence of Krylov subspace approximation to $\exp(A)v$ and upper bounds of Theorem 5.

Rational approximation and partial fraction expansion. When A is large, the methods of the previous sections can be employed to approximate $\exp(A)v$. In particular, using diagonal Padé or Chebyshev rational functions and their partial fraction expansion, one gets

$$\exp(A)v \approx \mathcal{N}_\nu(A)\mathcal{D}_\nu(A)^{-1}v = \omega_0 v + \sum_{j=1}^{\nu} \omega_j (A - \tau_j I)^{-1} v,$$

where the coefficients and the poles are pairwise complex conjugates. A recent discussion on the main properties of this approximation and on related references can be found in [47].

Application to the solution of Lyapunov matrix equations. The approximate solution of the Lyapunov matrix equation has been addressed in a large body of literature, in which both the sign function and the exponential function have played a leading role over the years. Here we focus on low-rank approximations, assuming that B has few columns and that it is full column rank.

The dependence of P (and Q) on the exponential has classically motivated the use of quadrature formulas for the approximation of the integral defining P (cf. (39)), together with the use of low degree polynomial or rational function approximations to $\exp(t\lambda)$. More precisely, setting $X(t) = \exp(tA)B$, P could be approximated as

$$P(\tau) = \int_0^\tau X(t)X(t)^{\mathrm{H}} dt,$$

for some $\tau \geq 0$. Then, by using a quadrature formula with discretization points t_i and weights δ_i, the integral in $[0,\tau]$ is approximated as $P(\tau) \approx \sum_{i=1}^k X(t_i)\delta_i X(t_i)^{\mathrm{H}}$. The practical effectiveness of the approach depends on the value of τ, but most notably on the quadrature formula used, under the constraint that all δ_i be positive, to ensure that $P(\tau)$ is positive semidefinite; see [59] for some experiments using different quadrature formulas.

The procedure above determines a low-rank approximation to the corresponding Gramian since the number of columns of B is small. An alternative approach that bypasses the integral formulation within the low-rank framework, is obtained by reducing the problem dimension. If an approximation to $\exp(tA)B$ is available as $x_k = V_k \exp(tH_k)E$, where E and V_k are defined so that $B = V_k E$, then

$$P_k = V_k \int_0^\infty \exp(tH_k)EE^T \exp(tH_k^{\mathrm{H}})dt\, V_k^{\mathrm{H}} =: V_k G_k V_k^{\mathrm{H}}.$$

If H_k is stable, Theorem 2 ensures that G_k is the solution to the following small dimensional Lyapunov equation:

$$H_k G + G H_k^{\mathrm{H}} + EE^T = 0. \tag{40}$$

This derivation highlights the theoretical role of the exponential in the approximation procedure. However, one can obtain G_k by directly solving the small matrix equation, by means of methods that exploit matrix factorizations [3, 30].

The following result sheds light onto the reduction process performed by this approximation; see, e.g., [40, 59].

Proposition 3. *Let the columns of* V_k, *with* $V_k^{\mathrm{H}} V_k = I_k$, *span* $K_k(A, B) = \mathrm{span}\{B, AB, \ldots, A^{k-1}B\}$. *The approximate solution* $P_k = V_k G_k V_k^{\mathrm{H}}$ *where* G_k *solves (40) is the result of a Galerkin process onto the space* $K_k(A, B)$.

Proof. Let V_k be a matrix whose orthonormal columns span $K_k(A, B)$. Let $R_k = AP_k + P_k A^{\mathrm{H}} + BB^{\mathrm{H}}$ be the residual associated with $P_k = V_k G_k V_k^{\mathrm{H}}$ for some G_k and let $H_k = V_k^{\mathrm{H}} A V_k$. A Galerkin process imposes the following orthogonality condition on the residual[1]

$$V_k^{\mathrm{H}} R_k V_k = 0.$$

Expanding R_k and using $V_k^{\mathrm{H}} V_k = I$, we obtain

$$\begin{aligned} V_k^{\mathrm{H}} A V_k G_k V_k^{\mathrm{H}} + G_k V_k^{\mathrm{H}} A^{\mathrm{H}} V_k + V_k^{\mathrm{H}} BB^{\mathrm{H}} V_k &= 0 \\ H_k G_k + G_k H_k + V_k^{\mathrm{H}} BB^{\mathrm{H}} V_k &= 0. \end{aligned}$$

Recalling that $B = V_k E$, the result follows. □

Other methods have been proposed to approximately solve large-scale Lyapunov equations; see [28, 45, 56] and references therein.

5 The Matrix Sign Function

In this section we discuss methods for the matrix sign function, with the sign function on $\mathbb{C}$ defined as

$$\mathrm{sign}(z) = \begin{cases} +1 \text{ if } \Re(z) > 0, \\ -1 \text{ if } \Re(z) < 0. \end{cases} \tag{41}$$

We do not define $\mathrm{sign}(z)$ on the imaginary axis where, anyway, it is not continuous. Outside the imaginary axis, sign is infinitely often differentiable, so that $\mathrm{sign}(A)$ is defined as long as the matrix $A \in \mathbb{C}^{n \times n}$ has no eigenvalues on the imaginary axis. We first recall a few application problems where the sign function is commonly employed.

5.1 Motivation

The *algebraic Riccati equation* arises in control theory as a very fundamental system to be solved in order to compute, for example, certain observers or stabilizers, see [4], [44]. It is a quadratic matrix equation of the form

$$G + A^{\mathrm{H}} X + XA - XFX = 0, \tag{42}$$

[1] This "two–sided" condition can be derived by first defining the matrix inner product $\langle X, Y \rangle = \mathrm{tr}(XY^{\mathrm{H}})$ and then imposing $\langle R_k, P_k \rangle = 0$ for any $P_k = V_k G V_k^{\mathrm{H}}$ with $G \in \mathbb{R}^{k \times k}$.

where $A, F, G \in \mathbb{R}^{n\times n}$ and F and G are symmetric and positive definite. One aims at finding a symmetric positive definite and stabilizing solution X, i.e. the spectrum of $A-FX$ should lie in $\mathbb{C}^-$. The quadratic equation (42) can be linearized by turning it into a system of doubled size as

$$K := \begin{bmatrix} A^{\mathrm{H}} & G \\ F & -A \end{bmatrix} = \begin{bmatrix} X & -I \\ I & 0 \end{bmatrix} \cdot \begin{bmatrix} -(A-FX) & -F \\ 0 & (A-FX)^{\mathrm{H}} \end{bmatrix} \cdot \begin{bmatrix} X & -I \\ I & 0 \end{bmatrix}^{-1}. \tag{43}$$

If we assume that X is a stabilizing solution (such a solution exists under mild conditions, see [4,44]), a standard approach is to use the matrix sign function to compute X. We have $\mathrm{sign}(-(A-FX)) = I$ and $\mathrm{sign}(A-FX) = -I$. Therefore,

$$\mathrm{sign} \begin{bmatrix} -(A-FX) & -F \\ 0 & (A-FX)^{\mathrm{H}} \end{bmatrix} = \begin{bmatrix} I & Z \\ 0 & -I \end{bmatrix}, \quad Z \in \mathbb{R}^{n\times n} \tag{44}$$

and we see that

$$\mathrm{sign}(K) - I = \begin{bmatrix} X & -I \\ I & 0 \end{bmatrix} \cdot \begin{bmatrix} 0 & Z \\ 0 & -2I \end{bmatrix} \cdot \begin{bmatrix} X & -I \\ I & 0 \end{bmatrix}^{-1}. \tag{45}$$

Split $\mathrm{sign}(K) - I$ vertically in its middle as $[M|N]$, move the inverse matrix to the left-hand side in (45) and then equate the first halves, to get

$$MX = -N. \tag{46}$$

This is an overdetermined, but consistent linear system for X, and by working with the second half blocks, it can be shown that M is full column rank. Therefore, the procedure above outlines a method to derive a stabilizing solution X to (42) by means of the sign function.

As discussed in section 4.3, the Lyapunov equation

$$A^{\mathrm{H}}X + XA + C^{\mathrm{H}}C = 0, \ \text{ where } A, C \in \mathbb{R}^{n\times n} \tag{47}$$

also arises in control theory. It was already shown in [58] that the (Hermitian) solution X is the (2,1) block of the sign-function of a matrix of twice the dimension, that is

$$\begin{bmatrix} 0 & 0 \\ X & I \end{bmatrix} = \frac{1}{2}\left(I + \mathrm{sign}\left(\begin{bmatrix} A & 0 \\ C^{\mathrm{H}}C & -A^{\mathrm{H}} \end{bmatrix}\right)\right). \tag{48}$$

This follows in a way similar to what we presented for the algebraic Riccati equation; see also [5] for a generalization.

The matrix sign function also appears in the modelling (and subsequent simulation) of complex physical systems. One example is given by the so-called overlap fermions of lattice quantum chromodynamics [54], where one has to solve linear systems of the form

$$(I + \Gamma_5 \mathrm{sign}(Q))x = b. \tag{49}$$

Here Γ_5 is a simple permutation matrix and Q is a huge, sparse, complex Hermitian matrix representing a nearest neighbour coupling on a regular 4-dimensional grid with 12 variables per grid point. Note that here we may situate ourselves in an order reduction context, since if we solve (49) with some iterative method, the basic operation will be to compute matrix-vector products, i.e. we need the action $\operatorname{sign}(Q)v$ rather than $\operatorname{sign}(Q)$ itself.

5.2 Matrix Methods

A detailed survey on methods to compute the whole matrix $\operatorname{sign}(A)$ is given in [43], see also [2]. We shortly describe the most important ones.

The Newton iteration to solve $z^2 - 1 = 0$ converges to $+1$ for all starting values in the right half plane, and to -1 for all those from $\mathbb{C}^-$. According to (24), the corresponding matrix iteration reads

$$S_{k+1} = \frac{1}{2}\left(S_k + S_k^{-1}\right), \quad \text{where } S_0 = A. \tag{50}$$

Although the convergence is global and asymptotically quadratic, it can be quite slow in the presence of large eigenvalues or eigenvalues with a small real part. Therefore, several accelerating scaling strategies have been proposed [42], for example by using the determinant [8], i.e.

$$S_{k+1} = \frac{1}{2}\left((c_k S_k) + \frac{1}{c_k} S_k^{-1}\right), \quad \text{with } c_k = \det(S_k). \tag{51}$$

Note that $\det(S_k)$ is easily available if S_k is inverted using the LU-factorization. An alternative which avoids the computation of inverses is the Schulz iteration, obtained as Newton's method for $z^{-2} - 1 = 0$, which yields

$$S_{k+1} = \frac{1}{2} \cdot S_k \cdot \left(3I - S_k^2\right), \quad S_0 = A. \tag{52}$$

This iteration is guaranteed to converge only if $\|I - A^2\| < 1$ (in an arbitrary operator norm).

In [41], several other iterations were derived, based on Padé approximations of the function $(1-z)^{-1/2}$. They have the form

$$S_{k+1} = S_k \cdot \mathcal{N}_{\mu\nu}(S_k^2) \cdot \mathcal{D}_{\mu\nu}(S_k^2)^{-1}, \quad S_0 = A. \tag{53}$$

For $\mu = 2p, \nu = 2p - 1$, an alternative representation is

$$S_{k+1} = \left((I + S_k)^{2p} + (I - S_k)^{2p}\right) \cdot \left((I + S_k)^{2p} - (I - S_k)^{2p}\right)^{-1}. \tag{54}$$

In this case, the coefficients of the partial fraction expansion are explicitly known, giving the equivalent representation

$$S_{k+1} = \frac{1}{p} \cdot S_k \cdot \sum_{i=1}^{p} \frac{1}{\xi_i} \left(S_k^2 + \alpha_i I\right)^{-1}, \quad S_0 = A, \tag{55}$$

$$\text{with } \xi_i = \frac{1}{2}\left(1 + \cos\frac{(2i-1)\pi}{2p}\right), \quad \alpha_i^2 = \frac{1}{\xi_i} - 1, \quad i = 1, \dots, p.$$

Interestingly, ℓ steps of the iteration for parameter p are equivalent to one step with parameter $p\ell$. The following global convergence result on these iterations was proved in [41].

Theorem 6. *If A has no eigenvalues on the imaginary axis, the iteration (54) converges to* $\operatorname{sign}(A)$. *Moreover, one has*

$$(\operatorname{sign}(A) - S_k)(\operatorname{sign}(A) + S_k)^{-1} = \left((\operatorname{sign}(A) - A)(\operatorname{sign}(A) + A)^{-1}\right)^{(2p)^k}, \tag{56}$$

which, in the case that A is diagonalizable, gives

$$\|(\operatorname{sign}(A) - S_k)(\operatorname{sign}(A) + S_k)^{-1}\| \le \|T\| \cdot \|T^{-1}\| \cdot \left(\max_{\lambda \in \operatorname{spec}(A)} \frac{\operatorname{sign}(\lambda) - \lambda}{\operatorname{sign}(\lambda) + \lambda}\right)^{2p^k}. \tag{57}$$

5.3 Krylov Subspace Approximations

We now look at Krylov subspace approximations for

$$\operatorname{sign}(A)v, \quad v \in \mathbb{C}^n \tag{58}$$

with special emphasis on A Hermitian. The Krylov subspace projection approach from (32) gives

$$\operatorname{sign}(A)v \approx V_k \operatorname{sign}(H_k) e_1 \cdot \|v\|. \tag{59}$$

If one monitors the approximation error in this approach as a function of k, the dimension of the Krylov subspace, one usually observes a non-monotone, jig-saw like behaviour. This is particularly so for Hermitian indefinite matrices, where the real eigenvalues lie to the left and to the right of the origin. This can be explained by the fact, formulated in Proposition 2, that the Krylov subspace approximation is given as $p_{k-1}(A)v$ where p_{k-1} is the degree $k-1$ polynomial interpolating at the Ritz values. But the Ritz values can get arbitrarily close to 0 (or even vanish), even though the spectrum of A may be well separated from 0, then producing a (relatively) large error in the computed approximation. A Ritz value close to 0 is likely to occur if $k-1$ is odd, so the approximation has a tendency to degrade every other step. A remedy to this phenomenon is to use the polynomial that interpolates A at the *harmonic* Ritz values, since these can be shown to be as well separated from zero as $\operatorname{spec}(A)$. Computationally, this can be done using the same Arnoldi recurrence (31) as before, but applying a simple rank-one modification to H_k before computing its sign function. Details are given in [67].

An alternative is to use the identity

$$\operatorname{sign}(z) = z \cdot (z^2)^{-\frac{1}{2}}, \tag{60}$$

then use the Krylov subspace approach on the squared matrix A^2 to approximate

$$\left.\begin{array}{l}(A^2)^{-\frac{1}{2}}\, v \approx V_k(H_k)^{-\frac{1}{2}} e_1 \cdot \|v\| =: y_k, \\ \operatorname{sign}(A)v \approx x_k = Ay_k.\end{array}\right\} \tag{61}$$

Note that in (61) the matrix H_k represents the projection of A^2 (not of A !), onto the Krylov subspace $K_k(A^2, b)$. Interestingly, this is one of the special cases when explicitly generating the space $K_k(A^2, b)$ is more effective than using $K_k(A, b)$; see [68] for a general analysis of cases when using the latter is computationally more advantageous.

It is also remarkable that, in case that A is Hermitian, it is possible to give *a posteriori* error bounds on the quality of the approximation x_k as formulated in the following theorem taken from [67].

Theorem 7. *Let A be Hermitian and non-singular. Then x_k from (61) satisfies*

$$\|\operatorname{sign}(A)v - x_k\|_2 \le \|r_k\|_2 \le 2\kappa \left(\frac{\kappa - 1}{\kappa + 1}\right)^k \cdot \|v\|_2, \tag{62}$$

where $\kappa \equiv \|A\|_2\|A^{-1}\|_2$ and r_k is the residual in the k-th step of the CG method applied to the system $A^2x = v$ (with initial residual v, i.e. initial zero guess).

The residual norms $\|r_k\|$ need not be computed via the CG method since they can be obtained at almost no cost from the entries of the matrix H_k in the Lanczos recursion (31). This can be seen as follows: Since A^2 is Hermitian, H_k is Hermitian and tridiagonal. If p_{k-1} is the degree $k-1$ polynomial expressing the k-th Lanczos vector v_k as $v_k = p_{k-1}(A^2)v$, the Lanczos recursion gives $h_{k+1,k}p_k(z) = (z - h_{k,k}) \cdot p_{k-1}(z) - h_{k-1,k}p_{k-2}(z)$. On the other hand, it can be shown that $r_k = \sigma v_{k+1}$ with $\|v_{k+1}\| = 1$ for some scalar σ (see [61, Proposition 6.20]), and since $r_k = q_k(A^2)v$ for some polynomial q_k of degree k satisfying $q_k(0) = 1$, it must be $q_k = p_k/p_k(0)$, so that $r_k = p_k(A^2)v/p_k(0) = v_{k+1}/p_k(0)$. Therefore, along with the Lanczos process we just have to evaluate the recursion $p_k(0) = -[h_{k,k} \cdot p_{k-1}(0) + h_{k-1,k}p_{k-2}(0)]/h_{k+1,k}$ to obtain $\|r_k\| = 1/|p_k(0)|$.

We do not know of comparable error bounds for the other two approaches outlined earlier ((59) and its variant using harmonic Ritz values). Note also that x_k from (61) satisfies $x_k = Ap_{k-1}(A^2)v$, where $q(z) = z \cdot p_{k-1}(z^2)$ is an odd polynomial, that is $q(-z) = -q(z)$, of degree $2k-1$ in z. This is a restriction as compared to the other two approaches where we do not enforce any symmetry on the interpolating polynomials. However, this restriction will most probably have an effect only if the spectrum of A is very unsymmetric with respect to the origin. Our computational experience in simulations from lattice QCD indicates that x_k is actually the best of the three approximations discussed so far. Since, in addition, x_k comes with a bound of the true error norm, we definitely favor this approach.

5.4 Partial Fraction Expansions

If we have sufficient information on the eigensystem of A available, we can use (57) to estimate a number $p \in \mathbb{N}$ such that the first iterate from (55) already gives a sufficiently good approximation to the sign function. As discussed in section 3.6, we can then approximate

$$\mathrm{sign}(A)v \approx A \cdot \sum_{i=1}^{p} \frac{1}{p\xi_i} \widetilde{x}^{(j)}, \tag{63}$$

with $\widetilde{x}^{(j)}$ an approximate solution of the linear system

$$(A^2 + \alpha_j I)x^{(j)} = v, \qquad j = 1, \ldots, p. \tag{64}$$

We already discussed in section 3.6 how we can make efficient use of the shifted nature of these systems when solving them with standard Krylov subspace methods.

A particularly important situation arises when A is Hermitian and the intervals $[-b, -a] \cup [c, d]$, with $0 < a \leq b, 0 < c \leq d$, containing the eigenvalues of A are available. Under these hypotheses, Zolotarev explicitly derived the best rational approximation in the Chebyshev sense; see [57]. The next theorem states this result for $[-b, -a] = -[c, d]$. The key point is that for fixed $\mu = 2p - 1, \nu = 2p$, finding the optimal rational approximation $\mathcal{R}_{\mu\nu}(z) = \mathcal{N}_{\mu\nu}(z)/\mathcal{D}_{\mu\nu}(z)$ to the sign function on $[-b, -a] \cup [a, b]$ is equivalent to finding the best such rational approximation $\mathcal{S}_{p-1,p}(z) = \mathcal{N}_{p-1,p}(z)/\mathcal{D}_{p-1,p}(z)$ in *relative sense* to the inverse square root on $[1, (b/a)^2]$. The two functions are then related via $\mathcal{R}_{2p-1,2p}(z) = az \cdot \mathcal{S}_{p-1,p}(az)$.

Proposition 4. *Let* $\mathcal{R}_{2p-1,2p}(z) = \mathcal{N}_{2p-1,2p}(z)/\mathcal{D}_{2p-1,2p}(z)$ *be the Chebyshev best approximation to* $\mathrm{sign}(z)$ *on the set* $[-b, -a] \cup [a, b]$, *i.e. the function which minimizes*

$$\max_{a<|z|<b} |\mathrm{sign}(z) - \widetilde{\mathcal{R}}_{2p-1,2p}(z)| \tag{65}$$

over all rational functions $\widetilde{\mathcal{R}}_{2p-1,2p}(z) = \widetilde{\mathcal{N}}_{2p-1,2p}(z)/\widetilde{\mathcal{D}}_{2p-1,2p}(z)$. *Then the factored form of* $\mathcal{R}_{2p-1,2p}$ *is given by*

$$\mathcal{R}_{2p-1,2p}(z) = az \cdot \mathcal{S}_{p-1,p}((az)^2) \;\; \text{with} \;\; \mathcal{S}_{p-1,p}(z) = D \frac{\prod_{i=1}^{p-1}(z + c_{2i})}{\prod_{i=1}^{p}(z + c_{2i-1})}, \tag{66}$$

where

$$c_i = \frac{\mathrm{sn}^2\left(iK/(2p); \sqrt{1-(b/a)^2}\right)}{1 - \mathrm{sn}^2\left(iK/(2p); \sqrt{1-(b/a)^2}\right)},$$

K is the complete elliptic integral, sn *is the Jacobi elliptic function, and D is uniquely determined by the condition*

$$\max_{z \in [1,(b/a)^2]} \left(1 - \sqrt{z}\mathcal{S}_{p-1,p}(z)\right) = - \min_{z \in [1,(b/a)^2]} \left(1 - \sqrt{z}\mathcal{S}_{p-1,p}(z)\right).$$

Table 1. Number of poles necessary to achieve accuracy of 0.01

b/a	(55)	Zolotarev
200	19	5
1000	42	6

For a given number of poles, the Zolotarev approximation is much more accurate than that of the rational approximation (55) and is therefore to be preferred. This is illustrated in Table 1, taken form [67]. However, the usc of the Zolotarev approximation is restricted to Hermitian matrices for which lower and upper bounds (a and b, resp.) on the moduli of the eigenvalues are known.

As a final point, let us again assume that A is Hermitian and that we approximate $\text{sign}(A)$ by some rational approximation $\mathcal{R}(A)$ with $\mathcal{R}$ having a partial fraction expansion of the form

$$\mathcal{R}(z) = \sum_{j=1}^{p} \omega_j \frac{z}{z^2 + \alpha_j}, \quad \omega_j \geq 0, \quad \alpha_j \geq 0, \quad j = 1, \ldots, p. \tag{67}$$

Note that this is the case for the Zolotarev approximation from Proposition 4 as well as for the Padé approximations from (55). In order to compute $\mathcal{R}(A)v$, let us assume that we use the (shifted) CG-method to simultaneously solve $(A^2 + \alpha_j I)x^{(j)} = v$ for all j of interest, so that we get CG-iterates $x_k^{(j)}$ with residual $r_k^{(j)} = v - (A^2 + \alpha_j I)x_k^{(j)}$. Then the following estimate holds [67].

Proposition 5. *Let $g_j > 0$ be such that $\sum_{j=1}^{p} g_j = 1$ and $\varepsilon > 0$. If the CG iteration for system j is stopped at step k_j in which the residual satisfies*

$$\|r_{k_j}^{(j)}\|_2 \leq \varepsilon g_j \frac{\sqrt{\alpha_j}}{\omega_j}, \tag{68}$$

then

$$\| \mathcal{R}(A)v - \sum_{j=1}^{p} \omega_j x_{k_j}^{(j)} \|_2 \leq \varepsilon. \tag{69}$$

This proposition formulates a computationally feasible stopping criterion. If we also know the approximation accuracy of the rational approximation, i.e. if we have an information of the kind

$$\max_{z \in \text{spec}(A)} |\mathcal{R}(z) - \text{sign}(z)| \leq \varepsilon_2, \tag{70}$$

then we know that

$$\| \text{sign}(A)v - \sum_{j=1}^{p} \omega_j x_{k_j}^{(j)} \|_2 \leq \varepsilon + \varepsilon_2. \tag{71}$$

This fact is in agreement with the discussion on rational approximation of section 3.6.

Acknowledgement

We thank Daniel Kressner, Marlis Hochbruch and Henk A. van der Vorst for helpful comments on an earlier version of this chapter.

References

1. A. C. Antoulas. *Approximation of large-scale Dynamical Systems*. Advances in Design and Control. SIAM, Philadelphia, 2005.
2. Z. Bai and J. Demmel. Using the matrix sign function to compute invariant subspaces. *SIAM J. Matrix Anal. Appl.*, 19(1):205–225, 1998.
3. R. H. Bartels and G. W. Stewart. Algorithm 432: Solution of the Matrix Equation AX+XB=C. *Comm. of the ACM*, 15(9):820–826, 1972.
4. P. Benner. Control Theory. In Leslie Hogben (editor), 'Handbook of Linear Algebra', Chapter 57, Chapman & Hall/CRC, 2006.
5. P. Benner and E.S. Quintana-Orti. Solving stable generalized Lyapunov equations with the matrix sign function. *Numer. Algorithms*, 20:75–100, 1999.
6. D. A. Bini, N. J. Higham, and B. Meini. Algorithms for the matrix pth root. *Numerical Algorithms*, 39(4):349–378, 2005.
7. P. B. Borwein. Rational approximations with real poles to e^{-x} and x^n. *Journal of Approximation Theory*, 38:279–283, 1983.
8. R. Byers. Solving the algebraic Riccati equation with the matrix sign function. *Linear Algebra Appl.*, 85:267–279, 1987.
9. A. J. Carpenter, A. Ruttan, and R. S. Varga. Extended numerical computations on the "1/9" conjecture in rational approximation theory. In P. R. Graves-Morris, E. B. Saff, and R. S. Varga, editors, *Rational approximation and interpolation. Proceedings of the United Kingdom-United States conference held at Tampa, Florida, December 12-16, 1983*, pages 503–511, Berlin, 1990. Springer-Verlag, Lecture Notes Math.
10. E. Celledoni and A. Iserles. Approximating the exponential from a Lie algebra to a Lie group. *Mathematics of Computation*, 69(232):1457–1480, 2000.
11. E. Celledoni and I. Moret. A Krylov projection method for systems of ODEs. *Applied Num. Math.*, 24:365–378, 1997.
12. W. J. Cody, G. Meinardus, and R. S. Varga. Chebyshev rational approximations to e^{-x} in $[0, +\infty)$ and applications to heat-conduction problems. *J. Approx. Theory*, 2(1):50–65, March 1969.
13. M. J. Corless and A. E. Frazho. *Linear systems and control - An operator perspective.* Pure and Applied Mathematics. Marcel Dekker, New York - Basel, 2003.
14. Ph. I. Davies and N. J. Higham. Computing $f(A)b$ for matrix functions f. In *QCD and numerical analysis III*, volume 47 of *Lect. Notes Comput. Sci. Eng.*, pages 15–24. Springer, Berlin, 2005.
15. P. J. Davis and P. Rabinowitz. *Methods of Numerical Integration*. Academic Press, London, 2nd edition edition, 1984.
16. J. W. Demmel. *Applied numerical linear algebra.* SIAM, Philadelphia, 1997.
17. V. Druskin, A. Greenbaum, and L. Knizhnerman. Using nonorthogonal Lanczos vectors in the computation of matrix functions. *SIAM J. Sci. Comput.*, 19(1):38–54, 1998.
18. V. Druskin and L. Knizhnerman. Two polynomial methods of calculating functions of symmetric matrices. *U.S.S.R. Comput. Math. Math. Phys.*, 29:112–121, 1989.

19. V. Druskin and L. Knizhnerman. Extended Krylov subspaces: approximation of the matrix square root and related functions. *SIAM J. Matrix Analysis and Appl.*, 19(3):755–771, 1998.
20. M. Eiermann and O. Ernst. A restarted Krylov subspace method for the evaluation of matrix functions. *SIAM J. Numer. Anal.*, 44:2481–2504, 2006.
21. R. W. Freund. Solution of shifted linear systems by quasi-minimal residual iterations. In L. Reichel et al., editor, *Numerical linear algebra. Proceedings of the conference in numerical linear algebra and scientific computation, Kent, OH, USA, March 13-14, 1992*, pages 101–121. Walter de Gruyter, Berlin, 1993.
22. A. Frommer. BiCGStab(ℓ) for families of shifted linear systems. *Computing*, 70:87–109, 2003.
23. A. Frommer and U. Glässner. Restarted GMRES for shifted linear systems. *SIAM J. Sci. Comput.*, 19(1):15–26, 1998.
24. A. Frommer and P. Maass. Fast CG-based methods for Tikhonov–Phillips regularization. *SIAM J. Sci. Comput.*, 20(5):1831–1850, 1999.
25. E. Gallopoulos and Y. Saad. Efficient solution of parabolic equations by Krylov approximation methods. *SIAM J. Sci. Stat. Comput.*, 13(5):1236–1264, 1992.
26. W. Gautschi. *Numerical Analysis. An Introduction*. Birkhäuser, Boston, 1997.
27. G. Golub and C. F. Van Loan. *Matrix computations*. The Johns Hopkins Univ. Press, Baltimore, 3rd edition, 1996.
28. S. Gugercin, D. C. Sorensen, and A. C. Antoulas. A modified low-rank Smith method for large-scale Lyapunov equations. *Numer. Algorithms*, 32:27–55, 2003.
29. E. Hairer, C. Lubich, and G. Wanner. *Geometric numerical integration. Structure-preserving algorithms for ordinary differential equations*, volume 31 of *Springer Series in Computational Mathematics*. Springer, Berlin, 2002.
30. S. J. Hammarling. Numerical solution of the stable, non-negative definite Lyapunov equation. *IMA J. Numer. Anal.*, 2:303–323, 1982.
31. N. J. Higham. Newton's method for the matrix square root. *Math. Comp.*, 46(174): 537–549, April 1986.
32. N. J. Higham. Stable iterations for the matrix square root. *Numer. Algorithms*, 15(2): 227–242, 1997.
33. N. J. Higham. The Scaling and Squaring Method for the Matrix Exponential Revisited. *SIAM J. Matrix Analysis and Appl.*, 26(4):1179–1193, 2005.
34. N. J. Higham. *Functions of Matrices: Theory and Computation*. SIAM, Philadelphia, 2008.
35. M. Hochbruck and C. Lubich. On Krylov subspace approximations to the matrix exponential operator. *SIAM J. Numer. Anal.*, 34(5):1911–1925, 1997.
36. M. Hochbruck, C. Lubich, and H. Selhofer. Exponential integrators for large systems of differential equations. *SIAM J. Sci. Comput.*, 19(5):1552–1574, 1998.
37. M. Hochbruck and J. van den Eshof. Preconditioning Lanczos approximations to the matrix exponential. *SIAM J. Sci. Comput.*, 27(4):1438–1457, 2006.
38. R. A. Horn and C. R. Johnson. *Topics in matrix analysis. 1st paperback ed. with corrections*. Cambridge University Press, Cambridge, 1994.
39. A. Iserles, H. Z. Munthe-Kaas, S. P. Nørsett, and A. Zanna. Lie-group methods. *Acta Numerica*, 9:215–365, 2000.
40. I. M. Jaimoukha and E. M. Kasenally. Krylov subspace methods for solving large Lyapunov equations. *SIAM J. Numer. Anal.*, 31(1):227–251, Feb. 1994.
41. C. Kenney and A. J. Laub. Rational iterative methods for the matrix sign function. *SIAM J. Matrix Anal. Appl.*, 12(2):273–291, 1991.

42. C. Kenney and A. J. Laub. On scaling Newton's method for polar decomposition and the matrix sign function. *SIAM J. Matrix Anal. Appl.*, 13(3):688–706, 1992.
43. C. S. Kenney and A. J. Laub. The matrix sign function. *IEEE Trans. Autom. Control*, 40(8):1330–1348, 1995.
44. P. Lancaster and L. Rodman. *The Algebraic Riccati Equation*. Oxford University Press, Oxford, 2nd edition, 1995.
45. J.-R. Li and J. White. Low-Rank solutions of Lyapunov equations. *SIAM Review*, 46(4):693–713, 2004.
46. C. Van Loan. The sensitivity of the matrix exponential. *SIAM J. Numer. Anal.*, 14(6): 971–981, 1977.
47. L. Lopez and V. Simoncini. Analysis of projection methods for rational function approximation to the matrix exponential. *SIAM J. Numer. Anal.*, 44(2):613 – 635, 2006.
48. L. Lopez and V. Simoncini. Preserving geometric properties of the exponential matrix by block Krylov subspace methods. *BIT, Numerical Mathematics*, 46(4):813–830, 2006.
49. The MathWorks, Inc. *MATLAB 7*, September 2004.
50. C. Moler and C. Van Loan. Nineteen dubious ways to compute the exponential of a matrix, twenty-five years later. *SIAM Review*, 45(1):3–49, 2003.
51. I. Moret and P. Novati. An interpolatory approximation of the matrix exponential based on Faber polynomials. *J. Comput. and Applied Math.*, 131:361–380, 2001.
52. I. Moret and P. Novati. RD-rational approximations of the matrix exponential. *BIT, Numerical Mathematics*, 44(3):595–615, 2004.
53. I. Moret and P. Novati. Interpolating functions of matrices on zeros of quasi-kernel polynomials. *Numer. Linear Algebra Appl.*, 11(4):337–353, 2005.
54. H. Neuberger. The overlap Dirac operator. In *Numerical Challenges in Lattice Quantum Chromodynamics*, volume 15 of *Lect. Notes Comput. Sci. Eng.*, pages 1–17. Springer, Berlin, 2000.
55. S. P. Nørsett. Restricted Padè approximations to the exponential function. *SIAM J. Numer. Anal.*, 15(5):1008–1029, Oct. 1978.
56. T. Penzl. A cyclic low-rank Smith method for large sparse Lyapunov equations. *SIAM J. Sci. Comput.*, 21(4):1401–1418, 2000.
57. P. P. Petrushev and V. A. Popov. *Rational approximation of real functions*. Cambridge University Press, Cambridge, 1987.
58. J. D. Roberts. Linear model reduction and solution of the algebraic Riccati equation by use of the sign function. *Int. J. Control*, 32:677–687, 1980.
59. Y. Saad. Numerical solution of large Lyapunov equations. In M. A. Kaashoek, J. H. van Schuppen, and A. C. Ran, editors, *Signal Processing, Scattering, Operator Theory, and Numerical Methods. Proceedings of the international symposium MTNS-89, vol III*, pages 503–511, Boston, 1990. Birkhauser.
60. Y. Saad. Analysis of some Krylov subspace approximations to the matrix exponential operator. *SIAM J. Numer. Anal.*, 29:209–228, 1992.
61. Y. Saad. *Iterative Methods for Sparse Linear Systems*. The PWS Publishing Company, Boston, 1996. Second edition, SIAM, Philadelphia, 2003.
62. R. B. Sidje. Expokit: A Software Package for Computing Matrix Exponentials. *ACM Transactions on Math. Software*, 24(1):130–156, 1998.
63. V. Simoncini and D. B. Szyld. Recent computational developments in Krylov subspace methods for linear systems.
J. Numerical Linear Algebra with Appl., 14(1):1-59, 2007.
64. D. E. Stewart and T. S. Leyk. Error estimates for Krylov subspace approximations of matrix exponentials. *J. Comput. and Applied Math.*, 72:359–369, 1996.

65. H. Tal-Ezer. Spectral methods in time for parabolic problems. *SIAM J. Numer. Anal.*, 26:1–11, 1989.
66. L. N. Trefethen, J. A. C. Weideman, and T. Schmelzer. Talbot Quadratures and Rational Approximations. *BIT, Numerical Mathematics*, 46:653–670, 2006.
67. J. van den Eshof, A. Frommer, Th. Lippert, K. Schilling, and H.A. van der Vorst. Numerical methods for the QCD overlap operator. I: Sign-function and error bounds. *Comput. Phys. Commun.*, 146(2):203–224, 2002.
68. H.A. van der Vorst. An iterative solution method for solving $f(A)x = b$, using Krylov subspace information obtained for the symmetric positive definite matrix A. *J. Comput. Appl. Math.*, 18:249–263, 1987.
69. H.A. van der Vorst. Solution of $f(A)x = b$ with projection methods for the matrix A. In *Numerical Challenges in Lattice Quantum Chromodynamics*, volume 15 of *Lect. Notes Comput. Sci. Eng.*, pages 18–28. Springer, Berlin, 2000.
70. R. S. Varga. On higher order stable implicit methods for solving parabolic partial differential equations. *J. of Mathematics and Physics*, XL:220–231, 1961.
71. A. Zanna and H. Z. Munthe-Kaas. Generalized polar decompositions for the approximation of the matrix exponential. *SIAM J. Matrix Analysis and Appl.*, 23(3):840–862, 2002.

Model Reduction of Interconnected Systems

Antoine Vandendorpe[1] and Paul Van Dooren[2]

[1] Department of Mathematical Engineering, Université catholique de Louvain, Belgium
vandendorpe@csam.ucl.ac.be
[2] Department of Mathematical Engineering, Université catholique de Louvain, Belgium
vdooren@csam.ucl.ac.be

Summary. We consider a particular class of structured systems that can be modelled as a set of input/output subsystems that interconnect to each other, in the sense that outputs of some subsystems are inputs of other subsystems. Sometimes, it is important to preserve this structure in the reduced order system. Instead of reducing the entire system, it makes sense to reduce each subsystem (or a few of them) by taking into account its interconnection with the other subsystems in order to approximate the entire system in a so-called structured manner. The purpose of this paper is to present both Krylov-based and Gramian-based model reduction techniques that preserve the structure of the interconnections. Several structured model reduction techniques existing in the literature appear as special cases of our approach, permitting to unify and generalize the theory to some extent.

1 Introduction

Specialized model reduction techniques have been developed for various types of structured problems such as weighted model reduction, controller reduction and second order model reduction. Interconnected systems, also called aggregated systems, have been studied in the eighties [FB87] in the model reduction framework, but they have not received a lot of attention lately. This is in contrast with controller and weighted SVD-based model reduction techniques, which have been extensively studied [AL89, Enn84]. Controller reduction Krylov techniques have also been considered recently in [GBAG04]. It turns out that many structured systems can be modelled as particular cases of more general *interconnected* systems defined below (the behavioral approach [PW98] for interconnected systems is not considered here).

In this paper, we define an *interconnected system* as a linear system $G(s)$ composed of an interconnection of k sub-systems $T_i(s)$. Each subsystem is assumed to be a linear MIMO transfer function. Subsystem $T_i(s)$ has α_i inputs denoted by the vector a_i and β_i outputs denoted by the vector b_i:

$$b_i(s) = T_i(s)a_i(s). \tag{1}$$

Note that these inputs and outputs can also be viewed as internal variables of the interconnected system. The input $a_i(s)$ of each subsystem is a linear combination of

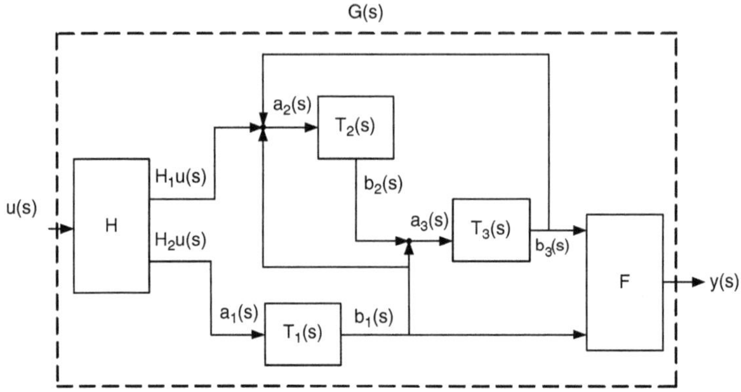

Fig. 1. Example of interconnected system

the outputs of all subsystems and of the external input $u(s) \in \mathbb{R}^m(s)$:

$$a_i(s) = H_i u(s) + \sum_{j=1}^{k} K_{i,j} b_j(s), \tag{2}$$

where $H_i \in \mathbb{R}^{\alpha_i \times m}$. The output $y(s) \in \mathbb{R}^p(s)$ of $G(s)$ is a linear function of the outputs of the subsystems:

$$y(s) = \sum_{i=1}^{k} F_i b_i(s), \tag{3}$$

with $F_i \in \mathbb{R}^{p \times \beta_i}$. Figure 1 gives an example of an interconnected system $G(s)$ composed of three subsystems.

We now introduce some notation in order to rewrite this in a block form. The matrix I_n denotes the identity matrix of size n and the matrix $0_{p,q}$ the $p \times q$ zero matrix. If $M_1, \ldots M_k$ is a set of matrices, then the matrix $diag\{M_1, \ldots, M_k\}$ denotes the block diagonal matrix

$$\begin{bmatrix} M_1 & & \\ & \ddots & \\ & & M_k \end{bmatrix}.$$

We also define $\alpha := \sum_{i=1}^{k} \alpha_i$ and $\beta := \sum_{i=1}^{k} \beta_i$. If the transfer functions $T_i(s) \in \mathbb{R}^{\beta_i \times \alpha_i}(s)$ are rational matrix function with real coefficients, then (1) can be rewritten as $b(s) = T(s)a(s)$, where

$$b(s) := \begin{bmatrix} b_1(s) \\ \vdots \\ b_k(s) \end{bmatrix}, \quad T(s) := diag\{T_1(s), \ldots, T_k(s)\}, \quad a(s) := \begin{bmatrix} a_1(s) \\ \vdots \\ a_k(s) \end{bmatrix}, \tag{4}$$

are respectively in $\mathbb{R}^{\beta}(s)$, $\mathbb{R}^{\beta\times\alpha}(s)$ and $\mathbb{R}^{\alpha}(s)$. If we also define $F \in \mathbb{R}^{p\times\beta}$, $K \in \mathbb{R}^{\alpha\times\beta}$ and $H \in \mathbb{R}^{\alpha\times m}$, as follows:

$$F := \begin{bmatrix} F_1 \dots F_k \end{bmatrix}, \quad K := \begin{bmatrix} K_{1,1} & \dots & K_{1,k} \\ \vdots & \ddots & \vdots \\ K_{k,1} & \dots & K_{k,k} \end{bmatrix}, \quad H := \begin{bmatrix} H_1 \\ \vdots \\ H_k \end{bmatrix}, \tag{5}$$

then (2), (3) can then be rewritten as follows:

$$a(s) = Hu(s) + Kb(s), \quad y(s) = Fb(s), \tag{6}$$

from which it easily follows that

$$y(s) = F(I_\beta - T(s)K)^{-1}T(s)Hu(s). \tag{7}$$

We assume that the Mc Millan degree of $T_i(s)$ is n_i and that (A_i, B_i, C_i, D_i) is a minimal state space realization of $T_i(s)$. If we define $n := \sum_{i=1}^{k} n_i$, then a realization for $T(s)$ is given by $C(sI_n - A)^{-1}B + D$ with

$$\begin{aligned} A &:= diag\{A_1, \dots, A_k\}, \quad B := diag\{B_1, \dots, B_k\}, \\ C &:= diag\{C_1, \dots, C_k\}, \quad D := diag\{D_1, \dots, D_k\}. \end{aligned} \tag{8}$$

In others words, $G(s) = F(I_\beta - T(s)K)^{-1}T(s)H$ and a state space realization of $G(s)$ is given by (A_G, B_G, C_G, D_G) (see for instance [ZDG96]), where

$$\begin{aligned} A_G &:= A + BK(I_\beta - DK)^{-1}C, \quad B_G := B(I_\alpha - KD)^{-1}H, \\ C_G &:= F(I_\beta - DK)^{-1}C, \quad D_G := FD(I_\alpha - KD)^{-1}H. \end{aligned} \tag{9}$$

If all the transfer functions are strictly proper, i.e. $D = 0$, the state space realization (9) of $G(s)$ reduces to:

$$A_G = A + BKC, \quad B_G = BH, \quad C_G = FC, \quad D_G = 0.$$

Let us finally remark that if all systems are connected in parallel, i.e. $K = 0$, then $G(s) = FT(s)H$.

The problem of *interconnected systems model reduction* proposed here consists in reducing some (e.g. one) of the subsystems $T_i(s)$ in order to approximate the global mapping from $u(s)$ to $y(s)$ and not the internal mappings from $a_i(s)$ to $b_i(s)$.

This paper is organized as follows. After some preliminary results, a Balanced Truncation framework for interconnected systems is derived in Section 2. Krylov model reduction techniques for interconnected systems are presented in Section 3. In Section 4, several connections with existing model reduction techniques for structured systems are given, and Section 5 contains some concluding remarks.

2 Interconnected Systems Balanced Truncation

We first recall the well-known Balanced Truncation method and emphasize their energetic interpretation. We then show how to extend Balanced Truncation to the so-called *Interconnected System Balanced Truncation*.

We consider a general transfer function $T(s) := C(sI_n - A)^{-1}B + D$ which corresponds to the linear system

$$\mathcal{S}\begin{cases} \dot{x}(t) = Ax(t) + Bu(t) \\ y(t) = Cx(t) + Du(t). \end{cases} \tag{10}$$

If the matrix A is Hurwitz, the controllability and observability Gramians, denoted respectively by P and Q are the unique solutions of the following equations

$$AP + PA^T + BB^T = 0\,, \quad A^TQ + QA + C^TC = 0.$$

If we apply an input $u(.) \in \mathcal{L}_2[-\infty, 0]$ to the system (10) for $t < 0$, the position of the state at time $t = 0$ (by assuming the zero initial condition $x(-\infty) = 0$) is a linear function of $u(t)$ given by the convolution

$$x(0) = \mathcal{C}_o(u(t)) := \int_{-\infty}^{0} e^{-At}Bu(t)dt.$$

By assuming that a zero input is applied to the system for $t > 0$, then for all $t \geq 0$, the output $y(.) \in \mathcal{L}_2[0, +\infty]$ of the system (10) is a linear function of $x(0)$, given by

$$y(t) = \mathcal{O}_b(x(0)) := Ce^{At}x(0).$$

The so-called controllability operator $\mathcal{C}_o : \mathcal{L}_2[-\infty, 0] \mapsto \mathbb{R}^n$ (mapping past inputs $u(.)$ to the present state) and observability operator $\mathcal{O}_b : \mathbb{R}^n \mapsto \mathcal{L}_2[0, +\infty]$ (mapping the present state to future outputs $y(.)$) have dual operators, respectively denoted by $\mathcal{C}_o^*$ and $\mathcal{O}_b^*$ (see [Ant05]).

A physical interpretation of the Gramians is the following. The controllability matrix arises from the following optimization problem. Let

$$J(v(t), a, b) := \int_a^b v(t)^T v(t)dt$$

be the *energy* of the vector function $v(t)$ in the interval $[a, b]$. Then [Glo84]

$$\min_{\mathcal{C}_o u(t)=x_0} J(u(t), -\infty, 0) = x_0^T P^{-1} x_0, \tag{11}$$

and, by duality, we have that

$$\min_{\mathcal{O}_b^* y(t)=x_0} J(y(t), -\infty, 0) = x_0^T Q^{-1} x_0. \tag{12}$$

Essential properties of the Gramians P and Q are as follows. First, under a coordinate transformation $x(t) = S\bar{x}(t)$, the new Gramians $\bar{P}$ and $\bar{Q}$ corresponding to the state-space realization $(\bar{A}, \bar{B}, \bar{C}) = (S^{-1}AS, S^{-1}B, CS)$ undergo the following (so-called *contragradient*) transformation:

$$\bar{P} = S^{-1}PS^{-T}, \quad \bar{Q} = S^T QS. \tag{13}$$

This implies that there exists a state-space realization $(A_{bal}, B_{bal}, C_{bal})$ of $T(s)$ such that the corresponding Gramians are equal and diagonal $\bar{P} = \bar{Q} = \Sigma$ [ZDG96]. Secondly, because these Gramians appear in the solutions of the optimization problems (11) and (12), they tell something about the energy that goes through the system, and more specifically, about the distribution of this energy among the state variables. The idea of the Balanced Truncation model reduction framework is to perform a state space transformation that yields equal and diagonal Gramians and to keep only the most controllable and observable states. If the original transfer function is stable, the reduced order transfer function is guaranteed to be stable and an a priori global error bound between both systems is available [Ant05].

If the standard balanced truncation technique is applied to the state space realization (A, B, C) (8) of an interconnected system, the structure of the subsystems is lost in the resulting reduced order transfer function. We show then how to preserve the structure in the balancing process. We first recall a basic lemma that will be used in the sequel.

Lemma 1. *Let $x_i \in \mathbb{R}^{n_i}$ and $M_{i,j} \in \mathbb{R}^{n_i \times n_j}$ for $1 \le i \le k$ and define*

$$x := \begin{bmatrix} x_1 \\ \vdots \\ x_k \end{bmatrix}, \quad M := \begin{bmatrix} M_{1,1} & \dots & M_{1,k} \\ \vdots & \ddots & \vdots \\ M_{k,1} & \dots & M_{k,k} \end{bmatrix}.$$

Assume M to be positive definite and consider the product

$$J(x, M) := x^T M^{-1} x.$$

Then, for any fixed $x_i \in \mathbb{R}^{n_i \times n_i}$,

$$J(x, M)_{x_j = 0, j \neq i} = x_i^T \left(M_{i,i} - M_{i,j} M_{j,j}^{-1} M_{j,i}\right)^{-1} x_i, \tag{14}$$

and

$$\min_{x_j, j \neq i} J(x, M) = x_i^T M_{i,i}^{-1} x_i. \tag{15}$$

Proof. Without loss of generality, let us assume that $i = 1$. For ease of notation, define $y := \begin{bmatrix} x_2^T & \dots & x_k^T \end{bmatrix}^T$ and $\begin{bmatrix} N_{1,1} & N_{1,2} \\ N_{1,2}^T & N_{2,2} \end{bmatrix} = N := M^{-1}$ with $N_{1,1} \in \mathbb{R}^{n_1 \times n_1}$. We obtain the following expression

$$J(x, M) = x_1^T N_{1,1} x_1 + 2x_1^T N_{1,2} y + y^T N_{2,2} y. \tag{16}$$

For $y = 0$ and using the Schur complement formula for the inverse of a matrix, we retrieve (14). In order to prove (15) we note that N is positive definite since M is positive definite. This implies that $N_{1,1}$ and $N_{2,2}$ are positive definite. $J(x, M)$ is a quadratic form and the Hessian of $J(x, M)$ with respect to y is equal to $N_{2,2}$. The minimum is then obtained by annihilating the gradient:

$$\frac{tialJ(x, M)}{tialy} = 2N_{1,2}^T x_1 + 2N_{2,2} y,$$

which is obtained for $y = -N_{2,2}^{-1} N_{1,2}^T x_1$ and yields

$$\min_y J(x, M) = x_1^T N_{1,1} x_1 - x_1^T N_{1,2} N_{2,2}^{-1} N_{1,2}^T x_1 = x_1^T M_{1,1}^{-1} x_1.$$

The last equality is again obtained by using the Schur complement formula. □

Let us now consider the controllability and observability Gramians of $G(s)$:

$$A_G P_G + P_G A_G^T + B_G B_G^T = 0, \quad A_G^T Q_G + Q_G A_G + C_G^T C_G = 0, \tag{17}$$

and let us partition them as follows:

$$P_G = \begin{bmatrix} P_{1,1} & \ldots & P_{1,k} \\ \vdots & \ddots & \vdots \\ P_{k,1} & \ldots & P_{k,k} \end{bmatrix}, \quad Q_G = \begin{bmatrix} Q_{1,1} & \ldots & Q_{1,k} \\ \vdots & \ddots & \vdots \\ Q_{k,1} & \ldots & Q_{k,k} \end{bmatrix}, \tag{18}$$

where $P_{i,j} \in \mathbb{R}^{n_i \times n_j}$. If we perform a state space transformation S_i to the state $x_i(t) = S_i \bar{x}_i(t)$ of each interconnected transfer function $T_i(s)$, we actually perform a state space transformation

$$S := diag\{S_1, \ldots, S_k\}$$

to the realization $(\bar{A}, \bar{B}, \bar{C}, \bar{D}) = (S^{-1}AS, S^{-1}B, CS, D)$ of $T(s)$. This, in turn, implies that $(\bar{A}_G, \bar{B}_G, \bar{C}_G, \bar{D}_G) = (S^{-1}A_G S, S^{-1}B_G, C_G S, D_G)$ and

$$(\bar{P}_G, \bar{Q}_G) = (S^{-1}P_G S^{-T}, S^T Q_G S),$$

i.e. they undergo a contragradient transformation. This implies that $(\bar{P}_{i,i}, \bar{Q}_{i,i}) = (S_i^{-1} P_{i,i} S_i^{-T}, S_i^T Q_{i,i} S_i)$, which is a contra-gradient transformation that only depends on the state space transformation on x_i, i.e. on the state space associated to $T_i(s)$.

Let us recall that the minimal past energy necessary to reach $x_i(0) = x_i$ for each $1 \leq i \leq k$ with the pair (A_G, B_G) is given by the expression

$$\begin{bmatrix} x_1^T & \ldots & x_k^T \end{bmatrix} P_G^{-1} \begin{bmatrix} x_1 \\ \vdots \\ x_k \end{bmatrix}. \tag{19}$$

The following result is then a consequence of Lemma 1.

Lemma 2. *With the preceding notation, the minimal past input energy*

$$J := \int_{-\infty}^{0} u(t)^T u(t) dt$$

needed to apply to the interconnected transfer function $G(s)$ in order that for subsystem i at time $t = 0$, $x_i(0) = x_i$ over all initial input condition $x_j(0), j \neq i$, is given by

$$x_i^T P_{i,i}^{-1} x_i.$$

Moreover, the minimal input needed in order that for subsystem i at time $t = 0$, $x_i(0) = x_i$ and that for all the other subsystems, $x_j(0) = 0, j \neq i$, is given by

$$x_i^T (P_G^{-1})_{i,i} x_i,$$

where $(P_G^{-1})_{i,i}$ is the i,i block of the inverse of P_G, and this block is equal to the inverse of the Schur complement of $P_{i,i}$.

Finally,

$$0 < P_{i,i}^{-1} \leq (P_G^{-1})_{i,i}. \tag{20}$$

Proof. The two first results are direct consequences of Lemma 1. Let us prove (20). For any nonzero vector x_i, the minimum energy necessary for subsystem i at time $t = 0$ to reach $x_i(0) = x_i$ over all initial input conditions $x_j(0), j \neq i$, cannot be larger than by imposing $x_j(0) = 0, j \neq i$. This implies that for any nonzero vector v,

$$v^T \left((P_G^{-1})_{i,i} - P_{i,i}^{-1} \right) v \geq 0.$$

□

Similar energy interpretations hold for the diagonal blocks of the observability matrix Q_G and of its inverse.

Because of Lemma 2, it makes sense to truncate the part of the state x_i of each subsystem $T_i(s)$ corresponding to the smallest eigenvalues of the product $P_{i,i}Q_{i,i}$. We can thus perform a block diagonal transformation in order to make the Gramians $P_{i,i}$ and $Q_{i,i}$ both equal and diagonal: $P_{i,i} = Q_{i,i} = \Sigma_i$. Then, we can truncate each subsystem $T_i(s)$ by deleting the states corresponding to the smallest eigenvalues of Σ_i. This is resumed in the following *Interconnected Systems Balanced Truncation* (ISBT) Algorithm. Let $(A_G, B_G, C_G, D_G) \sim G(s)$, where $G(s)$ is an interconnection of k subsystems

$$(A_i, B_i, C_i, D_i) \sim T_i(s),$$

of order n_i. In order to construct a reduced order system $\widehat{G}(s)$ while preserving the interconnections, proceed as follows.

ISBT Algorithm

1. Compute the Gramians P_G and Q_G satisfying (17).
2. For each subsystem $T_i(s)$ requiring an order reduction, perform the contragradient transformation S_i in order to make the Gramians $P_{i,i}$ and $Q_{i,i}$ equal and diagonal.

3. For each subsystem (A_i, B_i, C_i, D_i), keep only the space of states corresponding to the largest eigenvalues of $P_{i,i} = Q_{i,i} = \Sigma_i$, giving the reduced subsystems $\widehat{T}_i(s)$.
4. Define
$$\widehat{G}(s) = F(I_\beta - \widehat{T}(s)K)^{-1}\widehat{T}(s)H,$$
with $\widehat{T}(s) := diag\{\widehat{T}_i(s)\}$.

Remark 1. A variant of the ISBT Algorithm consists in performing a *balance and truncate* procedure for each subsystem $T_i(s)$ with respect to the Schur complements of $P_{i,i}$ and $Q_{i,i}$ instead of $P_{i,i}$ and $Q_{i,i}$. From Lemma 2, this corresponds to sorting the state-space of each system (A_i, B_i, C_i) with respect to the optimization problem $\min_u allelu(t)allel^2$ such that $x_i(0) = x_i$ and $x_j(0) = 0$ for $j \neq i$. Mixed strategies are also possible (see for instance [VA03] in the Controller Order Reduction framework).

It should be mentioned that a related balanced truncation approach for second order systems can be found in [MS96, CLVV06].

A main criticism concerning the ISBT Algorithm is that the reduced order system is not guaranteed to be stable. If all the subsystems $T_i(s)$ are stable, it is possible to impose all the subsystems $\widehat{T}_i(s)$ to remain stable by following a technique similar to that described in [WSL99]. Let us consider the $(1, 1)$ block of P_G and Q_G, i.e. $P_{1,1}$ and $Q_{1,1}$. These Gramians are positive definite because P_G and Q_G are assumed to be positive definite (here, $G(s)$ is assumed stable and (A_G, B_G, C_G, D_G) is a minimal realization). From (17), $P_{1,1}$ and $Q_{1,1}$ satisfy the Lyapunov equation

$$A_1 P_{1,1} + P_{1,1} A_1 + X = 0, \quad A_1^T Q_{1,1} + Q_{1,1} A_1 + Y = 0$$

where the symmetric matrices X and Y are not necessary positive definite. If one modifies X and Y to positive semi-definite matrices $\bar{B}\bar{B}^T$ and $\bar{C}^T\bar{C}$, one is guaranteed to obtain a stable reduced system $\widehat{T}_1(s)$. The main criticism about this technique is that the energetic interpretation of the modified Gramians is lost.

3 Krylov Techniques for Interconnected Systems

Krylov subspaces appear naturally in interpolation-based model reduction techniques. Let us recall that for any matrix M, $Im(X)$ is the space spanned by the columns of M.

Definition 1. *Let $A \in \mathbb{R}^{n\times n}$ and $B \in \mathbb{R}^{n\times m}$. The Krylov matrix $K_k(A, B) \in \mathbb{R}^{n\times km}$ is defined as follows*

$$K_k(A, B) := \begin{bmatrix} B \; AB \; \ldots \; A^{k-1}B \end{bmatrix}.$$

The subspace spanned by the columns of $K_k(A, B)$ is denoted by $\mathcal{K}_k(A, B)$.

Krylov techniques have already been considered in the literature for particular cases of structured systems. See for instance [SA86] in the controller reduction framework, or [SC91] in the second-order model reduction framework. This last case has been revisited recently in [Fre05] and [VV04]. But, to our knowledge, it is the first time they are studied in the general framework of *Interconnected Systems*.

The problem is the following. If one projects the state-space realizations (A_i, B_i, C_i) of the interconnected transfer functions $T_i(s)$ with projecting matrices Z_i, V_i derived from Krylov subspaces, this yields reduced-order transfer functions $\widehat{T}_i(s)$ that satisfy interpolation conditions with respect to $T_i(s)$; what are then the resulting relations between $\widehat{G}(s)$ and $G(s)$?

If one imposes the same interpolation conditions for every pair of subsystems $T_i(s)$ and $\widehat{T}_i(s)$, then the same interpolation conditions hold between the block diagonal transfer functions $T(s)$ and $\widehat{T}(s)$ as well. Let us investigate what this implies for $G(s)$ and $\widehat{G}(s)$. Assume that

$$(\widehat{A}, \widehat{B}, \widehat{C}) = (Z^T AV, Z^T B, CV)$$

such that $Z^T V = I$ and

$$\mathcal{K}_k\left((\lambda I - A)^{-1}, (\lambda I - A)^{-1}B\right) \subseteq Im(V).$$

In such a case, it is well known that [VS87, Gri97] $\widehat{T}(s) := \widehat{C}(sI - \widehat{A})^{-1}\widehat{B} + D$ interpolates $T(s) := C(sI - A)^{-1}B + D$ at $s = \lambda$ up to the k first derivatives. Concerning $G(s)$, the matrices F, K, D and H are unchanged, from which it easily follows that

$$\widehat{G}(s) = C_G V(sI - Z^T A_G V)^{-1} Z^T B_G + D_G.$$

It can easily be proved recursively that

$$\mathcal{K}_k\left(A_G, B_G\right) = \mathcal{K}_k\left(A + BK(I - DK)^{-1}C, B(I - KD)^{-1}H\right) \subseteq \mathcal{K}_k\left(A, B\right),$$

and it turns out that such a result holds for arbitrary interpolation points in the complex plane, as shown in the following lemma.

Lemma 3. *Let $\lambda \in \mathbb{C}$ be a point that is neither an eigenvalue of A nor an eigenvalue of A_G (defined in* (9)*). Then*

$$\mathcal{K}_k\left((\lambda I - A_G)^{-1}, (\lambda I - A_G)^{-1}B_G\right) \subseteq \mathcal{K}_k\left((\lambda I - A)^{-1}, (\lambda I - A)^{-1}B\right), \quad (21)$$

$$\mathcal{K}_k\left((\lambda I - A_G)^{-T}, (\lambda I - A_G)^{-T}C_G^T\right) \subseteq \mathcal{K}_k\left((\lambda I - A)^{-T}, (\lambda I - A)^{-T}C^T\right). \quad (22)$$

Proof. Only (21) will be proved. An analog proof can be given for (22). First, let us prove that the column space of $(\lambda I - A_G)^{-1}B_G$ is included in the column space of $(\lambda I - A)^{-1}B$. In order to simplify the notation, let us define the following matrices

$$M := (\lambda I_n - A)^{-1}B, \quad X := K(I_\beta - DK)^{-1}C, \quad G := (I_\alpha - KD)^{-1}H. \quad (23)$$

From the identity $(I - MX)^{-1}M = M(I - XM)^{-1}$, it then follows that

$$\begin{aligned}(\lambda I - A_G)^{-1}B_G &= (\lambda I - A - BX)^{-1} \\ BG &= (I - MX)^{-1} \\ MG &= M\,(I - XM)^{-1}\,G.\end{aligned}$$

This clearly implies that the column space of $(\lambda I - A_G)^{-1}B_G$ is included in the column space of $M = (\lambda I - A)^{-1}B$. Let us assume that

$$\mathcal{K}_{k-1}\left((\lambda I - A_G)^{-1}, (\lambda I - A_G)^{-1}B_G\right) \subseteq \mathcal{K}_{k-1}\left((\lambda I - A)^{-1}, (\lambda I - A)^{-1}B\right),$$

and prove that this implies that

$$\mathcal{K}_{k}\left((\lambda I - A_G)^{-1}, (\lambda I - A_G)^{-1}B_G\right) \subseteq \mathcal{K}_{k}\left((\lambda I - A)^{-1}, (\lambda I - A)^{-1}B\right). \quad (24)$$

Since the image of $(\lambda I - A_G)^{-k+1}B_G$ belongs to $\mathcal{K}_{k-1}\big((\lambda I - A)^{-1}(\lambda I - A)^{-1}B\big)$, there exists a matrix Y such that

$$(\lambda I - A_G)^{-k+1}B_G = K_{k-1}\left((\lambda I - A)^{-1}, (\lambda I - A)^{-1}B\right)Y.$$

One obtains then that $(\lambda I - A_G)^{-k}B_G$ equals

$$(\lambda I - A_G)^{-1}(\lambda I - A_G)^{-k+1}B_G = \sum_{i=0}^{\infty}(MX)^i\,(\lambda I - A)^{-1}K_{k-1}\left((\lambda I - A)^{-1}, M\right)Y.$$

Note that

$$Im\left((\lambda I - A)^{-1}K_{k-1}\left((\lambda I - A)^{-1}, M\right)\right) \subseteq \mathcal{K}_k\left((\lambda I - A)^{-1}, M\right).$$

Moreover, for any integer $i > 0$, it is clear that

$$Im\left((MX)^i\right) \in Im(M).$$

This proves that (24) is satisfied. □

Thanks to the preceding lemma, there are at least two ways to project the subsystems $T_i(s)$ in order to satisfy a set of interpolation conditions using Krylov subspaces as follows.

Lemma 4. *Let $\lambda \in \mathbb{C}$ be neither a pole of $T(s)$ nor a pole of $G(s)$. Define*

$$V := \begin{bmatrix} V_1 \\ \vdots \\ V_k \end{bmatrix} \in \mathbb{C}^{n\times r},$$

such that $V_i \in \mathbb{C}^{n_i \times r}$. Assume that either

$$\mathcal{K}_k\left((\lambda I - A_G)^{-1}, (\lambda I - A_G)^{-1}B_G\right) \subseteq Im(V). \tag{25}$$

or

$$\mathcal{K}_k\left((\lambda I - A)^{-1}, (\lambda I - A)^{-1}B\right) \subseteq Im(V). \tag{26}$$

Construct matrices $Z_i \in \mathbb{C}^{n_i \times r}$ *such that* $Z_i^T V_i = I_r$. *Project each subsystem as follows:*

$$(\widehat{A}_i, \widehat{B}_i, \widehat{C}_i) := (Z_i^T A_i V_i, Z_i^T B_i, C_i V_i). \tag{27}$$

Then, $\widehat{G}(s)$ *interpolates* $G(s)$ *at* λ *up to the first* k *derivatives.*

Proof. First note that (26) implies (25) because of Lemma 3, and that (27) amounts to projecting (A, B, C) to $(\widehat{A}, \widehat{B}, \widehat{C}) := (\mathcal{Z}^T A\mathcal{V}, \mathcal{Z}^T B, C\mathcal{V})$ with

$$\mathcal{Z} := diag\{Z_1, \ldots, Z_k\}, \quad \mathcal{V} := diag\{V_1, \ldots, V_k\} \tag{28}$$

and hence also (A_G, B_G, C_G) to $(\widehat{A}_G, \widehat{B}_G, \widehat{C}_G) := (\mathcal{Z}^T A_G\mathcal{V}, \mathcal{Z}^T B_G, C_G\mathcal{V})$. The interpolation property then follows from $\mathcal{Z}^T\mathcal{V} = I$ and

$$\mathcal{K}_k\left((\lambda I - A_G)^{-1}, (\lambda I - A_G)^{-1}B_G\right) \subseteq Im(V) \subseteq Im(\mathcal{V}), \tag{29}$$

which concludes the proof. □

In some contexts, such as controller reduction or weighted model reduction, one does not construct a reduced order transfer function $\widehat{G}(s)$ by projecting the state spaces of all the subsystems (A_i, B_i, C_i) but one may choose to project only some or one of the subsystems. Let us consider this last possibility.

Corollary 1. *Under the assumptions* (26) *or* (25) *of Lemma 4,* $\widehat{G}(s)$ *interpolates* $G(s)$ *at* λ *up to the first* k *derivatives even if only one subsystem* i *is projected according to* (27) *and all the other subsystems are kept unchanged.*

Proof. This corresponds to $(\widehat{A}_G, \widehat{B}_G, \widehat{C}_G) := (\mathcal{Z}^T A_G\mathcal{V}, \mathcal{Z}^T B_G, C_G\mathcal{V})$ with

$$\mathcal{Z} := diag\{I_{\sum_{j=1}^{i-1} n_j}, Z_i, I_{\sum_{j=i+1}^{k} n_j}\}, \quad \mathcal{V} := diag\{I_{\sum_{j=1}^{i-1} n_j}, V_i, I_{\sum_{j=i+1}^{k} n_j}\} \tag{30}$$

Again we have $\mathcal{Z}^T\mathcal{V} = I$ and $Im(V) \subseteq Im(\mathcal{V})$, which concludes the proof. □

Remark 2. Krylov techniques have recently been generalized for MIMO systems with the *tangential interpolation* framework [GVV04]. It is also possible to project the subsystems $T_i(s)$ in such a way that the reduced interconnected transfer function $\widehat{G}(s)$ satisfies a set of tangential interpolation conditions with respect to the original interconnected transfer function $G(s)$, but special care must be taken. Indeed, Lemma 3 is generically not true anymore for generalized Krylov subspaces corresponding to tangential interpolation conditions. In other words, the column space of the matrix

$$\mathcal{K}_k\left((\lambda I - A_G)^{-1}B_G, (\lambda I - A_G)^{-1}, Y\right) :=$$

$$\left[(\lambda I - A_G)^{-1}B_G \dots (\lambda I - A_G)^{-k}B_G\right] \begin{bmatrix} y_0 & \cdots & y_{k-1} \\ & \ddots & \vdots \\ & & y_0 \end{bmatrix}$$

is in general not contained in the column space of the matrix

$$\mathcal{K}_k\left((\lambda I - A)^{-1}B, (\lambda I - A)^{-1}, Y\right) :=$$

$$\left[(\lambda I - A)^{-1}B \dots (\lambda I - A)^{-k}B\right] \begin{bmatrix} y_0 & \cdots & y_{k-1} \\ & \ddots & \vdots \\ & & y_0 \end{bmatrix}.$$

In such a case, interchanging matrices (A_G, B_G, C_G) by (A, B, C), as done in Lemma 4 and Corollary 1 is not always permitted. Nevertheless, Lemma 4 and Corollary 1 can be extended to the tangential interpolation framework by projecting the state space realizations (A_i, B_i, C_i) with generalized Krylov subspaces of the form $\mathcal{K}_k\left((\lambda I - A_G)^{-1}B_G, (\lambda I - A_G)^{-1}, Y\right)$ and not of the form $\mathcal{K}_k\left((\lambda I - A)^{-1}B, (\lambda I - A)^{-1}, Y\right)$.

4 Examples of Structured Model Reduction Problems

As we will see in this section, many structured systems can be modelled as *interconnected systems*. Three well known structured systems are presented, namely *weighted* systems, *second-order* systems and *controlled* systems. For each of these specific cases one recovers well-known formulas. It turns out that several existing model reduction techniques for structured systems are particular cases of our ISBT Algorithm.

The preceding list is by no means exhaustive. For instance, because linear fractional transforms correspond to making a constant feedback to a part of the state, this can also be described by an interconnected system. Periodic systems are also a typical example of interconnected system that is not considered below.

Weighted Model Reduction

As a first example, let us consider the following *weighted* transfer function:

$$y(s) = W_{out}(s)T(s)W_{in}(s)u(s) := G(s)u(s).$$

Let (A_o, B_o, C_o, D_o), (A, B, C, D) and (A_i, B_i, C_i, D_i) be the state space realizations of respectively $W_{out}(s)$, $T(s)$ and $W_{in}(s)$, of respective order n_o, n and n_i. A state space realization (A_G, B_G, C_G, D_G) of $G(s)$ is given by

$$\left[\begin{array}{c|c} A_G & B_G \\ \hline C_G & D_G \end{array}\right] := \left[\begin{array}{ccc|c} A_o & B_oC & B_oDC_i & B_oDD_i \\ 0 & A & BC_i & BD_i \\ 0 & 0 & A_i & B_i \\ \hline C_o & D_oC & D_oDC_i & D_oDD_i \end{array}\right]. \tag{31}$$

The transfer function $G(s)$ corresponds to the *interconnected* system $\mathcal{S}$ with

$$\mathcal{S} : \begin{cases} b_1(s) = W_o(s)a_1(s),\ b_2(s) = T(s)a_2(s), \\ b_3(s) = W_i(s)a_3(s),\ y(s) = b_1(s), \\ a_1(s) = b_2(s),\ a_2(s) = b_3(s),\ a_3 = u(s) \end{cases},$$

and

$$H = \begin{bmatrix} 0 \\ 0 \\ I \end{bmatrix}, \quad K = \begin{bmatrix} 0 & I & 0 \\ 0 & 0 & I \\ 0 & 0 & 0 \end{bmatrix}, \quad F = \begin{bmatrix} I & 0 & 0 \end{bmatrix}.$$

A frequency weighted balanced reduction method was first introduced by Enns [Enn84, ZDG96]. Its strategy is the following. Note that Enns assumes that $D = 0$ (otherwise D can be added to $\widehat{T}(s)$).

ENNS Algorithm

1. Compute the Gramians P_G and Q_G satisfying (17) with (A_G, B_G, C_G, D_G) defined in (31).

2. Perform a state space transformation on (A, B, C) in order to obtain $P = Q = \Sigma$ diagonal, where P and Q are the diagonal blocs of P_G and Q_G corresponding to the $T(s)$:

$$P = \begin{bmatrix} 0_{n,n_o} & I_n & 0_{n,n_i} \end{bmatrix} P_G \begin{bmatrix} 0_{n_o,n} \\ I_n \\ 0_{n_i,n} \end{bmatrix}, \quad Q = \begin{bmatrix} 0_{n,n_o} & I_n & 0_{n,n_i} \end{bmatrix} Q_G \begin{bmatrix} 0_{n_o,n} \\ I_n \\ 0_{n_i,n} \end{bmatrix}. \tag{32}$$

3. Truncate (A, B, C) by keeping only the part of the state space corresponding to the largest eigenvalues of Σ.

It is clear the ENNS Algorithm is exactly the same as the ISBT Algorithm applied to weighted systems. As for the ISBT Algorithm, there is generally no known a priori error bound for the approximation error and the reduced order model is not guaranteed to be stable either.

There exists other weighted model reduction techniques. See for instance [WSL99] where an elegant error bound is derived.

A generalization of weighted systems are *cascaded systems*. If we assume that the interconnected systems are such that the input of $T_i(s)$ is the output of $T_{i+1}(s)$, we obtain a structure similar than for the weighted case. The matrix K has then the form

$$K = \begin{bmatrix} 0 \; I_{\beta_1} & & \\ & \ddots \; \ddots & \\ & & \ddots \; I_{\beta_{k-1}} \\ & & 0 \end{bmatrix}.$$

Second-Order systems

Second order systems arise naturally in many areas of engineering (see, for example, [Pre97, Rub70, WJ87]) with the following form:

$$\begin{cases} M\ddot{q}(t) + D\dot{q}(t) + Sq(t) = F_{in}\, u(t), \\ \qquad\qquad\qquad\qquad y(t) = F_{out}\, q(t). \end{cases} \tag{33}$$

We assume that $u(t) \in \mathbb{R}^m$, $y(t) \in \mathbb{R}^p$, $q(t) \in \mathbb{R}^n$, $F_{in} \in \mathbb{R}^{n\times m}$, $F_{out} \in \mathbb{R}^{p\times n}$, and $M, D, S \in \mathbb{R}^{n\times n}$ with M invertible. For mechanical systems the matrices M, D and S represent, respectively, the *mass* (or *inertia*), *damping* and *stiffness* matrices, $u(t)$ corresponds to the vector of *external* forces, F_{in} is the input distribution matrix, $y(\cdot)$ is the output measurement vector, F_{out} is the output measurement matrix, and $q(t)$ to the vector of *internal generalized coordinates.*

Second-Order systems can be seen as an interconnection of two subsystems as follows. For simplicity, the mass matrix M is assumed equal to the identity matrix. Define $T_1(s)$ and $T_2(s)$ corresponding to the following system:

$$\begin{cases} \dot{x}_1(t) = -Dx_1(t) - Sy_2(t) + F_{in}u(t) \\ y_1(t) = x_1(t) \end{cases},$$

$$\begin{cases} \dot{x}_2(t) = 0x_2(t) + y_1(t) \\ y_2(t) = x_2(t) \end{cases}. \tag{34}$$

From this, $y_1(s) := T_1(s)a_1(s) = (sI_n + D)^{-1}a_1(s)$ with $a_1(s) := u_1(t) - Sy_2(s)$ (with the convention $u_1(t) = F_{in}u(t)$) and $y_2(s) = F_{out}s^{-1}a_2(s) := T_2(s)a_2$ with $a_2(s) = y_1(s)$. Matrices F, H, K are given by

$$F := \begin{bmatrix} 0 \; F_{out} \end{bmatrix}, \quad H := \begin{bmatrix} F_{in} \\ 0 \end{bmatrix}, \quad K := \begin{bmatrix} 0 & -S \\ I & 0 \end{bmatrix}.$$

From the preceding definitions, one obtains

$$C = \begin{bmatrix} I & 0 \\ 0 & I \end{bmatrix}, \quad A = \begin{bmatrix} -D & 0 \\ 0 & 0 \end{bmatrix}, \quad B = \begin{bmatrix} I & 0 \\ 0 & I \end{bmatrix},$$

$$C_G = \begin{bmatrix} 0 \; F_{out} \end{bmatrix}, A_G = \begin{bmatrix} -D & -K \\ I & 0 \end{bmatrix}, B_G = \begin{bmatrix} F_{in} \\ 0 \end{bmatrix}.$$

The matrices (A_G, B_G, C_G) are clearly a state space realization of $F_{out}(s^2 I_n + Ds + S)^{-1}F_{in}$. It turns out that the Second-Order Balanced Truncation technique proposed in [CLVV06] is exactly the same as the Interconnected Balanced Truncation technique applied to $T_1(s)$ and $T_2(s)$. In general, systems of order k can be rewritten as an interconnection of k subsystems by generalizing the preceding ideas.

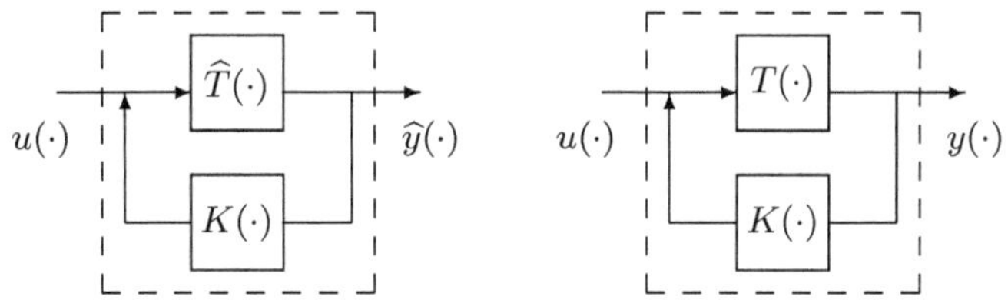

Fig. 2. Controller Order Reduction

Controller Order Reduction

The Controller Reduction problem (see Figure 2) introduced by Anderson and Liu [AL89] is the following. Most high-order linear plants $T(s)$ are controlled with a high order linear system $K(s)$. In order to model such *structured* systems by satisfying the computational constraints, it is sometimes needed to approximate either the plant, or the controller, or both systems by reduced order systems, denoted respectively by $\widehat{T}(s)$ and $\widehat{K}(s)$.

The objective of Controller Order Reduction is to find $\widehat{T}(s)$ and/or $\widehat{K}(s)$ that minimize the *structured* error $\|G(s) - \widehat{G}(s)\|$ with

$$G(s) := (I - T(s)K(s))^{-1}T(s), \quad \widehat{G}(s) := (I - \widehat{T}(s)\widehat{K}(s))^{-1}\widehat{T}(s). \qquad (35)$$

Balanced Truncation model reduction techniques have also been developed for this problem. Again, most of these techniques are very similar to the ISBT Algorithm. See for instance [VA03] for recent results. Depending on the choice of the pair of Gramians, it is possible to develop balancing strategies that ensure the stability of the reduced system, under certain assumptions [LC92].

5 Concluding Remarks

In this paper, general structure preserving model reduction techniques have been developed for interconnected systems, and this for both SVD-based and Krylov-based techniques. Of particular interest, the ISBT Algorithm is a generic tool for performing structured preserving balanced truncation. The advantage of studying model reduction techniques for general interconnected systems is twofold. Firstly, this permits to unify several model reduction techniques developed for weighted systems, controlled systems and second order systems in the same framework. Secondly, our approach permits to extend existing model reduction techniques for a large class of structured systems, namely those that can fit our definition of *interconnected* systems.

Acknowledgment

This paper presents research results of the NSF contract ACI-03-24944 and of the Belgian Network DYSCO (Dynamical Systems, Control, and Optimization), funded by the Interuniversity Attraction Poles Programme, initiated by the Belgian State, Science Policy Office. The scientific responsibility rests with its authors.

References

[AL89] Anderson, B.D.O., Liu, Y.: Controller reduction: concepts and approaches. IEEE Trans. Autom. Control, **34(8)**, 802–812 (1989)

[Ant05] Antoulas, A.C.: Approximation of Large-Scale Dynamical Systems. Siam Publications, Philadelphia (2005)

[CLVV06] Chahlaoui, Y., Lemonnier, D., Vandendorpe, A., Van Dooren, P.: Second-order balanced truncation. Linear Algebra and its Applications, **415**, 373–384 (2006)

[VS87] de Villemagne, C., Skelton, R.E.: Model reductions using a projection formulation. Internat. J. Control, **46(6)**, 2141–2169 (1987)

[Enn84] Enns, D.: Model reduction for control system design. PhD thesis, Stanford University, Stanford (1984)

[FB87] Feliachi, A., Bhurtun, C.:. Model reduction of large-scale interconnected systems. Int. J. Syst. Sci., **18**, 2249–2259 (1987)

[Fre05] Freund, R.W.: Padé-type model reduction of second-order and higher-order linear dynamical systems. In: Benner, P. et al. (eds) Dimension Reduction of Large-Scale Systems, Springer Verlag, LNCSE **45**, 191–223 (2005)

[GVV04] Gallivan, K., Vandendorpe, A., Van Dooren, P.: Model reduction of MIMO systems via tangential interpolation. SIAM Journal on Matrix Analysis and Applications, **26(2)**, 328–349 (2004)

[Glo84] Glover, K.D.: All optimal Hankel-norm approximation of linear multivariable systems and their L^∞-error bounds. Int. J. Control, **39(6)**, 1115–1193 (1984)

[Gri97] Grimme, E.: Krylov projection methods for model reduction. PhD Thesis, University of Illinois, Urbana-Champaign (1997)

[GBAG04] Gugercin, S., Beattie, C.A., Antoulas, A.C., Gildin, E.: Controller reduction by Krylov projection methods. In: De Moor, B. et al. (eds) Proceedings of 16th Symp. on the Mathematical Theory of Networks and Systems, Belgium (2004)

[LC92] Lin, C.A., Chiu, T.Y.: Model reduction via frequency weighted balanced realization. Control-Theory and Advanced Technology, **8(2)**, 341–351 (1992)

[MS96] Meyer, D.G., Srinivasan, S.: Balancing and model reduction for second-order form linear systems. IEEE Trans. Automat. Control, **41(11)**, 1632–1644 (1996)

[PW98] Polderman, J.W., Willems, J.C.: Introduction to Mathematical Systems Theory: a Behavioral Approach. Springer Verlag, Berlin (1998)

[Pre97] Preumont, A.: Vibration Control of Active Structures. Kluwer Academic Publishers, Dordrecht Boston London (1997)

[Rub70] Rubinstein, M.F.: Structural Systems - Statics, Dynamics and Stability. Prentice-Hall, Upper Saddle River (1970)

[SA86] Skelton, R.E., Anderson, B.D.O.: Q-Markov covariance equivalent realizations. Int. J. Control, **44**, 1477–1490 (1986)

[SC91] Su, T.J., Craig, R.R.Jr.: Model reduction and control of flexible structures using Krylov vectors. J. Guidance, **14(2)**, 260–267 (1991)

[VV04] Vandendorpe, A., Van Dooren, P.: Krylov techniques for model reduction of second-order systems. Technical Report CESAME 2004-07, Université catholique de Louvain (2004)

[VA03] Varga, A., Anderson, B.D.O.: Accuracy-enhancing methods for balancing-related frequency-weighted model and controller reduction. Automatica, **39(5)**, 919–927 (2003)

[WSL99] Wang, G., Sreeram, V., Liu, W.Q.: A new frequency-weighted balanced truncation method and an error bound. IEEE Trans. Autom. Control, **44(9)**, 1734–1737 (1999)

[WJ87] Weaver, W.Jr., Johnston, P.R.: Structural Dynamics by Finite Elements. Prentice-Hall, Upper Saddle River (1987)

[ZDG96] Zhou, K., Doyle, J.C., Glover, K.D.: Robust and Optimal Control. Prentice-Hall, Upper Saddle River (1996)

Quadratic Inverse Eigenvalue Problem and Its Applications to Model Updating — An Overview

Moody T. Chu**

Department of Mathematics, North Carolina State University, Raleigh, NC 27695-8205
chu@math.ncsu.edu

1 Introduction

Modeling is one of the most fundamental tools that we use to simulate the complex world. The goal of modeling is to come up with a representation that is simple enough for mathematical manipulation yet powerful enough for describing, inducing, and reasoning complicated phenomena. When modeling physical systems, the resulting mathematical models are sometimes of a very high order too expensive for simulation. One remedy is the notion of model reduction that assists in approximating very high order mathematical models with lower order models. As is evidenced in this collection, model reduction has been under extensive study and rapid development over the past few years with many physical and engineering applications. On the other hand, precise mathematical models of physical systems are hardly available in practice. Many factors, including inevitable disturbances to the measurement and imperfect characterization of the model, attribute to the inexactitude. Since the model reduction process begets only a partial effect of the original model, it is reasonable to expect that the reduced model might not be consonant with realistic data either. For various reasons, it often becomes necessary to *update* a primitive model to attain consistency with empirical results. This procedure of updating or revising an existing model is another essential ingredient for establishing an effective model. The emphasis of the following discussion is on the model updating of quadratic pencils.

The second order differential system

$$M\ddot{\mathbf{x}} + C\dot{\mathbf{x}} + K\mathbf{x} = f(t), \tag{1}$$

where $\mathbf{x} \in \mathbb{R}^n$ and $M, C, K \in \mathbb{R}^{n\times n}$, arises frequently in a wide scope of important applications, including applied mechanics, electrical oscillation, vibro-acoustics,

** This research was supported in part by the National Science Foundation under grants CCR-0204157 and DMS-0505880.

fluid mechanics, signal processing, and finite element discretization of PDEs. In most applications involving (1), specifications of the underlying physical system are embedded in the matrix coefficients M, C and K. It is well known that if

$$\mathbf{x}(t) = \mathbf{v}e^{\lambda t} \tag{2}$$

represents a fundamental solution to (1), then the scalar λ and the vector $\mathbf{v}$ must solve the *quadratic eigenvalue problem* (QEP)

$$(\lambda^2 M + \lambda C + K)\mathbf{v} = 0. \tag{3}$$

That is, characteristic behavior of the system (1) usually can be interpreted via the eigenvalues and eigenvectors of the system (3). Because of this connection, considerable efforts have be devoted to the QEP in the literature. Readers are referred to the treatise by Tisseur and Meerbergen [25] for a good survey of many applications, mathematical properties, and a variety of numerical techniques for the QEP.

Two aspects of the quadratic pencil associated with the model (1) deserve consideration. The process of analyzing and deriving the spectral information and, hence, inducing the dynamical behavior of a system from *a priori* known physical parameters such as mass, length, elasticity, inductance, capacitance, and so on is referred to as a *direct* problem. The *inverse* problem, in contrast, is to validate, determine, or estimate the parameters of the system according to its observed or expected behavior. The concern in the direct problem is to express the behavior in terms of the parameters whereas in the inverse problem the concern is to express the parameters in term of the behavior. The inverse problem is just as important as the direct problem in applications. The model updating problem can be regarded as a special case of the inverse eigenvalue problem.

The inverse eigenvalue problem is a diverse area full of research interests and activities. See the newly revised book by Gladwell [17], the review article [5], and the recently completed monograph by Chu and Golub [7] in which more than 460 references are collected. Among current development, the quadratic inverse eigenvalue problem (QIEP) is particularly more important and challenging with many unanswered questions. Depending on the applications, the term QIEP has been used in the literature to mean a rather wide range of diverse formulations. For instance, the QIEP studied by Ram and Elhay in [22] involves only symmetric tridiagonal matrix coefficients where *two* sets of eigenvalues are given. The QIEP studied by Starek and Inman in [24] is associated with nonproportional underdamped systems. Lancaster and Prells [20] considered the QIEP with symmetric and positive semi-definite damping C where *complete* information on eigenvalues and eigenvectors is given and all eigenvalues are simple and non-real. There are also works which utilize notions of feedback control to reassign the eigenstructure [10, 21]. The list goes on and on and can hardly be exhaustive.

In this article, we shall consider the QIEP under one common scenario, that is, the spectral information furnished is obtained from empirical data. In vibration industries, including aerospace, automobile, and manufacturing, through vibration tests where the excitation and the response of the structure at selected points are measured experimentally, there are identification techniques to extract a portion of eigenpair information from the measurements. However, the size of the system can be so large and complicated that it is not always possible to attain knowledge of the entire spectrum. While there is no reasonable analytical tool available to evaluate the entire spectral information, it is simply unwise to use experimental values of high natural frequencies to reconstruct a model. Additionally, it is often demanded, especially in structural design, that certain eigenvectors should also satisfy some specific conditions. A finite-element generated symmetric model therefore needs to be updated using only a few measured eigenvalues and eigenvectors [13, 14]. Furthermore, quantities related to high frequency terms in a *finite model* generally are susceptible to measurement errors due to the finite bandwidth of measuring devices. Spectral information, therefore, should not be used at its full extent. For these reasons, it might be more sensible to consider an inverse eigenvalue problem where only a *portion* of eigenvalues and eigenvectors is prescribed. Under these circumstances, the quadratic model updating problem (MUP) therefore can be formulated as follows:

> **(MUP)** Given a structured quadratic pencil (M_0, C_0, K_0) and a few of its associated eigenpairs $\{(\lambda_i, \mathbf{u}_i)\}_{i=1}^k$ with $k < 2n$, assume that new measured eigenpairs $\{(\sigma_i, \mathbf{y}_i)\}_{i=1}^k$ have been obtained. Update the pencil (M_0, C_0, K_0) to (M, C, K) *of the same structure* such that the subset $\{(\lambda_i, \mathbf{u}_i)\}_{i=1}^k$ is replaced by $\{(\sigma_i, \mathbf{y}_i)\}_{i=1}^k$ as k eigenpairs of (M, C, K).

2 Challenges

The MUP as stated above is of immense practical importance. However, there are considerable difficulties when solving a model updating problem. Many issues remain open for further research. We briefly outline three challenges below. We shall comment on current status of development for facing these challenges in later sections.

Structural Constraint. The structure imposed on a MUP depends inherently on the *connectivity* of the underlying physical system. The typical structure for a general mass-spring system, for example, is that the mass matrix M is diagonal, both the damping matrix C and the stiffness matrix K are symmetric and banded, M is positive definite ($M > 0$) and K is positive semi-definite ($K \geq 0$). As an illustration, the structure corresponding to the four-degree-of-freedom mass-spring system depicted in Figure 1 should be of the form where,

$$M = \begin{bmatrix} m_1 & 0 & 0 & 0 \\ 0 & m_2 & 0 & 0 \\ 0 & 0 & m_3 & 0 \\ 0 & 0 & 0 & m_4 \end{bmatrix}, \quad C = \begin{bmatrix} c_1 + c_2 & 0 & -c_2 & 0 \\ 0 & 0 & 0 & 0 \\ -c_2 & 0 & c_2 + c_3 & -c_3 \\ 0 & 0 & -c_3 & c_3 \end{bmatrix}, \tag{4}$$

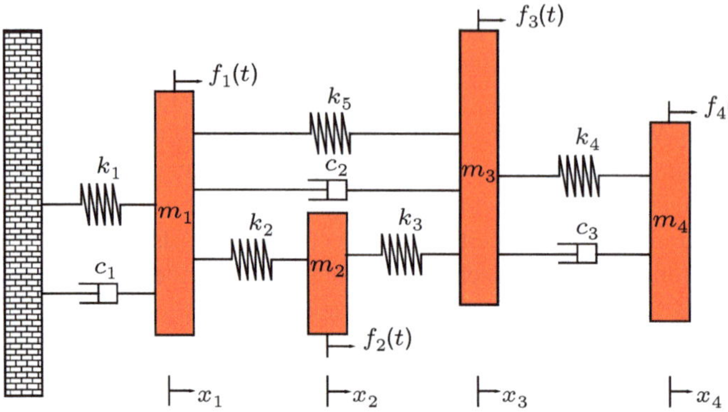

Fig. 1. A four-degree-of-freedom mass-spring system.

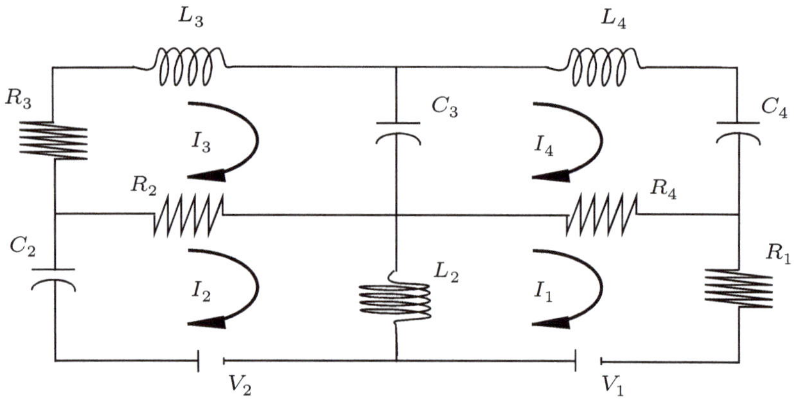

Fig. 2. An RLC electronic network.

$$K = \begin{bmatrix} k_1+k_2+k_5 & -k_2 & -k_5 & 0 \\ -k_2 & k_2+k_3 & -k_3 & 0 \\ -k_5 & -k_3 & k_3+k_4+k_5 & -k_4 \\ 0 & 0 & -k_4 & k_4 \end{bmatrix}. \tag{5}$$

In contrast, the structure associated with an electronic circuit may not be definite or even symmetric. As another illustration, the matrix coefficients in the differential system associated with the RLC network depicted in Figure 2 should have the following structure:

$$M = \begin{bmatrix} -L_2 & L_2 & 0 & 0 \\ L_2 & -L_2 & 0 & 0 \\ 0 & 0 & L_3 & 0 \\ 0 & 0 & 0 & L_4 \end{bmatrix}, C = \begin{bmatrix} 0 & R_2 & -R_2 & 0 \\ R_1+R_4 & 0 & 0 & -R_4 \\ 0 & -R_2 & R_2+R_3 & 0 \\ -R_4 & 0 & 0 & R_4 \end{bmatrix}, \tag{6}$$

$$K = \begin{bmatrix} 0 & \frac{1}{C_2} & 0 & 0 \\ 0 & 0 & 0 & 0 \\ 0 & 0 & \frac{1}{C_3} & -\frac{1}{C_3} \\ 0 & 0 & -\frac{1}{C_3} & \frac{1}{C_3} + \frac{1}{C_4} \end{bmatrix}, \tag{7}$$

For the sake of physical feasibility, the updated model usually is required to inherit the same connectivity as the original model. Since structured problems often results in special interrelationship within its eigenstructure, the observed measurement which often is contaminated with random noise may not be consistent with that innate structure. In other words, the structural constraint often severely limits whether a model could be updated.

Spurious Eigeninformation. An added challenge, known as the *no spill-over phenomenon* in the engineering literature, is that in updating an existing model it is often desirable that the current vibration parameters not related to the newly measured parameters should remain invariant. No spill-over is desirable either because these parameters have already been proven to be acceptable in the previous model and engineers do not wish to introduce new vibrations via updating or because engineers simply do not know of any information about these parameters. The MUP under such a circumstance therefore should be formulated as finding the updated model (M, C, K) so that not only he subset $\{(\lambda_i, \mathbf{u}_i)\}_{i=1}^k$ of the original model is replaced by newly measured $\{(\sigma_i, \mathbf{y}_i)\}_{i=1}^k$ as k eigenpairs of (M, C, K), but also the remaining $2n - k$ eigenpairs of (M, C, K), which often are unmeasurable and stay unknown, are the same as those of the original (M_0, C_0, K_0).

Minimal or Robust Modification. The solution to an MUP is not unique. The notion of optimizing the adjustment or the robustness therefore is highly plausible. Earlier work by Friswell, Inman and Pilkey [15] considers model updating by minimal changes of only the damping and the stiffness matrices. The work by Baruch [1], Bermann and Nagy [2], and Wei [26] concentrates only on undamped systems. More recently, the feedback control techniques have also been employed by Nichols and Kautsky [21] and Datta, Elhay, Ram, and Sarkissian [10–12] to manage the robustness.

Despite much effort, there does not seem to exist adequate theory or techniques thus far that can solve the MUP while addressing the aforementioned concerns. Existing methods have severe computational and engineering limitations, which restrict their usefulness in real applications. The purpose of this article is to provide an overview of this interesting topic with the hope of stimulating further studies toward its solution.

3 Quadratic Inverse Eigenvalue Problem

To answer whether a quadratic pencil can be updated, a more fundamental question is whether a quadratic pencil can have arbitrary k prescribed eigenpairs. For convenience, we adopt the notation that the diagonal matrix $\Lambda \in \mathbb{R}^{k \times k}$ represents the "eigenvalue matrix" of the quadratic pencil (3) in the sense that Λ is in real diagonal form with 2×2 blocks along the diagonal replacing the complex-conjugate pairs of eigenvalues originally there. Similarly, let $X \in \mathbb{R}^{n \times k}$ represent the "eigenvector matrix" in the sense that each pair of column vectors associated with a 2×2 block in Λ holds the real and the imaginary part of the original complex eigenvector. For the quadratic pencil (M, C, K) to have eigenstructure (Λ, X), it is clear that the relationship

$$MX\Lambda^2 + CX\Lambda + KX = 0_{n\times k} \tag{8}$$

must hold.

3.1 Self-Adjoint Pencils

At first glance, the relationship (8) is only a homogeneous linear system of nk algebraic equations. If there are no other constraints, the triplet (M, C, K) constitutes $3n^2$ unknowns. Since k is bounded above by $2n$, the system is well under-determined. It is intuitively true that the system should be solvable in general. The challenge is to characterize the solution in terms of the given (Λ, X). In this section, we discuss how a parametric representation can be obtained for (M, C, K) when these matrix coefficients are required to be symmetric.

To derive the parametric representation, observe that the matrix

$$\Omega := [I_k, \Lambda^\top, \Lambda^{2\top}] \in \mathbb{R}^{k\times 3k} \tag{9}$$

has a null space of dimension $2k$. Let columns of the matrix

$$\begin{bmatrix} U \\ T \\ S \end{bmatrix} \in \mathbb{R}^{3k\times 2k}, \tag{10}$$

where S, T and U are matrices in $\mathbb{R}^{k\times 2k}$, denote a basis of the null space of Ω. It is clear that once S and T are specified, then

$$U = -\Lambda^\top T - \Lambda^{2\top} S \tag{11}$$

is determined. Note that the system (8) can be written as

$$\Omega \begin{bmatrix} X^\top K \\ X^\top C \\ X^\top M \end{bmatrix} = 0_{k\times n}, \tag{12}$$

implying that there must exist a matrix $\Psi \in \mathbb{R}^{2k\times n}$ such that

$$\begin{bmatrix} U \\ T \\ S \end{bmatrix} \Psi = \begin{bmatrix} X^\top K \\ X^\top C \\ X^\top M \end{bmatrix}. \tag{13}$$

Since M, C and K are symmetric, the three matrices

$$A := S\Psi X, \tag{14}$$
$$B := T\Psi X, \tag{15}$$
$$F := U\Psi X \tag{16}$$

must also be symmetric in $\mathbb{R}^{k\times k}$. From (11) we know that F is specified once A and B are given. We shall use A and B to characterize the solution (M, C, K) to the QIEP associated with (Λ, X). It is important to note a critical relationship between A and B. Upon substituting (11) into (16) and using the fact that $F = F^\top$, we find that A and B are related by the equation:

$$\Lambda^\top B - B\Lambda = A\Lambda^2 - \Lambda^{2\top} A. \tag{17}$$

That is to say, not all entries in A or B are free. We shall exploit those entries which are free and establish a parametric representation of (M, C, K). Observe that each side of (17) represents a skew-symmetric matrix.

We begin with the case when $k = n$ and formulate the following result [8].

Theorem 1. *Given n distinct eigenvalues Λ and n linearly independent eigenvectors X both of which are closed under conjugation, let $A \in \mathbb{R}^{n\times n}$ be an arbitrary symmetric matrix and let B be a solution to the equation (17). Then the self-adjoint quadratic pencil with coefficients defined by*

$$M = X^{-\top} A X^{-1}, \tag{18}$$
$$C = X^{-\top} B X^{-1}, \tag{19}$$
$$K = -X^{-\top}\Lambda^\top(B + \Lambda^\top A)X^{-1}. \tag{20}$$

has the prescribed pair (X, Λ) as part of its eigenstructure.

Proof. The proof is straightforward. The relationship (13) implies that $M = X^{-\top} S\Psi$ for some $\Psi \in \mathbb{R}^{2n\times n}$. We also know from (14) that $A = S\Psi X$. Together, we can express M as $M = X^{-\top} A X^{-1}$. Similar arguments can be applied to C and K.

The choice of A gives rise to $\frac{n(n+1)}{2}$ free parameters. For each given $A \in \mathbb{R}^{n\times n}$, we need to see how B can be determined from the equation (17). Without loss of generality, we may assume that Λ is the diagonal matrix with $\ell \times \ell$ blocks,

$$\Lambda = \mathrm{diag}\{\lambda_1^{[2]}, \ldots, \lambda_\nu^{[2]}, \lambda_{\nu+1}, \ldots, \lambda_\ell\}, \tag{21}$$

where $\lambda_j^{[2]} = \begin{bmatrix} \alpha_j & \beta_j \\ \beta_j & \alpha_j \end{bmatrix} \in \mathbb{R}^{2\times 2}$, $\beta_j \neq 0$, if $j = 1, \ldots, \nu$; $\lambda_j \in \mathbb{R}$ if $j = \nu + 1, \ldots, \ell$; and $\ell + \nu = n$. Partition B into $\ell \times \ell$ blocks in such a way that, if the (i, j)-block is denoted by B_{ij}, then $\mathrm{diag}\{B_{11}, \ldots, B_{\ell\ell}\}$ has exactly the same structure as Λ. It is not difficult to see that the (i, j)-block of $\Lambda^\top B - B\Lambda$ is given by

$$\begin{cases} \lambda_i^\top B_{ij} - B_{ij}\lambda_j, & \text{if } \nu+1 \le i,j \le \ell, \\ (\lambda_i^{[2]})^\top B_{ij} - B_{ij}\lambda_j, & \text{if } 1 \le i \le \nu \text{ and } \nu+1 \le j \le \ell, \\ (\lambda_i^{[2]})^\top B_{ij} - B_{ij}(\lambda_j^{[2]}), & \text{if } 1 \le i,j \le \nu. \end{cases} \tag{22}$$

From a comparison with the corresponding blocks in $A\Lambda^2 - \Lambda^{2\top}A$ (cf. (17)), we draw the following conclusion. In the first case, B_{ij} is a scalar and is uniquely determined except that B_{ii} is free. In the second case, B_{ij} is a 2×1 block with all entries being uniquely determined. In the third case, if we write

$$B_{ij} = \begin{bmatrix} x & y \\ y & z \end{bmatrix}, \tag{23}$$

then

$$(\lambda_i^{[2]})^\top B_{ij} - B_{ij}(\lambda_j^{[2]}) = \begin{bmatrix} x(\alpha_i - \alpha_j) - y(\beta_i - \beta_j) & -z\beta_i - x\beta_j \\ x\beta_i + y(\alpha_i - \alpha_j) + z\beta_j & y(\beta_i - \beta_j) \end{bmatrix}. \tag{24}$$

It is clear that if $i = j$, then y is free and $x + z = 0$, still giving rise to two degrees of freedom. If $i \neq j$, the all entries of B_{ij} are uniquely determined. We conclude that $A \in \mathbb{R}^{n\times n}$ can be totally arbitrary and B is determined up to n free parameters. We thus have proved the following theorem.

Corollary 1. *The solutions (M, C, K) to the quadratic inverse eigenvalue problem with eigenstructure (X, Λ) as described in Theorem 1 form a subspace of dimensionality $\frac{n(n+3)}{2}$ in the product space $\mathbb{R}^{n\times n} \times \mathbb{R}^{n\times n} \times \mathbb{R}^{n\times n}$.*

It is worth mentioning that if A is selected to be symmetric and positive definite, then so is the leading coefficient M. Indeed, the above construction parameterizes all possible solutions.

We point out in passing that, in contrast to the construction described in Theorem 1, Kuo, Lin and Xu [18] have developed independently another parametrization for the solution (M, C, K). Let

$$\Omega := \operatorname{diag}\left\{ \begin{bmatrix} \xi_1 & \eta_1 \\ \eta_1 & -\xi_1 \end{bmatrix}, \ldots, \begin{bmatrix} \xi_\nu & \eta_\nu \\ \eta_\nu & -\xi_\nu \end{bmatrix}, \xi_{\nu+1}, \ldots, \xi_\kappa \right\}. \tag{25}$$

where $\xi_1, \ldots, \xi_\nu, \xi_{\nu+1}, \ldots, \xi_\kappa$ and $\eta_1, \ldots, \eta_\nu$ are arbitrary real numbers. Then the matrices defined by

$$M := \text{an arbitrary symmetric matrix}, \tag{26}$$

$$C := -\left(MX\Lambda X^{-1} + (X\Lambda X^{-1})^\top M + X^{-\top}\Omega X^{-1}\right), \tag{27}$$

$$K := (X\Lambda X^{-1})^\top M (X\Lambda X^{-1}) + X^{-\top}\Omega\Lambda X^{-1}, \tag{28}$$

also solves the QIEP associated with (Λ, X). It can been checked that these two ways of parametrization are equivalent, except that our approach is also able to handle the case $k > n$ which we now explore.

The case $k > n$ is a little bit more involved. It remains true from the relationships (13), (14), (15) and (16) that

$$A = S\Psi X = X^\top M X, \tag{29}$$
$$B = T\Psi X = X^\top C X, \tag{30}$$
$$F = U\Psi X = X^\top K X, \tag{31}$$

are symmetric even in the case $k > n$, but we cannot obtain a parametric representation of (M, C, K) from A and B directly because $X \in \mathbb{R}^{n\times k}$ with $k > n$ is no longer an injection transformation. To retrieve (M, C, K), we rewrite the eigenvectors as

$$X = [Z_1, Z_2], \tag{32}$$

where $Z_1 \in \mathbb{R}^{n\times n}$ and $Z_2 \in \mathbb{R}^{n\times(k-n)}$. Then we see that

$$A = \begin{bmatrix} A_{11} & A_{12} \\ A_{12}^\top & A_{22} \end{bmatrix} = \begin{bmatrix} Z_1^\top M Z_1 & Z_1^\top M Z_2 \\ Z_2^\top M Z_1 & Z_2^\top M Z_2 \end{bmatrix}, \tag{33}$$

where A_{ij}, $i, j = 1, 2$, are blocks with appropriate sizes. Thus, instead of using the matrix A as parameter, we should select a symmetric submatrix $A_{11} \in \mathbb{R}^{n\times n}$ arbitrarily and define

$$M = Z_1^{-\top} A_{11} Z_1^{-1}. \tag{34}$$

Once $M \in \mathbb{R}^{n\times n}$ is determined, the matrix $A \in \mathbb{R}^{k\times k}$ is completely specified. This selection gives rise to $\frac{n(n+1)}{2}$ degrees of freedom. There is no additional freedom in the choice of A.

With $A \in \mathbb{R}^{k\times k}$ specified, we next want to determine the matrix $B \in \mathbb{R}^{k\times k}$ based on the necessary condition (17). Write

$$B = \begin{bmatrix} B_{11} & B_{12} \\ B_{12}^\top & B_{22} \end{bmatrix} = \begin{bmatrix} Z_1^\top C Z_1 & Z_1^\top C Z_2 \\ Z_2^\top C Z_1 & Z_2^\top C Z_2 \end{bmatrix}. \tag{35}$$

Consider the B_{11} block first. Partition the given eigenvalues as

$$\Lambda = \operatorname{diag}\{\Upsilon_1, \Upsilon_2\} \tag{36}$$

where $\Upsilon \in \mathbb{R}^{n\times n}$ and $\Upsilon_2 \in \mathbb{R}^{(k-n)\times(k-n)}$. We note that B_{11} and A_{11} satisfy a relationship

$$\Upsilon_1^\top B_{11} - B_{11}\Upsilon_1 = A_{11}\Upsilon_1^2 - \Upsilon_1^{2\top} A_{11}, \tag{37}$$

that is similar to (17). The same argument used earlier for the case when $k = n$ can be applied and we conclude that the submatrix B_{11} can be completely determined up to n free parameters. It follows that a symmetric matrix

$$C = Z_1^{-\top} B_{11} Z_1^{-1} \tag{38}$$

can be determined and, hence, it appears that the matrix B is completely determined up to n free parameters.

The point is that there are additional limitations on the choice of A and B in the case $k > n$. The very same C defined by (38) should also equate the two sides of equation (17) for the $(1,2)$ and $(2,2)$ blocks, respectively. These blocks involve more than n equations to be satisfied. Thus, B_{11} and consequently A_{11} must be special. We have to go back to modify the selection of A_{11}. In other words, the n free parameters in B_{11} and the matrix A_{11} must be further restricted so that the remaining part of B also satisfies (17). To that end, we observe that if we define

$$W := Z_1^{-1} Z_2, \tag{39}$$

then it follows that

$$A = \begin{bmatrix} A_{11} & A_{11}W \\ W^\top A_{11} & W^\top A_{11} W \end{bmatrix},$$

$$B = \begin{bmatrix} B_{11} & B_{11}W \\ W^\top B_{11} & W^\top B_{11} W \end{bmatrix}.$$

Expressing the equation (17) in block form, we obtain (37) and the following two equations:

$$\Upsilon_1^\top B_{11} W - B_{11} W \Upsilon_2 = A_{11} W \Upsilon_2^2 - \Upsilon_1^{2\top} A_{11} W, \tag{40}$$

$$\Upsilon_2^\top W^\top B_{11} W - W^\top B_{11} W \Upsilon_1 = W^\top A_{11} W \Upsilon_2^2 - \Upsilon_2^{2\top} W^\top A_{11} W. \tag{41}$$

Post-multiplying (37) by W and subtracting (40), we obtain an equivalent relationship:

$$A_{11}\Upsilon_1^2 W + B_{11}\Upsilon_1 W = A_{11} W \Upsilon_2^2 + B_{11} W \Upsilon_2. \tag{42}$$

It follows that

$$\begin{aligned} W^\top(A_{11}W\Upsilon_2^2 + B_{11}W\Upsilon_2) &= W^\top(A_{11}\Upsilon_1^2 W + B_{11}\Upsilon_1 W) \\ &= (W^\top A_{11}\Upsilon_1^2 + W^\top B_{11}\Upsilon_1)W \\ &= (\Upsilon_2^\top W^\top B_{11} + \Upsilon_2^{2\top} W^\top A_{11})W, \end{aligned}$$

which is precisely (41). The final equality follows from taking the transpose of equation (40). We have just proved that if we can solve the two equations (37) and (40), then the third equation (41) is automatically solved. We have indicated earlier that any given A_{11} will determine B_{11} through (37) up to n free parameters. Thus, it only remains to choose the n free parameters in B_{11} and the $n \times n$ symmetric matrix A_{11} to satisfy the $n(k-n)$ linear equations imposed by (40). In total there are

$$\frac{n(n+1)}{2} + n - n(k-n) = \frac{3n(n+1)}{2} - nk \tag{43}$$

degrees of freedom. For nontrivial solutions, it is clear that we need $k < \frac{3(n+1)}{2}$.

Finally, we discuss the case when $k < n$. If less than n eigenpairs (X, Λ) are given, we can solve the inverse eigenvalue problem by embedding this eigeninformation in a larger set of n eigenpairs. In particular, we expand $X \in \mathbb{R}^{n\times k}$ to

$$\widehat{X} := [X, \widetilde{X}] \in \mathbb{R}^{n\times n}, \tag{44}$$

where $\widetilde{X} \in \mathbb{R}^{n\times(n-k)}$ is arbitrary under the condition that $\widehat{X}$ is nonsingular. Caution should be taken when counting the degrees of freedom. We should consider the columns in $\widetilde{X}$ as being *normalized* since, otherwise, a normalization factor would have been added to the arbitrariness of A. With this normalization in mind, the expansion of eigenvectors involves additional $(n-1)(n-k)$ degrees of freedom. We then expand $\Lambda \in \mathbb{R}^{k\times k}$ to

$$\widehat{\Lambda} := \operatorname{diag}\{\Lambda, \widetilde{\Lambda}\}, \tag{45}$$

where $\widehat{\Lambda}$ is a diagonal matrix with distinct eigenvalues. This expansion of eigenvalues gives rise to another $n-k$ degrees of freedom. With $(\widehat{X}, \widehat{\Lambda})$ playing the role of (X, Λ) in Theorem 1, we can now construct the coefficient matrices M, C and K according to the formulas (18), (19) and (20), respectively. Recall that A is taken as an arbitrary symmetric matrix in $\mathbb{R}^{n\times n}$ and B, though depending on $\widehat{\Lambda}$ through the relationship (17), maintains n degrees of freedom. We conclude that the solutions to the QIEP with $k < n$ form a subspace of dimensionality $\frac{n(n+3)}{2} + n(n-k)$. Note that this embedding approach characterizes the solution (M, C, K) via the parametrization (18), (19) and (20) which in *nonlinear* in terms of A, B, $\widetilde{X}$ and $\widetilde{\Lambda}$.

We end this section with the following summarizing theorem.

Theorem 2. *Assume $1 \le k < \frac{3(n+1)}{2}$. Let (Λ, X) represent k arbitrarily prescribed eigenpairs (Λ, X) which are closed under conjugation. The self-adjoint quadratic inverse eigenvalue problem associated with (Λ, X) is generally solvable. The solutions form a subspace of dimension $\frac{3n(n+1)}{2} - nk$. The maximal allowable number of prescribed eigenpairs is given by (50).*

3.2 Structured Pencils

Thus far, the only structure laid upon the QIEP is the symmetry, in which case we have shown its solvability. However, it is important to note that algebraic solvability does not necessarily imply physical feasibility. Physical feasibility means, for

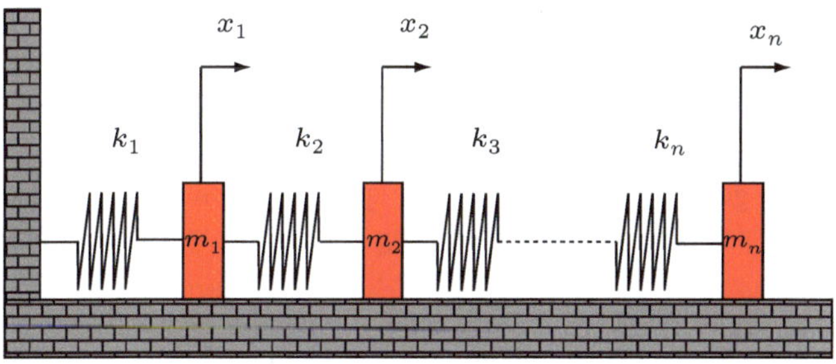

Fig. 3. An undamped mass-spring system.

example, that the special matrix structure resulting from the underlying connectivity must hold or that the physical parameters must be nonnegative. These additional constraints make the QIEP much more interesting but harder to solve. There does not seem to exist reported research in this direction. We believe that the issue of solvability probably is problem dependent and will have to be analyzed case by case. For demonstration purpose, we shall discuss only one setting in this section.

Consider the serially linked, undamped mass-spring system depicted in Figure 3, which can be used to model many other physical systems, including a vibrating beam, a composite pendulum, or a string with beads. The corresponding quadratic pencil $\lambda^2 M + K$ has the structure

$$M = \begin{bmatrix} m_1 & 0 & \dots & 0 \\ 0 & m_2 & & \\ \vdots & & \ddots & \\ 0 & & & m_n \end{bmatrix}, \; K = \begin{bmatrix} k_1+k_2 & -k_2 & 0 & \dots & 0 & 0 \\ -k_2 & k_2+k_3 & -k_3 & & & 0 \\ 0 & -k_3 & k_3+k_4 & & & \\ \vdots & & & \ddots & & \vdots \\ 0 & & & & k_{n-1}+k_n & -k_n \\ 0 & & & & -k_n & k_n \end{bmatrix}. \tag{46}$$

The inverse eigenvalue problem would imply to find positive values for the masses $m_1, \dots, m_n$ and spring constants $k_1, \dots, k_n$ from prescribed eigeninformation. A typical approach in the literature has been to recast the quadratic pencil as a linear pencil $\mu I + J$ with a Jacobi matrix $J = M^{-1/2} K M^{-1/2}$. A classical theory has been that *two sets of eigenvalues* can uniquely solve the corresponding Jacobi inverse eigenvalue problem [7, Section 4.2]. What can be said if the system is to be reconstructed from eigenpairs?

Each eigenpair provides n equations. Imposing two eigenpairs generally will lead to the trivial algebraic solution in such a system, unless the prescribed eigenpairs satisfy some additional internal relationship. So we ask the even more fundamental question of constructing the system with one prescribed eigenpair $(i\beta, \mathbf{x})$ where $i = \sqrt{-1}$, $\beta \in \mathbb{R}$ and $\mathbf{x} \in \mathbb{R}^n$.

Denote $\mathbf{x} = [x_1, \dots, x_n]^\top$ and $x_0 = 0$. It is not difficult to see that the recursive relationship,

$$k_n = \frac{\beta^2 m_n x_n}{x_n - x_{n-1}}, \tag{47}$$

$$k_i = \frac{\beta^2 m_i x_i + k_{i+1}(x_{i+1} - x_i)}{x_i - x_{i-1}}, \quad i = n-1, \dots, 1, \tag{48}$$

must hold. Our goal is to find a positive approximation for m_i, which then defines a positive value for k_i. The following is a *make-or-break* algorithm for the construction [4].

The above algorithm appears naive since it only checks a few signs, but its simplicity is in fact closely related to the classical Courant Nodal Line Theorem [27]. Roughly speaking, it is known in the literature that critical information about a vibration system can be recovered from places where *nothing happens*. These places are

Given an arbitrary eigenpair $(i\beta, \mathbf{x})$, assume the normalization $x_n = 1$ and $m_n = 1$. The following steps either construct the masses $m_1, \ldots, m_{n-1}$ and spring constants $k_1, \ldots, k_n$, all positive, for the pencil $\lambda^2 M + K$, or determine that such a system with the prescribed eigenpair does not exist.

```
1. initialization:
       s_n = 0.9;                                   (mass decreasing factor)
       s_p = 1.1;                                   (mass increasing factor)
       η = β^2;
2. if x_{n-1} < 1,
       k_n = β^2/(1 - x_{n-1});                     (use formula (47))
   else
       return                                       (inconsistent eigenvector)
   end
3. for i from n - 1 to 2, do
   a) ρ = η/x_i;
   b) if x_{i-1} < x_i,
          if ρ > 0,
              if x_i < 0,
                  return                            (inconsistent eigenvector)
              else
                  m_i = 1;                          (any m_i > 0 will be fine)
              end
          else
              if x_i < 0,
                  m_i = -s_n ρ/β^2;                 (need 0 < m_i < -ρ/β^2)
              else
                  m_i = -s_p ρ/β^2;                 (need m_i > -ρ/β^2)
              end
          end
      else
          if ρ > 0,
              if x_i < 0,
                  m_i = 1;
              else
                  return                            (inconsistent eigenvector)
              end
          else
              if x_i < 0,
                  m_i = -s_p ρ/β^2;
              else
                  m_i = -s_n ρ/β^2;
              end
          end
      end
   c) η = η + β^2 m_i x_i;
   d) k_i = η/(x_i - x_{i-1});                      (use formula (48))
4. finale:
   a) rho = η/x_1;
   b) if ρ > 0,
          m_1 = 1;
      else
          m_1 = -s_p ρ/β^2;
      end
   c) η = η + β^2 m_1 x_1;
   d) k_1 = η/x_1;                                  (use formula (48))
```

referred to as the nodal lines. Courant's theorem gives a right count of the number of nodal lines. We shall not elaborate the particulars here. Readers are referred to the paper [9, 27] for more detailed discussion. In short, the effect of the algorithm is based on the following result, whose proof can be found in [4]. We believe that the necessary and sufficient conditions on the specified eigenvector is lucid but elegant.

Theorem 3. *A given vector* $\mathbf{x} = [x_1, \ldots, x_n]^\top \in \mathbb{R}^n$ *with distinct entries is an eigenvector of a quadratic pencil* $\lambda^2 M + K$ *with the structure specified in (46) if and only if* $x_n(x_n - x_{n-1}) > 0$ *and the signs of the triplets* $(x_{i+1} - x_i, x_i, x_i - x_{i-1})$ *for* $i = 2, \ldots, n-1$ *are not* $(+, +, -)$ *nor* $(-, -, +)$. *Furthermore, if* $\mathbf{x}$ *is feasible and if there are* τ *changes of signs when going through* x_1 *to* x_n, *then* $\mathbf{x}$ *is the* τ*-th eigenvector of the pencil regardless how the masses* m_i *are defined.*

For damped systems and other types of connectivity or RLC configurations, the resulting pencil structure will be different. It is likely that the conditions for solvability will also vary. This is a wide open area for further research.

4 Spill-Over Phenomenon

Recall that a model updating with no spill-over is mathematically equivalent to a QIEP with a *complete set* of prescribed eigenpairs (Λ, X) where we partition Λ and X as

$$\Lambda = \operatorname{diag}\{\Sigma, \Lambda_2\}, \quad X = [Y, X_2], \tag{49}$$

with the pair $(\Sigma, Y) \in \mathbb{R}^{k\times k} \times \mathbb{R}^{n\times k}$ representing the portion of eigenstructure that has been modified and (Λ_2, X_2) corresponding to the inert portion of eigenstructure in the original model which should not been changed (and perhaps is even not known). Ideally, we prefer to see no spill-over in the model updating. But can this be achieved? If not, to what extent do we know about the spurious eigenstructure brought in by the updating?

In the case when symmetry is required, we have seen that the additional constraint of symmetry imposes an upper bound on the number k of prescribed eigenpair. The maximal allowable number k_{max} of prescribed eigenpairs is given by

$$k_{max} = \begin{cases} 3\ell + 1, & \text{if } n = 2\ell, \\ 3\ell + 2, & \text{if } n = 2\ell + 1. \end{cases} \tag{50}$$

As a consequence, the remaining $2n - k_{max}$ eigenpairs of a quadratic pencil *cannot* be arbitrarily assigned anymore. That is to say, if $n \geq 3$ and if the updating intends to replace k_{max} original eigenpairs by newly measured data, then with probability one the phenomenon of spill-over will occur. The following example from [8] illustrates this.

Consider the case when $n = 3$. A quadratic pencil generally allows six eigenpairs. Suppose that five of them are prescribed by

$$\Sigma = \mathrm{diag}\{1,2,3,5,8\}, \quad Y = \begin{bmatrix} 1 & 0 & 0 & 2 & -1 \\ 0 & 1 & 0 & -2 & 0 \\ 0 & 0 & 1 & 2 & 2 \end{bmatrix}. \tag{51}$$

This is a case where $k = k_{max} = 5$. By Theorem 2, the solution has three degrees of freedom. We find that the solution to the QIEP can be represented as

$$M = \begin{bmatrix} s & -s+4u & u \\ -s+4u & t & -\frac{7}{10}s+\frac{14}{5}u \\ u & -\frac{7}{10}s+\frac{14}{5}u & -\frac{3}{10}u+\frac{7}{10}s \end{bmatrix},$$

$$C = \begin{bmatrix} -9s+10u & 3s-12u & -4u \\ 3s-12u & -\frac{27}{5}s+\frac{108}{5}u-7t & \frac{7}{2}s-14u \\ -4u & \frac{7}{2}s-14u & \frac{34}{5}u-\frac{77}{10}s \end{bmatrix},$$

$$K = -\begin{bmatrix} -8s+10u & 2s-8u & -3u \\ 2s-8u & -\frac{54}{5}s+\frac{216}{5}u-10t & \frac{21}{5}s-\frac{84}{5}u \\ -3u & \frac{21}{5}s-\frac{84}{5}u & \frac{177}{10}u-\frac{84}{5}s \end{bmatrix}.$$

It can be shown that

$$\det(M) = -\frac{1}{100}(7s-10u)\left(272u^2-136su-10tu-10ts+17s^2\right). \tag{52}$$

Obviously, we can choose s, t and u so that $\det(M) > 0$. Indeed, the three parameters can be chosen to make the other two principal minors of M positive so that M is positive definite. We also find that the sixth eigenvalue is given by

$$\lambda_6 = -2\frac{52u^2+37s^2-161su+40st-35tu}{17s^2-136su-10tu+272u^2-10st} \tag{53}$$

while its corresponding eigenvector is given by

$$\mathbf{x}_6 = \left[\frac{2}{5}\frac{9s-36u+5t}{7s-10u}, 1, \frac{2}{5}\frac{9s-36u+5t}{7s-10u}\right]^\top. \tag{54}$$

It is clear that the sixth eigenvector $\mathbf{x}_6$ cannot be arbitrarily assigned and, hence, no spill-over *cannot* be guaranteed.

On the other hand, suppose $k = n$ eigenpairs have been modified. Then according to the construction specified in Theorem 1, we can derive the following result.

Theorem 4. *Given n distinct eigenvalues Σ and n linearly independent eigenvectors Y both of which are closed under conjugation, construct (M, C, K) as in Theorem 1 with A and B as parameters. Then the corresponding self-adjoint quadratic pencil can be factorized as*

$$\begin{aligned} \lambda^2 M + \Sigma C + K &= Y^{-\top}\left(\lambda I_n - \Sigma^\top\right)\left(B + (\lambda I_n + \Sigma^\top)A\right)Y^{-1} \\ &= Y^{-\top}\left(B + A(\lambda I_n + \Sigma)\right)(\lambda I_n - \Sigma)\,Y^{-1}. \end{aligned} \tag{55}$$

It is interesting to note from Theorem 4 that the remaining eigenvalues are the same as the eigenvalues of the linear pencil $\lambda A + B + A\Sigma$. Since the entire matrix A and (diagonal) part of B are free, there is room to impose additional eigeninformation to the pencil. In [18], for instance, it has been argued that *additional n eigenvalues* could be arbitrarily specified. This fulfills only partially the no spill-over phenomenon. In our context where we require that eigenvalues and eigenvectors are prescribed in pairs, we believe that spill-over phenomenon is inevitable except in the undamped case. In undamped case, note that the quadratic pencil $\lambda^2 M + K$ can be reduced to a linear pencil $\mu M + K$ with $\mu = \lambda^2$. The following result for a self-adjoint linear pencil is proved in [8].

Theorem 5. *A self-adjoint linear pencil $\mu M + K$ can have arbitrary eigenstructure with n distinct eigenvalues and linearly independent eigenvectors. Indeed, given an eigenstructure (Λ, X) in $\mathbb{R}^{n\times n} \times \mathbb{R}^{n\times n}$, the solutions (M, K) form a subspace of dimensionality n in the product space $\mathbb{R}^{n\times n} \times \mathbb{R}^{n\times n}$ and can be parameterized by the diagonal matrix Γ via the relationships,*

$$M = X^{-\top}\Gamma X^{-1}, \tag{56}$$

$$K = -X^{-\top}\Gamma\Lambda X^{-1}. \tag{57}$$

5 Least Squares Update

Inverse eigenvalue problems generally are ill-posed. Any measure of sensitivity or robustness of a solution to perturbations must be designed by taking several factors into consideration [21]. One such attempt is to require that the updating is made with minimal changes [15]. The model updating problem can then be formulated as an optimization problem:

$$\text{minimize} \quad \frac{1}{2}\left(\|M - M_0\|_F^2 + \|C - C_0\|_F^2 + \|K - K_0\|_F^2\right), \tag{58}$$

$$\text{subject to} \quad MY\Sigma^2 + CY\Sigma + KY = 0, \quad M, C \text{ and } K \text{ symmetric}, \tag{59}$$

where $(\Sigma, Y) \in \mathbb{R}^{k\times k} \times \mathbb{R}^{n\times k}$ are the newly measured eigenpairs. Note that the above formulation is actually a quadratic programming problem for which many techniques are available. See, for example, [16]. In principle, the least squares model updating problem can be handled by standard optimization procedures, provided the feasible set is known to be nonempty.

Of course, advantage can be taken of the special features of the problem so that the quantities needed for numerical computation are calculated with minimal effort. For example, since the problem involves only linear equality constraints, the projected gradient and the projected Hessian can be calculated formally in terms of the null space of the $k \times 3n$ matrix $[X^\top, \Lambda^\top X^\top, \Lambda^{2\top} X^\top]$. Another approach is to utilize the parametric representation of (M, C, K) and rewrite the objective function as an unconstrained optimization in terms of the free parameters. Readers are referred to [19] for an implementation that uses a similar but different parametrization specified in [18].

Structured model updating problems can be formulated in a similar way except that the matrices (M,C,K) in (59) are restricted to the specified structures. If the matrices are further required to be nonnegative, then we also have bounded constraints. As we have indicated, we can solve currently only a few structured QIEPs, if they are solvable at all, by numerical algorithms. That is to say, when solving a structured least squares model updating problem, a feasible candidate (M,C,K) can be identified only through point-to-point calculation. This would make it very hard to find the optimal solution. Again, we believe that this is an area open for further research.

6 Conclusions

Model updating so as to attain consistent spectral property with empirical data is an essential ingredient for establishing an effective model. In this chapter, we presented an overview on this subject by briefly addressing three important issues involved in model updating — that we have to satisfy the structural constraint for physical feasibility, that we prefer to see that no spurious modes are introduced into the range of the frequency range of interest, and that we want to keep the modifications minimal. Before we are able to determine whether some updating can be achieved, a more fundamental question is to solve the quadratic inverse eigenvalue problem when a set of eigenpairs is prescribed. For this problem we are able to provide a parametric representation of the solution to the QIEP if only symmetry is required of the matrices involved. We demonstrated an algorithmic approach for an undamped QIEP when the structure and nonnegativity are to be maintained, but a general solution procedure is not available yet. From inspection of the dimension of the solution space of the QIEP, we conclude that the spill-over phenomenon is unavoidable. We pointed out many open questions that deserve further study.

References

1. M. Baruch, Optimization procedure to correct stiffness and flexibility matrices using vibration data, AIAA Journal, 16(1978) 1208–1210.
2. A. Berman and E. J. Nagy, Improvement of a large analytical model using test data, AIAA Journal, 21(1983) 1168–1173.
3. J. Carvalho, State Estimation and Finite-Element Model Updating for Vibrating Systems, Ph.D. Dissertation, Northern Illinois University, 2002.
4. M. T. Chu, N. Del Buono and B. Yu, Structured quadratic inverse eigenvalue problems, I. Serially linked systems, SIAM J. Sci. Comput., 29(2007), 2668–2685.
5. M. T. Chu and G. H. Golub, Structured inverse eigenvalue problems, Acta Numerica, 11(2002), 1–71.

6. M. T. Chu, Y.-C. Kuo and W.-W. Lin, On inverse quadratic eigenvalue problems with partially prescribed eigenstructure, SIAM J. Matrix Anal. Appl., 25(2004), 995–1020.
7. M. T. Chu and G. H. Golub, Inverse Eigenvalue Problesm: Theory, Algorithms, and Applications, Oxford University Press, 2005.
8. M. T. Chu, B. N. Datta, W.-W. Lin and S.-F. Xu, The spill-over phenomenon in quadratic model updating, AIAA. J., to appear, 2008.
9. R. Courant and D. Hilbert, Methods of Mathematical Physics, Vol. 1, Interscience, New York, 1956.
10. B. N. Datta, S. Elhay, Y. M. Ram, and D. R. Sarkissian, Partial eigenstructure assignment for the quadratic pencil, Journal of Sound and Vibration, 230(2000), 101–110.
11. B. N. Datta and D. R. Sarkissian, Theory and computations of some inverse eigenvalue problems for the quadratic pencil, in Contemporary Mathematics, Volume on "Structured Matrices in Operator Theory, Control, and Signal and Image Processing", 2001, 221–240.
12. B. N. Datta, Finite element model updating, eigenstructure assignment and eigenvalue embedding techniques for vibrating systems, mechanical systems and signal processing, Special Volume on "Vibration Control", 16(2002), 83–96.
13. D. J. Ewins, Adjustment or updating of models, Sādhanā, 25(2000), 235–245.
14. M. I. Friswell and J. E. Mottershed, Finite Element Model Updating in Structural Dynamics, Kluwer Academic Publishers, Boston, 1995.
15. M. I. Friswell, D. J. Inman, and D. F. Pilkey, The direct updating of damping and stiffness matrices, AIAA Journal, 36(1998), 491–493.
16. P. E. Gill, W. Murray, and M. H. Wright, Practical Optimization, Academic Press, London, UK, 1981.
17. G. M. L. Gladwell, Inverse Problems in Vibration, Martinus Nijhoff, Dordrecht, Netherlands, 1986, 2nd ed., 2004.
18. Y.-C. Chen, W.-W Lin and S.-F. Xu, Solution of the partially described inverse quadratic eigenvalue problem, SIAM J. Matrix Analysis Appl., 29(2006), 33–53.
19. Y.-C. Kuo, W.-W Lin and S.-F. Xu, A new model updating methods for quadratic eigenvalue problems, preprint, National Tsinghua University, Taiwan, 2005.
20. P. Lancaster and U. Prells, Inverse problems for damped vibration systems, J. Sound Vibration, 283(2005), 891–914.
21. N. K. Nichols and J. Kautsky, Robust eigenstructure assignment in quadratic matrix polynomials: nonsingular case, SIAM J. Matri Anal. Appl., 23(2001), 77–102.
22. Y. M. Ram and S. Elhay An inverse eigenvalue problem for the symmetric tridiagonal quadratic pencil with application to damped oscillatory systems, SIAM J. Appl. Math., 56(1996), 232–244.
23. W. H. A. Schilders and H. A. van der Vorst , Model order reduction, coupled problems and optimzation, http://www.lorentzcenter.nl/lc/web/2005/160/info.php3?wsid=160.
24. L. Starek and D. J. Inman, Symmetric inverse eigenvalue vibration problem and its applications, Mechanial Systems and Signal Process, 15(2001), 11–29.
25. F. Tisseur and K. Meerbergen, The quadratic eigenvalue problem, SIAM Review, 43(2001), 235–286.
26. F.-S. Wei, Mass and stiffness interaction effects in analytical model modification, AIAA Journal, 28(1990) 1686–1688.
27. H.-M. Zhu, Courant's Nodal Line Theorem and Its Discrete Counterparts, Ph. D. Dissertation, University of Waterloo, Waterloo, ONtario, Canada, 2000.

Data-Driven Model Order Reduction Using Orthonormal Vector Fitting

Dirk Deschrijver and Tom Dhaene

Ghent University, Sint Pietersnieuwstraat 41, 9000 Gent, Belgium
{dirk.deschrijver,tom.dhaene}@intec.ugent.be

Accurate frequency-domain macromodels are becoming increasingly important for the design, study and optimization of complex physical systems. These macromodels approximate the complex frequency-dependent input-output behaviour of broadband multi-port systems in the frequency domain by rational functions [28]. Unfortunately, due to the complexity of the physical systems under study and the dense discretization required for accurately modelling their behaviour, the rational or state-space macromodels may lead to unmanageable levels of storage and computational requirements. Therefore, Model Order Reduction (MOR) methods can be applied to build a model of reduced size, which captures the dynamics of the larger model as closely as possible.

Orthonormal Vector Fitting (OVF) [5, 9] is an identification method, which is typically used to approximate simulated or measured frequency responses by an analytic function. In this chapter, it is shown that the OVF method can also be seen as a *data-driven* MOR method. Rather than reducing the dimensions of the state-space matrices of a model directly (*model-based* MOR), this technique is used to build a new state-space model with a reduced model complexity based on input-output data. The goal of this algorithm is to parameterize a rational transfer function, such that its spectral behaviour matches the response of the larger model in a least-squares sense.

Most available identification methods suffer poor numerical conditioning for large state-space dimensions or broad frequency ranges. The OVF method tackles these issues by combining the benefits of a Sanathanan-Koerner iteration [32] and a well-chosen set of orthonormal rational basis functions. It is shown that the method is applicable to reduce systems with a large amount of poles. The method does not preserve passivity by default, however several techniques are available to enforce a desired physical behaviour in a post-processing step [12, 14].

1 Identification Problem

1.1 Goal

The major goal of the algorithm is to identify the mapping between the inputs and outputs of a complex system by an analytic model of reduced size. For continuous-time linear-time-invariant (LTI) systems in the frequency domain, this reduces to finding a rational transfer function

$$R(s) = \frac{N(s)}{D(s)} = \frac{\sum_{n=0}^{N} N_n \phi_n(s)}{\sum_{d=0}^{D} D_d \phi_d(s)} \qquad s = i2\pi f \tag{1}$$

which approximates the spectral response of a system over some predefined frequency range of interest $[f_{min}, f_{max}]$. The spectral behaviour is characterized by a set of frequency-domain data samples $(s_k, H(s_k))$, $\forall k = 0, ..., K$, which are obtained by evaluating the state-space matrices of the large model. N_n and D_d are the real-valued system parameters which need to be estimated, and N and D represent the order of numerator and denominator respectively. In practice, N and D are chosen to be much smaller than the order of the large model. A dense frequency sweep is required in many situations, so the amount of available data samples can be quite numerous. Therefore, numerically stable fitting techniques are required which estimate the model coefficients in a least-squares sense [10].

1.2 Non-linearity of the Estimator

Rational least-squares approximation is essentially a non-linear problem, and corresponds to minimizing the following cost function [29]

$$\arg \min_{N_n, D_d} \sum_{k=0}^{K} \left| H(s_k) - \frac{N(s_k)}{D(s_k)} \right|^2 = \arg \min_{N_n, D_d} \sum_{k=0}^{K} \frac{|D(s_k)H(s_k) - N(s_k)|^2}{|D(s_k)|^2} . \tag{2}$$

Due to its non-linear nature, it can be hard to estimate the system parameters in a fast and accurate way. In many papers, e.g. [34], this difficulty is avoided by assuming that a-priori knowledge about the poles is available. In this case, the non-linear problem reduces to a linear problem since the denominator parameters are assumed to be known. In practice, however, this situation is often not a realistic one. Another possible option is the use of non-linear optimization techniques, such as Newton-Gauss type algorithms, in order to minimize (2). A known drawback of these methods, is that the solutions may converge to local minima, even when Levenberg-Marquardt algorithms are used to extend the region of convergence [22, 25].

In [2], it was proposed to minimize Levi's linearized cost function [19, 23]

$$\arg \min_{N_n, D_d} \sum_{k=0}^{K} |D(s_k)H(s_k) - N(s_k)|^2 . \tag{3}$$

This formulation basically reduces to (2), if the weighting factor $1/\left|D(s_k)\right|^2$ is set equal to one for all frequencies s_k. Clearly, this weighting will bias the fitted transfer function, and this often results in poor low-frequency fits, due to an undesired overemphasis of high-frequency errors.

In this chapter, the use of a Sanathanan-Koerner iteration is advocated [32]. First, an estimate of the poles is obtained by minimizing Levi's linearized cost function. Given this initial (iteration step 0) or previous (iteration step $t-1$) estimate of the poles, the model parameters of the next iteration step are calculated by minimizing the weighted linear cost function

$$\arg \min_{N_n^{(t)}, D_d^{(t)}} \left(\sum_{k=0}^{K} \frac{\left|D^{(t)}(s_k)H(s_k) - N^{(t)}(s_k)\right|^2}{\left|D^{(t-1)}(s_k)\right|^2} \right). \tag{4}$$

By analyzing the gradients of the error criterion, it is straightforward to show that this method generates solutions that don't converge asymptotically to the solution of (2) either, even though the error criterion itself tends asymptotically to the fundamental least squares criterion [35]. In practice, however, this approach often gives favorable results for sufficiently small modelling errors. The interested reader is hereby referred to an excellent survey [29].

1.3 Choice of Basisfunctions

To solve the identification problem, (4) reduces naturally to a linear set of least-squares equations, which needs to be solved with sufficient accuracy.

Suppose that $\mathbf{H} = diag(H(s_0), ..., H(s_K))$, $w_k = [D^{(t-1)}(s_k)]^{-1}$, and $\Phi_{0:X}$ is defined as

$$\Phi_{0:X} = \begin{pmatrix} w_0\phi_0(s_0) & ... & w_0\phi_X(s_0) \\ ... & ... & ... \\ w_K\phi_0(s_K) & ... & w_K\phi_X(s_K) \end{pmatrix}, \tag{5}$$

then the least-squares solution of $\mathbf{Vx} = \mathbf{b}$ can be calculated to estimate the parameter vector $\mathbf{x}$, provided that $\mathbf{V}$, $\mathbf{x}$ and $\mathbf{b}$ are defined as ($D_0 = 1$)

$$\mathbf{V} = \begin{pmatrix} \Re e\left(\Phi_{0:N} \; -\mathbf{H}\Phi_{1:D}\right) \\ \Im m\left(\Phi_{0:N} \; -\mathbf{H}\Phi_{1:D}\right) \end{pmatrix}, \quad \mathbf{b} = \begin{pmatrix} \Re e(\mathbf{H}\Phi_0) \\ \Im m(\mathbf{H}\Phi_0) \end{pmatrix} \tag{6}$$

$$\mathbf{x} = (N_0^{(t)} \; ... \; N_N^{(t)} \; D_1^{(t)} \; ... \; D_D^{(t)})^T . \tag{7}$$

Each equation is split in its real and imaginary part to enforce the poles and zeros to be real, or to occur in complex conjugate pairs (under the assumption that the basis functions $\phi(s)$ are real-valued as well). This ensures that the coefficients of the transfer function are real, and that no imaginary terms occur in the time-domain.

It becomes clear that the accuracy of the parameter vector $\mathbf{x}$, and the numerical conditioning of this problem is highly dependent on the structure of the system equations. If the basisfunctions $\phi(s)$ are chosen to be a monomial power series basis $(1,s,s^2,...)$, the matrix Φ will be a Vandermonde matrix which is notoriously ill-conditioned. Adcock and Potter [1] suggested the use of polynomials which are orthogonal with respect to a continuous inner product, such as Chebyshev polynomials, as basis functions. The large variation of the Chebyshev polynomials with increase in order makes it possible to downsize the effects of ill-conditioning. On the other hand, Richardson and Formenti [30] proposed the use of Forsythe polynomials which are orthonormal with respect to a discrete inner product, defined by the normal equations of the estimator. This implies that a different set of basis functions is used for numerator and denominator. Rolain et al. [31] have shown that a basis transformation from the Forsythe polynomials to a different, arbitrary polynomial basis results in an inferior conditioning of $\mathbf{V}^T\mathbf{V}$. Hence, the Forsythe polynomial basis is optimal in a sense that there doesn't exist any other polynomial basis resulting in a better conditioned form of the normal equations.

2 Vector Fitting

2.1 Model Representation

Quite recently, Gustavsen and Semlyen [13] proposed the use of partial fractions as basis functions for the numerator and denominator

$$R(s) = \frac{N(s)}{D(s)} = \frac{\sum_{p=1}^{P} c_p \phi_p(s)}{1 + \sum_{p=1}^{P} \tilde{c}_p \phi_p(s)} = \frac{\sum_{p=1}^{P} \frac{c_p}{s+a_p}}{1 + \sum_{p=1}^{P} \frac{\tilde{c}_p}{s+a_p}}, \tag{8}$$

provided that c_p and $\tilde{c}_p$ represent the residues, and $-a_p$ are a set of prescribed poles. The denominator has an additional basisfunction which equals the constant value 1. Its coefficient can be fixed to one, since numerator and denominator can be divided by the same constant value without loss of generality. Other non-triviality constraints are also possible [16]. Given the constraint that the poles of the numerator and denominator expression of (8) are the same, it's easy to see that these basis functions are complete, in a sense that they can approximate any strictly proper transfer function with distinct poles arbitrarily well. To approximate systems which require a proper or improper transfer function, an optional constant and linear term can be added to the numerator expression.

2.2 Parameterization of the Transfer Function

In the first iteration, Levi's cost function is applied, which results that $\mathbf{V}$ becomes

$$\mathbf{V} = \begin{pmatrix} \Re e\left(\Phi_{1:P} \; -H\Phi_{1:P}\right) \\ \Im m\left(\Phi_{1:P} \; -H\Phi_{1:P}\right) \end{pmatrix} \tag{9}$$

and that the parameter vector consist of unknown residues

$$\mathbf{x} = (c_1 \dots c_P \, \widetilde{c}_1 \dots \widetilde{c}_P)^T \, . \tag{10}$$

The Φ matrix is then a Cauchy matrix, which makes the system equations often well-conditioned if the prescribed poles are well-chosen. To make sure that the transfer function has real-valued coefficients, a linear combination of $\phi_p(s)$ and $\phi_{p+1}(s)$ is formed to make the residues $c_{p+1} = c_p^*$ complex conjugate if the poles $-a_{p+1} = -a_p^*$. This way, two basis functions of the following form are obtained

$$\phi_p(s) = \frac{1}{(s + a_p)} + \frac{1}{(s + a_{p+1})} \tag{11}$$

$$\phi_{p+1}(s) = \frac{i}{(s + a_p)} - \frac{i}{(s + a_{p+1})} \, . \tag{12}$$

This causes the corresponding elements in the solution vector to become equal to $\Re e(c_p)$, $\Im m(c_p)$ and $\Re e(\widetilde{c}_p)$, $\Im m(\widetilde{c}_p)$.

In successive iterations, a Sanathanan-Koerner iteration can be applied. In theory, one could use the denominator of the previous iteration as an inverse weighting to the system equations. The Vector Fitting technique is different, in a sense that weighting is performed implicitly by pole-relocation without weighting. The implicit weighting was found to be more robust if poles need to be relocated over long distances [8]. More details about this procedure are described in Appendix A.

As suggested in [13] and [17], the poles of the basis functions are optimally selected as complex conjugate pairs on a vertical or skew line, close to the imaginary axis. Due to the iterative behaviour of the SK-iteration, the prescribed poles are relocated until the poles converge in such way that the minimization of the SK cost function is converged. In general, this happens quite fast (i.e. <3 iterations). When poles are chosen too far to the left in the complex plane, the real part of the poles dominates the matrix entries, which deteriorates the numerical conditioning. However, even when the initial poles are inappropriately chosen, the algorithm succeeds in minimizing (4), at the expense of additional iterations.

After parameterization of $\mathbf{x}$, (8) can be simplified by cancelling out common poles. This means that the zeros of the denominator expression become the poles of the final transfer function. Calculating the zeros can easily be done, as shown in the following section.

2.3 Calculation of Transfer Function Poles

The minimal LTI state-space realization

$$\begin{aligned} sX(s) &= \mathbf{A}X(s) + \mathbf{B}U(s) \\ Y(s) &= \mathbf{C}X(s) + \mathbf{D}U(s) \end{aligned} \tag{13}$$

of the denominator

$$D(s) = \sum_{p=1}^{P} \frac{\widetilde{c}_p}{s + a_p} + 1 \tag{14}$$

can be obtained by a parallel connection (initially $\mathbf{A}$,$\mathbf{B}$,$\mathbf{C} = \emptyset$ and $\mathbf{D} = 0$)

$$\mathbf{A} = \begin{pmatrix} \mathbf{A} & 0 \\ 0 & \mathbf{A}_p \end{pmatrix}, \mathbf{B} = \begin{pmatrix} \mathbf{B} \\ \mathbf{B}_p \end{pmatrix} \tag{15}$$

$$\mathbf{C} = \begin{pmatrix} \mathbf{C} & \mathbf{C}_p \end{pmatrix}, \mathbf{D} = \mathbf{D} + \mathbf{D}_p \tag{16}$$

of the minimal state space realizations $(\mathbf{A}_p$,$\mathbf{B}_p$,$\mathbf{C}_p$,$\mathbf{D}_p)$ of each simple fraction, with

$$\mathbf{A}_p = -a_p,\ \mathbf{B}_p = 1,\ \mathbf{C}_p = \widetilde{c}_p,\ \mathbf{D}_p = 0 \tag{17}$$

provided that $-a_p$ is real. If $-a_p$ and $-a_{p+1}$ constitute a complex conjugate pair of poles (i.e. $-a_{p+1} = -a_p^*$), the corresponding state space realization of the linear combination is given as

$$\mathbf{A}_p = \begin{pmatrix} \Re e(-a_p) & \Im m(-a_p) \\ -\Im m(-a_p) & \Re e(-a_p) \end{pmatrix}, \mathbf{B}_p = \begin{pmatrix} 2 \\ 0 \end{pmatrix}$$
$$\mathbf{C}_p = \begin{pmatrix} \Re e(\widetilde{c}_p) & \Im m(\widetilde{c}_p) \end{pmatrix}, \mathbf{D}_p = 0\ . \tag{18}$$

Afterwards, the constant term 1 of (14) can simply be added to the scalar $\mathbf{D}$. This transformation makes the state-space realization of $D(s)$

$$D(s) = \mathbf{C}(sI - \mathbf{A})^{-1}\mathbf{B} + \mathbf{D} \tag{19}$$

real-valued, such that the poles and zeros occur as complex conjugate pairs. The zeros of (19) can then be solved by calculating the eigenvalues of $\mathbf{A}$-$\mathbf{BC}$. After simplification of (8), these eigenvalues will become the relocated poles of the transfer function

$$-a_p = eig(\mathbf{A} - \mathbf{BC}) \tag{20}$$

and this procedure can be repeated iteratively ($t = 1, ..., T$) until the minimization of the SK-cost function is converged.

2.4 Identification of the Residues

Once the final poles $-a_p^{(T)}$ are identified, the corresponding residues θ_p can be solved as a linear problem

$$\arg\min_{\theta_p} \sum_{k=0}^{K} \left| H(s_k) - \left(\sum_{p=1}^{P} \frac{\theta_p}{s_k + a_p^{(T)}} \right) \right|^2 . \tag{21}$$

This technique was called "Vector Fitting" [13], and it has been widely applied to many modelling problems within power systems, high-speed interconnection structures, electronic packages and microwave systems.

3 Orthonormal Vector Fitting

3.1 Orthonormalization of Partial Fraction Basis

Instead of using the partial fractions as rational basis functions, it was shown that orthonormal rational basis functions can lead to significant improvements in numerical conditioning [6, 7, 26]. A straightforward way to calculate an orthonormal basis, is to apply a Gram-Schmidt procedure on the partial fractions [2, 21, 27]. Hence, orthonormal rational functions $\phi_p(s)$ are obtained, which are in fact linear combinations of the partial fractions, of the form

$$\phi_p(s) = \frac{Q_p(s)}{\prod_{j=1}^{p}(s + a_j)} \tag{22}$$

for $p = 1, ..., P$ and $Q_p(s)$ an arbitrary polynomial of order $p - 1$, such that

$$\langle \phi_m(s), \phi_n(s) \rangle = \delta_{mn} \tag{23}$$

with $1 \leq m, n \leq P$. If the inner product is defined as

$$\langle \phi_m(s), \phi_n(s) \rangle = \frac{1}{2\pi i} \int\limits_{i\mathbb{R}} \phi_m(s) \phi_n^*(s) ds \tag{24}$$

then the $Q_p(s)$ polynomial can be determined by imposing the orthonormality conditions on the basis functions. As an example, consider the construction of the first function $\phi_1(s)$.

$$\langle \phi_1(s), \phi_1(s) \rangle = \frac{1}{2\pi i} \int\limits_{i\mathbb{R}} \phi_1(s) \phi_1^*(s) ds \tag{25}$$

$$= \frac{1}{2\pi i} \int\limits_{i\mathbb{R}} \frac{|\gamma_1|^2}{(s + a_1)(-s + a_1^*)} ds \tag{26}$$

$$= \frac{|\gamma_1|^2}{a_1 + a_1^*} \tag{27}$$

To normalize $\phi_1(s)$, $Q_1(s) = \gamma_1$ must equal $\kappa_1 \sqrt{2\Re e(a_1)}$, where κ_1 is an arbitrary unimodular complex number. $\phi_1(s)$ is then obtained as

$$\phi_1(s) = \kappa_1 \sqrt{2\Re e(a_1)} \frac{1}{s + a_1} . \tag{28}$$

Now consider the construction of the second function $\phi_2(s)$. First of all, $\phi_2(s)$ must be orthogonal to $\phi_1(s)$

$$\langle \phi_1(s), \phi_2(s) \rangle = \frac{1}{2\pi i} \int\limits_{i\mathbb{R}} \phi_1(s) \phi_2^*(s) ds = 0 \tag{29}$$

which implies that $\phi_2^*(s)$ must vanish for $s = -a_1$. Therefore $Q_2(s) = \gamma_2(s - a_1^*)$. This constant γ_2 is determined by imposing the normalization condition

$$\begin{aligned}
&\langle \phi_2(s), \phi_2(s) \rangle \\
&= \frac{1}{2\pi i} \int_{i\mathbb{R}} \frac{\gamma_2(s - a_1^*)}{(s + a_1)(s + a_2)} \frac{\gamma_2^*(-s - a_1)}{(-s + a_1^*)(-s + a_2^*)} ds \qquad (30) \\
&= \frac{1}{2\pi i} \int_{i\mathbb{R}} \frac{|\gamma_2|^2}{(s + a_2)(-s + a_2^*)} ds = \frac{|\gamma_2|^2}{a_2 + a_2^*} . \qquad (31)
\end{aligned}$$

Clearly, it follows that $\gamma_2 = \kappa_2 \sqrt{2\Re e(a_2)}$, where κ_2 is an arbitrary unimodular complex number. So, $\phi_2(s)$ is then given by

$$\phi_2(s) = \kappa_2 \sqrt{2\Re e(a_2)} \frac{s - a_1^*}{(s + a_1)(s + a_2)} . \qquad (32)$$

Similarly continuing this approach, the general polynomials are obtained

$$\phi_p(s) = \kappa_p \sqrt{2\Re e(a_p)} \left(\prod_{j=1}^{p-1} \frac{s - a_j^*}{s + a_j} \right) \frac{1}{s + a_p} . \qquad (33)$$

This basis originates from the discrete-time Takenaka-Malmquist basis [24, 33], and has later been transformed to the continuous time domain. It is a generalization of the Laguerre basis [4], where all poles $\{-a_p\}$ are the same real number, and the 2-parameter Kautz bases [20] where all poles $\{-a_p, -a_{p+1}\}$ are the same complex conjugate pair with $-a_p^* = -a_{p+1}$. A theoretical analysis of these basis functions is well-described in literature. The interested reader is referred to [18] which gives an excellent survey.

To make sure that the transfer function has real-valued coefficients, a linear combination of $\phi_p(s)$ and $\phi_{p+1}(s)$ is formed which can be made real-valued if the poles are real or occur in a complex conjugate pair. This way, two orthonormal functions of the following form are obtained

$$\phi_p(s) = \gamma_p \left(\prod_{j=1}^{p-1} \frac{s - a_j^*}{s + a_j} \right) \frac{s - x}{(s + a_p)(s + a_{p+1})} \qquad (34)$$

$$\phi_{p+1}(s) = \gamma_{p+1} \left(\prod_{j=1}^{p-1} \frac{s - a_j^*}{s + a_j} \right) \frac{s - y}{(s + a_p)(s + a_{p+1})} \qquad (35)$$

with real $\gamma_p, \gamma_{p+1}, x$ and y. To impose the orthogonality,

$$\langle \phi_p(s), \phi_{p+1}(s) \rangle = \gamma_p \gamma_{p+1} \frac{xy + a_p a_{p+1}}{2(a_p + a_{p+1}) a_p a_{p+1}} = 0 \qquad (36)$$

x and y are set to be $\sqrt{a_p a_{p+1}} = |a_p|$ and $-\sqrt{a_p a_{p+1}} = -|a_p|$ respectively. Similarly, γ_p and γ_{p+1} are set to $\sqrt{a_p + a_{p+1}} = \sqrt{2\Re e(a_p)}$. Note that this choice is not unique, and that other possibilities exist. Note also that the orthonormalization of the basis functions is done analytically instead of numerically, so it doesn't require any additional computation time.

3.2 Calculation of Transfer Function Poles

The minimal continuous-time LTI state-space realization

$$sX(s) = \mathbf{A}X(s) + \mathbf{B}U(s) \tag{37}$$

$$Y(s) = \mathbf{C}X(s) + \mathbf{D}U(s) \tag{38}$$

of the denominator

$$D(s) = 1 + \sum_{p=1}^{P} \tilde{c}_p \phi_p(s) \tag{39}$$

can then be calculated, by cascading the minimal state-space realization of smaller, first and second order sections [11]

$$\frac{s - a_1^*}{s + a_1} \rightarrow \frac{s - a_2^*}{s + a_2} \rightarrow \ldots \rightarrow \frac{s - a_{P-1}^*}{s + a_{P-1}} \rightarrow \frac{1}{s + a_P}\,. \tag{40}$$

The minimal state-space realization $(\mathbf{A}_p,\mathbf{B}_p,\mathbf{C}_p,\mathbf{D}_p)$ of the all-pass function

$$\frac{Y_p(s)}{U_p(s)} = \frac{s - a_p^*}{s + a_p} \tag{41}$$

for $p = 1, \ldots, P-1$ is given as

$$\mathbf{A}_p = -a_p,\ \mathbf{B}_p = 1,\ \mathbf{C}_p = 2\Re e(-a_p),\ \mathbf{D}_p = 1 \tag{42}$$

and the minimal state-space realization $(\mathbf{A}_p,\mathbf{B}_p,\mathbf{C}_p,\mathbf{D}_p)$ of the low-pass function

$$\frac{Y_p(s)}{U_p(s)} = \frac{1}{s + a_p} \tag{43}$$

is given as

$$\mathbf{A}_p = -a_p,\ \mathbf{B}_p = 1,\ \mathbf{C}_p = 1,\ \mathbf{D}_p = 0 \tag{44}$$

for $p = P$. Then the minimal state-space realization of the compound system (40) is obtained as the cascade construction

$$\mathbf{A} = \begin{bmatrix} \mathbf{A}_1 & 0 & ... & 0 \\ \mathbf{B}_2\mathbf{C}_1 & \mathbf{A}_2 & ... & 0 \\ \mathbf{B}_3\mathbf{D}_2\mathbf{C}_1 & \mathbf{B}_3\mathbf{C}_2 & ... & 0 \\ \mathbf{B}_4\mathbf{D}_3\mathbf{D}_2\mathbf{C}_1 & \mathbf{B}_4\mathbf{D}_3\mathbf{C}_2 & ... & 0 \\ ... & ... & ... & ... \\ \mathbf{B}_P\mathbf{D}_{P-1}...\mathbf{D}_2\mathbf{C}_1 & \mathbf{B}_P\mathbf{D}_{P-1}...\mathbf{D}_3\mathbf{C}_2 & ... & \mathbf{A}_P \end{bmatrix}$$

$$\mathbf{B} = \begin{bmatrix} \mathbf{B}_1 \\ \mathbf{B}_2\mathbf{D}_1 \\ \mathbf{B}_3\mathbf{D}_2\mathbf{D}_1 \\ \mathbf{B}_4\mathbf{D}_3\mathbf{D}_2\mathbf{D}_1 \\ ... \\ \mathbf{B}_P\mathbf{D}_{P-1}...\mathbf{D}_1 \end{bmatrix}, C = \begin{bmatrix} \mathbf{D}_P...\mathbf{D}_2\mathbf{C}_1 \\ \mathbf{D}_P...\mathbf{D}_3\mathbf{C}_2 \\ ... \\ \mathbf{C}_P \end{bmatrix}^T \tag{45}$$

$$\mathbf{D} = \mathbf{D}_P...\mathbf{D}_1$$

of the smaller state space models, with $y_p(t) = u_{p+1}(t)$.

The state matrix $\mathbf{A}$ and the input vector $\mathbf{B}$ are build such that the states contain exactly the unnormalized basis functions. The output vector $\mathbf{C}$ and scalar $\mathbf{D}$ are chosen to obtain the denominator expression (39), by compensating for the coefficients $\widetilde{c}_p$ and normalization constant $\sqrt{2\Re e(a_p)}$ in the vector $\mathbf{C}$, and setting the scalar $\mathbf{D}$ equal to the constant value 1. The following real-valued state space realization is obtained

$$\mathbf{A}_{P\times P} = \begin{bmatrix} -a_1 & 0 & 0 & ... & 0 \\ 2\Re e(-a_1) & -a_2 & 0 & ... & 0 \\ 2\Re e(-a_1) & 2\Re e(-a_2) & -a_3 & ... & 0 \\ 2\Re e(-a_1) & 2\Re e(-a_2) & 2\Re e(-a_3) & ... & 0 \\ ... & ... & ... & ... & ... \\ 2\Re e(-a_1) & 2\Re e(-a_2) & 2\Re e(-a_3) & ... & -a_P \end{bmatrix}$$

$$\mathbf{B}_{1\times P} = \begin{bmatrix} 1 \\ 1 \\ ... \\ 1 \end{bmatrix}, \mathbf{C}_{P\times 1} = \begin{bmatrix} \widetilde{c}_1\sqrt{2\Re e(a_1)} \\ \widetilde{c}_2\sqrt{2\Re e(a_2)} \\ ... \\ \widetilde{c}_P\sqrt{2\Re e(a_P)} \end{bmatrix}^T \tag{46}$$

$$\mathbf{D}_{1\times 1} = 1$$

provided that the poles $-a_p$ are real.

If $-a_p$ and $-a_{p+1}$ constitute a complex conjugate pair of poles (i.e. $-a_{p+1} = -a_p^*$), a real-valued state-space realization is obtained by replacing

$$\frac{s-a_p^*}{s+a_p} \rightarrow \frac{s-a_{p+1}^*}{s+a_{p+1}} \tag{47}$$

in the cascade scheme (40) by

$$\frac{(s-a_p^*)(s-a_{p+1}^*)}{(s+a_p)(s+a_{p+1})} = 1 + \frac{4\Re e(-a_p)s}{(s+a_p)(s+a_p^*)}. \tag{48}$$

This corresponds to replacing

$$\begin{pmatrix} -a_p & 0 \\ 2\Re e(-a_p) & -a_{p+1} \end{pmatrix} \tag{49}$$

in the state matrix A, by

$$\begin{pmatrix} \Re e(-a_p) & \Re e(-a_p) - |a_p| \\ \Re e(-a_p) + |a_p| & \Re e(-a_p) \end{pmatrix}. \tag{50}$$

The other state space matrices remain unchanged. Appendix B describes this transformation in more detail. Again, the zeros of the denominator are calculated by solving the eigenvalues of $\mathbf{A}$-$\mathbf{BC}$. These eigenvalues replace the set of prescribed poles, and the procedure is repeated iteratively ($t = 1, ..., T$) until the minimization of the SK-cost function is converged.

Once the final poles are identified, the residues can be solved as a linear problem using the partial fraction basis (21). The orthonormal basis functions can also be applied if stability of the poles is enforced. Both representations can easily be realized to state-space as was shown before.

4 Example

As an example, the technique is illustrated on a dense model of an atmospheric storm track (eady), which is obtained from the NICONET benchmark dataset collection [3]. Based on the state-space matrices of the large model (598×598), the frequency response is densely calculated over the frequency range of interest [10^{-1},10^2] and shown in Fig. 1.

First, a prescribed set of complex conjugate starting poles is chosen as was proposed by [13]

$$-a_p = -\alpha + \beta i, -a_{p+1} = -\alpha - \beta i \tag{51}$$

$$\alpha = \beta/100 \tag{52}$$

with imaginary parts β logarithmically spaced over the frequency range of interest. The amount of poles is chosen in terms of the desired reduction. In this example, it was chosen to be 54, in order to have an RMS error which corresponds to the order of 10^{-8}.

The weighted linear cost function (4) is solved using the orthonormal rational basis functions (33), (34), (35) and an estimate for the residues c_p and $\widetilde{c}_p$ is obtained. Using the residues $\widetilde{c}_p$ and the poles $-a_p$, the minimal state-space realization ($\mathbf{A}$,$\mathbf{B}$,$\mathbf{C}$,$\mathbf{D}$) of the denominator $D(s)$ (39) is calculated. From this state-space

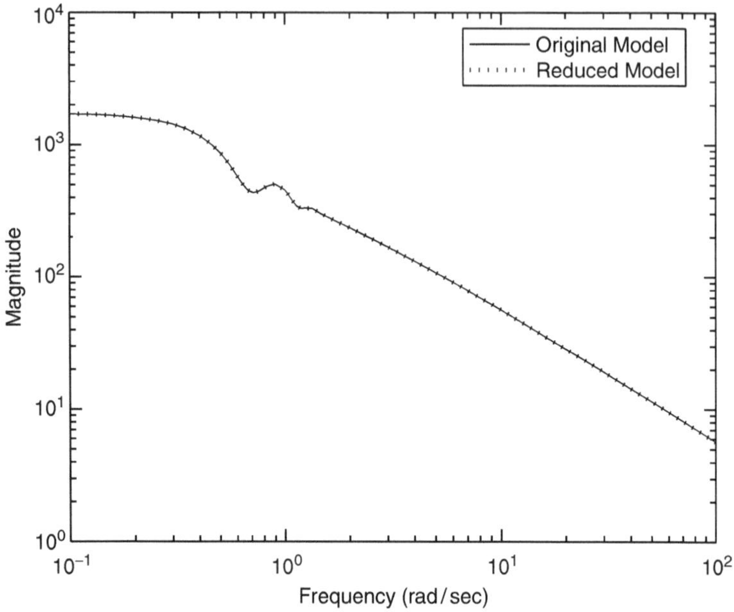

Fig. 1. Frequency response : original and reduced model

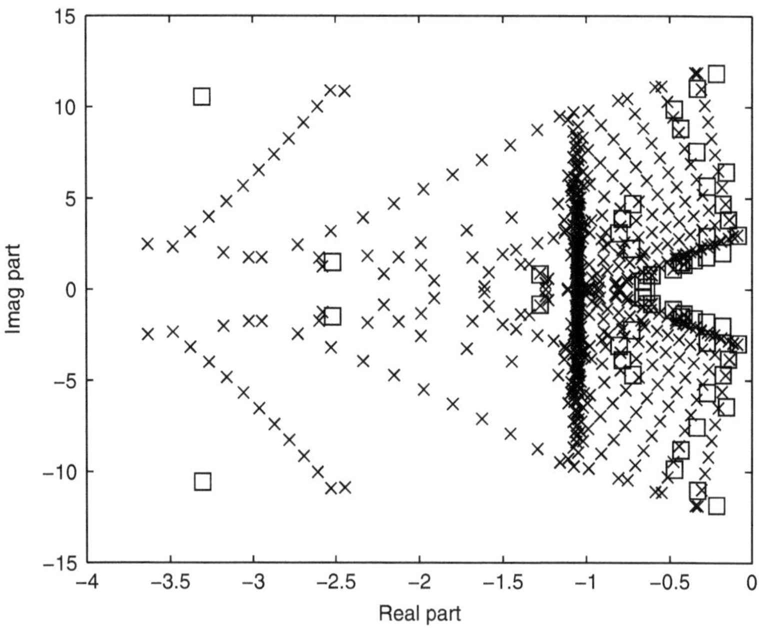

Fig. 2. Poles of original (×) and reduced (□) model

model, the poles of the transfer function are calculated by solving the eigenvalues of **A**-**BC**. These poles are chosen as new starting poles, and the method iterates until the poles are converged to their optimal location. Once the poles are known, the residues of the transfer function can be estimated as a linear problem. The final accuracy of the model in terms of RMS error is 5.61×10^{-8}.

As is shown in Fig. 1, no visual difference can be observed between the frequency response of the original and reduced system. The poles of the original model and the reduced model are shown in Fig. 2. The OVF method can also be extended to fit systems with multiple ports. It is noted that the extension is completely analogous to the Matrix Fitting algorithm [15].

5 Conclusion

This paper shows that Orthonormal Vector Fitting can be useful to reduce the state space dimensions of large circuit models. First, the spectral response of the large model is calculated, and then the OVF algorithm is used to approximate the data with a model of reduced size. It was shown that the method is quite robust, even when the original system has a large amount of poles. The method extends in a natural way to multi-port systems (not shown in this paper). The reduced model is represented as state-space realization, which can easily be converted e.g. to an RLCG circuit.

6 Acknowledgements

This work was supported by the Fund for Scientific Research Flanders (FWO Vlaanderen).

A Sanathanan-Koerner Iteration

The least-squares SK-cost function is defined as

$$\arg \min_{N_n^{(t)}, D_d^{(t)}} \left(\sum_{k=0}^{K} \left| \frac{1}{D^{(t-1)}(s_k)} \right|^2 \left| D^{(t)}(s_k) H(s_k) - N^{(t)}(s_k) \right|^2 \right) . \tag{53}$$

If the basis functions are chosen as partial fractions, based on a prescribed set of poles $-a_1, ..., -a_P$, then it follows that

$$N^{(t)}(s) = \sum_{p=1}^{P} \frac{c_p^{(t)}}{s + a_p} = \frac{\prod_{p=1}^{P-1}(s + z_{p,n}^{(t)})}{\prod_{p-1}^{P}(s + a_p)} \tag{54}$$

$$D^{(t)}(s) = \sum_{p=1}^{P} \frac{\tilde{c}_p^{(t)}}{s + a_p} + \tilde{c}_0^{(t)} = \frac{\prod_{p=1}^{P}(s + z_{p,d}^{(t)})}{\prod_{p=1}^{P}(s + a_p)} . \tag{55}$$

The denominator has an additional basisfunction, which equals the constant value 1. In the first iteration step ($t = 0$), Levi's linearization is applied to obtain a first guess of the denominator $(D^{(-1)}(s) = 1)$

$$\arg \min_{N_n^{(t)}, D_d^{(t)}} \left(\sum_{k=0}^{K} \left| \frac{1}{D^{(t-1)}(s_k)} \right|^2 \left| D^{(t)}(s_k) H(s_k) - N^{(t)}(s_k) \right|^2 \right) \tag{56}$$

$$= \arg \min_{\widetilde{c}_p^{(0)}, c_p^{(0)}} \left(\sum_{k=0}^{K} \left| \left(\sum_{p=1}^{P} \frac{\widetilde{c}_p^{(0)}}{s_k + a_p} + \widetilde{c}_0^{(0)} \right) H(s_k) - \sum_{p=1}^{P} \frac{c_p^{(0)}}{s_k + a_p} \right|^2 \right). \tag{57}$$

This reduces to solving the following set of least-squares equations, for all complex frequencies s

$$\left(\sum_{p=1}^{P} \frac{\widetilde{c}_p^{(0)}}{s + a_p} + \widetilde{c}_0^{(0)} \right) H(s) - \sum_{p=1}^{P} \frac{c_p^{(0)}}{s + a_p} = 0\,. \tag{58}$$

One coefficient of the rational function, e.g. $\widetilde{c}_0^{(0)}$, can be fixed to unity, since numerator and denominator can be divided by the same complex value without loss of generality. So, (58) is equivalent to

$$\sum_{p=1}^{P} \frac{c_p^{(0)}}{s + a_p} - \sum_{p=1}^{P} \frac{\widetilde{c}_p^{(0)}}{s + a_p} H(s) = H(s)\,. \tag{59}$$

Once the parameters $c_p^{(0)}$ and $\widetilde{c}_p^{(0)}$ are estimated, $N^{(0)}(s)$ and $D^{(0)}(s)$ are known (54-55). It's straightforward to calculate $z_{p,n}^{(0)}$ and $z_{p,d}^{(0)}$ in a robust way, by solving the eigenvalue problem (20). In practice, only $z_{p,d}^{(0)}$ is needed.

Now, the Sanathanan-Koerner linearization can be applied for iteration step $t = 1, ..., T$

$$\arg \min_{N_n^{(t)}, D_d^{(t)}} \left(\sum_{k=0}^{K} \left| \frac{1}{D^{(t-1)}(s_k)} \right|^2 \left| D^{(t)}(s_k) H(s_k) - N^{(t)}(s_k) \right|^2 \right) \tag{60}$$

$$= \arg \min_{\widetilde{c}_p^{(t)}, c_p^{(t)}} \left(\sum_{k=0}^{K} \left| \frac{\prod_{p=1}^{P} (s_k + a_p)}{\prod_{p=1}^{P} (s_k + z_{p,d}^{(t-1)})} \right|^2 \left| \left(\sum_{p=1}^{P} \frac{\widetilde{c}_p^{(t)}}{s_k + a_p} + \widetilde{c}_0^{(t)} \right) \right.\right.$$

$$\left.\left. \times\ H(s_k) - \sum_{p=1}^{P} \frac{c_p^{(t)}}{s_k + a_p} \right|^2 \right) \tag{61}$$

$$= \arg \min_{z_{p,d}^{(t)}, z_{p,n}^{(t)}} \left(\sum_{k=0}^{K} \left| \frac{\prod_{p=1}^{P}(s_k + a_p)}{\prod_{p=1}^{P}(s_k + z_{p,d}^{(t-1)})} \right|^2 \left| \left(\frac{\prod_{p=1}^{P}(s_k + z_{p,d}^{(t)})}{\prod_{p=1}^{P}(s_k + a_p)} \right) \right. \right.$$

$$\left. \left. \times H(s_k) - \frac{\prod_{p=1}^{P-1}(s_k + z_{p,n}^{(t)})}{\prod_{p=1}^{P}(s_k + a_p)} \right|^2 \right) \tag{62}$$

$$= \arg \min_{z_{p,d}^{(t)}, z_{p,n}^{(t)}} \left(\sum_{k=0}^{K} \left| \left(\frac{\prod_{p=1}^{P}(s_k + z_{p,d}^{(t)})}{\prod_{p=1}^{P}(s_k + z_{p,d}^{(t-1)})} \right) \right. \right.$$

$$\left. \left. \times H(s_k) - \frac{\prod_{p=1}^{P-1}(s_k + z_{p,n}^{(t)})}{\prod_{p=1}^{P}(s_k + z_{p,d}^{(t-1)})} \right|^2 \right) \tag{63}$$

$$= \arg \min_{\widetilde{d}_p^{(t)}, d_p^{(t)}} \left(\sum_{k=0}^{K} \left| \left(\sum_{p=1}^{P} \frac{\widetilde{d}_p^{(t)}}{s_k + z_{p,d}^{(t-1)}} + \widetilde{d}_0^{(t)} \right) \right. \right.$$

$$\left. \left. \times H(s_k) - \sum_{p=1}^{P} \frac{d_p^{(t)}}{s_k + z_{p,d}^{(t-1)}} \right|^2 \right) . \tag{64}$$

When the classical SK-iteration is used, one can solve the coefficients $c_p^{(t)}$ and $\widetilde{c}_p^{(t)}$ of $N^{(t)}$ and $D^{(t)}$ if a weighting is applied to each row of the system equations (explicit weighting). The Vector Fitting performs this weighting implicitly, by calculating the coefficients $d_p^{(t)}$ and $\widetilde{d}_p^{(t)}$ of $N^{(t)}/D^{(t-1)}$ and $D^{(t)}(s)/D^{(t-1)}$ instead (without an explicit weighting). In successive iterations ($t > 0$), the coefficients $\widetilde{d}_p^{(t)}$ of $D^{(t)}/D^{(t-1)}$ are then used to calculate the poles, which does not pose a problem, as the zeros of $D^{(t)}$ and $D^{(t)}/D^{(t-1)}$ are the same. It is noted, however, that the poles of the basis functions of $N^{(t)}(s)$ and $D^{(t)}(s)$ remain unchanged, and cancel out in each iteration (62).

B Real-Valued State Space

This appendix describes how the real-valued state-space realization of

$$\frac{(s - a_p^*)(s - a_{p+1}^*)}{(s + a_p)(s + a_{p+1})} = 1 + \frac{4\Re e(-a_p)s}{(s + a_p)(s + a_p^*)} \tag{65}$$

can be obtained.

Define the state matrix $\mathbf{A}$ and input vector $\mathbf{B}$ as

$$\mathbf{A} = \begin{pmatrix} \mathbf{A}_{11} & \mathbf{A}_{12} \\ \mathbf{A}_{21} & \mathbf{A}_{22} \end{pmatrix}, \mathbf{B} = \begin{pmatrix} \mathbf{B}_1 \\ \mathbf{B}_2 \end{pmatrix} \tag{66}$$

where $\mathbf{A}_{ij}$ and $\mathbf{B}_i$ are the scalar elements of the matrix and the vector respectively. Capitals are used to avoid notational confusion between the poles and the entries of the state matrix.

A first constraint on the entries, is that the poles of (65), $\{-a_p, -a_p^*\}$, must equal the eigenvalues of $\mathbf{A}$. More specifically, the transfer function from the input $U_1(s)$ and $U_2(s)$ to the states $X_1(s)$ and $X_2(s)$ respectively, must satisfy

$$\frac{X_1(s)}{U_1(s)} = \frac{s - |a_p|}{(s + a_p)(s + a_p^*)} \tag{67}$$

$$\frac{X_2(s)}{U_2(s)} = \frac{s + |a_p|}{(s + a_p)(s + a_p^*)} \, . \tag{68}$$

The input-to-state transfer function is given by

$$\frac{X(s)}{U(s)} = (sI - \mathbf{A})^{-1}\mathbf{B} \tag{69}$$

$$= \frac{\begin{pmatrix} s - \mathbf{A}_{22} & \mathbf{A}_{12} \\ \mathbf{A}_{21} & s - \mathbf{A}_{11} \end{pmatrix}}{(s - \mathbf{A}_{11})(s - \mathbf{A}_{22}) - \mathbf{A}_{12}\mathbf{A}_{21}} \mathbf{B} \, , \tag{70}$$

so

$$\frac{X_1(s)}{U_1(s)} = \frac{(s - \mathbf{A}_{22})\mathbf{B}_1 + \mathbf{A}_{12}\mathbf{B}_2}{(s - \mathbf{A}_{11})(s - \mathbf{A}_{22}) - \mathbf{A}_{12}\mathbf{A}_{21}} \tag{71}$$

and

$$\frac{X_2(s)}{U_2(s)} = \frac{\mathbf{A}_{21}\mathbf{B}_1 + (s - \mathbf{A}_{11})\mathbf{B}_2}{(s - \mathbf{A}_{11})(s - \mathbf{A}_{22}) - \mathbf{A}_{12}\mathbf{A}_{21}} \tag{72}$$

By equating the numerators of (67) to (71), (68) to (72), and applying some basic linear algebra, the following constraints are easily obtained

$$\mathbf{B}_1 = 1 \tag{73}$$

$$\mathbf{B}_2 = 1 \tag{74}$$

$$-\mathbf{A}_{22} + \mathbf{A}_{12} = -|a_p| \tag{75}$$

$$\mathbf{A}_{21} - \mathbf{A}_{11} = |a_p| \tag{76}$$

which determine the input vector $\mathbf{B}$ completely. Unfortunately, the elements of the state matrix $\mathbf{A}$ are still ambiguous.

By equating the denominators, it follows that

$$s^2 - (\mathbf{A}_{11} + \mathbf{A}_{22})s + (\mathbf{A}_{11}\mathbf{A}_{22} + \mathbf{A}_{12}\mathbf{A}_{21}) \tag{77}$$

$$= s^2 + (a_p + a_p^*)s + a_p a_p^* \tag{78}$$

so

$$\mathbf{A}_{11} + \mathbf{A}_{22} = -2\Re e(a_p) \tag{79}$$

$$\mathbf{A}_{11}\mathbf{A}_{22} - \mathbf{A}_{12}\mathbf{A}_{21} = |a_p|^2 \, . \tag{80}$$

Combining (80) with (75) and (76) gives

$$\mathbf{A}_{11} - \mathbf{A}_{12} = |a_p| \,. \tag{81}$$

Using (75) and (81)

$$\mathbf{A}_{11} - \mathbf{A}_{22} = 0 \tag{82}$$

$$\mathbf{A}_{11} = \mathbf{A}_{22} \,. \tag{83}$$

Obviously from (79) and (83), it results that

$$\mathbf{A}_{11} = \Re e(-a_p) \tag{84}$$

$$\mathbf{A}_{22} = \Re e(-a_p) \,. \tag{85}$$

Combining this with (75) and (76), it follows that

$$\mathbf{A}_{12} = \Re e(-a_p) - |a_p| \tag{86}$$

$$\mathbf{A}_{21} = \Re e(-a_p) + |a_p| \tag{87}$$

which determines **A** uniquely.

Verifying that the eigenvalues of **A** are actually equal to $-a_p$ and $-a_p^*$ is trivial. Now, **C** and **D** can easily be formed to obtain (65)

$$\mathbf{C} = \begin{pmatrix} 2\Re e(-a_p) & 2\Re e(-a_p) \end{pmatrix}, \mathbf{D} = 1 \,. \tag{88}$$

References

[1] Adcock J., Potter, R.: A Frequency Domain Curve Fitting Algorithm with Improved Accuracy, Proceedings 3rd International Modal Analysis Conference, 1, 541–547 (1985).

[2] Akcay, H., Ninness, B., Orthonormal Basis Functions for Modelling Continuous-Time Systems, Signal Processing, **77 (1)**, 261–274 (1999).

[3] Chahlaoui, Y., Van Dooren, P.: A Collection of Benchmark Examples for Model Reduction of Linear Time Invariant Dynamical Systems, SLICOT Working Note 2002-2 (2002).

[4] Clement, P.R.: Laguerre Functions in Signal Analysis and Parameter Identification, Journal of the Franklin Institute, **313**, 85–95 (1982).

[5] Deschrijver, D., Dhaene, T., Rational Modeling of Spectral Data using Orthonormal Vector Fitting, 9th IEEE Workshop on Signal Propagation on Interconnects, 111–114 (2005).

[6] Deschrijver, D., Dhaene, T.: Broadband Macromodelling of Passive Components using Orthonormal Vector Fitting, IEE Electronics Letters, **41 (21)**, 1160–1161 (2005).

[7] Deschrijver, D., Dhaene, T.: Parametric Identification of Frequency Domain Systems Using Orthonormal Rational Bases, 14th IFAC Symposium on System Identification, 837–847 (2006).

[8] Deschrijver, D., Gustavsen, B., Dhaene, T.: Advancements in Iterative Methods for Rational Approximation in the Frequency Domain, IEEE Transactions on Power Delivery, **22 (3)**, 1633–1642 (2007).

[9] Deschrijver, D., Haegeman, B., Dhaene, T.: Orthonormal Vector Fitting : A Robust Macromodeling Tool for Rational Approximation of Frequency Domain Responses, IEEE Transactions on Advanced Packaging, **30 (2)**, 216–225 (2007).

[10] Golub, G.H., Van Loan, C.F.: Matrix computations - Third Edition, London, The Johns Hopkins University Press (1996).

[11] Gomez, J.C.: Analysis of Dynamic System Identification using Rational Orthonormal Bases, Phd Thesis, University of Newcastle (1998).

[12] Grivet-Talocia, S.: Passivity Enforcement via Perturbation of Hamiltonian Matrices, IEEE Transactions on Circuits and Systems 1: Fundamental Theory and Applications, **51 (9)**, 1755–1769 (2004).

[13] Gustavsen, B., Semlyen, A.: Rational Approximation of Frequency Domain Responses by Vector Fitting, IEEE Transactions on Power Delivery, **14 (3)**, 1052–1061 (1999).

[14] Gustavsen, B., Semlyen, A.: Enforcing Passivity for Admittance Matrices approximated by Rational Functions, IEEE Transactions on Power Systems, **16 (1)**, 97–104 (2001).

[15] Gustavsen, B.: Computer Code for Rational Approximation of Frequency Dependent Admittance Matrices, IEEE Transactions on Power Delivery, **17 (4)**, 1093–1098 (2002).

[16] Gustavsen, B.: Improving the Pole Relocating Properties of Vector Fitting, IEEE Transactions on Power Delivery, **21 (3)**, 1587–1592 (2006).

[17] Hendrickx, W., Deschrijver, D., Dhaene, T.: Some Remarks on the Vector Fitting Iteration, Post-Conference Proceedings of ECMI 2004, Mathematics in Industry, Springer-Verlag, 134–138 (2006).

[18] Heuberger, P.S.C., Van Den Hof, P.M.J., Wahlberg B.: Modelling and Identification with Rational Orthogonal Basis Functions, London, Springer-Verlag (2005).

[19] Kalman, R.E., Design of a Self Optimizing Control System, Transactions ASME, **80**, 468–478 (1958).

[20] Kautz, W.H.: Transient Synthesis in the Time-Domain, IRE Transactions on Circuit Theory, **1**, 29–39 (1954).

[21] Knockaert, L.: On Orthonormal Muntz-Laguerre Filters, IEEE Transactions on Signal Processing, **49 (4)**, 790–793 (2001).

[22] Levenberg, K., A Method for the Solution of Certain Problems in Least Squares, Quarterly Journal on Applied Mathematics, **2**, 164–168 (1944).

[23] Levi, E.C.: Complex Curve Fitting, IEEE Transactions on Automatic Control, AC-4, 37–43 (1959).

[24] Malmquist, F.: Sur la Détermination d'une Classe de Fonctions Analytiques par leurs Valeurs dans un Ensemble donné de Points, Compte Rendus de Sixieme Congres de Mathematiciens Scandinaves, 253–259 (1926).

[25] Marquardt D.: An Algorithm for Least-Squares Estimation of Nonlinear Parameters, SIAM Journal on Applied Mathematics, **11**, 431–441 (1963).

[26] Ninness, B., Gibson S., Weller, S.R.: Practical Aspects of using Orthonormal System Parameterisations in Estimation Problems, 12th IFAC Symposium on System Identification (2000).

[27] Oliveira e Silva, T.: Rational Orthonormal Functions on the Unit Circle and on the Imaginary Axis, with Applications in System Identification (2005).

[28] R. Pintelon and J. Schoukens, System Identification : A Frequency Domain Approach, Piscataway NJ (USA), IEEE Press (2000).

[29] Pintelon, R., Guillaume, P., Rolain, Y., Schoukens, J., Hamme, H.V.: Parametric Identification of Transfer Functions in the Frequency Domain - a Survey, IEEE Transactions on Automatic Control, **39 (11)**, 2245–2260 (1994).

[30] Richardson, M., Formenti, D.L.: Parameter Estimation from Frequency Response Measurements using Rational Fraction Polynomials, Proceedings 1st International Modal Analysis Conference, 167–182 (1982).

[31] Rolain, Y., Pintelon, R., Xu, K.Q., Vold, H.: Best Conditioned Parametric Identification of Transfer Function Models in the Frequency Domain, IEEE Transactions on Automatic Control, AC-40 **(11)**, 1954–1960 (1995).

[32] Sanathanan, C.K., Koerner, J., Transfer Function Synthesis as a Ratio of Two Complex Polynomials, IEEE Transactions on Automatic Control, AC-8, 56–58 (1963).

[33] Takenaka, S.: On the Orthogonal Functions and a New Formula of Interpolation, Japanese Journal of Mathematics, 129–145 (1925).

[34] Wahlberg, B., Makila, P., On Approximation of Stable Linear Dynamical Systems using Laguerre and Kautz Functions, Automatica, **32**, 693–708 (1996).

[35] Whitfield, A.H.: Asymptotic Behavior of Transfer Function Synthesis Methods, International Journal of Control, 45, 1083–1092 (1987).

Model-Order Reduction of High-Speed Interconnects Using Integrated Congruence Transform

Emad Gad[1], Michel Nakhla[2], and Ram Achar[2]

[1] School of Information Technology and Engineering (SITE), Univ. of Ottawa, Ottawa, ON, Canada, K1N 6N5
egad@site.uottawa.ca.
[2] Dept. of Electronics, Carleton Univ., Ottawa, ON, Canada, K1S 5B6
{msn,achar}@doe.carleton.ca

1 High-Speed Interconnects and Its Effects on Signal Propagation

With the rapid developments in Very Large Scale Integration (VLSI) technology at both the chip and package level, the operating frequencies are quickly reaching the vicinity of GHz and switching times are getting to the sub-nano second levels. The ever increasing quest for high-speed applications has placed higher demands on interconnect performance and highlighted the previously negligible effects of interconnects such as ringing, signal delay, distortion, reflections and crosstalk. As depicted by Figure 1, interconnects can exist at various levels of design hierarchy such as on-chip, packaging structures, multichip modules, printed circuit boards and backplanes. In addition, the trend in the VLSI industry towards miniature designs, low power consumption and increased integration of analog circuits with digital blocks has further complicated the issue of signal integrity analysis. It is predicted that interconnects will be responsible for majority of signal degradation in high-speed systems [1]. High-speed interconnect problems are not always handled appropriately by the conventional circuit simulators, such as SPICE [2]. If not considered during the design stage, these interconnect effects can cause logic glitches which render a fabricated digital circuit inoperable, or they can distort an analog signal such that it fails to meet specifications. Since extra iterations in the design cycle are costly, accurate prediction of these effects is a necessity in high-speed designs. Hence it becomes extremely important for designers to simulate the entire design along with interconnect subcircuits as efficiently as possible while retaining the accuracy of simulation.

Speaking on a broader perspective, a "high-speed interconnect" is the one in which the time taken by the propagating signal to travel between its end points can not be neglected. An obvious factor which influences this definition is the physical extent of the interconnect, where the longer the interconnect, the more time the signal takes to travel between its end points. Smoothness of signal propagation suffers once the line becomes long enough for signals rise/fall times to roughly match its propagation time through the line. Then the interconnect electrically isolates the

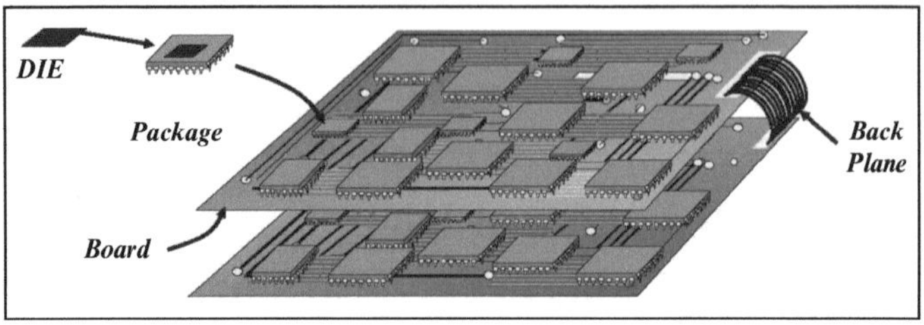

Fig. 1. Electrical interconnects are encountered at all levels of design hierarchy.

driver from the receivers, which no longer function directly as loads to the driver. Instead, within the time of signals transition between its high and low voltage levels, the impedance of interconnect becomes the load for the driver and also the input impedance to the receivers [1]. This leads to various transmission line effects, such as reflections, overshoot, undershoot, crosstalk and modeling of these effects needs the blending of EM and circuit theory.

Alternatively, the term 'high-speed' can be defined in terms of the frequency content of the signal. At low frequencies an ordinary wire, in other words, an interconnect, will effectively short two connected circuits. However, this is not the case at higher frequencies. The same wire, which is so effective at lower frequencies for connection purposes, has too many inductive/capacitive effects to function as a short at higher frequencies. Faster clock speeds and sharper slew rates tend to add more and more high-frequency contents. An important criterion used for classifying interconnects is the *electrical length* of an interconnect. An interconnect is considered to be "*electrically short*", if at the highest operating frequency of interest, the interconnect length is physically shorter than approximately one-tenth of the wave-length (i.e., length of interconnect/$\lambda \approx 0.1$, $\lambda = v/f$). Else the interconnect is referred to as "*electrically long*" [1]. In most digital applications, the desired highest operating frequency (which corresponds to the minimum wavelength) of interest is governed by the rise/fall time of the propagating signal. For example, the energy spectrum of a trapezoidal pulse is spread over an infinite frequency range, however, most of the signal energy is concentrated near the low frequency region and decreases rapidly with increase in frequency. Hence ignoring the high-frequency components of the spectrum above a maximum frequency, $f_{\max}$, will not seriously alter the overall signal shape. Consequently, for all practical purposes, the width of the spectrum can be assumed to be finite. In other words, the signal energy of interest is assumed to be contained in the major lobes of the spectrum and the relationship between the desired $f_{\max}$ and $1/f_r$, the rise/fall time of the signal, can be expressed as [3,4]

$$f_{\max} \approx 0.35/f_r \tag{1}$$

This implies that, for example, for a rise time of 0.1ns, the maximum frequency of interest is approximately 3GHz or the minimum wave-length of interest is 10cms. In some cases the limit can be more conservatively set as $f_{\max} \approx 1/f_r$ [5].

In summary, the primary factors which influence the decision that, "whether high-speed signal distortion effects should be considered", are interconnect length, cross-sectional dimensions, signal slew rate and the clock-speed. Other factors which also should be considered are logic levels, dielectric material and conductor resistance. Electrically short interconnects can be represented by lumped models where as electrically long interconnects need distributed or full-wave models.

High-speed effects influencing a signal propagating on an interconnect could be multifold, such as delay, rise time degradation, attenuation, crosstalk, skin effect, overshoots, undershoots, ringing, and reflection. The following list presents a more detailed account of these high-speed effects (which are also known as transmission line effects).

- **Propagation Delay.** A signal traversing from one end of a transmission line to the other end takes a finite amount of time; in other words, it experiences a certain amount of delay (T_d). In addition, the signal may encounter rise time degradation, where the rise time at the receiver end (t_R) is larger than the rise time at the source end (t_r). Rise-time degradation further adds to the overall delay experienced by the signal, as it influences the maximum and minimum attainable logic levels between the switching intervals.
- **Attenuation.** The signal through an interconnect may suffer attenuation, due to ohmic or conductance losses. Ohmic losses are more pronounced at higher frequencies due to the uneven current distributions. Conductance losses are proportional to the dielectric loss factor of the dielectric material and are also a function of the frequency. If the losses are high, the signals may not retain the specified logic levels during the transit through an interconnect and may result in false switching of digital circuitry.
- **Signal Reflection and Ringing.** Signal reflection and the associated ringing can severely distort signal propagation at higher frequencies. The prime cause of reflection-related signal degradation is the discontinuity in characteristic impedance of the transmitting line. Such a discontinuity can be either distributed or lumped in nature. In the case of distributed discontinuity, the impedance variation on a line takes place over a certain length. For example, this can occur due to the change in the medium along the length of the signal trace, which may have to traverse several layers on a printed circuit board (impedance may not be well controlled from layer to layer). Following are some common causes of discontinuities: connectors between card-to-board, cable-to-card, leads between chip and chip carriers, or between card wiring and chip carriers, long vias, orthogonal wiring, flip-chip soldier balls, wire bonds, and redistribution lines, etc.
  Another major contributor to the reflection related signal degradation is the impedance mismatch between the line characteristic impedance and source/ terminating impedances. In general, undershoots occur when the terminating impedance is less than the characteristic impedance of the interconnect. Overshoots occur when the terminating impedance is larger than the characteristic impedance of the line. As described earlier, the undershoots, overshoots, and the ringing experienced by the signal increases with the delay of the interconnect.

- **Crosstalk.** Crosstalk refers to the interaction between signals that are propagating on various lines in the system. An analogy of crosstalk could be the interference from other lines while talking on the phone. Crosstalk is mainly due to the dense wiring required by compact and high-performance systems. High-density and closely laid interconnects result in electromagnetic coupling between signal lines. The active signal energy is coupled to the quiet line through both mutual capacitance and inductances, resulting in noise voltages and currents. This may lead to inadvertent switching and system malfunctioning. Crosstalk is a major constraint while routing in high-speed designs. By its very nature, crosstalk analysis involves systems of two or more conductors. Such systems are studied on the basis of dominant propagating modes. System behavior in response to any general excitation is then a linear combination of modal responses.

2 Time-Domain Macromodeling of High-Speed Interconnects

In the context of high-speed interconnects, the term "Time-Domain Macromodeling" typically refers to the task of incorporating interconnects in commercial circuit simulators to simulate the effect of the aforementioned phenomena on signal propagation. A typical circuit simulator works by representing the circuit, which is usually described in a text file, in a concise mathematical form, whereupon a number of appropriate numerical analysis techniques can be invoked to simulate its behavior under different stimulus conditions. The next subsection presents a brief background on the techniques adopted by commercial simulators to represent general circuits in the mathematical domain.

2.1 Formulation of Circuits with Lumped Elements

The presence of nonlinear elements in virtually all circuit designs mandates that the natural domain for mathematically describing general circuits is the time-domain. A widely adopted time-domain formulation is known as the Modified Nodal Analysis or MNA approach [6]. Using the MNA formulation, a general circuit with lumped elements, such as resistors, inductors, capacitors, etc., can be described by the following system of Differential Algebraic Equations (DAEs),

$$\mathbf{C}\frac{\mathrm{d}\mathbf{x}(t)}{\mathrm{d}t} + \mathbf{G}\mathbf{x}(t) + \mathbf{f}(\mathbf{x}(t)) = \mathbf{b}(t) \tag{2}$$

where

- $\mathbf{C}, \mathbf{G} \in \mathbb{R}^{N\times N}$ are real matrices describing the memory and memoryless elements in the network, respectively;
- $\mathbf{x}(t) \in \mathbb{R}^{N}$ is a vector of node voltage waveforms appended by waveforms of currents in independent-voltage sources and inductors currents;
- $\mathbf{b}(t) \in \mathbb{R}^{N}$ is a vector of voltage and current waveforms of independent voltage and current sources representing the external stimulus in the circuits;

- $\mathbf{f}(\mathbf{x}(t)) \in \mathbb{R}^N$ is a vector whose entries are scalar nonlinear functions $N \to 1$ that represent the nonlinear elements in the circuit, and
- N is the total number of variables in the MNA formulation.

To simulate the time-domain behavior of the circuit variables in $\mathbf{x}(t)$, various time marching techniques can be used. Examples of these techniques are the Trapezoidal rule (TR) or its high-order variants such as the Back-Differentiation Formulas (BDF) [7].

Constructing the MNA formulation is usually done on elements basis, where each element contributes in a prescribed manner to a specific set of entries in the matrices $\mathbf{C}$ or $\mathbf{G}$, or to the vectors $\mathbf{f}(\mathbf{x}(t))$ or $\mathbf{b}(t)$. For example, a capacitor with capacitance of C_1 Farads connected between nodes i and j is represented in the MNA formulation by adding C_1 to the (i, i) and (j, j) entries and subtracting C_1 from the (i, j) and (j, i) entries. Representation of each lumped circuit element in the MNA formulation is best described by the notion of stamp or stencil, where an element stamp defines precisely the contribution that this element leaves on the MNA formulation. A compiled list of stamps for different circuit elements as well as the method by which these stamps have been derived can be found in [7]. Figure 2 shows a sample of stencils used for some lumped passive elements and independent

Element	Circuit Schematic	Stamp
Resistor	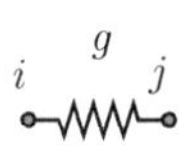	$\begin{array}{c|ccc} & i & & j \\ \hline i & g & \cdots & -g \\ & \vdots & \cdots & \vdots \\ j & -g & \cdots & g \end{array}$
Capacitor	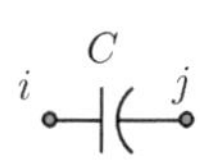	$\begin{array}{c|ccc} & i & & j \\ \hline i & C & \cdots & -C \\ & \vdots & \cdots & \vdots \\ j & -C & \cdots & C \end{array}$
Inductor	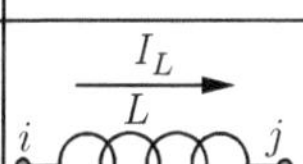	$\begin{array}{c|cccc} & i & & j & \\ \hline i & 0 & \cdots & 0 & 1 \\ & \vdots & \cdots & \vdots & 0 \\ j & 0 & \cdots & 0 & -1 \\ & 1 & 0 & -1 & -sL \end{array}$
Voltage Source	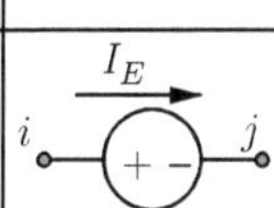	$\begin{array}{c|cccc} & i & & j & \\ \hline i & 0 & \cdots & 0 & 1 \\ & \vdots & \cdots & \vdots & 0 \\ j & 0 & \cdots & 0 & -1 \\ & 1 & 0 & -1 & 0 \end{array} \quad \begin{bmatrix} \\ \\ \\ E \end{bmatrix}$
Current Source	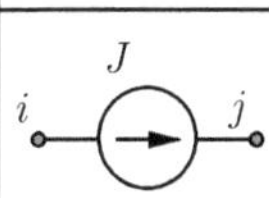	$\begin{array}{c|c} i & J \\ & \vdots \\ j & -J \end{array}$

Fig. 2. Component Stamps in MNA formulation.

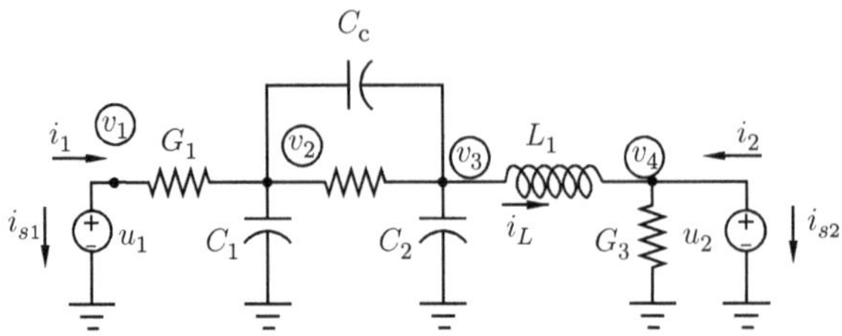

Fig. 3. A linear circuit used as an example to illustrate the MNA approach.

sources. To further illustrate the idea of using elements stamps to construct the MNA matrices, we consider the circuit shown in Figure 3.

Using the MNA formulation, we get

$$\mathbf{G} = \begin{bmatrix} G_1 & -G_1 & 0 & 0 & 0 & 1 & 0 \\ -G_1 & G_1+G_2 & -G_2 & 0 & 0 & 0 & 0 \\ 0 & -G_2 & G_2 & 0 & 1 & 0 & 0 \\ 0 & 0 & 0 & G_3 & -1 & 0 & 1 \\ 0 & 0 & -1 & 1 & & & \\ -1 & 0 & 0 & 0 & & \mathbf{0} & \\ 0 & 0 & 0 & -1 & & & \end{bmatrix} \tag{3}$$

$$\mathbf{C} = \begin{bmatrix} 0 & 0 & 0 & 0 & & & \\ 0 & C_c+C_1 & -C_c & 0 & & \mathbf{0} & \\ 0 & -C_c & C_2+C_c & 0 & & & \\ 0 & 0 & 0 & 0 & & & \\ & & & & L_1 & 0 & 0 \\ & \mathbf{0} & & & 0 & 0 & 0 \\ & & & & 0 & 0 & 0 \end{bmatrix} \tag{4}$$

$$\mathbf{b}(t) = \begin{bmatrix} 0 & 0 \\ 0 & 0 \\ 0 & 0 \\ 0 & 0 \\ 0 & 0 \\ -1 & 0 \\ 0 & -1 \end{bmatrix} \begin{bmatrix} u_1(t) \\ u_2(t) \end{bmatrix}, \quad \mathbf{x}(t) = \begin{bmatrix} v_1(t) \\ v_2(t) \\ v_3(t) \\ v_4(t) \\ i_L(t) \\ i_{S_1}(t) \\ i_{S_2}(t) \end{bmatrix} \tag{5}$$

Notice that this circuit does not have nonlinear elements and therefore the vector mapping $\mathbf{f}(\mathbf{x}(t))$ is not present in this circuit formulation.

2.2 Incorporating High-Speed Interconnects in Circuit Formulation

In order to be included in a general circuit simulator, a high-speed interconnect subcircuit must have a well defined systematic approach that enables such a subcircuit to be included in the MNA time-domain formulation. This section describes briefly the main bottleneck that makes deriving such an approach a challenging task. Figure 4

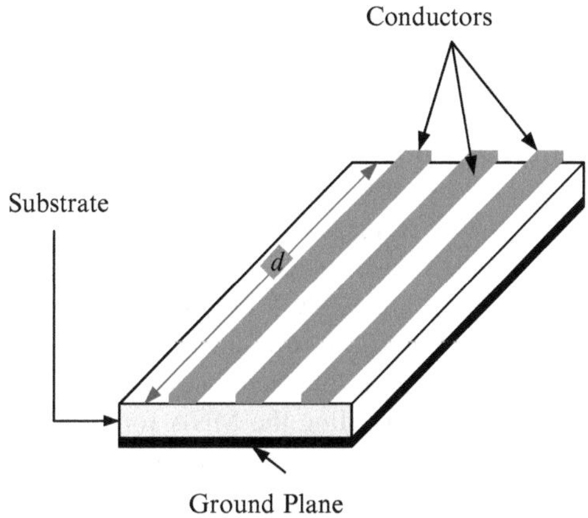

Fig. 4. A physical representation for a high-speed interconnect structure.

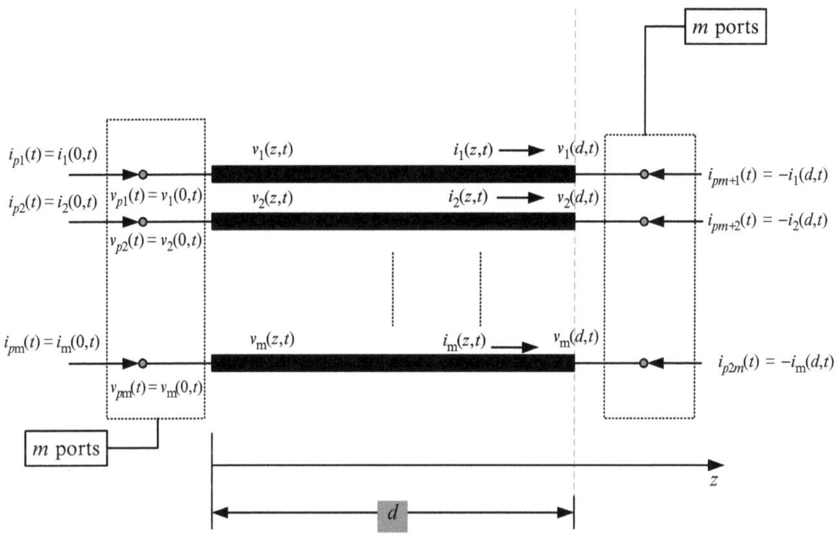

Fig. 5. A physical representation for a high-speed interconnect structure.

shows a physical description of an interconnect structure. A schematic representation of the interconnect as a Multi-conductor Transmission Line (MTL) with distributed voltages and currents is presented in Figure 5. In this figure, notice that an interconnect with m conductors can be viewed as multiport network with $2m$ ports.

Including a high-speed interconnect in the circuit formulation presents a particular difficulty. The main difficulty that arises thereof stems from the fact that interconnects in general are distributed structures (as opposed to lumped)[I] whose

[I] In the current context, the term "distributed structure" typically refers to an electrical device, in which the relation between its electrical parameters (e.g., voltage, current, charge, etc.) involves one or more spatial variables such as the (x, y, z) space coordinates.

underlying physics are naturally derived in the frequency- or the Laplace-domain. Furthermore, this description always comes with a complex dependence on the Laplace variable s which makes obtaining the equivalent time-domain form a cumbersome task. In fact, to obtain a straightforward time-domain representation for interconnects, one would have to use a set of Partial Differential Equations (PDEs), which involves derivatives with respect to the spatial variable, z, that represents the point of observation along the interconnect. Unfortunately, using PDEs to derive a stamp for the interconnect is not a feasible solution.

One possible way to overcome this difficulty is based on direct computation of a time-domain representation from the frequency-domain using convolution [8]. This approach usually requires amending the MNA formulation with an additional convolution-based term. In that case the MNA formulation takes on the following form,

$$\mathbf{C}\frac{\mathrm{d}\mathbf{x}(t)}{\mathrm{d}t} + \mathbf{G}\mathbf{x}(t) + \mathbf{f}(\mathbf{x}(t)) + \sum_{k=1}^{K}\mathbf{D}_k \underbrace{\int_0^t \mathbf{Y}_k(t-\tau)\mathbf{D}_k^T\mathbf{x}(\tau)\mathrm{d}\tau}_{\text{Convolution term}} = \mathbf{b}(t) \tag{6}$$

where

- $\mathbf{Y}_k(t) \in \mathbb{R}^{2m_k \times 2m_k}$ is a matrix whose (i, j) entry represents the current waveform at port i in response to an impulse voltage $\delta(t)$ applied at port j for the k-th interconnect, while maintaining the voltages at all other ports at zero;
- $\mathbf{D}_k \in \mathbb{R}^{N \times 2m_k}$ is a selector mapping, with elements in $\{0, \pm 1\}$ that maps the currents at the ports of the k-th interconnect subnetwork to the variables space of the rest of the circuit;
- m_k is the number of conductors in the k-th interconnect subnetwork, and
- K is the total number of interconnect subnetworks in the circuit.

Nonetheless, numerical solution of DAEs systems with a recursive convolution term, such as (6), is not an easy task. The main goal of this chapter is to present advanced approaches that rely on the concept of Model-Order Reduction (MOR) to represent the high-speed interconnects in general circuit formulation while maintaining the numerically desirable format of (2).

The next subsection shows that high-speed interconnects are best represented by Multi-conductor Transmission Lines (MTL). This presentation will further clarify the hurdles encountered in trying to include high-speed interconnects in circuit simulators.

2.3 High-Speed Interconnects as MTLs

transmission line characteristics are in general described by Telegraphers equations (TE). TE is usually derived by discretizing the line into infinitesimal sections of length Δz and assuming uniform per-unit length (p.u.l.) parameters of resistance (R), inductance (L), conductance (G) and capacitance (C). Each section then includes a resistance $R\Delta z$, inductance $L\Delta z$, conductance $G\Delta z$ and capacitance $C\Delta z$. Using Kirchoffs current and voltage laws, one can write [9]

$$v(z+\Delta z,t)=v(z,t)-R\Delta z i(z,t)-L\Delta z\frac{\partial}{\partial t}i(z,t) \tag{7}$$

or

$$\frac{y(z+\Delta z,t)-v(z,t)}{\Delta z}=-Ri(z,t)-L\frac{\partial}{\partial t}i(z,t) \tag{8}$$

Taking the limit $\Delta z\to 0$, one gets

$$\frac{\partial}{\partial z}v(z,t)=-Ri(z,t)-L\frac{\partial}{\partial t}i(z,t) \tag{9}$$

Similarly, we can obtain the second transmission line equation in the form:

$$\frac{\partial}{\partial z}i(z,t)=-Gv(z,t)-C\frac{\partial}{\partial t}v(z,t) \tag{10}$$

Equations (9) and (10) can be generalized to a Multiconductor Transmission Line (MTL), with m conductors as follows,

$$\frac{\partial}{\partial z}\begin{pmatrix}\mathbf{v}(z,t)\\ \mathbf{i}(z,t)\end{pmatrix}=-\begin{pmatrix}\mathbf{0} & \mathbf{R}\\ \mathbf{G} & \mathbf{0}\end{pmatrix}\begin{pmatrix}\mathbf{i}(z,t)\\ \mathbf{v}(z,t)\end{pmatrix}-\begin{pmatrix}\mathbf{0} & \mathbf{L}\\ \mathbf{C} & \mathbf{0}\end{pmatrix}\frac{\partial}{\partial t}\begin{pmatrix}\mathbf{i}(z,t)\\ \mathbf{v}(z,t)\end{pmatrix} \tag{11}$$

where $\mathbf{i}(z,t)$ and $\mathbf{v}(z,t)\in\mathbb{R}^N$ represent the currents and voltages at a given point along the MTL, and $\mathbf{R},\mathbf{G},\mathbf{L}$ and $\mathbf{C}\in\mathbb{R}^{N\times N}$ are p.u.l. parameter *matrices* of the MTL. In the Laplace- (frequency-) domain, (11) takes the following form,

$$\frac{\partial}{\partial z}\begin{bmatrix}\mathbf{V}(z,s)\\ \mathbf{I}(z,s)\end{bmatrix}=-\begin{bmatrix}\mathbf{0} & \mathbf{R}+s\mathbf{L}\\ \mathbf{G}+s\mathbf{C} & \mathbf{0}\end{bmatrix}\begin{bmatrix}\mathbf{V}(z,s)\\ \mathbf{I}(z,s)\end{bmatrix} \tag{12}$$

The above derivations for MTL's assumes that the currents in the conductors are distributed uniformly throughout the cross section of the conductors. This assumption, however, is valid for low frequencies, and results in having the p.u.l. matrices constant and independent of the frequency variable s. However, as the operating frequency increases, the current distribution gets uneven and it starts getting concentrated more and more near the surface or edges of the conductor. This phenomenon can be categorized as follows: skin, edge and proximity effects [9]. The skin effect causes the current to concentrate in a thin layer near the conductor surface and this reduces the effective cross-section available for signal propagation. This leads to an increase in the resistance to signal propagation and other related effects [10]. The edge effect causes the current to concentrate near the sharp edges of the conductor. The proximity effect causes the current to concentrate in the sections of ground plane that are close to the signal conductor. To account for these effects, modelling based on frequency-dependent p.u.l. parameters may be necessary, and hence matrices $\mathbf{R},\mathbf{L},\mathbf{G}$ and $\mathbf{C}$ will be functions of s. Under this condition, restoring the TE to a set of time-domain PDEs of the form (11) becomes a difficult task.

Nonuniformity of the MTL is also one more aspect that should be taken into account to describe more general interconnect structures. For example, if the spacing between the conductors or their cross sectional areas are not constant, but vary along the spatial variable z, the p.u.l parameters can not be considered as z-independent, and the TE will have to be modified to account for this nonuniformity. The MTL is referred to in this case as Nonuniform MTL (NMTL).

To take into account nonuniformity and the frequency-dependence of the MTL, the TE would have to be written in the following form

$$\frac{\partial}{\partial z}\begin{bmatrix}\mathbf{V}(z,s)\\ \mathbf{I}(z,s)\end{bmatrix} = -\begin{bmatrix}\mathbf{0} & \mathbf{R}(s,z)+s\mathbf{L}(s,z)\\ \mathbf{G}(s,z)+s\mathbf{C}(s,z) & \mathbf{0}\end{bmatrix}\begin{bmatrix}\mathbf{V}(z,s)\\ \mathbf{I}(z,s)\end{bmatrix} \quad (13)$$

The above formulation shows clearly the difficulties involved in trying to deduce a stamp for the MTL (or the NMTL in general), relating terminal voltages $(\mathbf{v}(0,t),\mathbf{v}(d,t))$ and terminal currents $(\mathbf{i}(0,t),\mathbf{i}(d,t))$, which can be incorporated in the MNA time-domain formulation of a general circuit, in a similar manner to other lumped elements. One possible approach that may be used to overcome this difficulty is based on finding a closed-form solution for the TE. This is achievable only in the case of uniform MTL, where a *frequency*-domain stamp relating the terminal voltages and currents is obtained as follows,

$$\begin{bmatrix}\mathbf{V}(d,s)\\ \mathbf{I}(d,s)\end{bmatrix} = \mathrm{e}^{(\mathbf{D}(\mathbf{s})+s\mathbf{E}(s))}\begin{bmatrix}\mathbf{V}(0,s)\\ \mathbf{I}(0,s)\end{bmatrix} \quad (14)$$

with

$$\mathbf{D}(s) = \begin{bmatrix}\mathbf{0} & \mathbf{R}(s)\\ \mathbf{G}(s) & \mathbf{0}\end{bmatrix}, \quad \mathbf{E}(s) = \begin{bmatrix}\mathbf{0} & \mathbf{L}(s)\\ \mathbf{G}(s) & \mathbf{0}\end{bmatrix}, \quad (15)$$

and $\mathrm{e}^{(\mathbf{D}(\mathbf{s})+s\mathbf{E}(s))}$ is the matrix exponential function. Nonetheless, this stamp does not have a direct representation in the time-domain, and its incorporation in the time-domain formulation will necessitate introducing the convolution term. To avoid using convolution in the time-domain, special techniques have been introduced in the literature. Examples of these techniques are the Matrix Rational Approximation (MRA) [11, 12] of the exponential function, and delay-based approaches such as Method of Characteristics (MoC) [13] and the DEPACT algorithm [14]. This approach, however, still does not address the general case of NMTL.

2.4 Time-Domain Macromodeling Based on Discretization

Discretization techniques represent a very straightforward approach to overcome the above difficulties and incorporate high-speed interconnects in circuit simulators. The basic idea in these techniques is to divide the line into segments of length Δz, chosen to be small fraction of the smallest wavelength in the driving signal. If each of these segments is electrically small, and the p.u.l. remains constant in $z_1 < z < z_1 + \Delta z$, then each segment can be replaced by a model with lumped circuit elements. Generally, the lumped structures used to discretize an MTL contain the series elements

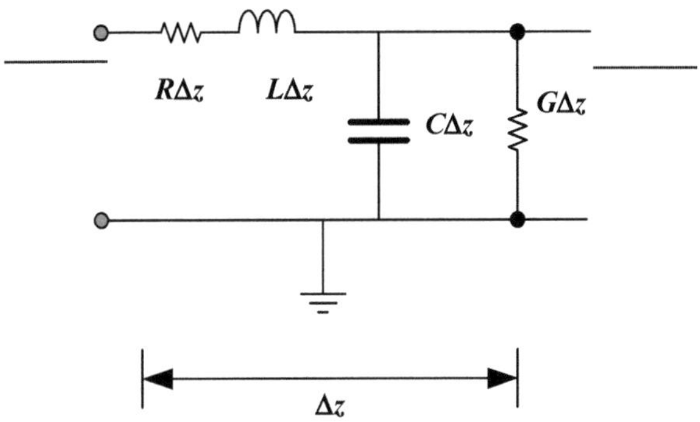

Fig. 6. Modeling a segment of single-conductor transmission line using lumped circuit elements.

$L\Delta z$ and $R\Delta z$, and the shunt elements $G\Delta z$ and $C\Delta z$, where, for simplicity, we assumed that interconnect has only one conductor $m = 1$, in which case the p.u.l. parameter matrices reduce to scalar values. A schematic representation for the model of such a segment is shown in Figure 6.

It is of practical interest to know how many of these segments are required to reasonably approximate the MTL. For illustration consider a lossless line, i.e. $R = G = 0$, with only LC elements which can be viewed as a low pass filter. For a reasonable approximation, this filter must pass at least some multiples of the highest frequency $f_{\max}$ of the propagating signal (say ten times, $f_0 \geq 10 f_{\max}$). In order to relate these parameters, we make use of the 3-dB passband frequency of the LC filter given by [3]

$$f_0 = \frac{1}{\pi\sqrt{LdCd}} = \frac{1}{\pi\tau d} \tag{16}$$

where d is the length of the line and $\tau = \sqrt{LC}$ represents the delay p.u.l. From (1), we have $f_{\max} = 0.35/t_r$, and using (16), we can express the relation $f_0 \geq 10 f_{\max}$ in terms of the delay of the line and the rise time as $1/(\pi\tau d) \geq 10 \times 0.35/tr$ or $t_r \geq 3.5(\pi\tau d) \approx 10\tau d$. In other words, the delay allowed per segment is approximately $t_r/10$. Hence the total number of segments (P) needed to represent the a total delay of τd is given by,

$$P = \tau d/(t_r/10) = 10\tau d/t_r \tag{17}$$

In the case of RLC segments, in addition to satisfying (16), the series resistance of each segment must also be accounted for. As an example, consider a digital signal with rise time $t_r = 0.2$ns propagating on a lossless wire of length 10 cm with a p.u.l. delay of 70.7 ps, which can be represented by a distributed model with p.u.l parameters of $L = 5$nH/cm and $C = 1$pF/cm. If the same transmission line were to be represented by lumped segments, one needs $P \approx 35$ sections.

One of the major drawbacks of the above approach is that it requires a large number of sections especially for circuits with high operating speeds and sharper rise times. This leads to large circuit sizes and the simulation becomes CPU inefficient.

The next section presents an approach for handling this problem based on the concept of Model-Order Reduction (MOR).

3 Time-Domain Macromodeling Through MOR

Interest in MOR techniques for circuit simulation grew out of the difficulties encountered in handling the large lumped circuits arising upon discretization of high-speed interconnects. Application of MOR in circuit simulation can be classified into two main categories: explicit moment matching MOR [15, 16] and projection-based MOR techniques [17, 18]. The latter category enjoys an important advantage over the former one, namely, it can guarantee the passivity of the resulting macromodel. Passivity of the macromodel is a crucial requirement for guaranteeing the numerical stability of the time-domain simulation of the whole circuit since a non-passive macromodel can cause the circuit to become unstable when formulated in the time-domain. We consider here two projection-based approaches. The first approach, discussed in Section 3.1, is used to reduce the large circuits that arise from the discretization. The second approach, discussed in Section 3.2, is a projection technique that reduces the high-speed interconnect without going through the discretization step.

3.1 MOR based on Congruence Transform (MOR-CT)

Given that large circuits resulting from the discretization contain only *linear lumped* elements, the MNA formulation for these circuits is given by,

$$\mathbf{C}_k \frac{\mathrm{d}\mathbf{x}_k(t)}{dt} + \mathbf{G}_k\mathbf{x}_k(t) = \mathbf{B}_k\mathbf{u}(t) \tag{18}$$

where the matrices $\mathbf{C}_k$ and $\mathbf{G}_k$ contain the memory and memoryless elements, respectively, and are similar to the ones introduced earlier in Section 2.1, while the subscript k is only used to emphasize that these matrices contain *only* the lumped circuit elements arising from discretizing the k-th interconnect subnetwork[II]. The left-side of (18), which represents the driving stimulus at the ports of the subnetwork, is described here by a matrix $\mathbf{B}_k \in \mathbb{R}^{N_k \times 2m_k}$ and vector of driving voltages $\mathbf{u}(t) \in \mathbb{R}^{2m_k}$, were N_k is the number of variables in the MNA formulation of the k-th subnetwork, and m_k is its number of conductors. One can show that the currents at the $2m_k$ ports can be given by [19]

$$\mathbf{i}_k(t) = \mathbf{B}_k^T\mathbf{x}_k(t) \tag{19}$$

The matrix of Y-parameters for the network is given by

$$\mathbf{Y}(s) = \mathbf{B}_k^T\left(\mathbf{G}_k + s\mathbf{C}_k\right)^{-1}\mathbf{B}_k \tag{20}$$

[II] As before, $k = 1, 2, \cdots, K$.

The main objective in projection-based MOR algorithm is to construct a reduced model whose Y-parameters matrix approximates the Y-parameters matrix of the original system, but with matrices ($\widehat{\mathbf{C}}_k$ and $\widehat{\mathbf{G}}_k$) that have smaller sizes than that of the original ones ($\mathbf{C}_k$ and $\mathbf{G}_k$). This approximation is carried out by ensuring the first derivatives of the Y-parameters matrix for the reduced model, denoted here by $\widehat{\mathbf{Y}}(s)$, match the same first derivatives of $\mathbf{Y}(s)$ at certain points in the Laplace-domain.

Before explaining how that part is performed, it would beneficial to highlight the physical interpretation of the Y-parameters matrix, which can be stated as follows. The r-th column of the Y-parameters matrix of a $2m_k$-port network represents the currents at the port of this subnetwork, when an impulse voltage is applied at the r-th port, while keeping all other ports voltages at zero value. In other words, obtaining the r-th column of the Y-parameters matrix is carried out by substituting for $\mathbf{U}_k(s)$, the Laplace-domain version of the port voltages $\mathbf{u}_k(t)$, with the r-th column of the identity matrix $\mathbf{e}_r$ and solving the system in (18) in the Laplace-domain,

$$(s\mathbf{C}_k + \mathbf{G}_k)\,\mathbf{X}_{k,r}(s) = \mathbf{B}_k\mathbf{e}_r \tag{21}$$

for $\mathbf{X}_{k,r}(s)$ at all possible values of $s \in \mathbb{C}$. The r-th column is then obtained from the product $\mathbf{B}_k^T\mathbf{X}_{k,r}(s)$. For the purpose of later usage we define the matrix $\boldsymbol{\mathcal{X}}_k$ as a matrix in $\mathbb{C}^{N_k \times 2m_k}$, whose $2m_k$ columns are the solutions of (21) for $r = 1, \cdots, 2m_k$. Hence,

$$(s\mathbf{C}_k + \mathbf{G}_k)\,\boldsymbol{\mathcal{X}}_k(s) = \mathbf{B}_k \tag{22}$$

The following theorem states succinctly how to form the reduced system and its approximative relation to the original system.

Theorem 1. *Consider the system defined as follows,*

$$\widehat{\mathbf{C}}_k\frac{\widehat{\mathbf{x}}(t)}{dt} + \widehat{\mathbf{G}}\widehat{\mathbf{x}}(t) = \widehat{\mathbf{B}}_k\mathbf{u}(t) \tag{23}$$

where

$$\widehat{\mathbf{C}}_k = \boldsymbol{\mathcal{Q}}_k^T\mathbf{C}_k\boldsymbol{\mathcal{Q}}_k, \quad \widehat{\mathbf{G}}_k = \boldsymbol{\mathcal{Q}}_k^T\mathbf{G}_k\boldsymbol{\mathcal{Q}}_k, \quad \widehat{\mathbf{B}}_k = \boldsymbol{\mathcal{Q}}_k^T\mathbf{B}_k \tag{24}$$

and $\boldsymbol{\mathcal{Q}}$ is a matrix in $\mathbb{R}^{N_k \times h}$. Then the Y-parameters matrix for the above system (referred to, henceforth, as the reduced system) which is given by

$$\widehat{\mathbf{Y}}(s) = \widehat{\mathbf{B}}^T\left(s\widehat{\mathbf{C}}_k + \widehat{\mathbf{G}}_k\right)^{-1}\widehat{\mathbf{B}} \tag{25}$$

will have its first, say $q_0, \cdots, q_H$, "matrix-valued" derivatives computed, respectively, at points $s_0, \cdots, s_H$ identical to those derivatives of the original system (20) at the same points if the columns of the matrix $\boldsymbol{\mathcal{Q}}$ are orthonormal and span the subspace of the first $q_0, \cdots, q_H$ matrix-valued derivatives of $\boldsymbol{\mathcal{X}}_k(s)$ at $s = s_0, \cdots, s_H$.

A proof of the above theorem can be found in [19]. Theorem 1 prescribes precisely how to construct the reduced system. According to this theorem, the central step to be performed here is the computation of an orthonormal basis for the subspace of the first q derivatives of $\boldsymbol{\mathcal{X}}_k(s)$ at $s = s_0, \cdots, s_H$. The derivatives of $\boldsymbol{\mathcal{X}}_k(s)$, at say

$s = s_0$ can be computed by first writing $\boldsymbol{\mathcal{X}}_k(s)$ in a Taylor series expansion around $s = s_0$,

$$\boldsymbol{\mathcal{X}}(s) = \sum_{i=0}^{\infty} \boldsymbol{\mathcal{U}}^{(i)} (s - s_0)^i \tag{26}$$

then substituting from (26) in (22) and equating similar powers of s to obtain the following recursive relation for $\boldsymbol{\mathcal{U}}^{(i)}$

$$\mathbf{A}(s_0)\boldsymbol{\mathcal{U}}_k^{(0)} = \mathbf{B}_k \tag{27}$$

$$\mathbf{A}(s_0)\boldsymbol{\mathcal{U}}_k^{(i)} = -\mathbf{C}_k \mathbf{U}_k^{(i-1)} \tag{28}$$

where $\mathbf{A}(s_0) = \mathbf{G}_k + s_0\mathbf{C}_k$. Note here that $\boldsymbol{\mathcal{U}}_k^{(i)}$ represent the i-th moment of $\boldsymbol{\mathcal{X}}(s)$ at $s = s_0$. Given that moments are only a scaled version of the derivatives, i.e.,

$$\boldsymbol{\mathcal{U}}_k^{(i)} = \frac{1}{i!} \left. \frac{\mathrm{d}\boldsymbol{\mathcal{X}}_k(s)}{\mathrm{d}s} \right|_{s=s_0} \tag{29}$$

then an orthonormal basis for the moments will also span the subspace of the derivatives. Once such an orthonormal basis has been made available, then obtaining the reduced-order system is done through the matrix projection operations shown by (24).

There are various ways to construct an orthonormal basis for a set of vectors [20]. Perhaps the simplest method is that based on the Modified-Gram Schmidt (MGS) process. Unfortunately, a direct application of the MGS process here on the matrices $\boldsymbol{\mathcal{U}}^{(i)}, i = 0, \cdots, q$ usually leads to numerical problems that degrade the accuracy of the reduced-order system. A direct application here means that the moments $\boldsymbol{\mathcal{U}}^{(i)}, i = 0, \cdots, q$ are first computed recursively through (27) and (28), then MGS is invoked to compute a spanning orthonormal basis for them. The numerical difficulty that arises from that approach is a result from getting the roundoff error[III], which is inadvertently incurred while computing low order moments, amplified in computing higher-order moments. Such an error causes the high-order moments to be totally inaccurate which makes their contributions in matching higher-order derivatives of the original system practically unnoticeable.

Such a problem is addressed using the block form of the Arnoldi algorithm [20]. The Arnoldi algorithm is essentially an adaptation of the MGS for obtaining an orthonormal basis for a set of vectors related recursively as shown in (28), however, its basic numerical advantage is that it does not require computing the moments explicitly to construct the spanning orthonormal basis.

A test example is given here to illustrate the numerical advantage of using the Arnoldi algorithm over the explicit approach based on the direct application of MGS to $\boldsymbol{\mathcal{U}}^{(i)}$. In this test, a transmission line network of 3 conductors has been approximated using a suitable number of lumped RLC sections. However, instead of using the Arnoldi algorithm to compute the orthonormal basis $\boldsymbol{\mathcal{Q}}$, the moments $\boldsymbol{\mathcal{U}}^{(i)}$ were

[III] This is a very small error that results from using a machine finite-precision storage.

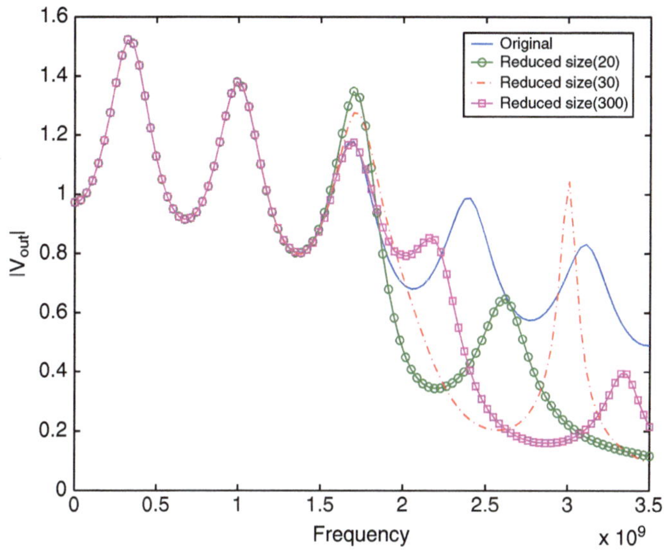

Fig. 7. MOR-CT using with *explicit* moment computation.

first computed *explicitly* and then the Householder algorithm[IV] [20] was used to generate a spanning orthonormal basis for them. Figure 7 shows the performance of the reduced system under these conditions for reduced systems of sizes 20, 30, and 300, respectively. As can be seen from Figure 7, using high-order moments could not help in enhancing the accuracy of the reduced system. On the other hand, Figure 8 shows the results obtained from using the Arnoldi algorithm to construct a reduced system of size 35. It is evident here that implicit usage of the higher order moments via the Arnoldi algorithm has succeeded in capturing the entire frequency range of interest with better accuracy.

Figures 9 and 10 depict a pseudo-code representation for the Arnoldi algorithm. The main algorithm in Figure 9 runs the block version of the Arnoldi process which calls the ORTHOGONALIZE procedure (shown in Figure 10) to perform an orthogonalization using the Modified Gram Schmidt (MGS) process on the input vectors. Note that the mappings used in the pseudocode, $\langle \cdot, \cdot \rangle$ and $\| \cdot \|$, denote the classical inner-product and norm mappings of vectors, i.e.,

$$\langle \mathbf{u}, \mathbf{v} \rangle = \mathbf{u}^T \mathbf{v} \tag{30}$$

$$\|\mathbf{u}\| = \mathbf{u}^T \mathbf{u} \tag{31}$$

It is to be noted that the pseudocode given here constructs an orthonormal basis for the derivatives of $\boldsymbol{\mathcal{X}}_k(s)$ at $s = s_0$, and therefore the reduced system matches the first r derivatives of the original system at $s = s_0$ only. Nonetheless, generalization

[IV] The Householder algorithm was adopted here since it has better orthogonalization properties than the MGS.

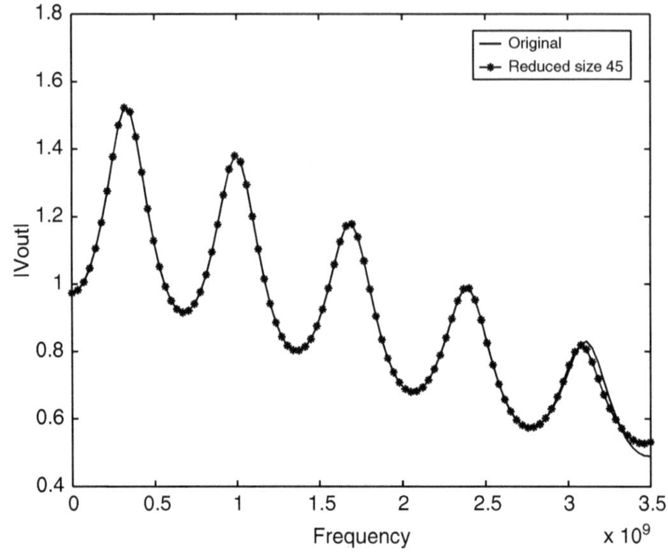

Fig. 8. MOR-CT with *implicit* moments usage matching, where the orthonormal basis $\mathcal{Q}$ was constructed via the Arnoldi algorithm.

Algorithm 6: Computing $\mathcal{Q}_k$

---

input: $\mathbf{G}_k, \mathbf{C}_k, \mathbf{B}_k, h, m_k$
output: $\mathcal{Q}_k$
$r \leftarrow \lceil \frac{h}{2m_k} \rceil$
Solve: $\mathbf{G}_k \widetilde{\mathbf{Q}}_0 = \mathbf{B}_k$
$\mathbf{Q}_0 \leftarrow$ ORTHOGONALIZE($\widetilde{\mathbf{Q}}_0$)
begin
  for $m \leftarrow 1$ **to** $r - 1$ **do**
    Solve: $\mathbf{G}_k \widetilde{\mathbf{Q}}_m = -\mathbf{C}_k \mathbf{Q}_{m-1}$
    for $v = 0 \leftarrow$ **to** $m - 1$ **do**
      $\widetilde{\mathbf{Q}}_m \leftarrow \widetilde{\mathbf{Q}}_m - \mathbf{Q}_v \left\langle \mathbf{Q}_v, \widetilde{\mathbf{Q}}_m \right\rangle$
    $\mathbf{Q}_m \leftarrow$ORTHOGONALIZE($\widetilde{\mathbf{Q}}_m$)
  $\mathcal{Q}_k \leftarrow [\mathbf{Q}_0, \cdots, \mathbf{Q}_{r-1}]$
  Truncate: $\mathcal{Q}_k$ *to* h *columns only.*
  return $\mathcal{Q}_k$
end

Fig. 9. Pseudocode description for the Arnoldi process.

of this pseudocode to make it produce a basis that can be used in matching derivatives at other points should be straightforward.

Procedure : ORTHOGONALIZE

input: A set of vectors $\mathbf{U}_i \in \mathbb{R}^L, i = 0, \cdots, l-1$
output: An orthogonal basis $\mathbf{Q}$ for $\mathbf{U}_i$
$\mathbf{q}_0 \leftarrow \mathbf{U}_0 / \|\mathbf{U}_0\|$
begin
    for $k \leftarrow 1$ **to** $l-1$ **do**
        $\widetilde{\mathbf{q}}_k \leftarrow \mathbf{U}_k$
        for $h \leftarrow 0$ **to** $k-1$ **do**
            $\widetilde{\mathbf{q}}_k \leftarrow \widetilde{\mathbf{q}}_k - \langle \widetilde{\mathbf{q}}_k, \mathbf{q}_h \rangle \mathbf{q}_h$
        $\mathbf{q}_k \leftarrow \widetilde{\mathbf{q}}_k / \|\widetilde{\mathbf{q}}_k\|$
    $\mathbf{Q} \leftarrow [\mathbf{q}_0, \cdots \mathbf{q}_{l-1}]$
    return Q
end

Fig. 10. Pseudocode description for the Arnoldi process (*Cont.*). The ORTHOGONALIZE is given based on the Modified Gram Schmidt process.

3.2 MOR based on Integrated Congruence Transform (MOR-ICT)

The main objective in using the ICT approach is to carry the idea of MOR directly to the interconnect structure in its original distributed form represented by the TE without going through the discretization step. There are a number of advantages to using the ICT approach. The first advantage is that an MTL is a distributed structure by nature, and a discretization into lumped circuit remains only an approximation. Another drawback in the discretization-based MOR-CT, which is handled naturally in MOR-ICT, is the underlying assumption that the p.u.l. parameter matrices are independent of the spatial variable z within each segment. As noted earlier, this assumption does not take into account the case of NMTL in which the p.u.l parameter matrices are continuous functions of z. The idea of using the MOR-ICT was first pioneered in [21] and further developed later in [22] to handle NMTL with an improved numerical accuracy. It was also used in [23] to address the issue of sensitivity analysis. The concept of MOR-ICT was employed in [24] to obtain reduced-order models for linear periodically time-varying systems.

Describing MOR-ICT is best approached through highlighting its main similarities and differences with MOR-CT. In the description of MOR-CT, it was obvious that the pivotal operation in constructing the reduced-order system is the computation of an orthonormal basis for the moments of the internal system states, $\boldsymbol{\mathcal{X}}_k(s)$. In MOR-ICT, a similar operation is also carried out. The only difference in this situation is that the distributed voltages and currents, $\mathbf{v}(z,t)$ and $\mathbf{i}(z,t), 0 \leq z \leq d$ play the role of internal system states, where an orthonormal basis for their derivatives subspace is being sought. To state this idea more precisely, we put (13) in the following form,

$$\mathbf{T}\frac{\partial \mathbf{X}(z,s)}{\partial z} = -\left(\mathbf{N}(z) + s\mathbf{M}(z)\right)\mathbf{X}(z,s) \tag{32}$$

where[V],

$$\mathbf{N}(z) = \begin{bmatrix} \mathbf{R}(z) & \mathbf{0} \\ \mathbf{0} & \mathbf{G}(z) \end{bmatrix}, \qquad \mathbf{M}(z) = \begin{bmatrix} \mathbf{L}(z) & \mathbf{0} \\ \mathbf{0} & \mathbf{C}(z) \end{bmatrix}$$

$$\mathbf{T} = \begin{bmatrix} \mathbf{0} & \mathbf{J}_m \\ \mathbf{J}_m & \mathbf{0} \end{bmatrix}, \qquad \mathbf{X}(z,s) = \begin{bmatrix} \mathbf{I}(z,s) \\ \mathbf{V}(z,s) \end{bmatrix} \tag{33}$$

and $\mathbf{J}_m$ is an $m \times m$ identity matrix. Notice that in the above formulation of TE, the dependence of the p.u.l. matrices on the Laplace variable s was omitted to simplify presenting the basic approach. This issue will be later revisited once the main ideas have been established.

The moments of $\mathbf{X}(z, s)$ are in fact the coefficients of a Taylor series expansion around an arbitrary point, s_0, in the Laplace-domain,

$$\mathbf{X}(z,s) = \sum_{i=0}^{\infty} \mathbf{U}^{(i)}(z)\,(s - s_0)^i \tag{34}$$

It is obvious here that the moments in this case, $\mathbf{U}^{(i)}(z)$, are z-dependent vectors, where $0 \leq z \leq d$. This is to be contrasted with the moments encountered in MOR-CT which were constant vectors. Therefore, MOR-ICT moments can not be considered as elements in an Euclidian space but should be treated as elements belonging to Hilbert space [25]. In fact, this seemingly slight departure from MOR-CT impacts various issues in constructing the reduced-order model. More specifically, we will have to revisit the following three issues.

- Computing the moments,
- computing the orthonormal basis, and
- constructing the reduced-order system.

The remainder of this subsection is dedicated to examining these issues in more depth.

Computation of Moments in Hilbert Space

Computing the moments $\mathbf{U}^{(i)}(z)$ is typically approached by substituting (34) into (32) and equating like powers of s. This leads to the following recursive system of differential equations,

$$\mathbf{T}\frac{\mathrm{d}\mathbf{U}^{(0)}(z)}{\mathrm{d}z} = -\left(\mathbf{N}(z) + s_0\mathbf{M}(z)\right)\mathbf{U}^{(0)}(z) \tag{35}$$

$$\mathbf{T}\frac{\mathrm{d}\mathbf{U}^{(i)}(z)}{\mathrm{d}z} = -\left(\mathbf{N}(z) + s_0\mathbf{M}(z)\right)\mathbf{U}^{(i)}(z) - \mathbf{M}(z)\mathbf{U}^{(i-1)}(z) \tag{36}$$

Thus moments computations can proceed by first solving (35) for $\mathbf{U}^{(0)}(z)$ and then using (36) to solve for high-order moments $\mathbf{U}^{(i)}(z), i > 0$. The above equations,

[V] To simplify the mathematical notations, the subscript k is dropped here.

however, are differential equations and their solutions can be only approached as either an Initial Value Problem (IVP) or a (BVP). In fact, these equations are treated here as BVP. The immediate question that arises here is "What are the appropriate Boundary Conditions (BC) that need to be imposed?". To address this point, it is important here to recall that in MOR-CT, the moments of the internal state variables were computed under excitation conditions corresponding to computing the columns of the Y-parameters matrix. The same rationale is also used in MOR-ICT, where the BC are utilized to enforce these same excitations conditions. To further elaborate on this point, we partition the vectors $\mathbf{U}^{(0)}(z)$ and $\mathbf{U}^{(i)}(z)$ as follows,

$$\mathbf{U}^{(0)}(z)=\begin{bmatrix}\mathbf{U}_I^{(0)}(z)\\ \mathbf{U}_V^{(0)}(z)\end{bmatrix},\quad \mathbf{U}^{(i)}(z)=\begin{bmatrix}\mathbf{U}_I^{(i)}(z)\\ \mathbf{U}_V^{(i)}(z)\end{bmatrix},\tag{37}$$

where $\mathbf{U}_I^{(0)}(z)\in\mathbb{C}^m$ and $\mathbf{U}_V^{(0)}(z)\in\mathbb{C}^m$ (or $\mathbf{U}_I^{(i)}(z)\in\mathbb{C}^m$ and $\mathbf{U}_V^{(i)}(z)\in\mathbb{C}^m$) correspond to the zero-order[VI] (or high-order) moment of the currents and voltages at $s=s_0$, respectively. A BC for $\mathbf{U}^{(0)}(z)$ that enforces an excitation condition corresponding to the first column of the Y-Parameters matrix would have to satisfy,

$$\mathbf{U}_V^{(0)}(0)=\begin{bmatrix}1\\0\\\vdots\\0\end{bmatrix},\quad \mathbf{U}_V^{(0)}(d)=\begin{bmatrix}0\\0\\\vdots\\0\end{bmatrix}\tag{38}$$

where the unity in the first entry in $\mathbf{U}_V^{(0)}(0)$ represents the Laplace-domain version of the impulse voltage applied at port 1, while the rest of the zeros in $\mathbf{U}^{(0)}(0)$ and $\mathbf{U}^{(0)}(d)$ reflect the fact that all other ports are kept at zero voltages (i.e., short-circuited). The BC that enforce excitation conditions corresponding to other columns of the Y-Parameters matrix can be deduced in an analogous manner. Hence, if we use $\boldsymbol{\mathcal{U}}^{(0)}(z)\in\mathbb{C}^{2m\times 2m}$ to denote the set of $2m$ solutions of (37) under $2m$ BC that enforce the excitation conditions corresponding to all columns of the Y-parameters matrix, then $\boldsymbol{\mathcal{U}}^{(0)}(z)$ is the solution of the following BVP,

$$\begin{aligned}\left(\mathbf{T}\tfrac{\mathrm{d}}{\mathrm{d}z}+\mathbf{N}(z)+s_0\mathbf{M}(z)\right)\overbrace{\begin{bmatrix}\boldsymbol{\mathcal{U}}_I^{(0)}(z)\\ \boldsymbol{\mathcal{U}}_V^{(0)}(z)\end{bmatrix}}^{\boldsymbol{\mathcal{U}}^{(0)}(z)}&=\mathbf{0},\\ \boldsymbol{\mathcal{U}}_V^{(0)}(0)&=\underbrace{\begin{bmatrix}\mathbf{J}_m\ \mathbf{0}\end{bmatrix}}_{2m\text{ BC at }z=0}\\ \boldsymbol{\mathcal{U}}_V^{(0)}(d)&=\underbrace{\begin{bmatrix}\mathbf{0}\ \mathbf{J}_m\end{bmatrix}}_{2m\text{ BC at }z=d}\end{aligned}\tag{39}$$

[VI] Note that the zero-order moment of $u(x)$ at $x=x_0$ is actually $u(x_0)$.

The set of BC for higher-order moments may be found by noting that all higher-order derivatives of the Laplace-domain of an impulse function are identical to zero. Therefore, $\boldsymbol{\mathcal{U}}^{(i)}(z)$ result as a solution to the following BVP,

$$\left(\mathbf{T}\frac{\mathrm{d}}{\mathrm{d}z} + \mathbf{N}(z) + s_0\mathbf{M}(z)\right)\overbrace{\begin{bmatrix}\boldsymbol{\mathcal{U}}_I^{(i)}(z)\\ \boldsymbol{\mathcal{U}}_V^{(i)}(z)\end{bmatrix}}^{\boldsymbol{\mathcal{U}}^{(i)}(z)} = -\mathbf{M}(z)\overbrace{\begin{bmatrix}\boldsymbol{\mathcal{U}}_I^{(i-1)}(z)\\ \boldsymbol{\mathcal{U}}_V^{(i-1)}(z)\end{bmatrix}}^{\boldsymbol{\mathcal{U}}^{(i-1)}(z)},$$

$$\boldsymbol{\mathcal{U}}_V^{(i)}(0) = \underbrace{\begin{bmatrix}\mathbf{0}\;\mathbf{0}\end{bmatrix}}_{2m \text{ BC at } z=0} \tag{40}$$

$$\boldsymbol{\mathcal{U}}_V^{(i)}(d) = \underbrace{\begin{bmatrix}\mathbf{0}\;\mathbf{0}\end{bmatrix}}_{2m \text{ BC at } z=d}$$

Note here that the above BVPs do not specify the boundary values for the derivatives of the currents. In fact, that issue as well as prescribing an efficient solution mechanism for (39) and (40) will be expounded later when the focus shifts to addressing NMTL.

Computing the Orthonormal Basis

The problem of computing an orthonormal basis for the first q moments $\boldsymbol{\mathcal{U}}^{(i)}(z)$, $0 \leq i < q$ is easily addressed once those moments are treated as elements of a Hilbert space, which is denoted in the context of this chapter by $\boldsymbol{\mathcal{L}}(0,d)$. A typical method to generate an orthonormal basis for a set of elements in a general linear space is the MGS process [20]. Using MGS to generate the orthonormal basis for $\mathbf{U}^{(i)}(z)$, however, requires adopting the proper definitions for the "inner-product" and "norm" mappings on the space $\boldsymbol{\mathcal{L}}(0,d)$. For this purpose, the following two mappings have been adopted

$$\langle \mathbf{u}(z)|\mathbf{v}(z)\rangle = \int_0^d \mathbf{u}(z)^T\mathbf{v}(z)\mathrm{d}z \tag{41}$$

$$\|\mathbf{u}(z)\| = \int_0^d \mathbf{u}(z)^T\mathbf{u}(z)\mathrm{d}z \tag{42}$$

where $\mathbf{u}(z)$ and $\mathbf{v}(z) \in \boldsymbol{\mathcal{L}}(0,d)$. Once $\boldsymbol{\mathcal{U}}^{(i)}(z), 0 \leq i < q$ have been computed with the proper boundary conditions, the above two mappings can be used to construct a spanning orthonormal basis, denoted here by $\boldsymbol{\mathcal{Q}}(z)$.

Constructing the Reduced Order System

Given that an orthonormal basis for the first q moments $\boldsymbol{\mathcal{U}}^{(i)}(z)$ has been computed as shown above, we can now proceed to construct the reduced-order system. To this end, $\mathbf{X}(z,s)$ in (32) is replaced by another set of variables through the following change of variables

$$\mathbf{X}(z,s) \leftarrow \mathcal{Q}(z)\widehat{\mathbf{X}}(s) \tag{43}$$

where $\widehat{\mathbf{X}}(s) : \mathbb{C} \rightarrow \mathbb{C}^{q\times 2m}$ is a vector-valued mapping, and q is the number of columns in $\mathcal{Q}(z)$.[VII] (32) is then premultiplied by $\mathcal{Q}(z)^T$ and then integrated to yield

$$\left(\widehat{\mathbf{T}} + \widehat{\mathbf{N}}_1 + s\widehat{\mathbf{M}}\right)\widehat{\mathbf{X}}(s) = \mathbf{0} \tag{44}$$

where

$$\begin{aligned}
\widehat{\mathbf{M}} &= \int_0^d \mathcal{Q}(z)^T \mathbf{M}(z)\mathcal{Q}(z)\mathrm{d}z \\
\widehat{\mathbf{N}}_1 &= \int_0^d \mathcal{Q}(z)^T \mathbf{N}(z)\mathcal{Q}(z)\mathrm{d}z \\
\widehat{\mathbf{T}} &= \int_0^d \mathcal{Q}(z)^T \mathbf{T}\frac{\mathrm{d}\mathcal{Q}(z)}{\mathrm{d}z}\mathrm{d}z
\end{aligned} \tag{45}$$

The goal of the following steps is to derive a relation between the terminal voltages $\mathbf{V}(0,s), \mathbf{V}(d,s)$ and terminal currents $\mathbf{I}(0,s), \mathbf{I}(d,s)$. The voltages at the terminals of the TL network are obtained from,

$$\mathbf{V}(s) = \begin{bmatrix}\mathbf{V}(0,s)\\ \mathbf{V}(d,s)\end{bmatrix} = \begin{bmatrix}\mathcal{Q}_V(0,s)\\ \mathcal{Q}_V(d,s)\end{bmatrix}\widehat{\mathbf{X}}(s) \tag{46}$$

where the subscript V above denotes only that portion of the orthonormal basis corresponding to the voltage variables. Defining the matrix $\mathbf{P}$ as

$$\mathbf{P} = \left(\mathcal{Q}_I(d)\right)^T \mathcal{Q}_V(d) - \left(\mathcal{Q}_I(0)\right)^T \mathcal{Q}_V(0) \tag{47}$$

where the subscript i in (47) means that only the portion of $\mathcal{Q}(z)$ corresponding to the current variables is considered. $\mathbf{V}(s)$ can then be obtained using

$$-\mathbf{P}\widehat{\mathbf{X}}(s) = \widehat{\mathbf{b}}\mathbf{V}(s) \tag{48}$$

where

$$\widehat{\mathbf{b}} = \begin{bmatrix}\mathcal{Q}_I(0)\\ -\mathcal{Q}_I(d)\end{bmatrix}^T \tag{49}$$

and the currents at the terminals of the network can be described using $\mathcal{Q}(z)$

$$\mathbf{I}(s) = \begin{bmatrix}\mathcal{Q}_I(0)\\ -\mathcal{Q}_I(d)\end{bmatrix}\widehat{\mathbf{X}}(s) = \widehat{\mathbf{b}}^T\widehat{\mathbf{X}}(s) \tag{50}$$

Let $\hat{\mathbf{N}}_2 = \hat{\mathbf{T}} - \mathbf{P}$, and substitute in (44) using (48),

[VII] In fact, if $\mathcal{Q}(z)$ is constructed from M moments at $s = s_0$ then $q = 2mM$, if s_0 is real and $q = 4mM$ if s_0 is complex.

$$\underbrace{(\widehat{\mathbf{N}}_1 + \widehat{\mathbf{N}}_2}_{\widehat{\mathbf{N}}} + s\widehat{\mathbf{M}})\widehat{\mathbf{X}}(s) = \widehat{\mathbf{b}}\mathbf{V}(s) \tag{51}$$

Hence the relation between the terminal voltages and currents of the TL network in the frequency-domain are obtained from

$$\left(\widehat{\mathbf{N}} + s\widehat{\mathbf{M}}\right)\widehat{\mathbf{X}}(s) = \widehat{\mathbf{b}}\mathbf{V}(s) \tag{52}$$

$$\mathbf{I}(s) = \widehat{\mathbf{b}}^T\widehat{\mathbf{X}}(s) \tag{53}$$

A key advantage of the above constitutive relation for the MTL is that it can be represented in the time-domain as a set of ODEs. Thus it enables general nonlinear circuits with MTL to be formulated in the desirable form of DAEs. Consider for example the circuit shown in Figure 11. Using the reduced-order system constructed as described above, this circuit can be represented in the time-domain using (MNA) formulation as follows,

$$\begin{aligned}\begin{bmatrix} \mathbf{G} & \mathbf{A}\widehat{\mathbf{b}}^T \\ -\widehat{\mathbf{b}}\mathbf{A}^T & \widehat{\mathbf{N}} \end{bmatrix}\begin{bmatrix} v_1 \\ v_2 \\ \widehat{\mathbf{x}}(t) \end{bmatrix} + \begin{bmatrix} \mathbf{C} & \mathbf{0} \\ \mathbf{0} & \widehat{\mathbf{M}} \end{bmatrix}\begin{bmatrix} \frac{dv_1}{dt} \\ \frac{dv_2}{dt} \\ \frac{d\widehat{\mathbf{x}}(t)}{dt} \end{bmatrix} \\ + \begin{bmatrix} 0 \\ I_o\left(\exp\left(v_2/V_T\right) - 1\right) \\ \mathbf{0} \end{bmatrix} = \begin{bmatrix} J(t) \\ 0 \\ \mathbf{0} \end{bmatrix}\end{aligned} \tag{54}$$

where the matrices $\mathbf{G}$ and $\mathbf{C} \in \mathbb{R}^{2\times 2}$ are given by

$$\mathbf{G} = \begin{bmatrix} \frac{1}{R_1} & 0 \\ 0 & \frac{1}{R_2} \end{bmatrix}, \quad \mathbf{C} = \begin{bmatrix} C & 0 \\ 0 & 0 \end{bmatrix} \tag{55}$$

and $\widehat{\mathbf{N}}, \widehat{\mathbf{M}} \in \mathbb{R}^{q\times q}$, $\widehat{\mathbf{b}} \in \mathbb{R}^{q\times 2}$ are obtained using the MOR-ICT as shown above, with q being the size of the reduced system. $\mathbf{A}$ in (54) is an incidence matrix that maps the currents at the terminals of the TL to the nodes of the network. In the case of the above example, $\mathbf{A}$ is a 2×2 identity matrix.

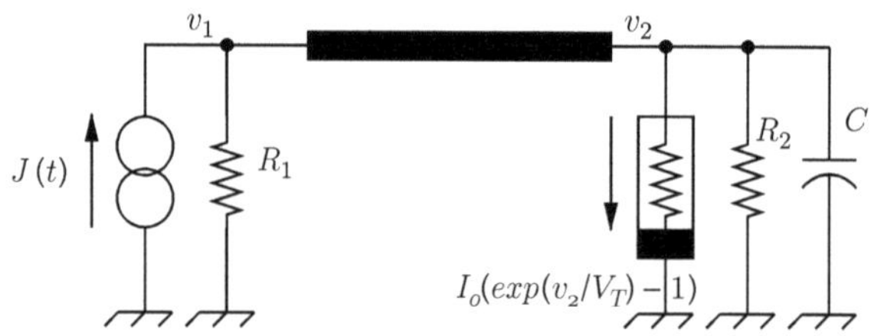

Fig. 11. An example of a circuit containing a TL with nonlinear termination.

Equivalence between the Reduced System and High-Speed Interconnect

Let $\mathbf{Y}_j(s)$ be a column vector defined by

$$\mathbf{Y}_j(s) = \left[I_1(0,s)\; I_2(0,s) \cdots I_m(0,s)\; -I_1(d,s)\; -I_2(d,s)\; \cdots -I_m(d,s) \right]^T \tag{56}$$

where $I_k(0,s)$ and $I_k(d,s)$ are, in respective order, the near- and far-end currents at the k-th conductor when the terminal voltages, $\mathbf{V}(s)$,

$$\mathbf{V}(s) = \left[V_1(0,s)\; V_2(0,s) \cdots V_m(0,s)\; V(d,s)\; V_2(d,s) \cdots V_m(d,s) \right]^T \tag{57}$$

are equal to $\mathbf{e}_j$. Thus the Y-parameters matrix for the distributed high-speed interconnect is given, column-wise, by

$$\mathbf{Y}(s) = \left[\mathbf{Y}_1(s)\; \mathbf{Y}_2(s) \cdots \mathbf{Y}_{2m}(s) \right] \tag{58}$$

The following theorem describes the relation between the reduced system and the high-speed interconnect represented by the TE.

Theorem 2. *Denote by $\boldsymbol{\mathcal{U}}^{(i)}(z, s_h)$ where $0 \le i \le q_h$, $0 \le h \le H$ the moments of the matrix-valued distributed voltages and currents on the high-speed interconnect, computed at $s = s_h$, and let $\boldsymbol{\mathcal{Q}}(z)$ be an orthonormal basis for their Hilbert subspace. Then*

$$\mathbf{Y}^{(m)}(s_h) = \left(\widehat{\mathbf{G}} + s_h \widehat{\mathbf{C}} \right)^{-m} \left(\widehat{\mathbf{G}} \right)^{-1} \widehat{\mathbf{b}}, \quad 0 \le m \le q_h, \quad 0 \le h \le H \tag{59}$$

where $\mathbf{Y}^{(m)}(s_h)$ is the mth moment of $\mathbf{Y}(s)$ computed at $s = s_h$.

Handling Frequency-Dependent p.u.l. Parameter Matrices

The analysis presented here is focused on the treatment of a single-conductor TLs. However, generalization to MTL is straightforward and will be outlined briefly at the end of this section. For the case of a single-conductor TL, the p.u.l parameter matrices reduce to scalar functions of s, and can be modelled using scalar rational function approximations,

$$Z(s) \equiv R(s) + sL(s) \approx \widetilde{Z}(s) = \frac{f_k s^k + \cdots + f_0}{b_{k-1} s^{k-1} + \cdots + b_0} \tag{60}$$

$$Y(s) \equiv G(s) + sC(s) \approx \widetilde{Y}(s) = \frac{g_\ell s^\ell + \cdots + g_0}{c_{\ell-1} s^{\ell-1} + \cdots + c_0} \tag{61}$$

However, to guarantee that the reduced system is passive, the fitting rational function needs to be a positive-real function. Several approaches have been proposed to achieve that through synthesizing the rational function into a network of passive circuit components such as RLCK elements [26–28]. For the case of a single-conductor TL, the synthesized network is a single-port network whose driving point impedance

$\widetilde{Z}(s)$ (or admittance, $\widetilde{Y}(s)$) approximates $Z(s)$ (or $Y(s)$). This fact enables applying circuit formulation techniques, such as the MNA approach described earlier, to write $\widetilde{Z}(s)$ in the following form,

$$\widetilde{Z}(s) = \mathbf{b}_Z^T \left(\mathcal{G}_Z + s\mathcal{C}_Z\right)^{-1} \mathbf{b}_Z \tag{62}$$

where $\mathbf{b}_Z$ is a vector that contains "1" in its first component and "0" otherwise. $\mathcal{G}_Z$ and $\mathcal{C}_Z$ are matrices composed from the "stamps" contributed by the RLCK components. In general, these matrices have the following structures

$$\mathcal{G}_Z = \begin{bmatrix} \mathcal{N}_Z & \mathcal{E}_Z \\ -\mathcal{E}_Z^T & \mathbf{0} \end{bmatrix}, \qquad \mathcal{C}_Z = \begin{bmatrix} \mathcal{Q}_Z & \mathbf{0} \\ \mathbf{0} & \mathcal{P}_Z \end{bmatrix} \tag{63}$$

where $\mathcal{N}_Z$, $\mathcal{Q}_Z$, and $\mathcal{P}_Z$ are symmetric nonnegative definite matrices that contain, in respective order, the stamps of resistive, capacitive and inductive elements. $\mathcal{E}_Z$ serves as a mapping operator with elements $\mathcal{E}_{z,ij} \in \{\pm 1, 0\}$.

An analogous argument is made to show that $\widetilde{Y}(s)$ can be put in a similar form as follows,

$$\widetilde{Y}(s) = \mathbf{b}_Y^T \left(\mathcal{G}_Y + s\mathcal{C}_Y\right)^{-1} \mathbf{b}_Y \tag{64}$$

where the matrices $\mathcal{G}_Y$ and $\mathbf{C}_Y$ have the same structure as that of $\mathcal{G}_Z$ and $\mathcal{C}_Z$, and $\mathbf{b}_Y$ is a vector with a single unity entry and "0" otherwise entries. We illustrate the nature of the constituent matrices using a simple example.

As an example, assume that the s-dependency of $\widetilde{Z}(s)$ and $\widetilde{Y}(s)$ can be captured by two networks as shown in Fig. 12. For that particular case and using the MNA formulation, we can reduce the matrices in (62) and (64) to be,

$$\mathcal{G}_Z = \begin{bmatrix} 1/R_1 & -1/R_1 & 0 & 0 \\ -1/R_1 & 1/R_1 & 0 & 1 \\ 0 & 0 & 1/R_2 & -1 \\ 0 & -1 & 1 & 0 \end{bmatrix}, \quad \mathcal{C}_Z = \begin{bmatrix} 0 & 0 & 0 & 0 \\ 0 & 0 & 0 & 0 \\ 0 & 0 & C_1 & 0 \\ 0 & 0 & 0 & L_1 \end{bmatrix}$$

$$\mathcal{G}_Y = \begin{bmatrix} 1/R_1 & 0 & -1 \\ 0 & 1/R_2 & 0 \\ 1 & 0 & 0 \end{bmatrix}, \quad \mathcal{C}_Y = \begin{bmatrix} C_1 & -C_1 & 0 \\ -C_1 & C_1 + C_2 & 0 \\ 0 & 0 & 0 \end{bmatrix}$$

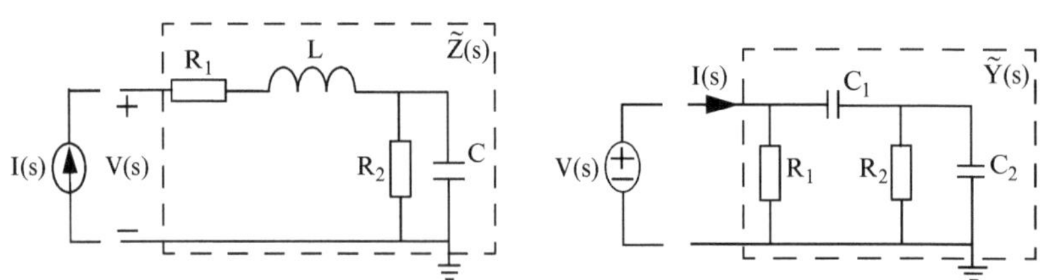

(a) An example of a FD-PUL impedance network

(b) An example of a FD-PUL admittance network

Fig. 12. Illustration of MNA formulation for equivalent $\widetilde{Z}(s)$ and $\widetilde{Y}(s)$ networks.

while the vectors $\mathbf{b}_Z$ and $\mathbf{b}_Y$ would be given by,

$$\mathbf{b}_Z = [1 \quad 0 \quad 0 \quad 0]^T, \qquad \mathbf{b}_Y = [0 \quad 0 \quad 1]^T$$

Thus far we have established that $R(s) + sL(s)$ and $G(s) + sC(s)$ can be approximated through $\widetilde{Z}(s)$ and $\widetilde{Y}(s)$, whose structures are shown in (62) and (64), respectively. The rationale used to represent both $\widetilde{Z}(s)$ and $\widetilde{Y}(s)$ for single conductor TLs can be generalized and carried to the case of multi-conductor TLs. In that case, $\mathbf{b}_Z$ and $\mathbf{b}_Y$ will become rectangular matrices having a number of columns equal to the number of conductors in the TL, with a single identity entry in each column and "0" entry otherwise.

Assuming that the dimensions of the corresponding $\mathcal{G}_Z$ (and $\mathcal{C}_Z$) and $\mathcal{G}_Y$ (and $\mathcal{C}_Y$) are given respectively by n_Z, n_Y, we then proceed to show how to incorporate these approximate network functions into the TEs. To this end, we introduce the set of auxiliary variables $\mathbf{V}_Z(z,s)$ and $\mathbf{I}_Y(z,s)$ which are defined as follows,

$$(\mathcal{G}_Z + s\mathcal{C}_Z)\,\mathbf{V}_Z(z,s) = \mathbf{b}_Z\mathbf{I}(z,s) \tag{65}$$

$$(\mathcal{G}_Y + s\mathcal{C}_Y)\,\mathbf{I}_Y(z,s) = \mathbf{b}_Y\mathbf{V}(z,s) \tag{66}$$

where $\mathbf{I}(z,s)$ and $\mathbf{V}(z,s)$ are, respectively, the currents and voltages at any point z on the line, as introduced earlier in (12). Substituting in the Telegraphers equations in (32) for $\mathbf{R}(s) + s\mathbf{L}(s)$ and $\mathbf{G}(s) + s\mathbf{C}(s)$ using the approximate $\widetilde{\mathbf{Z}}(s)$ and $\widetilde{\mathbf{Y}}(s)$ given, respectively, by (62) and (64) yields the following set of mixed DAEs,

$$\widetilde{\mathbf{T}}\frac{\partial}{\partial z}\widetilde{\mathbf{X}}(z,s) = -\left(\widetilde{\mathbf{N}} + s\widetilde{\mathbf{M}}\right)\widetilde{\mathbf{X}}(z,s) \tag{67}$$

where,

$$\widetilde{\mathbf{T}} = \begin{bmatrix} 0 & \mathbf{J}_m & 0 & 0 \\ \mathbf{J}_m & 0 & 0 & 0 \\ 0 & 0 & 0 & 0 \\ 0 & 0 & 0 & 0 \end{bmatrix}, \widetilde{\mathbf{N}} = \begin{bmatrix} 0 & 0 & \mathbf{b}_Z^T & 0 \\ 0 & 0 & 0 & \mathbf{b}_Y^T \\ -\mathbf{b}_Z & 0 & \mathcal{G}_Z & 0 \\ 0 & -\mathbf{b}_Y & 0 & \mathcal{G}_Y \end{bmatrix}, \widetilde{\mathbf{M}} = \begin{bmatrix} 0 & 0 & 0 & 0 \\ 0 & 0 & 0 & 0 \\ 0 & 0 & \mathcal{C}_Z & 0 \\ 0 & 0 & 0 & \mathcal{C}_Y \end{bmatrix},$$

$$\widetilde{\mathbf{X}}(z,s) = \begin{bmatrix} \mathbf{I}(z,s) \\ \mathbf{V}(z,s) \\ \mathbf{V}_Z(z,s) \\ \mathbf{I}_Y(z,s) \end{bmatrix} \tag{68}$$

It is important to highlight that the new equations relating the currents and voltages in (67) form a mixed set of DAEs. It is easy to see here that the new formulation enables adapting the basic ideas of MOR-ICT developed previously for the frequency independent case to the frequency-dependent p.u.l.

3.3 MOR-ICT for Nonuniform MTL

This subsection sheds more light on further implementation issues that improve the accuracy of MOR-ICT and make it more practically applicable to Nonuniform MTL (NMTL).

Basis Construction without Explicit Moments Computations

Similar to the case of MOR-CT, the approach used in MOR-ICT to construct $\mathcal{Q}(z)$ is of great importance in maintaining robust numerical performance for the reduced system. The approach based on MGS described earlier in Section 3.2, however, requires computing the block moments $\mathcal{U}^{(i)}(z)$ explicitly before constructing their spanning orthonormal basis. As with the case of MOR-CT, this process involves a recursive computation which introduces a mechanism by which small roundoff error in low-order moments gets amplified in computing high-order moments. It would therefore be desirable to have an algorithm similar to the Arnoldi algorithm to generate $\mathcal{Q}(z)$ but without computing the underlying moments explicitly.

Fortunately, this is possible since the block moments $\mathcal{U}^{(i)}(z)$ are related through a linear operator. This may be observed through putting (39) and (40) in the following form,

$$\mathcal{D}\mathcal{U}^{(i)}(z) = \mathbf{0} \tag{69}$$

$$\mathcal{D}\mathcal{U}^{(i)}(z) = -\mathbf{M}(z)\mathcal{U}^{(i-1)}(z) \tag{70}$$

where $\mathcal{D} = \mathbf{T}\frac{\mathrm{d}}{\mathrm{d}z} + \mathbf{N}(z) + s_0\mathbf{M}(z)$. This fact enables adapting the MGS presented earlier to an Arnoldi-like algorithm [22]. Figure 13 describes a general pseudocode representation for the required procedure.

Solution of BVPs

We now turn to the issue of solving the BVPs in (39) and (40), which are needed in Steps 4 and 14, respectively. For conciseness, these BVP are reproduced next,

$$\left(\mathbf{T}\frac{\mathrm{d}}{\mathrm{d}z} + \mathbf{N}(z) + s_0\mathbf{M}(z)\right)\widetilde{\mathcal{U}}^{(0)}(z) = \mathbf{0} \tag{71}$$

subject to BC,

$$\widetilde{\mathcal{U}}_V^{(0)}(0) = \left[\mathbf{J}_m\ \mathbf{0}\right] \tag{72}$$

$$\widetilde{\mathcal{U}}_V^{(0)}(d) = \left[\mathbf{0}\ \mathbf{J}_m\right] \tag{73}$$

and

$$\left(\mathbf{T}\frac{\mathrm{d}}{\mathrm{d}z} + \mathbf{N}(z) + s_0\mathbf{M}(z)\right)\widetilde{\mathcal{U}}^{(i)}(z) = -\mathbf{M}(z)\mathcal{V}_{i-1}(z) \tag{74}$$

subject to BC,

$$\widetilde{\mathcal{U}}_V^{(i)}(0) = \left[\mathbf{0}\ \mathbf{0}\right] \tag{75}$$

$$\widetilde{\mathcal{U}}_V^{(i)}(d) = \left[\mathbf{0}\ \mathbf{0}\right] \tag{76}$$

where we note that both of $\widetilde{\mathcal{U}}^{(0)}(z)$ and $\widetilde{\mathcal{U}}^{(i)}(z)$, are given by

Algorithm 8: Implicit-Basis-Computation $\mathcal{Q}(z)$

Inputs: $\{\mathbf{M}(z), \mathbf{N}(z)\}, \{\lambda_1, \cdots, \lambda_H\}, \{q_1, \cdots, q_H\}$

$\{\{\mathbf{M}(z)$ and $\mathbf{N}(z)\}$ are matrices constructed using p.u.l.matrices by (33).
$\{\lambda_1, \cdots, \lambda_H\}, \{q_1, \cdots, q_H\}$ define frequency points and # of moments at each point.
$\}$

Output: $\mathcal{Q}(z)$

$\mathcal{Q}(z) \leftarrow \phi$ {Initialize to an empty set.} **begin**

- **for** $h \leftarrow 1$ **to** H **do**
  - $\mathcal{R}(z) \leftarrow \phi$ {Initialize to the empty set.} *Solve:* $\left(\mathbf{T}\frac{\mathrm{d}}{\mathrm{d}z} + \mathbf{N}(z) + \lambda_h \mathbf{M}(z)\right)\widetilde{\mathcal{U}}^{(0)}(z) = \mathbf{0}$, BC $\widetilde{\mathcal{U}}_V^{(0)}(0) = [\mathbf{J}_m\ \mathbf{0}]$, $\widetilde{\mathcal{U}}_V^{(0)}(d) = [\mathbf{0}\ \mathbf{J}_m]$
  - **if** λ_h *is real* **then**
    - $\mathcal{R}_0(z) \leftarrow$ HilbertSpaceOrtho($\widetilde{\mathcal{U}}^{(0)}(z)$,*1*)
  - **else**
    - $\mathcal{R}_0^{\Re}(z) \leftarrow$ HilbertSpaceOrtho($\Re\left\{\widetilde{\mathcal{U}}^{(0)}(z)\right\}$,*1*)
    - $\widetilde{\mathcal{R}}_0^{\Im}(z) \leftarrow \Im\left\{\widetilde{\mathcal{U}}^{(0)}(z)\right\} - \mathcal{R}_0^{\Re}(z)\left\langle \mathcal{R}_0^{\Re}(z) | \Im\left\{\widetilde{\mathcal{U}}^{(0)}(z)\right\}\right\rangle$
    - $\mathcal{R}_0^{\Im}(z) \leftarrow$ HilbertSpaceOrtho($\widetilde{\mathcal{R}}_0^{\Im}(z)$,*1*)
    - $\mathcal{R}_0(z) \leftarrow \{\mathcal{R}_0^{\Re}(z) \bigcup \mathcal{R}_0^{\Im}(z)\}$
  - **for** $l = 1 \leftarrow$ **to** $q_h - 1$ **do**
    - **if** (λ_h *is real)* **then** $\left\{\mathcal{V}_{l-1}(z) \leftarrow \mathcal{R}_{l-1}(z)\right\}$ **else** $\left\{\mathcal{V}_{l-1}(z) \leftarrow \mathcal{R}_{l-1}^{\Im}(z)\right\}$
    - *Solve:* $\left(\mathbf{T}\frac{\mathrm{d}}{\mathrm{d}z} + \mathbf{N}(z) + \lambda_h \mathbf{M}(z)\right)\widetilde{\mathcal{U}}^{(l)}(z) = -\mathbf{M}(z)\mathcal{V}_{l-1}(z)$, BC $\widetilde{\mathcal{U}}_V^{(l)}(0) = [\mathbf{0}\ \mathbf{0}]$, $\widetilde{\mathcal{U}}_V^{(l)}(d) = [\mathbf{0}\ \mathbf{0}]$
    - **if** λ_h *is real* **then**
      - $\widetilde{\mathcal{R}}_l(z) \leftarrow \widetilde{\mathcal{U}}^{(l)}(z)$
      - **for** $p = 0 \leftarrow$ **to** $l - 1$ **do**
        - $\widetilde{\mathcal{R}}_l(z) \leftarrow \widetilde{\mathcal{R}}_l(z) - \mathcal{R}_p(z)\left\langle \widetilde{\mathcal{R}}_l(z) | \mathcal{R}_p(z)\right\rangle$;
      - $\mathcal{R}_l(z) \leftarrow$ HilbertSpaceOrtho($\widetilde{\mathcal{R}}_l(z)$,*1*)
    - **else**
      - $\widetilde{\mathcal{R}}_l^{\Re}(z) \leftarrow \Re\left\{\widetilde{\mathcal{U}}^{(l)}(z)\right\}$; $\widetilde{\mathcal{R}}_l^{\Im}(z) \leftarrow \Im\left\{\widetilde{\mathcal{U}}^{(l)}(z)\right\}$
      - **for** $p = 0 \leftarrow$ **to** $l - 1$ **do**
        - $\widetilde{\mathcal{R}}_l^{\Re}(z) \leftarrow \widetilde{\mathcal{R}}_l^{\Re}(z) - \mathcal{R}_p^{\Re}(z)\left\langle \widetilde{\mathcal{R}}_l^{\Re}(z) | \mathcal{R}_p^{\Re}(z)\right\rangle$;
          $\widetilde{\mathcal{R}}_l^{\Re}(z) \leftarrow \widetilde{\mathcal{R}}_l^{\Re}(z) - \mathcal{R}_p^{\Im}(z)\left\langle \widetilde{\mathcal{R}}_l^{\Re}(z) | \mathcal{R}_p^{\Im}(z)\right\rangle$
        - $\widetilde{\mathcal{R}}_l^{\Im}(z) \leftarrow \widetilde{\mathcal{R}}_l^{\Im}(z) - \mathcal{R}_p^{\Re}(z)\left\langle \widetilde{\mathcal{R}}_l^{\Im}(z) | \mathcal{R}_p^{\Re}(z)\right\rangle$;
          $\widetilde{\mathcal{R}}_l^{\Im}(z) \leftarrow \widetilde{\mathcal{R}}_l^{\Im}(z) - \mathcal{R}_p^{\Im}(z)\left\langle \widetilde{\mathcal{R}}_l^{\Im}(z) | \mathcal{R}_p^{\Im}(z)\right\rangle$
      - $\mathcal{R}_l^{\Re}(z) \leftarrow$ HilbertSpaceOrtho($\widetilde{\mathcal{R}}_l^{\Re}(z)$,*1*)
      - $\widetilde{\mathcal{R}}_l^{\Im}(z) \leftarrow \widetilde{\mathcal{R}}_l^{\Im}(z) - \mathcal{R}_l^{\Re}(z)\left\langle \widetilde{\mathcal{R}}_l^{\Im}(z) | \mathcal{R}_l^{\Re}(z)\right\rangle$
      - $\mathcal{R}_l^{\Im}(z) \leftarrow$ HilbertSpaceOrtho($\widetilde{\mathcal{R}}_l^{\Im}(z)$,*1*)
      - $\mathcal{R}_l(z) \leftarrow \{\mathcal{R}_l^{\Re}(z) \bigcup \mathcal{R}_l^{\Im}(z)\}$
    - $\mathcal{R}(z) \leftarrow \{\mathcal{R}(z) \bigcup \mathcal{R}_l(z)\}$
  - $\mathcal{Q}(z) \leftarrow \{\mathcal{Q}(z) \bigcup \mathcal{R}(z)\}$

end

return $\mathcal{Q}(z)$

Fig. 13. Pseudocode for Arnoldi-like algorithm to construct a basis for the Hilbert subspace of $\mathcal{U}^{(i)}(z)$.

$$\widetilde{\mathcal{U}}^{(0)}(z) = \begin{bmatrix} \widetilde{\mathcal{U}}_I^{(0)}(z) \\ \widetilde{\mathcal{U}}_V^{(0)}(z) \end{bmatrix}, \quad \widetilde{\mathcal{U}}^{(i)}(z) = \begin{bmatrix} \widetilde{\mathcal{U}}_I^{(i)}(z) \\ \widetilde{\mathcal{U}}_V^{(i)}(z) \end{bmatrix}, \tag{77}$$

It should be stressed here that the BC shown above are given only partially, where the portions corresponding to the moments of currents variables are left unspecified. The basic idea used to solve (74) hinges upon using the partial BC and the concept of State-Transition Matrix (STM) to compute the corresponding full set of Initial Conditions (IC), and subsequently enabling both problems to be approached as IVP.

Computation of IC. Consider first the BVP (71). This problem is a homogenous ODE, whose solution at any z can be obtained from,

$$\widetilde{\mathcal{U}}^{(0)}(z) = \mathbf{\Phi}(0, z)\widetilde{\mathcal{U}}^{(0)}(0) \tag{78}$$

where $\mathbf{\Phi}(0, z) \in \mathbb{C}^{2m\times 2m}$ is the STM of the system [29]. Partitioning $\mathbf{\Phi}(0, z)$ into four equally-sized block matrices in $\mathbb{C}^{m\times m}$, and substituting $z = d$ in (78) yields,

$$\widetilde{\mathcal{U}}^{(0)}(d) = \begin{bmatrix} \mathbf{\Phi}_{11}(0, d) & \mathbf{\Phi}_{12}(0, d) \\ \mathbf{\Phi}_{21}(0, d) & \mathbf{\Phi}_{22}(0, d) \end{bmatrix} \begin{bmatrix} \widetilde{\mathcal{U}}_I^{(0)}(0) \\ \widetilde{\mathcal{U}}_V^{(0)}(0) \end{bmatrix} \tag{79}$$

Using the BC given in (72) and (73), (79) becomes

$$\begin{bmatrix} \widetilde{\mathcal{U}}_I^{(0)}(d) \\ \mathbf{0}\ \mathbf{J}_m \end{bmatrix} = \begin{bmatrix} \mathbf{\Phi}_{11}(0, d) & \mathbf{\Phi}_{12}(0, d) \\ \mathbf{\Phi}_{21}(0, d) & \mathbf{\Phi}_{22}(0, d) \end{bmatrix} \begin{bmatrix} \widetilde{\mathcal{U}}_I^{(0)}(0) \\ \mathbf{J}_m\ \mathbf{0} \end{bmatrix} \tag{80}$$

Solving for $\widetilde{\mathcal{U}}_I^{(0)}(0)$ results in,

$$\widetilde{\mathcal{U}}_I^{(0)}(0) = (\mathbf{\Phi}_{21}(0, d))^{-1} \left[-\mathbf{\Phi}_{22}(0, d)\ \mathbf{J}_m \right] \tag{81}$$

Computing the IC for the currents portion in BVP (74) is approached in a similar manner. The complete solution for this BVP can be written in terms of STM and the product integral [29],

$$\widetilde{\mathcal{U}}^{(i)}(z) = \mathbf{\Phi}(0, z)\widetilde{\mathcal{U}}^{(i)}(0) + \underbrace{\int_0^z -\mathbf{\Phi}(\mathbf{0}, \boldsymbol{\tau})\mathrm{M}(\tau)\mathcal{V}_{i-1}(\tau)\mathrm{d}\tau}_{\widetilde{\mathcal{W}}_i(z)} \tag{82}$$

Note that the integral term represents a matrix-shaped solution in $\mathbb{C}^{2m\times 2m}$ for (74) under *zero* IC. We denote that term by $\widetilde{\mathcal{W}}_i(z)$, and assume that it is readily available. Substituting $z = d$ in (82) and using the BC specified in (75) and (76) yield the following equation

$$\begin{bmatrix} \widetilde{\mathcal{U}}^{(i)}(d) \\ \mathbf{0}\ \mathbf{0} \end{bmatrix} = \begin{bmatrix} \mathbf{\Phi}_{11}(0, d) & \mathbf{\Phi}_{12}(0, d) \\ \mathbf{\Phi}_{21}(0, d) & \mathbf{\Phi}_{22}(0, d) \end{bmatrix} \begin{bmatrix} \widetilde{\mathcal{U}}_I^{(i)}(0) \\ \mathbf{0}\ \mathbf{0} \end{bmatrix} + \begin{bmatrix} \widetilde{\mathcal{W}}_{i,1}(d) \\ \widetilde{\mathcal{W}}_{i,2}(d) \end{bmatrix} \tag{83}$$

Solving for $\widetilde{\mathcal{U}}_I^{(i)}(0)$ gives,

$$\widetilde{\mathcal{U}}_I^{(i)}(0) = -\left(\mathbf{\Phi}_{21}(0,d)\right)^{-1}\widetilde{\mathcal{W}}_{i,2}(d) \tag{84}$$

Hence, from the IC (81) and (84), and using (78) and (82), one can construct solutions for the BVPs (71) and (74), provided that the zero IC response for (74), $\widetilde{\mathcal{W}}_i(z)$ and the STM, $\mathbf{\Phi}(0,z)$ can be easily computed. While computing these terms can be easily obtained in the case of uniform MTL due to the availability of closed-form solution [12], it becomes significantly challenging for the case of nonuniform MTL since numerical solution becomes the sole option. For example, it is well known that solving thc systcm of ODE in (71) or (74) suffers an ill-conditioning problem [30]. This ill-conditioning results from the existence of a dichotomy between a dominant and dominated solutions. The existence and characterization of such spaces have been reported in [31,32]. The result of having such a dichotomy is that the dominant solution leads to instability during the course of numerical integration resulting in numerical "singularities" [33], especially for long lines. We present next a robust technique to obtain these two terms without running into these difficulties.

Computing zero IC response of (74), $\widetilde{\mathcal{W}}_i(z)$. Notice that the presence of a non-zero IC for $\widetilde{\mathcal{U}}^{(i)}(z)$ can be always taken into account by adding a forcing term on the right side of (74) as follows,

$$\left(\mathbf{T}\frac{\mathrm{d}}{\mathrm{d}z} + \mathbf{N}(z) + s_0\mathbf{M}(z)\right)\widetilde{\mathcal{U}}^{(i)}(z) = -\mathbf{M}(z)\mathcal{V}_{i-1}(z) + \underbrace{\mathcal{U}^{(i)}(0)\,\delta(z)}_{\text{IC}} \tag{85}$$

and treating the resulting problem as a zero IC problem. Since we are interested in the solution $\widetilde{\mathcal{W}}_i(z)$ under zero IC, then the second term on the right side vanishes, and $\widetilde{\mathcal{W}}_i(z)$ results as solution to the following problem

$$\left(\mathbf{T}\frac{\mathrm{d}}{\mathrm{d}z} + \mathbf{N}(z) + s_0\mathbf{M}(z)\right)\widetilde{\mathcal{W}}_i(z) = -\mathbf{M}(z)\mathcal{V}_{i-1}(z) \tag{86}$$

with zero IC. Assuming that both $\mathbf{N}(z)$ and $\mathbf{M}(z)$ are smooth functions of z, then computing $\widetilde{\mathcal{W}}_i(z)$ in (86) proceeds by first representing all the z-dependent quantities in (86) as a summation of $H+1$ Chebyshev polynomials, $T_h(\bar{z})$, of the first kind [34],

$$\begin{aligned}
\mathbf{N}(z) &= \sum_{h=0}^{H}\mathbf{N}_h T_h(\bar{z}) \\
\mathbf{M}(z) &= \sum_{h=0}^{H}\mathbf{M}_h T_h(\bar{z}) \\
\widetilde{\mathcal{W}}_i(z) &= \sum_{h=0}^{H}\Xi_{\widetilde{\mathcal{W}},h} T_h(\bar{z}) \\
\mathcal{V}_{i-1}(z) &= \sum_{h=0}^{H}\Xi_{\mathcal{V},h} T_h(\bar{\bar{z}})
\end{aligned} \tag{87}$$

where $\bar{z} = \frac{2}{d}z - 1$. Substituting from (87) into (86), taking the integral of both sides from -1 to $\bar{z}$, while employing the following relations,

$$\begin{aligned}
T_m(\bar{z})T_n(\bar{z}) &= \frac{1}{2}(T_{m+n}(\bar{z}) + T_{|m-n|}(\bar{z})) \\
\int_{-1}^{\bar{z}} T_0(\xi)\mathrm{d}\xi &= T_0(\bar{z}) + T_1(\bar{z}) \\
\int_{-1}^{\bar{z}} T_1(\xi)\mathrm{d}\xi &= \frac{1}{4}(T_2(\bar{z}) - T_0(\bar{z})) \\
\int_{-1}^{\bar{z}} T_h(\xi)\mathrm{d}\xi &= \frac{1}{2}\left(\frac{T_{h+1}(\bar{z})}{h+1} + \frac{T_{h-1}(\bar{z})}{h-1}\right) + \frac{(-1)^{h+1}}{h^2-1}
\end{aligned}$$

along with the orthonormality property of Chebyshev polynomials yields,

$$\left[\boldsymbol{\mathcal{A}} + \frac{2}{d}\left(\mathbf{J}_{H+1} \otimes \mathbf{T}\right)\right] \widetilde{\boldsymbol{\Lambda}}_i = -\boldsymbol{\mathcal{A}}_M \boldsymbol{\Lambda}_{i-1} \tag{88}$$

where

$$\begin{aligned}
\boldsymbol{\mathcal{A}} &= \frac{1}{2}(\boldsymbol{\mathcal{T}}_1 + \boldsymbol{\mathcal{T}}_2 + \boldsymbol{\mathcal{T}}_3) \\
\boldsymbol{\mathcal{A}}_M &= \frac{1}{2}(\boldsymbol{\mathcal{T}}_{M1} + \boldsymbol{\mathcal{T}}_{M2} + \boldsymbol{\mathcal{T}}_{M3}) \\
\boldsymbol{\Lambda}_{i-1} &= \left[\boldsymbol{\Xi}_{\mathcal{V},H}{}^T \cdots \boldsymbol{\Xi}_{\mathcal{V},0}{}^T\right]^T \\
\widetilde{\boldsymbol{\Lambda}}_i &= \left[\boldsymbol{\Xi}_{\widetilde{\mathcal{W}},H}{}^T \cdots \boldsymbol{\Xi}_{\widetilde{\mathcal{W}},0}{}^T\right]^T \\
\mathbf{J}_{H+1} &\rightarrow (H+1) \times (H+1) \text{ identity matrix.}
\end{aligned} \tag{89}$$

Matrices $\boldsymbol{\mathcal{T}}_1$, $\boldsymbol{\mathcal{T}}_2$ and $\boldsymbol{\mathcal{T}}_3$ are defined in (90)-(92), with $\mathbf{D}_h = \mathbf{N}_h + s_0\mathbf{M}_h$. Matrices $\boldsymbol{\mathcal{T}}_{M1}$, $\boldsymbol{\mathcal{T}}_{M2}$ and $\boldsymbol{\mathcal{T}}_{M3}$ have a similar structures to $\boldsymbol{\mathcal{T}}_1$, $\boldsymbol{\mathcal{T}}_2$ and $\boldsymbol{\mathcal{T}}_3$ except that $\mathbf{D}_h$ is substituted by $\mathbf{M}_h$.

$$\boldsymbol{\mathcal{T}}_1 = \begin{bmatrix}
0 & \cdots & 0 & \frac{1}{2H}D_H & \frac{1}{2H}D_{H-1} \\
0 & \cdots & \frac{1}{2(H-1)}D_H & \frac{1}{2(H-1)}D_{H-1} & \frac{1}{2(H-1)}(D_{H-2} - D_H) \\
\vdots & \ddots & \vdots & \vdots & \vdots \\
\vdots & \ddots & D_2 - \frac{1}{2}D_4 & D_1 - \frac{1}{2}D_3 & D_0 - \frac{1}{2}D_2 \\
D_H & \cdots & \cdots & \begin{pmatrix} D_1 - \frac{1}{4}D_2 + \\ \sum_{h=2}^{H-1} D_h \frac{(-1)^{h+1}}{h^2-1} \end{pmatrix} & \begin{pmatrix} D_0 - \frac{1}{4}D_1 + \\ \sum_{h=2}^{H} D_h \frac{(-1)^{h+1}}{h^2-1} \end{pmatrix}
\end{bmatrix} \tag{90}$$

$$\mathcal{T}_2 = \begin{bmatrix} \frac{1}{2H}D_1 & 0 & \cdots & \cdots & 0 & 0 \\ \frac{1}{2(H-1)}(D_2 - D_0) & \frac{1}{2(H-1)}D1 & 0 & \cdots & 0 & 0 \\ \vdots & \cdots & \cdots & \ddots & \vdots & \vdots \\ \frac{1}{4}(D_{H-1} - D_{H-3}) & \cdots & \frac{1}{4}D_1 & \frac{1}{4}D_0 & 0 & \\ -\frac{1}{2}D_{H-2} & \cdots & -\frac{1}{2}D_1 & -\frac{1}{2}D_0 & 0 & 0 \\ \left(\begin{array}{c} -\frac{1}{4}D_{H-1}+ \\ \sum_{h=2}^{H} D_{H-h}\frac{(-1)^{h+1}}{h^2-1} \end{array}\right) & \cdots & \cdots & -\frac{1}{4}D_1 - \frac{1}{3}D_0 & -\frac{1}{4}D_0 & 0 \end{bmatrix} \tag{91}$$

$$\mathcal{T}_3 = \begin{bmatrix} 0 & \cdots & \cdots & \frac{1}{2H}(D_{H-2} - D_H) & \frac{1}{2H}D_{H-1} \\ 0 & \ddots & \ddots & \cdots & \vdots \\ \vdots & -\frac{1}{4}D_0 & -\frac{1}{4}D_1 & \frac{1}{4}(D_0 - D_2) & \frac{1}{4}(D_1 - D_3) \\ 0 & 0 & -\frac{1}{2}D_0 & -\frac{1}{2}D_1 & D_0 - \frac{1}{2}D_2 \\ D_0 & \cdots & \sum_{h=2}^{H} D_{h-2}\frac{(-1)^{h+1}}{h^2-1} & \left(\begin{array}{c} -\frac{1}{4}D_0+ \\ \sum_{h=2}^{H} D_{h-1}\frac{(-1)^{h+1}}{h^2-1} \end{array}\right) & \left(\begin{array}{c} D_0 - \frac{1}{4}D_1+ \\ \sum_{h=2}^{H} D_h\frac{(-1)^{h+1}}{h^2-1} \end{array}\right) \end{bmatrix} \tag{92}$$

The operator $\otimes$ in (88) denotes the Kronecker product.

Computing the STM, $\boldsymbol{\Phi}(\mathbf{0}, \mathbf{z})$. Given that the STM, $\boldsymbol{\Phi}(0, z)$, is the matrix-valued solution of the homogeneous differential equations, i.e. the solution obtained in the absence of any forcing terms, with an IC given by a unity matrix of the same size [29], then considering (86), $\boldsymbol{\Phi}(0, z)$ becomes the solution of the following system

$$\left(\mathbf{T}\frac{\mathrm{d}}{\mathrm{d}z} + \mathbf{N}(z) + s_0\mathbf{M}(z)\right)\boldsymbol{\Phi}(0, z) = \mathbf{J}_{2m}\delta(z) \tag{93}$$

Expressing $\boldsymbol{\Phi}(0, z)$ in Chebyshev series with $H + 1$ polynomials,

$$\boldsymbol{\Phi}(0, z) = \sum_{h=0}^{H} \boldsymbol{\Xi}_{\boldsymbol{\Phi},h} T_h(\bar{z}) \tag{94}$$

and using the identities given above yields, upon integration, the following linear system in the coefficients $\boldsymbol{\Xi}_{\boldsymbol{\Phi},h}$

$$\left[\boldsymbol{\mathcal{A}} + \frac{2}{d}(\mathbf{J}_{H+1} \otimes \mathbf{T})\right]\widetilde{\boldsymbol{\Lambda}}_{\boldsymbol{\Phi}} = \boldsymbol{\Upsilon}_0 \tag{95}$$

where

$$\widetilde{\boldsymbol{\Lambda}}_{\boldsymbol{\Phi}} = \left[\boldsymbol{\Xi}_{\boldsymbol{\Phi},H}^T\ \boldsymbol{\Xi}_{\boldsymbol{\Phi},H-1}^T \cdots \boldsymbol{\Xi}_{\boldsymbol{\Phi},0}^T\right]^T \tag{96}$$

and

$$\boldsymbol{\Upsilon}_0 = \left[\underbrace{\mathbf{0}\ \mathbf{0} \cdots \mathbf{0}}_{H\text{matrices}} \quad \mathbf{J}_{2m}\right]^T \tag{97}$$

Hence, by solving (88) and (95) one can represent $\widetilde{\mathcal{W}}_i(z)$ and the STM $\mathbf{\Phi}(0, z)$ as a series of Chebyshev polynomials. This result enables expressing the solutions to the BVP (71) and (74) in the same form. More specifically, $\widetilde{\mathcal{U}}^{(0)}(z)$ and $\widetilde{\mathcal{U}}^{(i)}(z)$ can be written as,

$$\widetilde{\mathcal{U}}^{(0)}(z) = \sum_{h=0}^{H} \left(\Xi_{\mathbf{\Phi},h} \widetilde{\mathcal{U}}^{(0)}(0) \right) T_h(\bar{z}) \tag{98}$$

and

$$\widetilde{\mathcal{U}}^{(i)}(z) = \sum_{h=0}^{H} \left(\Xi_{\mathbf{\Phi},h} \widetilde{\mathcal{U}}^{(0)}(0) + \widetilde{\Xi}_{\widetilde{\mathcal{W}},i} \right) T_h(\bar{z}) \tag{99}$$

Modified Inner-Product and Norm Mappings

It is possible to take advantage of the fact that solutions to the BVPs (71) and (74) are represented a Chebyshev series through adopting a slightly different version of the inner-product and norm mappings in (41) and (42), respectively. More specifically, we choose $\langle \cdot | \cdot \rangle_w$ and $\langle \cdot \rangle_w$ to be obtained via a *weighted* integration given by

$$\langle \mathbf{u}(z) | \mathbf{v}(z) \rangle = \int_{-1}^{1} \mathbf{u}(z)^T \mathbf{v}(z) w(\bar{z}) \mathrm{d}\bar{z} \tag{100}$$

$$\langle \mathbf{u}(z) \rangle = \int_{-1}^{1} \mathbf{u}(z)^T \mathbf{u}(z) w(\bar{z}) \mathrm{d}\bar{z} \tag{101}$$

to define the inner-product and norm mappings for any $\mathbf{u}(z)$ and $\mathbf{v}(z) \in \mathcal{L}(0, d)$, respectively, where $\bar{z} = (2/d)z - 1$ and $w(\bar{z}) = 1/\sqrt{1 - \bar{z}^2})$. Under these mappings, if $\mathbf{u}(z)$ and $\mathbf{v}(z)$ happen to be represented by a series of Chebyshev polynomials of the first kind,

$$\mathbf{u}(z) = \sum_{h=0}^{H} \mathbf{U}_h T_h(2z/d - 1) \quad \mathbf{v}(z) = \sum_{h=0}^{H} \mathbf{V}_h T_h(2z/d - 1) \tag{102}$$

then (100) and (101) reduce to

$$\langle \mathbf{u}(z) | \mathbf{v}(z) \rangle = \pi {\mathbf{V}_0}^T \mathbf{U}_0 + \frac{\pi}{2} \sum_{h=1}^{H} {\mathbf{V}_h}^T \mathbf{U}_h \tag{103}$$

$$\langle \mathbf{u}(z) \rangle = \pi {\mathbf{U}_0}^T \mathbf{U}_0 + \frac{\pi}{2} \sum_{h=1}^{H} {\mathbf{U}_h}^T \mathbf{U}_h \tag{104}$$

In other words, the inner-product (norm) mappings defined over the Hilbert space $\mathcal{L}(0, d)$ become equivalent to a series of $H + 1$ classical inner-products defined over Euclidian spaces. This is a result of the orthonormality property of the Chebyshev polynomials [34], which is expressed as follows,

$$\int_{-1}^{1} T_i(\bar{z}) T_j(\bar{z}) \mathrm{d}\bar{z} = \begin{cases} \pi & i = j = 0 \\ \frac{\pi}{2} & i = j \neq 0 \\ 0 & i \neq j \end{cases} \tag{105}$$

4 Numerical Computations

4.1 Example 1

The objective of this example is to show a numerical comparison between using the moments explicitly [21] and implicitly (through the algorithm of Section 3.3) in constructing the orthogonal basis $\mathcal{Q}(z)$ used in obtaining the reduced-order model. For that purpose, the proposed algorithm was implemented to obtain a reduced-order model for a uniform 3-conductor TL network of length $d = 10cm$. Figure 14 shows a graphical comparison between the exact response and the response obtained from the proposed algorithm after running it for $k = 65$ iterations. The exact response for Y_{11} was computed by first using the exponential matrix expression [12] to deduce the H-parameters. The H-parameters were then converted to the Y-parameters. Also shown on the same graph, the response obtained by using 65 moments *explicitly* to construct the basis. It is clear that for the same size of the reduced system, the one based on the proposed algorithm could match up double the frequency range matched through using the moments explicitly.

4.2 Example 2

A 2-conductor tapered TL, as shown in Figure 15, has been considered for this example. The line p.u.l parameters are listed in Table 1. Figure 16 shows the $|Y_{11}|$

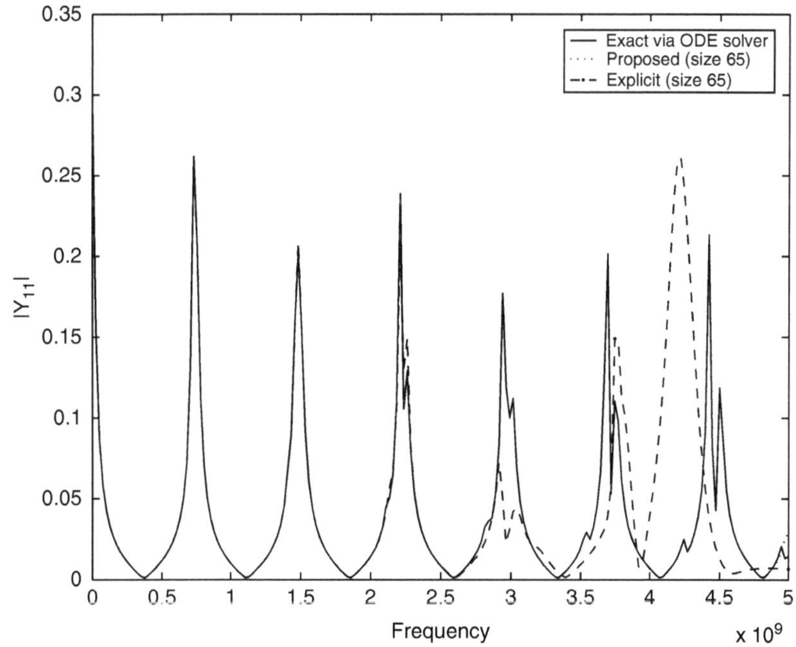

Fig. 14. A comparison for the $|Y_{11}|$ parameter between the proposed algorithm the algorithm in [21] using explicit moments computation.

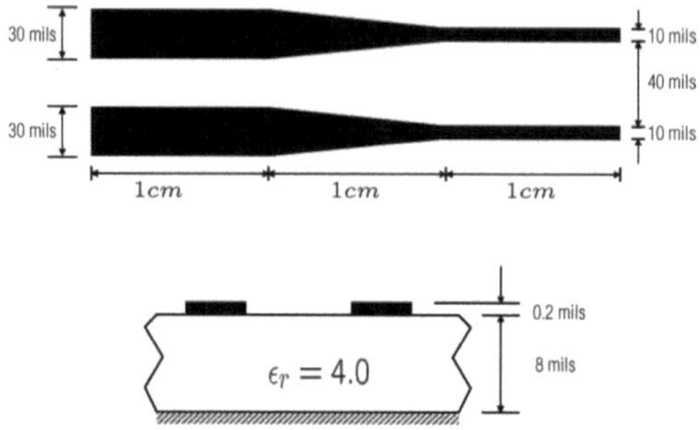

Fig. 15. Prototype chip interconnect.

Table 1. Parameters for the prototype chip interconnect (Example 2)

x(cm)	$C11 = C12$ (pF/cm)	$C12 = C21$ (pF/cm)	$L11 = L22$ (nF/cm)	$L12 = L21$ (nF/cm)
0.0-1.0	1.84	-0.090	1.96	0.23
1.14	1.76	-0.073	2.04	0.22
1.29	1.60	-0.050	2.22	0.20
1.43	1.44	-0.035	2.43	0.19
1.57	1.28	-0.024	2.69	0.17
1.71	1.12	-0.017	3.01	0.17
1.86	0.96	-0.012	3.44	0.16
2.0-3.0	0.880	-0.009	3.71	0.16

parameter as computed by the ICT algorithm and compares it with the exact solution obtained from a Matlab ODE solver used to solve the Telegraphers equations. The TL network was then embedded in a circuit as shown in Figure 17. Figure 18 shows the transient time-domain response due to an input signal of rise/fall time of 0.1 nSec at the active and quiescent lines, respectively. The results obtained are found to be in a very good agreement with those obtained using Fast Fourier Transform FFT.

4.3 Example 3

The MOR-ICT algorithm was used to simulate a nonuniform TL network (with $d = 50$cm) consisting of 3 conductors. First the exact frequency-domain response of the TL network was obtained by solving the Telegraphers ODEs in (32). The results obtained from that solution were then used to deduce the Y-parameters of the TL network. These results were then compared with those obtained via a reduced-order system representation for the TL network. Here the reduced-order system was constructed by first choosing a single expansion point at $s_0 = i2 * \pi(1.75$ GHz$)$, and the proposed orthogonalization procedure described in Section 3.3 was run for a total of 35 iterations to generate the orthogonal basis $\mathcal{Q}(z)$. Figure 19 shows the comparison between the exact approach resulting from solving the Telegraphers ODE

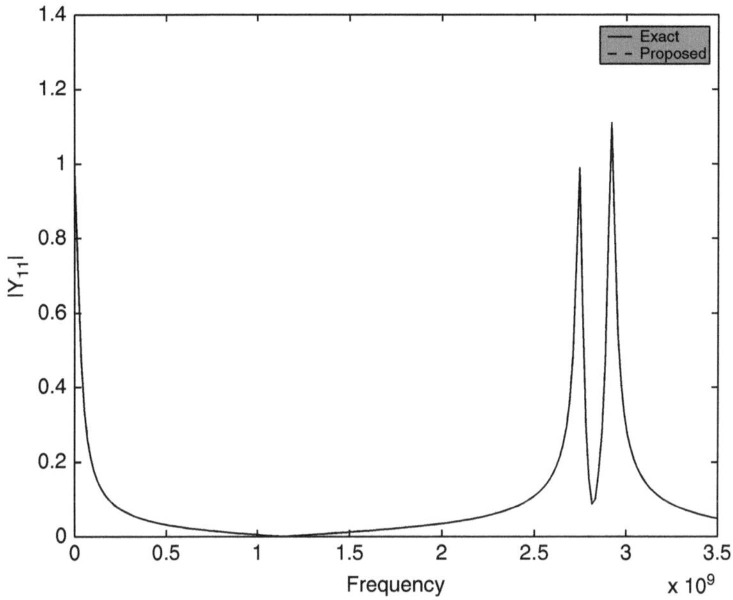

Fig. 16. A comparison between the exact and the proposed approach in computing the $Y_{1,1}$ for the TL network of Figure 15. Note that the two results are almost indistinguishable.

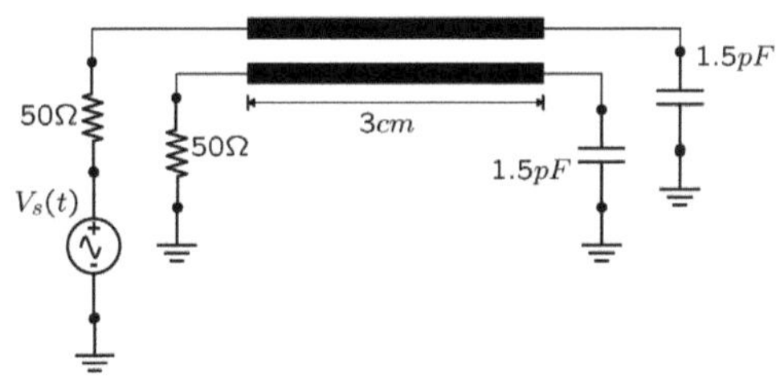

Fig. 17. Schematic representation for the circuit of Example 2.

system directly and the reduced-order system constructed using the orthogonal basis. Figure 20 depicts the relative error in percent.

Figure 21 demonstrates a comparison for the time-domain response due to an input pulse of rise/fall time of 0.1 ns and a pulse-width of 8 ns. The comparison is between a lumped RLC approximation of the line and the proposed algorithm. The lumped RLC model is obtained by dividing the nonuniform line into 50 sections and considering each section as a uniform line. Each uniform section is then represented by a number of lumped RLC sections according to its p.u.l. delay. Simulating the lumped model required 693 sec of CPU time on a Pentium III machine, while simulating the reduced system lasted for only 5.9 Sec, resulting in a speed-up ratio of about 117 times.

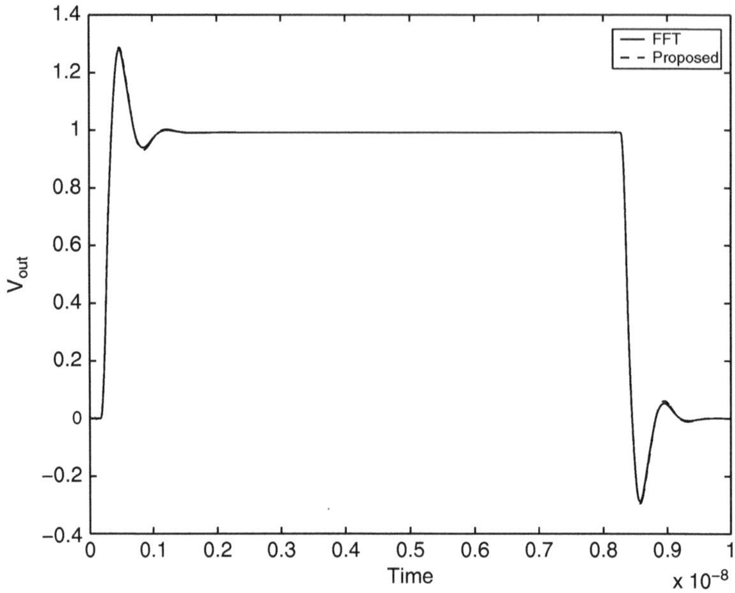

(a) Active line.

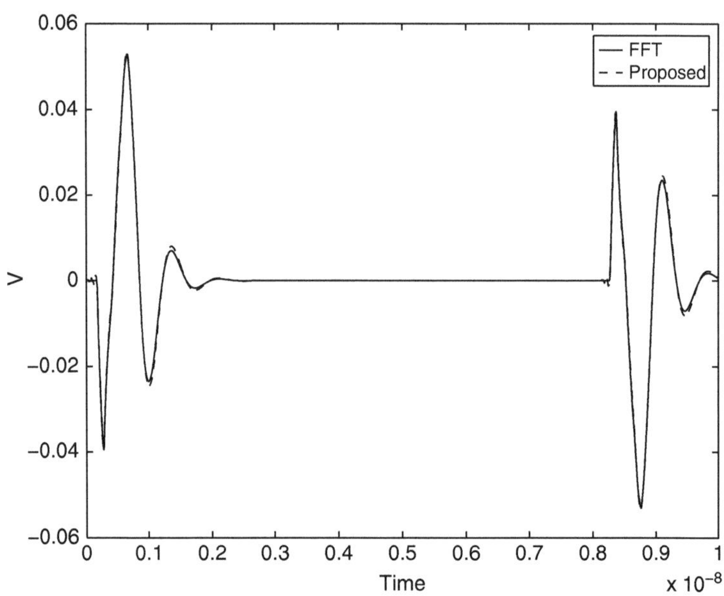

(b) Quiescent line.

Fig. 18. Time-domain Response for Example 2.

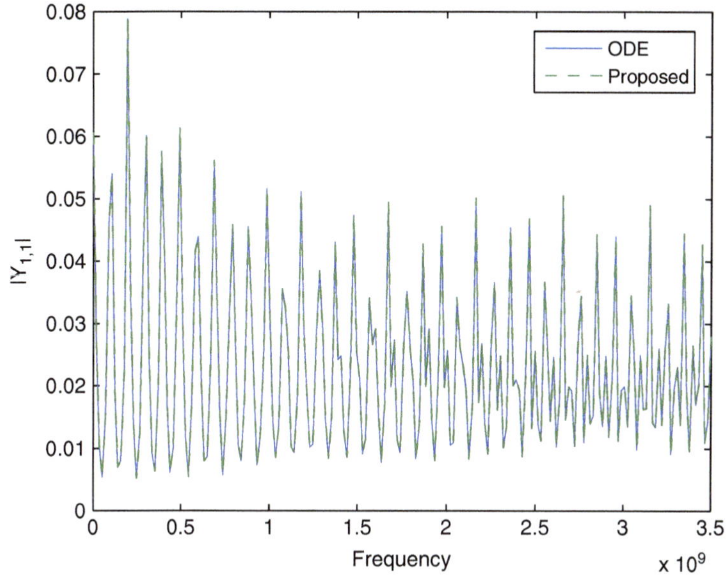

(a) Frequency Response at $Y_{1,1}$

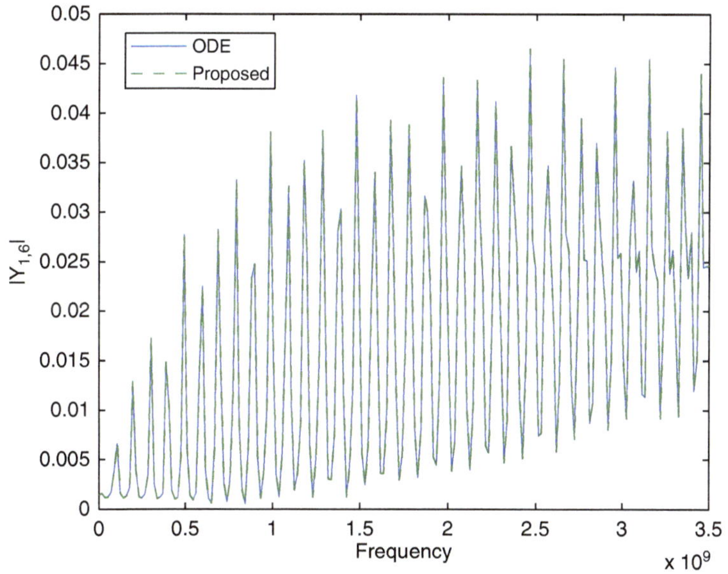

(b) Frequency Response at $Y_{1,6}$

Fig. 19. Numerical results for a 50 cm nonuniform transmission line

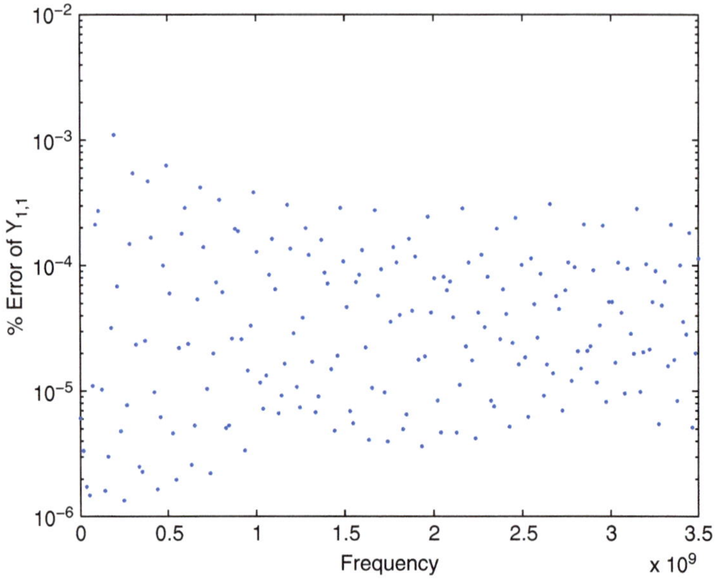

(a) Relative % Error of $Y_{1,1}$

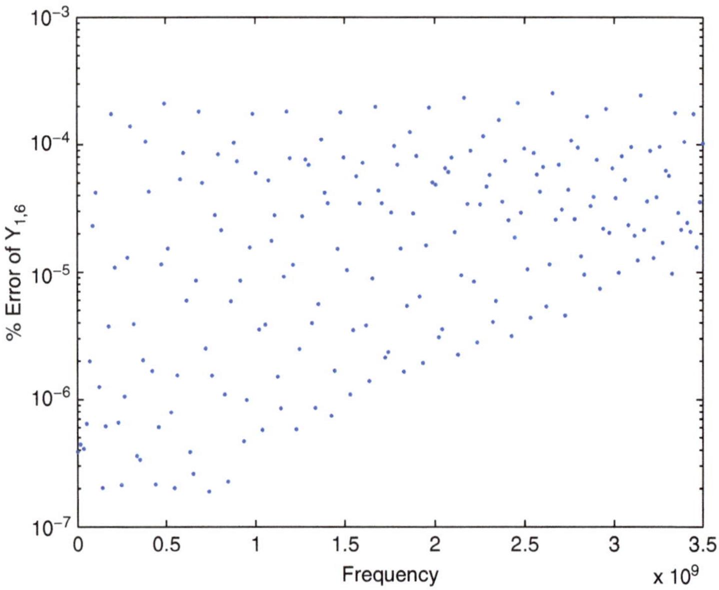

(b) Relative % Error of $Y_{1,6}$

Fig. 20. Relative percentage error in the a sample of the Y-paremeters of a 50 cm nonuniform transmission line

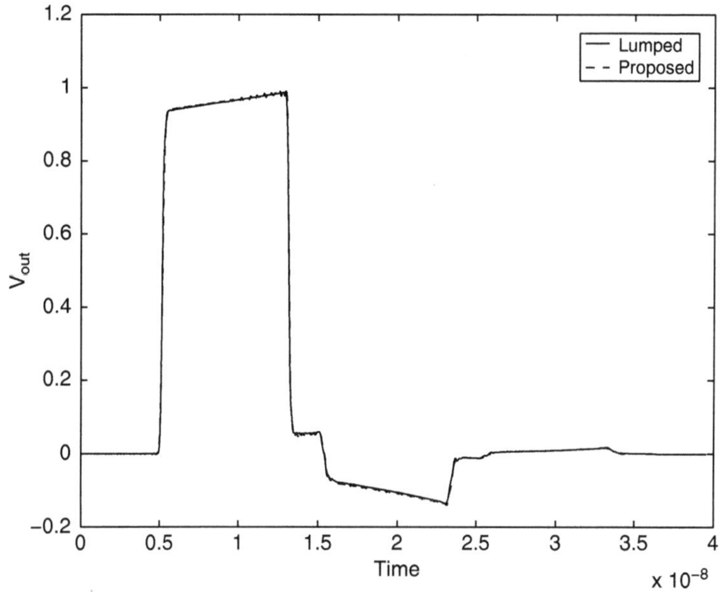

Fig. 21. Numerical results for a 50 cm nonuniform transmission line

5 Conclusion

This chapter described the application of Hilbert space-based MOR in the transient time-domain analysis of distributed MTL networks in circuit simulation environments. The basic idea relies on projecting the Telegraphers equations describing the line on the subspace of the system moments which are defined as the z-dependent scaled derivatives of the voltages and currents along the line. The basic algorithm was further refined numerically through employing the idea of implicit moment computation. Also a special procedure based on using the Chebyshev polynomials was described to gear the spectrum of potential applications more towards the general case of NMTL. Several numerical experiments were presented to validate the described techniques.

References

1. H. B. Bakoglu, *Circuits, interconnects and packaging for VLSI.* Reading, MA: Addison-Wesley, 1990.
2. T. L. Quarles, "The SPICE3 implementation guide," tech. rep., University of California, Berkeley, 1989.
3. H. W. Johnson and M. Graham, *High-Speed Digital Design: A Handbook of Black Magic.* Prentice Hall, 1993.
4. P. K. Poon, *Computer Circuits Electrical Design.* NJ: Prentice-Hall, 1995.

5. A. Cangellaris, S. Pasha, J. L. Prince, and M. Celik, "A new discrete transmission line model for passive model order reduction and macromodeling of high-speed interconnections," *IEEE Trans. Adv. Packag.*, vol. 22, pp. 356–364, Aug. 1999.
6. H. Chung-Wen, A. Ruehli, and P. Brennan, "The modified nodal approach to network analysis," vol. 22, pp. 504–509, Jun. 1975.
7. J. Vlach and K. Singhal, *Computer Methods for Circuit Analysis and Design.* New York: Van Nostrand Reinhold, 1983.
8. N. W. McLachlan, *Laplace Transforms and Their Applications to Differential Equations.* New York, Dover, 1962.
9. C. Paul, *Analysis of Multiconductor Transmission Lines.* New York: Wiley, 1994.
10. A. Deutsch, "Electrical characteristics of interconnections for high performance systems," *Proceedings of IEEE*, vol. 86, pp. 315–357, Feb. 1998.
11. A. Dounavis, R. Achar, and M. Nakhla, "A general class of passive macromodels for lossy multiconductor transmission lines," *IEEE Trans. Microwave Theory Tech.*, vol. 49, pp. 1686–1696, Oct. 2001.
12. A. Dounavis, X. Li, M. S. Nakhla, and R. Achar, "Passive closed-form transmission line model for general purpose circuit simulators," *IEEE Trans. Microwave Theory Tech.*, vol. 47, pp. 2450–2459, Dec. 1999.
13. S. Grivet-Talocia, H.-M. Huang, A. E. Ruehli, F. Canavero, and I. M. Elfadel, "Transient analysis of lossy transmission lines: An efficient approach based on the method of characteristics," *IEEE Trans. Adv. Packag.*, vol. 27, pp. 45–56, Feb. 2004.
14. N. M. Nakhla, A. Dounavis, R. Achar., and M. S. Nakhla, "DEPACT: Delay-extraction-based compact transmission-line macromodeling algorithm.," *IEEE Trans. Adv. Packag.*, vol. 28, pp. 13–23, Feb. 2005.
15. E. Chiprout and M. Nakhla, "Analysis of interconnect networks using complex frequency hopping (CFH)," *IEEE Trans. Computer-Aided Design of Integrated Circ. Sys.*, vol. 14, pp. 186–200, Feb. 1995.
16. L. T. Pillage and R. A. Rohrer, "Asymptotic waveform evaluation for timing analysis," *IEEE Trans. Computer-Aided Design of Integrated Circ. Sys.*, vol. 9, pp. 352–366, Apr. 1990.
17. R. W. Freund and P. Feldmann, "The SyMPVL algorithm and its applications to interconnect simulation," in *Proc. Intl. Conf. on Simulation of Semiconductor Processes and Devices*, (Sep.), pp. 113–116, 1997.
18. P. Feldmann and R. W. Freund, "Efficient linear circuit analysis by Padé approximation via Lanczos process," *IEEE Trans. Computer-Aided Design of Integrated Circ. Sys.*, vol. 14, pp. 639–649, May 1995.
19. A. Odabasioglu, M. Celik, and L. T. Pileggi, "PRIMA: passive reduced-order interconnect macromodeling algorithm," *IEEE Trans. Computer-Aided Design of Integrated Circ. Sys.*, vol. 17, pp. 645–654, Aug. 1998.
20. G. H. Golub and C. F. Van Loan, *Matrix Computations.* Johns Hopkins Press, 1989.
21. Q. Yu, J. M. L. Wang, and E. S. Kuh, "Passive multipoint moment matching model order reduction algorithm of multiport distributed interconnect networks," *IEEE Trans. Circ. Syst. I: Fundamental Theory and Applications*, vol. 46, pp. 140–160, Jan. 1999.
22. E. Gad and M. Nakhla, "Efficient simulation of nonuniform transmission lines using integrated congruence transform," *IEEE Trans. Very Large Scale Integration*, vol. 12, pp. 1307–1320, December 2004.
23. E. Gad and M. Nakhla, "Simulation and sensitivity analysis of nonuniform transmission lines using integrated congruence transform," *IEEE Trans. Adv. Packag.*, vol. 28, pp. 32–44, Feb. 2005.

24. E. Gad and M. Nakhla, "Efficient model reduction of linear periodically time-varying systems via compressed transient system function," *IEEE Trans. Circ. Syst. I: Fundamental Theory and Applications*, vol. 52, pp. 1188–1204, Jun 2005.
25. L. Debnath and P. Mikusinski, *Introduction to Hilbert Spaces with Applications*. Academic Press, Boston, MA, 1990.
26. B. Gustavsen and A. Semlyen, "Enforcing passivity for admittance matrices approximated by rational functions," *IEEE Trans. Power Syst.*, vol. 16, pp. 97–104, Feb. 2001.
27. B. Gustavsen and A. Semlyen, "Rational approximation of frequency responses by vector fitting," *IEEE Trans. Power Delivery*, vol. 14, pp. 1052–1061, July 1999.
28. A. Dounavis, R. Achar, and M. Nakhla, "Efficient passive circuit models for distributed networks with frequency-dependent parameters," *IEEE Trans. Adv. Packag.*, vol. 23, pp. 382–392, Aug. 2000.
29. H. D'Angelo, *Linear Time-Varying Systems: Analysis and Synthesis*. Boston, MA: Allyn and Bacon, 1970.
30. S. L. Manney, M. Nakhla, and Q.-j. Zhang, "Analysis of nonuniform, frequency-dependent high-speed interconnects using numerical inversion of laplace transform.," *IEEE Trans. Computer-Aided Design of Integrated Circ. Sys.*, vol. 13, pp. 1513–1525, Dec. 1994.
31. A. van Der Sluis, "Estimating the solutions of slowly varying recursions.," *SIAM J. Math. Anal.*, vol. 7, no. 5, pp. 662–695, 1976.
32. R. M. M. Mattheij, "Characterization of dominant and dominated solutions of linear recursions.," *Numerische Mathematik*, vol. 35, no. 421-442, 1980.
33. A. M. Ascher, R. M. M. Mattheij, and R. D. Russell, *Numerical solution of boundary value probelms.* Prentice Hall, 1988.
34. T. J. Rivlin, *Chebyshev Polynomials: From Approximation Theory to Algebra and Number Theory*. Pure and applied mathematics (Wiley-Interscience), New York: John Wiley, 1990.

Model Order Reduction for MEMS: Methodology and Computational Environment for Electro-Thermal Models

Tamara Bechtold[1], Evgenii B. Rudnyi[2], and Jan G. Korvink[3]

[1] NXP Semiconductors**, HTC48, 5656AE Eindhoven, The Netherlands
bechtold@nxp.com
[2] CAD-FEM GmbH Marktplatz 2,85567 Grafing, Germany
erudnyi@cadfem.de
[3] IMTEK, University of Freiburg, Georges-Koehler-Allee 102, D-79110, Germany
korvink@imtek.de

We present a methodology and computational environment for applying mathematical model order reduction (MOR) to electro-thermal MEMS[I]. MOR can successfully create dynamic compact thermal models of MEMS devices. It is currently possible to use software tool "MOR for ANSYS" (pronounced "more for ANSYS") to automatically create reduced order thermal models directly from ANSYS models with more than 500 000 degrees of freedom. Model order reduction is automatic and based on the implicit Pad approximation of the transfer function via the Arnoldi algorithm. After model reduction, one can visualize simulation results of the reduced model in Mathematica and can call the SLICOT library via the Mathlink interface in order to obtain mathematically optimal reduced models. Reduced models are easily convertible into hardware description language (HDL) form, and can be directly used for system-level simulation.

1 Introduction

The modeling of electro-thermal processes, e.g., Joule heating, becomes increasingly important during microsystems development [1]. With the decreasing size and growing complexity of micro-electronic systems, the power dissipation of integrated circuits has become a critical concern. The thermal influence upon the device caused by each transistor's self-heating and the thermal interaction with tightly placed neighboring devices cannot be neglected because excessive temperatures may cause

** The results presented here were performed at IMTEK, Freiburg

[I] MEMS traditionally stays for micro-electromechanical systems, although the term is increasingly being used even if different functionalities are employed models

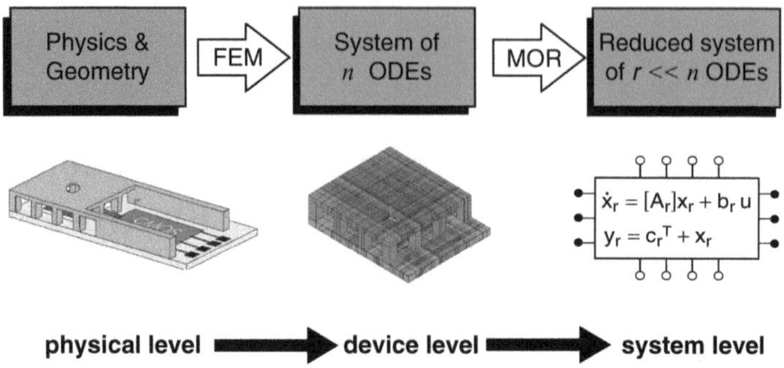

Fig. 1. Motivation for model order reduction.

the malfunction or even destruction of the device. Whereas Joule heating in microelectronics is a "parasitic" effect, some other devices like microsensors and microactuators use it (directly or indirectly) as a functioning principle [3]. In both cases, the engineer's task is to predict the temperature distribution for the given electrical input and the impact of the temperature on the device electronics in turn, i.e. to run an electro-thermal simulation. To go a step further, in each sequence of joint electro-thermal simulation the temperature field is computed on a discrete grid whose size easily exceeds 100 000 degrees of freedom (DOF), i.e. ordinary differential equations. Even though modern computers are able to handle this size of engineering problems, system-level simulation would become prohibitive if the full models were directly used. Hence, an efficient computational technique is needed. An alternative to "classical compact modeling", which is based on parametrization of equivalent thermal networks is mathematical model order reduction (MOR), which is formal, robust and can be made fully automated [4–7]. It is based on the formal conversion of the physical model, that is, governing partial differential equation to a low-dimensional ordinary differential equation (ODE) system. The intermediate level is a device level, which is a high dimensional ODE system (see Fig. 1). The first conversion of the physical to the device model is done via the finite element discretization. A second step, that is a conversion from the device to the system level simulation, can be efficiently performed via model order reduction.

2 Applications

Thermal simulation is an important issue in many engineering areas. In our work, we have focused on several hotplate-based MEMS devices (see Fig. 2). The pyrotechnic microthruster is based on the integration of solid fuel with a silicon micromachined structure [8]. The thermally tunable optical filter is a Fabry-Perot interferometer fabricated as a free-standing membrane [9]. The microhotplate gas sensor is supported by glass pillars emanating from a glass cap above the silicon wafer, which assures robust design and thermal isolation of the membrane from the surrounding wafer [10].

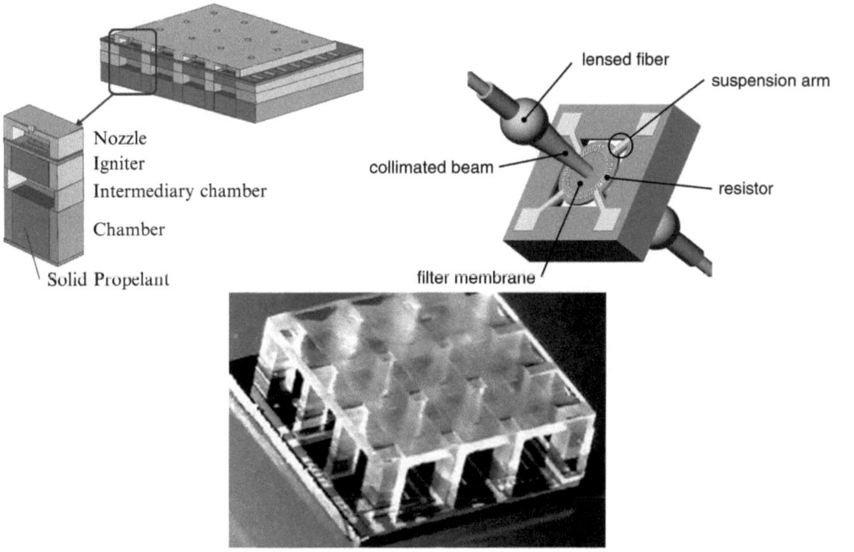

Fig. 2. MEMS case studies: microthruster (top left), gas sensor (top right), optical filter (bottom).

The heat transfer within each hotplate is described through the following equations:

$$\nabla \bullet (\kappa \nabla T) + Q - \rho C_p \frac{\partial T}{\partial t} = 0, Q = \mathbf{j}^2 R \tag{1}$$

where $\kappa(r)$ is the thermal conductivity in W/mK at the position r, $C_p(r)$ is the specific heat capacity (a material property that indicates the amount of energy a body stores for each degree increase in temperature, on a per unit mass basis) in J/kgK, $\rho(r,t)$ is the mass density in kg/m^3, is the temperature distribution and $Q(r,t)$ is the heat generation rate per unit volume in W/m^3. The engineer's task is to solve 1 for a thermal problem in question. Assuming that the heat generation is uniformly distributed within the heater, and that the system matrices are temperature independent around the working point, the finite element based spatial discretization of (1) leads to a large linear ODE system of the form:

$$\begin{aligned} C \cdot \dot{\mathbf{T}} + K \cdot \mathbf{T} &= F I^2(t) R(\mathbf{T}) \\ \mathbf{y} &= L^T \cdot \mathbf{T} \end{aligned} \tag{2}$$

where C and K are the global heat capacity and heat conductivity matrices, F is the load vector (matrix) and L is the output vector (matrix).

3 Model Order Reduction: Method and Numerical Results

As the number of equations in (2) is usually too high for a system-level simulation, MOR is performed and a new, reduced system of equations (of the same form as

(2)) is used to generate a system-level model. We use an Arnoldi reduction algorithm (with explicit projection of both system matrices as in [11]), which can be viewed as a projection, from the full space to the reduced Krylov-subspace:

$$K_r(A, b) := span(b, Ab, A^2b, \ldots, A^{r-1}b) \tag{3}$$

where $A = -K^{-1}C$ and $b = -K^{-1}F$. This projection is based on the transformation of the state vector $\mathbf{T}$ to the vector of generalized coordinates z, subjected to some small error ϵ:

$$\mathbf{T} = V \cdot \mathbf{z} + \epsilon \tag{4}$$

and the subsequent left hand side multiplication of (2) with V^T:

$$\begin{aligned} V^T C V \cdot \dot{\mathbf{z}} + V^T K V \cdot \mathbf{z} &= V^T F I^2(t) R(\mathbf{T}) \\ \mathbf{y} &= L^T V \cdot \mathbf{z} \end{aligned} \tag{5}$$

The transformation matrix $V \in \mathbb{R}^{n \times r}$, where $r << n$ are the dimensions of the reduced and the full system, respectively, is a direct output of the Arnoldi algorithm. The property of the Krylov-subspace (3) is such that the transfer function of (2) and the transfer function of (5) match the first r coefficients of their Taylor series around an arbitrarily chosen frequency. For our case studies the expansion frequency was set to zero in order to preserve the steady-state. The Taylor coefficients around zero frequency are called moments. The fact that neither input term nor the output matrix L take part in the order reduction, brings along two important properties of the Arnoldi algorithm, which distinguish it from other MOR methods: the approximation of the full output and the reduction of systems with the temperature dependent input power, both described in [12]. Using Arnoldi-based MOR we have reduced a gas sensor model with 73 955 ODEs to 10 ODEs and have implemented the reduced model in MAST, the hardware description language of the behavioral simulator SABER. Fig. 3 and Fig. 4 show the schematic structure of the implemented HDL model for the gas sensor and numerical simulation results of the full-scale model integrated in ANSYS, the reduced order model integrated in *Mathematica* and the MAST model integrated in SABER.

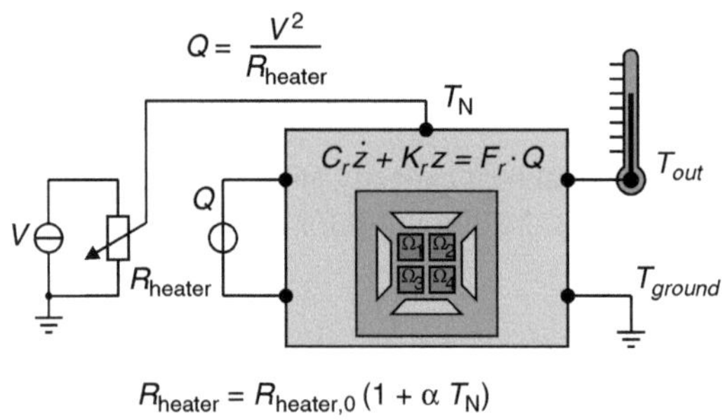

Fig. 3. HDL model structure containing back coupled temperature dependent heater.

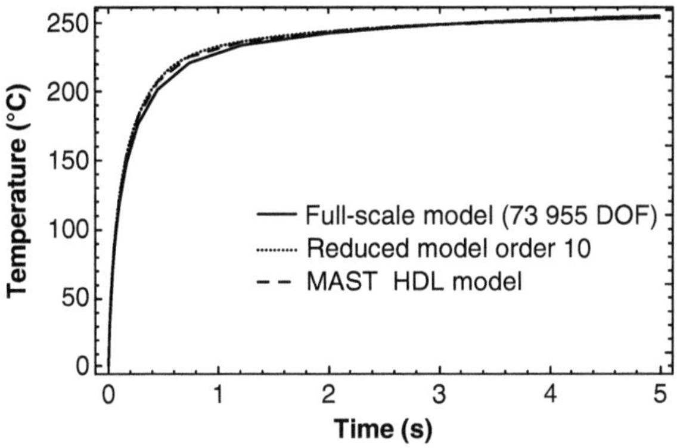

Fig. 4. Solution of the full-scale system (73 955 DOF) and of the reduced order 10 system in a central hotplate node of gas sensor device.

Our results show, that although the Arnoldi algorithm is presently limited to linear systems (see section 8 for the state-of-the-art on nonlinear MOR) and has no global error estimate (see section 5) it can already be used for highly effective modeling and simulation of electro-thermal MEMS devices. It works extremely well for heat transfer in 2D and 3D, which is consistent with observations of other researchers [13, 15].

4 Computational Environment

In order to use model reduction in practical work, it is necessary to develop a suitable computational environment. At present there is a number of tools that implement model reduction methods developed in control theory, e.g. MATLAB Control System Toolbox, MATCONTROL, MATRIXX, SLICOT etc. A good description of these packages can be found in [4]. Unfortunately, the computational complexity of theses methods is of $O(n^3)$ and their use is limited to at most a few thousand degrees of freedom. The software tool ROM Workbench, which has been developed within the EU project CODESTAR (see *http://www.imek.be/codestar*) is a MATLAB implementation of different MOR methods including Krylov and Laguerre based methods, asymptotic waveform evaluation and vector fitting. These methods are of an iterative nature and they can be used for higher dimensional systems. Still, the MATLAB implementation puts the limits to the dimensionality of the problem to solve. Another problem is to obtain system matrices for a particular engineering task. This turns out to be difficult, as the access to the system matrices is not considered important by software developers of commercial products.

Our solution was to employ the iterative Krylov-subspace method based on the Arnoldi process [11] and rely on C++ implementation to allow us to treat high dimensional problems. We have decided to target an ANSYS audience as many MEMS engineers use this commercial finite element software in practice. The result was a software tool "MOR for ANSYS" [16].

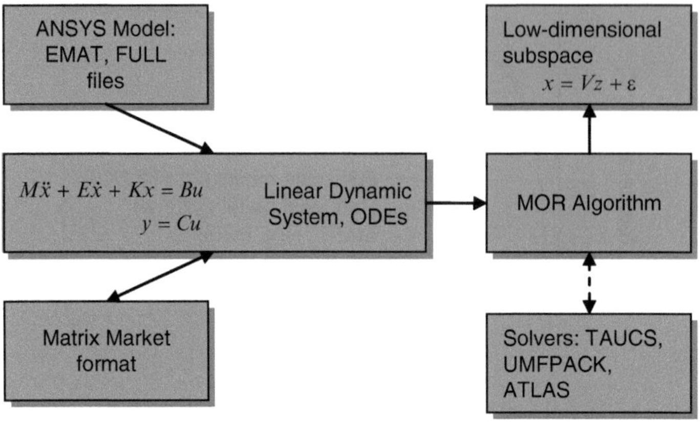

Fig. 5. Schematics of "MOR for ANSYS" environment.

The block-scheme of "MOR for ANSYS" is presented in Fig. 5. It is a command line tool that performs model reduction of ANSYS models directly, while allowing different options to construct a reduced model. Additionally, "MOR for ANSYS" can read the original dynamic system in the Matrix Market format specified in [15] that allows us to use it with other software packages than ANSYS as well. Specifying Matrix Market format to be the external format for the dynamic systems (2) and (5) offers large flexibility because the matrices in the Matrix Market format are written as an ASCII file and can be freely exchanged between different computer platforms. Furthermore, the software to read and write matrices in the Matrix Market format in C, Fortran and MATLAB is freely available at the Matrix Market site (*http://math.nist.gov/MatrixMarket/*). *Mathematica* supports it natively. "MOR for ANSYS" uses direct solvers TAUCS and UMPFACK based on the optimized BLAS made by ATLAS for efficient linear algebra. In our experience, on a computer with 4 Gb of RAM these solvers allow us to perform model reduction for finite element models up to 500 000 degrees of freedom.

The source code in C++ contains about 4000 lines (the half of it is for reading ANSYS binary files file.emat and file.full) and is available at *http://www.imtek.uni-freiburg.de/simulation/"MOR for ANSYS"/* under the GNU Public license.

As already mentioned, MATLAB can read matrices in the Matrix Market format and hence, one can read the reduced model into MATLAB/Simulink and postprocess it further. However, we have also developed a set of *Mathematica* functions within a package called Post4MOR (follow the link from "MOR for ANSYS" site), which enables the user to simulate the reduced model, visualize the results etc. Additionally, we have written a Mathlink interface to the SLICOT library in order to be able to use methods like the balanced truncation approximation for either direct reduction of smaller models or for consecutive MOR of the models already reduced with "MOR for ANSYS".

Our software environment allows an engineer to use model reduction for electro-thermal MEMS models as follows. After an ANSYS model is ready (the user must supply the meshed geometry), the dimension of the reduced system should be chosen in order to run "MOR for ANSYS". As already mentioned, afterwards it is possible to reduce the dimension of the reduced model even further by using the SLICOT library.

A tutorial to make model reduction for a thermal model of described microthruster device is available at the "MOR for ANSYS" site. It describes the MOR procedure step-by-step starting with an ANSYS model, performing reduction with "MOR for ANSYS" and then using *Mathematica* functions for postprocessing of the reduced model.

5 Error Estimation

As already mentioned, in order to apply Arnoldi-based model order reduction, the MEMS designer has to provide a discretized model (e.g. a finite element (FE) model) of the device and to specify which frequency band should be well approximated by the compact model. This is done by choosing one or more expansion points in the frequency domain. The next important step is to specify the desired order of the target reduced system. A key question is: which order of the reduced system do we need to select in order to achieve a desired accuracy. A reduced model is an approximation of the original large-scale model. Hence, the difference between the two can be characterized by some error norm. In order to automate the MOR process completely, one should be able to estimate this error as a function of the reduced model.s dimension. The automatic procedure from device-level to system-level modeling is schematically shown in Fig. 6.

Based on our numerical results, we propose three heuristic error indicators for the iterative model order reduction of electro-thermal MEMS models via the Arnoldi algorithm [12]. We first suggest a convergence criterion between two successive reduced models of order r and $r + 1$. We further propose to approximate a global

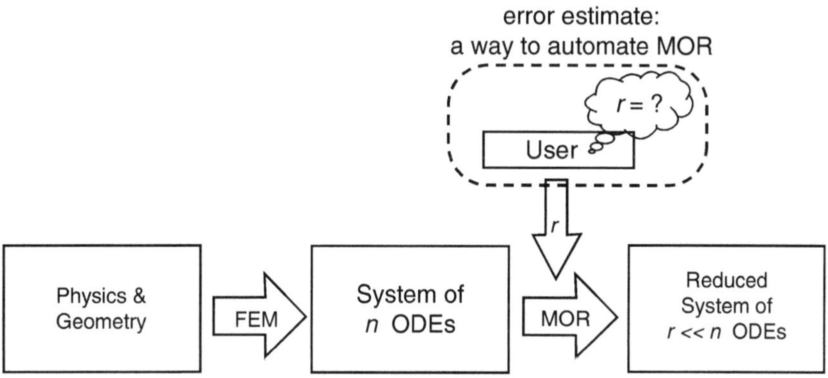

Fig. 6. Compact model extraction. Eliminating the need for user iteration makes the process fully automatic.

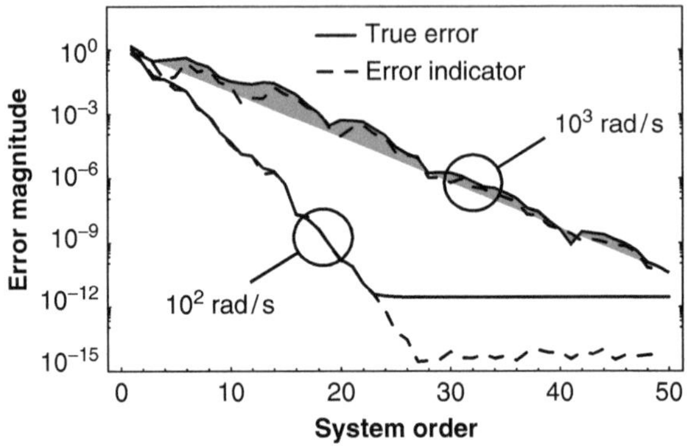

Fig. 7. Convergence of relative error between the two successive reduced models for the microthruster.

error bound provided by the exact control-theory methods and alternatively to employ sequential model order reduction, which is based on consecutively applying Arnoldi and control-theory methods (e.g. via calling SLICOT library). Fig. 7 shows the convergence of the relative error for the microthruster model for two different frequencies. The system order necessary to reach the convergence increases towards higher frequencies, as may be expected for the expansion around zero. It should be noted that for the frequencies far away from the expansion point oscillations may occur. Our observation is that the error indicator approximates the true error with high accuracy. The same observation has been recently made in [17].

6 Coupling of the Reduced Models

As MEMS are often composed of identical devices that are interconnected, array structures for example, it is desirable, especially with a large number of subsystems, to reduce each subsystem on its own and then to couple them back together. Hence, we seek a kind of compact thermal multiport representation which allows thermal fluxes to cross the boundaries and enables straightforward coupling to the next thermal multiport. The main problem thereby is that the thermal flow is not lumped by nature as, for example, the electrical flow is along metallic wire interconnects. The ratio of electrical conductivity of metals and that of insulators is of the order of 10^8. Hence, the electrical current flow takes place almost solely in metal paths. This is not the case with heat flow because the ratio of thermal conductivities in microtechnology is only of the order of 10^2. Therefore, it is unclear how to lump the thermal fluxes at shared surfaces between two finite element (FE) models in order to form the thermal ports (Fig. 8) which would serve to couple together several compact models. Indeed, very few works [18, 19] on how to couple compact thermal models are known to date.

Presently it is possible to use two methods for model order reduction of thermal MEMS arrays, Block Arnoldi and Guyan-based substructuring (available in

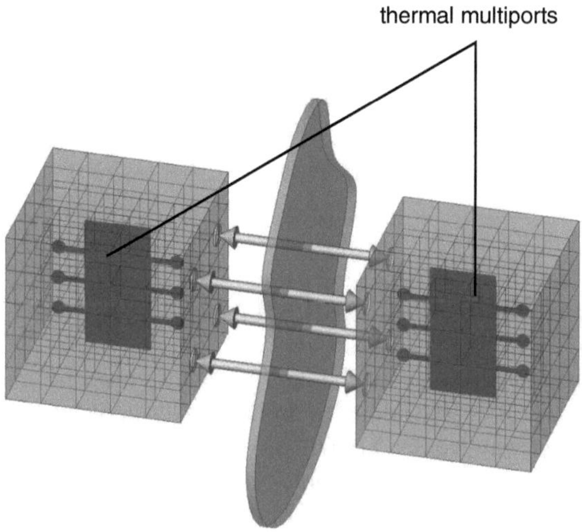

Fig. 8. Continuous thermal flux through the shared interface of two FE models. The goal is to model the "FE cubes" as thermal multiports i. e. to lump a flux.

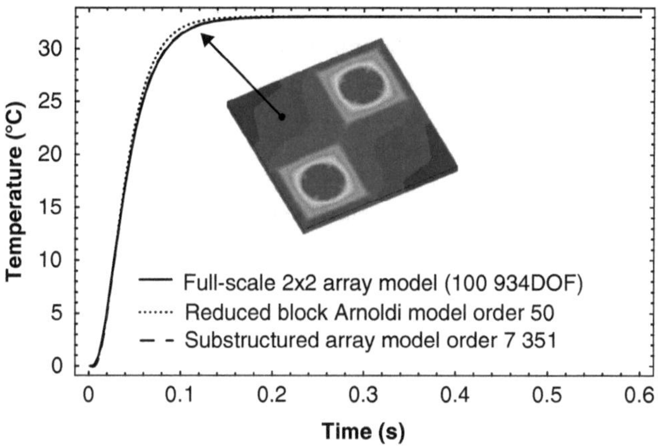

Fig. 9. Single output step responses of the full-scale and reduced order models created by block Arnoldi and Guyan-based substructuring.

ANSYS). The application of the block Arnoldi algorithm (classical Arnoldi which is suitable for multiple-input-multiple-output systems [11]) is straightforward. It reduces the entire array model and results in a much smaller reduced model than the substructuring. Its main disadvantage is that it does not scale well to a large number of devices within an array. Substructuring decouples the array model and physically preserves the shared nodes. This allows easy back-coupling of the reduced models, but results in unnecessarily large array model sizes. Both methods are described in [12]. Fig. 9 compares the step response of the substructured microhotplate test array with the step responses of the full-scale model and of the block Arnoldi reduced model.

In [12] we have also described a general technique of coupling two reduced thermal models gained by projection (e.g. when the Arnoldi algorithm is used). In such case the coupling is done via surface fluxes.

Another very promising approach is the structure preserving model order reduction [20–23], which seems to offer the possibility of directly reducing a device array without decoupling.

7 Model Order Reduction as a Fast Solver

An important engineering task is to build a validated model for each characterized novel MEMS device. It is possible to fit an RC-ladder network to the measured results [24], but in this way we don.t obtain a complete physical picture of the device. As in most MEMS applications the whole temperature field has to be known, so a more detailed FE model is required. Unfortunately, a common problem here is that the material properties of the employed thin film materials, like e.g. thermal conductivity κ and heat capacity c_p, strongly depend on fabrication conditions and may also be specific for the device under the test. In such case, it is possible to extract the material properties by fitting a parametrized FE model to a measured transient curve. However, the conventional optimization process is highly time consuming, because in each iteration a time integration of a full-scale model must be performed (see Fig. 10)

We suggest an alternative approach based on using model order reduction as a "fast solver". The right path in Fig. 10 shows that in each iteration of optimization loop, the suggested approach requires only the time integration of the reduced model (with less than 50 ODEs) and hence brings along an enormous saving in computational time, as the model reduction time is comparable with that for a static solution. By defining an objective function, which characterizes the difference between simulated and measured results, a data fitting cycle is performed.

Fig. 11 shows the flexible optimization environment coupled to MOR process. *Mathematica* is used for scripting, visualization and small size computations. Its function *eval* takes as arguments the fitting parameters κ and $c_p \cdot \rho$ and calls the external programs ANSYS (for rebuilding the FE model with changed material properties) and "MOR for ANSYS" (for creating a reduced model). It further integrates the reduced model and evaluates the objective function, which is defined as a quadratic error between the measured and the computed curves. Its value is transferred back to the DOT optimizer [26] which communicates with *Mathematica* via Mathlink (our implementation can be found at *http://evgenii.rudnyi.ru/soft/dot/*).

Fig. 12 shows the measured temperature response and the simulated temperature response for the reduced order 10 model of the optical filter device before the optimization. Fig. 13 shows further the measured and the simulated temperature response after 35 cycles of optimization.

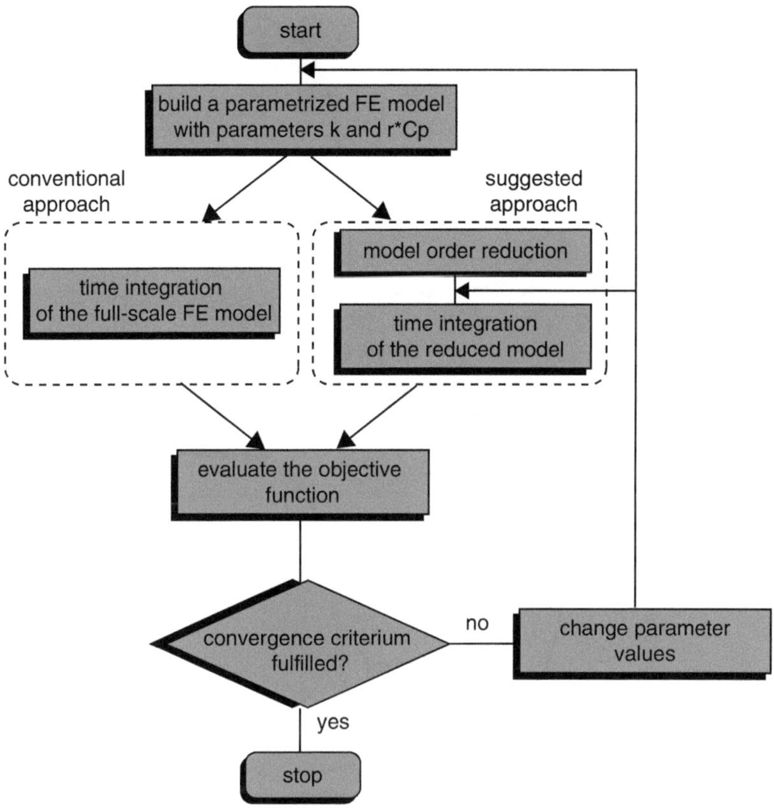

Fig. 10. Parameter extraction process for transient thermal problems via optimization and model order reduction [25].

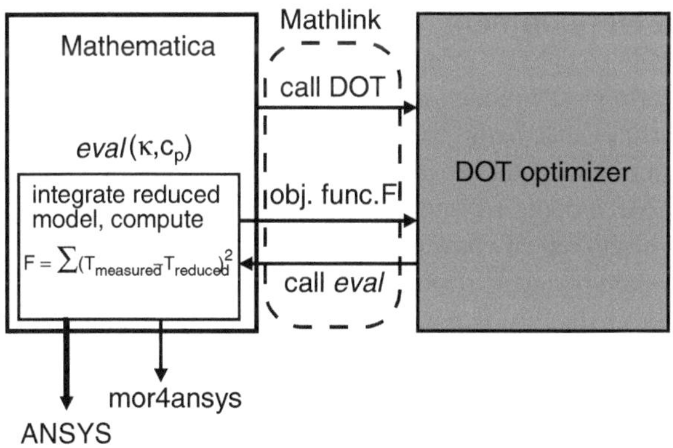

Fig. 11. Implementation of the reduced model optimization.

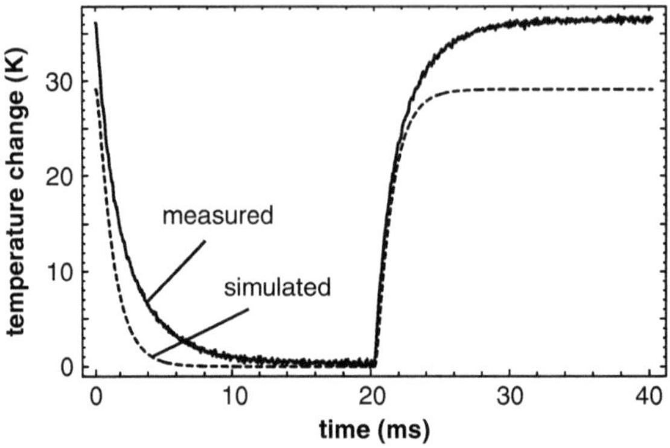

Fig. 12. Measured and simulated temperature response before optimization.

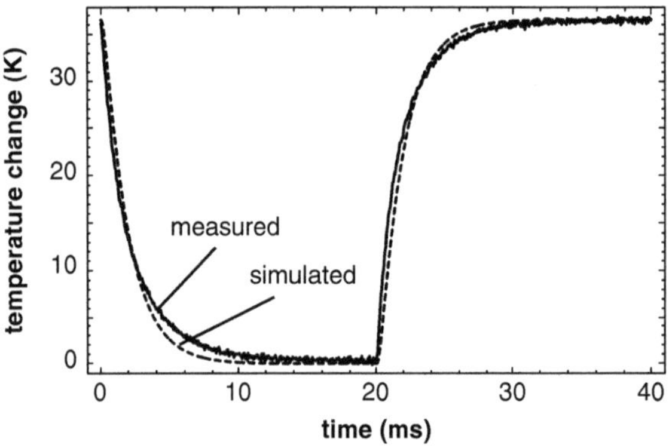

Fig. 13. Measured and simulated temperature response after 35 cycles of simulation.

8 Advanced Development

In its original form model reduction does not allow us to preserve parameters within the system matrices that naturally arise in many applications. For example, heat transfer (film) coefficients that are often used within convection boundary conditions for thermal MEMS models are within the heat conductivity matrix K and if they are to change, one has to repeat conventional model reduction again. Fortunately, a new development, called parametric model reduction, allows us to overcome this limit.

Parametric model reduction is based on generalization of moment matching. From a numerical point of view, a parameter within the system matrix is "similar" to the Laplace variable in the transfer function. If the transfer function of (2) is (without parameters) defined as:

$$G(s) = L^T(sC + K)^{-1}F \tag{6}$$

then, assuming linear dependence of K on the heat transfer coefficient h:

$$K(h) = K_0 + K_1 h \tag{7}$$

where K_0 is a constant part (see the definition of convection boundary condition for (1) in [27]), the transfer function (6) changes into:

$$G(s, h) = L^T (sC + K_0 + K_1 h)^{-1} F \tag{8}$$

As a result, the idea explored by several groups was to make multivariate expansions of the transfer function with respect to the Laplace variable s and parameters (e.g. h) which would be simultaneously preserved in symbolic form.
The problem of preserving film coefficients in symbolic form was approached by several groups [29, 30, 33, 34]. The numerical results showed that this problem can be solved in principle, i. e. one can change film coefficients in the reduced model over a very wide range of numerical values while approximating well the transient behavior of the original system.

In [28] and [35] the algorithm, which generates recursively all multivariate moments of G up to a chosen order, is described. However, the explicit computation of moments is numerically unstable and hence, in order to employ the method in practice, one has to perform orthogonalization. A novel method to do it in the case of a single parameter was presented in [29].The comparison with existing methods is given in [31]. In 2004 Codecasa et al [30] presented the algorithm with internal orthogonalization. In our view however, the most elegant solution for this problem was found recently in [32].

Of course, the main practical problem of parametric model reduction is that the number of mixed moments grows very rapidly. For example, if we choose to preserve four parameters then a reduced model, which contains all the first derivatives, has the dimension 6, a reduced model, which contains all second derivatives has the dimension 21, and a reduced model, which contains all third derivatives already has the dimension 56 (see Appendix F in [35]). At the same time, we may need derivatives of higher order than three to describe accurately transient behavior of the original model. A possible solution to ignore mixed moments is presented in [33, 36] and [34]. Additionally in [34], the local error control has been employed to automatically choose the number of moments to include into reduced model. The tutorial on parametric model reduction in respect to film coefficients is also available at the "MOR for ANSYS" site.

Another important aspect in model order reduction is to be able to deal with physically more realistic, nonlinear systems. At present, a leader among methods for nonlinear model reduction is proper orthogonal decomposition (POD) [37]. It uses simulation results of the original system ("snapshots") to build a low-dimensional subspace. It has been originally developed for fluid dynamics [38] but has recently been used for nonlinear heat transfer [39] as well. In [40] the application of the missing point estimation technique for speeding up the POD is presented.

There are recent results to generalize moment matching for weakly nonlinear systems [41, 42]. Under weakly nonlinear systems, one understands a system in which the nonlinear effects are limited to polynomial terms: quadratic, cubic, etc. The generalized moment matching has recently been used for the heat transfer with nonlinear film coefficient [43]. The difference with proper orthogonal decomposition is that a model reduction here employs system matrices directly.

An interesting idea to split a nonlinear trajectory into pieces, to build a linear reduced model for each piece and then to merge all reduced model has been explored in [44] and [45] (trajectory piecewise-linear approach). It has been already used in order to reduce a nonlinear thermal problem in [46].

9 Summary

We have developed a methodology and a software tool for applying mathematical model order reduction to the automatic generation of dynamic compact thermal models for MEMS. We have shown that with the Arnoldi algorithm it is possible to reduce linear thermal ODE systems of around 10^5 equations to orders between 20 and 50 with only a minimal loss of accuracy. This increases computational efficiency by more than 10 times in the case of a microhotplate gas sensor and in general reduces computational time to the time comparable with a single stationary solution of the original system.

Further advantages of Arnoldi-based reduction are the approximation of the complete output and the reduction of models with temperature dependent heating power. Its main disadvantage was the fact that no error estimate between the original and the reduced models exists. We have suggested three heuristic strategies for error estimation. At the present stage, the convergence of relative error and sequential model order reduction can be recommended for practical use. They are both straightforward to implement.

We have researched the possibilities for model order reduction of MEMS array structures. Presently, we are able to apply block Arnoldi and Guyan-based substructuring. Block Arnoldi can be recommended in cases of a moderate number of devices within an array. However, when the number of interconnected devices grows, both methods need alternatives.

We have further demonstrated the usability of model order reduction as a fast solver within an optimization loop. In electro-thermal MEMS design this can be used for the inverse thermal problem, i. e., the extraction of the material thermal parameters.

Computational environment "MOR for ANSYS", which has been developed at the university of Freiburg, Germany, offers the possibility for the efficient Krylov-based model order reduction of large scale (up to 500000 ODEs) MEMS models. It can be used either for ANSYS models directly or for the system matrices in Matrix Market format.

References

[1] M. N. Sabry, 11Dynamic Compact Thermal Models Used for Electronic design: A Review of Recent Progress", Proc. IPACK, pp. 1-17, (2003).

[2] S. D. Senturia, "Microsystem Design", Kluwer Academic Publishers, (2001).

[3] A. C. Antoulas, "Approximation of Large-Scale Dynamical Systems", Society for Industrial and Applied Mathematic, (2005).
[4] B. N. Datta, "Numerical Methods for Linear Control systems", Elsevier Incorporation, (2004).
[5] G. Obinata, B. D. O. Anderson, "Model Reduction for Control System Design", Springer, (2004).
[6] Z. Q. Qu, "Model Order Reduction Techniques with applications in finite element analysis", Springer, (2005).
[7] A. Varga, "Model reduction software in the SLICOT library", In: Applied and Computational Control, Signals and Circuits, Ed. B. Datta, KluwerAcademic Publishers, (2001).
[8] C. Rossi, "Micropropulsion for Space", Sensors Update, vol. 10, pp. 257-292, (2002).
[9] D. Hohlfeld, "Silicon based tunable optical filters", PhD Thesis, University of Freiburg, (2005).
[10] J. Wllenstein, H. Bttner, J. A. Plaza, C. Cane, Y. Min, H. L. Tuller, "A novel single chip thin film metal oxide array", Sensors and Actuators B: Chemical, 93(1-3), pp. 350-355, (2003).
[11] R. W. Freund, "Krylov-subspace methods for reduced-order modeling in circuit simulation", Journal of Computational and Applied Mathematics, Vol. 123, pp. 395-421, (2000).
[12] T. Bechtold, "Model Order Reduction of Electro-Thermal MEMS", PhD thesis, University of Freiburg, (2005).
[13] Y. J. Yang, C. Yu, "Extraction of heat-transfer macromodels for MEMS device", Journal of Micromechanics and Microengineering, vol. 14, pp. 587-596, (2004).
[14] L. Codecasa, D. D.Amore, P. Maffezzoni, "An Arnoldi Based Thermal Network Reduction Method for Electro-Thermal analysis", IEEE Transactions on Components and Packaging Technologies, 26(1), pp. 186 -192, (2003).
[15] J. G. Korvink, E. B. Rudnyi, "Oberwolfach Benchmark Collection", In: Benner, P., Mehrmann, V., Sorensen, D. (eds) Dimension Reduction of Large-Scale Systems, Lecture Notes in Computational Science and Engineering (LNCSE). Springer-Verlag, Berlin/Heidelberg, Germany, v. 45, p. 311-315, (2005).
[16] E. B. Rudnyi and J. G. Korvink, "Model Order Reduction for Large Scale Engineering Models Developed in ANSYS", Lecture Notes in Computer Science, v. 3732, pp. 349-356, (2005).
[17] P. Heres, "Robust and Efficient Krylov Subspace Methods for Model order Reduction", PhD thesis, Technical University of Eindhoven, (2005).
[18] D. Petit, R. Hachette, "Model Reduction in Linear Heat Conduction: Use of Interface Fluxes for the Numerical Coupling", International Journal of Heat and Mass Transfer, vol. 41, pp. 3177-3189, (1998).
[19] X. Guo, D. Celo, D. J. Walkey, T. Smy, "The Use of Constant Heat FlowPorts for Thermal Macro-Models", Proc. THERMICNIC, pp. 323-328, (2004).
[20] R. W. Freund, "SPRIM: Structure-preserving reduced-order interconnect macromodeling", IEEE/ACM ICCAD, (2004).
[21] H. Yu, L. He and S.X.D. Tan, "Block Structure Preserving Model Order Reduction", IEEE Behavioral Modeling and Simulation Wokshop pp. 1-6, (2005).
[22] A. Vanderdorpe, P. Van Dooren, "Model Rdeuction of Interconnected Systems", submitted to Automatica, 2005
[23] R. C. li, Z. Bai, "Structure-Preserving Model Reduction Using a Krylov Subspace Projection Formulation", Comm. Math. Sci., vol. 3(2), pp. 179-199, (2005).

[24] V. Szkely, M. Renz, "Thermal dynamics and the time constant domain", IEEE Trans. on Comp. Pack. Techn., 23(3), pp. 587-594, (2000).
[25] J. S. Han, E. B. Rudnyi, J. G. Korvink, "Efficient optimization of transient dynamic problems in MEMS devices using model order reduction", J. Micromech. Microeng., 15(4), pp. 822-832, (2005).
[26] DOT Users Manual version 4.20 (Colorado Springs, CO: Vanderplaats Research and Development), at http://www.vrand.com/DOT.html
[27] T. Bechtold, "Dynamic electro-thermal simulation of microsystems - a review", J. Micromech. Microeng., vol. 15, R17-R31, (2005).
[28] D. S. Weile and E. Michielssen, "Analysis of frequency selective surfaces using two-parameter generalized rational Krylov model-order reduction", IEEE Transactions on Antennas and Propagation, vol. 49, pp. 1539-1549, (2001).
[29] L. H. Feng, E. B. Rudnyi, J. G. Korvink, "Preserving the film coefficient as a parameter in the compact thermal model for fast electro-thermal simulation", IEEE Transactions on Computer-Aided Design of Integrated Circuits and Systems, v. 24, N 12, pp. 1838-1847, (2005).
[30] L. Codecasa, D. D'Amore, and P. Maffezzoni, "A novel approach for generating boundary condition independent compact dynamic thermal networks of packages", Proc. 10th International Workshop on Thermal Investigations of ICs and Systems (THERMINIC), pp. 305-310, (2004).
[31] L. H. Feng, "Parameter independent model order reduction", Mathematics and Computers in Simulation, v. 68, N 3, pp. 221-234, (2005).
[32] O. Farle, V. Hill, P. Ingelstrm, R. Dyczij-Edlinger, "Multi-parameter Polynomial Model Reduction Of Linear Finite Element Equation Systems", Proc. 5th MATHMOD, (2006).
[33] D. Celo, P. K. Gunupudi, R. Khazaka, D. J. Walkey, T. Smy, M. S. Nakhla, "Fast Simulation of Steady-State Temperature Distributions in Electronic Components Using Multidimensional Model Reduction", IEEE Transactions on Components and Packaging Technologies, vol. 28(1), pp. 70-79, (2005).
[34] E. B. Rudnyi, L. H. Feng, M. Salleras, S. Marco, J. G. Korvink, "Error Indicator to Automatically Generate Dynamic Compact Parametric Thermal Models", Proc. THERMINIC, (2005).
[35] L. Daniel, O. C. Siong, L. S. Chay, K. H. Lee, and J. White, "A Multiparameter Moment-Matching Model-Reduction Approach for Generating Geometrically Parameterized Interconnect Performance Models", IEEE Transactions on Computer-Aided Design of Integrated Circuits and Systems, vol. 23, pp. 678-693, (2004).
[36] P. K. Gunupudi, R. Khazaka, and M. Nakhla, "Analysis of transmissionline circuits using multidimensional model reduction techniques," IEEE Transactions on Advanced Packaging, vol. 25, pp. 174-180, (2002).
[37] P. Holmes, J. L. Lumley, and G. Berkooz, "Turbulence, coherent structures, dynamical systems, and symmetry", Cambridge University Press, (1996).
[38] D. J. Lucia, P. S. Beran, and W. A. Silva, "Reduced-order modeling: new approaches for computational physics", Progress in Aerospace Sciences, vol. 40, pp. 51-117, (2004).
[39] M. E. Kowalski and H. M. Jin, "Model-order reduction of nonlinear models of electromagnetic phased-array hyperthermia," IEEE Transactions on Biomedical Engineering, vol. 50, pp. 1243-1254, (2003).
[40] P. Astrid, A. Verhoeven, "Application of Least Squares MPE technique in the reduced order modeling of electrical circuits", Proc. MTNS, (2006).

[41] L. H. Feng, .Review of model order reduction methods for numerical simulation of nonlinear circuits", Applied Mathematics and Computation, vol. 167, N 1, pp. 576-591, (2005).

[42] J. R. Phillips, "Projection-based approaches for model reduction of weakly nonlinear, time-varying systems", IEEE Transactions on Computer-Aided Design of Integrated Circuits and Systems, 22, pp. 171-187, (2003).

[43] L. H. Feng, E. B. Rudnyi, J. G. Korvink, C. Bohm, T. Hauck, "Compact Electrothermal Model of Semiconductor Device with Nonlinear Convection Coefficient", Proc. EuroSimE, pp. 372-375, (2005).

[44] M. Rewienski, J. White, "A trajectory piecewise-linear approach to model order reduction and fast simulation of nonlinear circuits and micromachined devices", IEEE Transactions on Computer-Aided Design of Integrated Circuits and Systems, vol. 22, pp. 155-170, (2003).

[45] D. Vasilyev, M. Rewienski, and J. White, "Macromodel generation for bio MEMS components using a stabilized balanced truncation plus trajectory piecewise-linear approach", IEEE Transactions on Computer- Aided Design of Integrated Circuits and Systems, vol. 25, pp. 285-293, 2006.

[46] Y.-J. Yang, K.-Y. Shen, .Nonlinear heat-transfer macromodeling for MEMS thermal devices", "Journal of Micromechanics and Microengineering", vol. 15, pp. 408-418, (2005).

Model Order Reduction of Large RC Circuits

Nick P. van der Meijs

Delft University of Technology, Faculty of EEMCS, Delft, The Netherlands
N.P.vanderMeijs@tudelft.nl

1 Introduction

In this chapter, we will focus on direct techniques for reduction of RC circuits. Compared to iterative techniques, which are frequently (usually) based on subspace projection techniques, direct techniques are based on Gaussian Elimination or equivalent techniques.

Actually, Gaussian Elimination in linear algebra is the same as $Y\Delta$ transformation in circuit theory. The technique can be described in both ways, and we will switch between these frameworks as appropriate to gain an improved understanding, and to obtain further insights that can e.g. lead to more efficient implementations.

In this chapter, we will focus on the reduction of RC circuits. Formally, given a linear RC (sub-)circuit, let us define *port nodes* as a input or output nodes of the circuit. Typically, these are connected to the real inputs and outputs of the circuit or to the terminals of the active devices. Also, *internal nodes* are all the remaining nodes. Then, reduction aims at removing internal nodes and (resistive or capacitive) branches connecting them such that the result is simpler but still accurate enough. Port nodes typically should be preserved, although sometimes they can be merged without a large accuracy penalty.

While generalizations to circuits including inductance are sometimes possible, these may not always be very effective. This depends on the underlying method. Techniques which capitalize e.g. on a low-pass behavior of typical (parasitic) RC networks in integrated circuits are especially cumbersome to generalize. We will briefly mention such aspects when appropriate, but we will not go into details.

The close relationship between Gaussian Elimination and $Y\Delta$ transformation actually makes the so-called realization problem, which is certainly non-trivial with projection based MOR, less of an issue. This realization problem relates to the representation of the reduced circuit in the form of another, smaller, circuit rather than in the abstract form of merely one or more matrices that would describe the behavior of the reduced model but are not compatible with existing circuit simulators. Indeed, some straight-forward simplifications of Gaussian Elimination can actually be viewed as an RC-in RC-out technique. However, in other variations still some realization is required.

In this chapter, we will first present some background information on Gaussian elimination in Section 2. While this is exact for resistive circuits, the result can only be approximate for RC circuits (unless complex arithmetic is employed for a frequency-domain results with a fixed frequency). Therefore, we will study these approximation properties in Section 3, in the context of RC-in-RC-out techniques. In particular, we will show in Section 3.2 how $Y\Delta$ transformation can preserve the Elmore delay between all remaining nodes in the circuit, and in Section 3.3 how a *selective* node elimination procedure can maintain accuracy of the result in a user-specified frequency range $0 \ldots f_s$.

Subsequently, we will study in Section 4 two techniques for reduction that enable complete elimination of all internal (non-port) nodes while carrying some higher order information in the resulting admittances, which require a synthesis based realization procedure. Fortunately, this synthesis procedure is simpler than a standard synthesis procedure for arbitrary transfer functions, as it only needs to work on two-port admittances. Nevertheless, many of the same issues that normally arise in conjunction with the realizability issue, such as stability and passivity, also surface in this case. We will, however, only briefly touch upon this subject. Finally, we will study some issues related to actual efficient implementation of the proposed techniques in Section 5 and we conclude in Section 6.

2 Gaussian Elimination Background

2.1 Introduction

As noted in the introduction, we will first describe Gaussian Elimination for resistive networks, and illustrate the connection to $Y\Delta$ transformation. This is mainly for introductory purposes, on which the developments in subsequent sections are partially built. Also, we will present some high-level computational techniques that could be useful for matrix-based implementations. Furthermore, we will generalize these techniques to RC networks by representing these networks in the s domain.

2.2 Gaussian Elimination for Resistive Networks

In this section we will introduce the technique of Gaussian elimination, with only a limited view on model order reduction. In fact, in this section we will only consider resistive networks for which model reduction amounts to removing internal nodes (but the number of resistors can possibly increase). The result, however, is exact in the sense that the networks before and after the reduction when measured from the port nodes can't be distinguished. In the following sections, we will consider RC circuits and this exactness property will not hold anymore in general, although it can hold for a fixed frequency.

Gaussian elimination typically works on the (modified) admittance matrix formulation, where the lower triangular part of the matrix is made zero in a column-by-column fashion. It is equivalent to $Y\Delta$ transformations in a circuit, where the 'hub' node in a star network of resistances is removed and the resistances are replaced by a full graph of resistances (a clique) on the 'spoke' nodes of the star.

Understanding the relation between Gaussian elimination and $Y\Delta$ transformation allows to switch between representations as convenient. For example, $Y\Delta$ transformations are defined in the circuit theory domain, allowing very natural and efficient handling of sparsity and fill-in and of course good integration in circuit analysis tools. On the other hand, Gaussian elimination and matrix based approaches generally allow a different tool set for analysis and understanding. In particular, we will show that Gaussian climination allows an easy generalization towards elimination of groups of nodes using the so called Schur complement technique. This will offer some notational convenience, and also opens a route towards so-called 'sparse approximations'. In general, matrix based approaches also allow for more elegant but not necessarily computationally efficient prototype implementations in e.g. Matlab.

Usually, $Y\Delta$ transformations are explained as eliminating the common node in an Y (or star) circuit with 3 admittances, and replacing these 3 admittances by 3 others connecting the remaining nodes. Nevertheless, it is straightforward to generalize it to admittance structures of any degree. Consider Figure 1 where the circuit in (a) is replaced by the circuit in (b) with an admittance between each pair of remaining nodes. Such a circuit is called a clique circuit, it forms a full subgraph.

In (a), we have for the voltage of the star node, node x,

$$v_x = \frac{\sum_i v_i y_i}{\sum_i y_i}$$

which can be derived by writing the KCL equation for node x. Subsequently, we can write for the current entering node k, $i_k^Y = (v_k - v_x)y_k$,

$$i_k^Y = \frac{\sum_i (v_k - v_i) y_k y_i}{\sum_i y_i}. \tag{1}$$

In (b), we have for the current entering node k

$$i_k^\Delta = \sum_i i_{ki} = \sum_i (v_k - v_i) y_{ki} \tag{2}$$

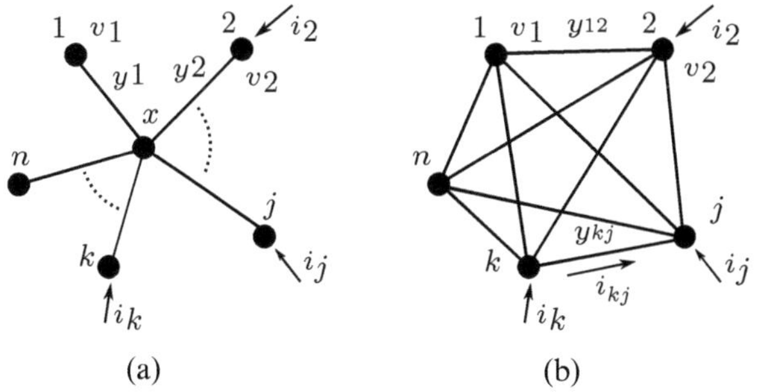

Fig. 1. Illustration of $Y\Delta$ transformation.

and we can identify i_k^Y and i_k^Δ by letting

$$y_{kj} = \frac{y_k y_j}{\sum_i y_i}. \tag{3}$$

That is, Equation (3) gives the admittance values of the Δ network in Figure 1(b) such that the currents entering the nodes in (b) are identical to those entering the corresponding nodes in (a). Thus, circuit (a) and (b) show exactly identical current-voltage relations and are indistinguishable from the outside. Circuit (b) has one node less compared to (a), and is in this sense a reduced circuit although the number of admittances generally has increased.

Note that in the above equations we have, for reasons of ease of notation and simplicity, adopted a single subscript notation (e.g. y_i) for resistances/admittances that are connected to the node to be deleted in the *original* network, while we have a double subscript notation (e.g. y_{kj}) for those that are to be connected to the former neighbors of the victim node after deletion. We will adhere to this notation throughout this chapter.

The equivalence of $Y\Delta$ transformation and Gaussian elimination can be illustrated with the following example. Consider the circuit in Figure 1 with $n = 3$. If we assume that node x is numbered first, the admittance matrix formulation looks as follows:

$$\begin{bmatrix} y_1+y_2+y_3 & -y_1 & -y_2 & -y_3 \\ -y_1 & y_1 & & \\ -y_2 & & y_2 & \\ -y_3 & & & y_3 \end{bmatrix} \begin{bmatrix} v_x \\ v_1 \\ v_2 \\ v_3 \end{bmatrix} = \begin{bmatrix} 0 \\ i_1 \\ i_2 \\ i_3 \end{bmatrix} \tag{4}$$

Eliminating v_x from the system to obtain a smaller system not depending on v_x (one step of Gaussian elimination) amounts to adding $y_2/(y_1+y_2+y_3)$ times the first row to the second row, and $y_3/(y_1+y_2+y_3)$ times the first row to the third row and similarly for row 4. Subsequently, the following system results:

$$\begin{bmatrix} y_1+y_2+y_3 & -y_1 & -y_2 & -y_3 \\ 0 & y_1 - \frac{y_1^2}{y_1+y_2+y_3} & \frac{-y_1y_2}{y_1+y_2+y_3} & \frac{-y_1y_4}{y_1+y_2+y_3} \\ 0 & \frac{-y_1y_2}{y_1+y_2+y_3} & y_2 - \frac{y_2^2}{y_1+y_2+y_3} & \frac{-y_2y_3}{y_1+y_2+y_3} \\ 0 & \frac{-y_1y_3}{y_1+y_2+y_3} & \frac{-y_2y_3}{y_1+y_2+y_3} & y_4 - \frac{y_4^2}{y_1+y_2+y_3} \end{bmatrix} \begin{bmatrix} v_x \\ v_1 \\ v_2 \\ v_3 \end{bmatrix} = \begin{bmatrix} 0 \\ i_1 \\ i_2 \\ i_3 \end{bmatrix} \tag{5}$$

The (3,2), (4,2) and (4,3) elements and their symmetrical counterparts precisely correspond to Equation (3), indeed confirming the equivalence between Gaussian elimination and $Y\Delta$ transformations.

It has to be noted that the right-hand side (RHS) in Equation (5) is identical to the RHS in Equation (4), since $i_1 = 0$ and adding multiples of i_1 as per the recipe of Gaussian elimination is not changing the RHS. Also, it is evident that so-called fill-in occurs in the matrix. The (i, j) entry of the matrix after elimination of, in this case, the first variable remains unchanged if and only if y_i or y_j is zero. Otherwise, an entry on position (i, j) and (j, i) is added if it was not already there. This illustrates that if in the original circuit the nodes i and j are both connected to the node to be eliminated, they are directly connected in the circuit after elimination. Thus, a clique (full subgraph) on the remaining nodes emerges.

The Schur complement technique for Gaussian elimination basically works as Gaussian elimination but then for block matrices. Consider the following partitioning of a system equivalent to Equation (4):

$$\begin{bmatrix} Y_{11} & Y_{12} \\ Y_{21} & Y_{22} \end{bmatrix} \begin{bmatrix} V_1 \\ V_2 \end{bmatrix} = \begin{bmatrix} 0 \\ I_2 \end{bmatrix}. \tag{6}$$

Here, the part to be eliminated corresponds to the Y_{11} block and is again put first in the matrix. The corresponding source-term in the RHS is zero, since we eliminate only nodes that are not connected to external sources. The elements in the Y_{22} block correspond to nodes to be preserved and the elements in the Y_{12} and Y_{21} blocks correspond to the admittances in the circuit connecting a node to be preserved to a node to be eliminated.

Now, we can write

$$Y_{11}V_1 + Y_{12}V_2 = 0$$

or

$$V_1 = -Y_{11}^{-1}Y_{12}V_2.$$

Using this equation for the second row in Equation (6) leads to

$$\left(Y_{22} - Y_{21}Y_{11}^{-1}Y_{12}\right) V_2 = I_2. \tag{7}$$

This equation represents a reduced system with only the nodes in the $(2,2)$ block remaining and trivially reduces to $Y\Delta$ transformation if there is only a single node in the $(1,1)$ block. The quantity between parenthesis in Equation (7) is called the Schur complement of Y_{11} in Y [5, 12]. A seminal presentation of this technique for admittance matrices is reference [14].

In general, Y is a sparse matrix. Also, the blocks Y_{12} and $Y_{21} = Y_{12}^T$ are sparse, since these blocks collect the admittances from the port nodes to the internal nodes, but generally by far not every internal node is incident to a port node. Upon appropriate re-ordering, the $Y_{21}Y_{11}^{-1}Y_{12}$ system looks as follows:

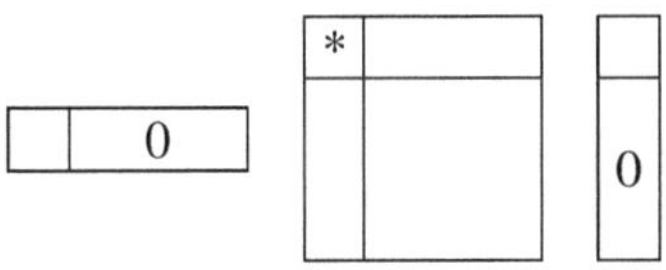

(8)

Hereby it follows that only the $*$ part of Y_{11}^{-1} is actually necssary. The other entries of Y_{11}^{-1} are unnecessary to compute.

An elegant way to exploit the fact that not all entries of Y_{11}^{-1} are necessary and thereby actually reducing the amount of computation was given by Clément [4]. Using the symmetry and positive definiteness of the system, Y_{11} can be Choleski decomposed as $Y_{11} = L_{11}L_{11}^T$ where L_{11} is lower-triangular. Hence, it follows that

$$
\begin{aligned}
Y_{21}Y_{11}^{-1}Y_{12} &= Y_{21}\left(L_{11}L_{11}^{T}\right)^{-1}Y_{12} \\
&= Y_{21}\left(L_{11}^{T}\right)^{-1}L_{11}^{-1}Y_{12} \\
&= Y_{21}\left(L_{11}^{-1}\right)^{T}L_{11}^{-1}Y_{12} \\
&= \left(L_{11}^{-1}Y_{12}\right)^{T}L_{11}^{-1}Y_{12} \\
&= P^{T}P \quad \text{with} \quad P = L_{11}^{-1}Y_{12}.
\end{aligned} \tag{9}
$$

Since L_{11}^{-1} has the same shape as L_{11}, the system for $P = L_{11}^{-1}Y_{12}$ has the following shape:

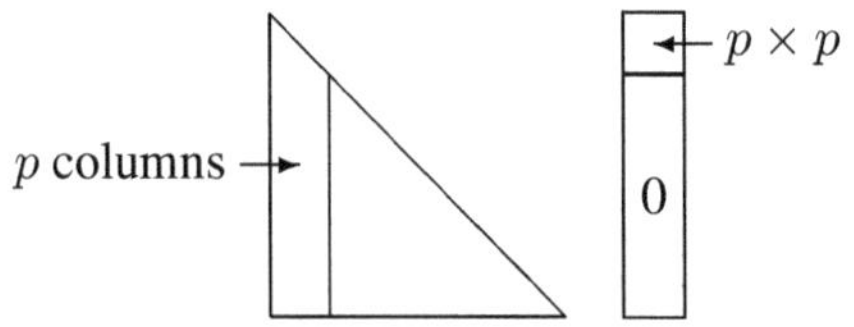

(10)

It is clear that now only the first p columns of L_{11}^{-1} are actually necssary, since the others are multiplied with the zero part of Y_{12}. Given L_{11}, the first columns of L_{11}^{-1} follow directly and efficiently from forward substitution.

Finally, let us show how the Schur complement can be computed from the LU factors of the blocks. This was shown recently in [2] but it can also be found in [11] which rephrases a result by [22]. Begin with partitioning the LU-decomposition of Y as

$$
Y = \begin{bmatrix} Y_{11} & Y_{12} \\ Y_{21} & Y_{22} \end{bmatrix} = \begin{bmatrix} L_{11} & 0 \\ L_{21} & L_{22} \end{bmatrix} \begin{bmatrix} U_{11} & U_{12} \\ 0 & U_{22} \end{bmatrix} \tag{11}
$$

(but note that L_{11} in this equation is not the same as L_{11} in (9)). By multiplication of the terms in the RHS we get

$$
Y = \begin{bmatrix} L_{11}U_{11} & L_{11}U_{12} \\ L_{21}U_{11} & L_{21}U_{12} + L_{22}U_{22}. \end{bmatrix}
$$

Now,

$$
\begin{aligned}
Y_{22} - Y_{21}Y_{11}^{-1}Y_{12} &= L_{21}U_{12} + L_{22}U_{22} - L_{21}U_{11}(L_{11}U_{11})^{-1}L_{11}U_{12} \\
&= L_{21}U_{12} + L_{22}U_{22} - L_{21}U_{11}U_{11}^{-1}L_{11}^{-1}L_{11}U_{12} \\
&= L_{22}U_{22}.
\end{aligned}
$$

Since Y_{11}^{-1} has lost sparsity compared to Y_{11}, using (7) directly is not efficient. However, the triangular factors in (9) and (11) are still sparse (although in general not as sparse as Y or Y_{11}) and these alternative formulations offer higher efficiency.

2.3 s-Domain Models for RC Networks

RC networks can be represented in the frequency or s domain using admittances. A resistor with value R will become an admittance $y(s) = 1/R$ and a capacitor with value C will become an admittance $y(s) = sC$. When the frequency is fixed, Gaussian elimination can produce an exact reduced model for that frequency by using complex arithmetic. Here, exact means that the reduced order model shows exactly accurate magnitude and phase behavior, but this is only the case for the fixed, a-priori specified, frequency. While this can sometimes be useful, it is often more interesting instead to investigate Gaussian elimination for a wider frequency range; typically we will be interested in the accuracy from DC to a pre-specified maximum frequency of interest.

When we leave s as a parameter, Gaussian elimination can generate admittances which are a rational function of s. Although theoretically such rational functions are still exact, it is cumbersome to work with them as their order typically becomes very high. However, truncation not only makes the result inexact but generally leads to issues w.r.t. stability and passivity. Also, such rational functions can not be directly represented using resistors and capacitors and are generally unsuitable for direct simulation. In that case, an RC network that approximately matches the reduced admittance model may be synthesized. Nevertheless, this is not a well-behaved procedure and actually introduces an approximation (of the synthesis) to an approximation (of the truncated reduced admittance model) and as such is less elegant.

While Gaussian elimination of $y(s)$ matrices produces rational functions of s, an alternative point of view is using moment expansions of $y(s)$ and such expansions can also be generated during Gaussian elimination. Again, truncation of these expansions leads to approximate reduced order representations. We will present and compare rational-function based and moment based elimination approaches in Section 4. Here we introduce an alternative moments-based point of view that we will use in Section 3 to show some of the properties of elimination. We thus take $y(s) = y_0 + y_1 s + y_2 s^2 + ...$ where y_0 is a DC admittance, y_1 is a capacitance and the higher order terms do not have a direct physical equivalent. Also, for direct physical relevance y_0 and y_1 should both be > 0. This is not guaranteed in general by Gaussian elimination but we will find that it is guaranteed for a number of important special cases.

Thus, we consider the moment expansion of the transfer function. We will only consider the single-input multiple-output case but because of linearity this can be easily generalized (superposition). For a single input, the transfer function can be expressed as a vector for the response of each (output) node. We will consider the impulse response, that is, the node voltages V in response to a unit source vector E being the Laplace transform of a Dirac impulse.

Hence, when considering the MNA formulation of an RC network, we would have

$$[G + sC]\, V = E \tag{12}$$

where E is independent of s as indicated above. Then, with a Taylor expansion of V around $s = 0$, we get

$$[G + sC]\left[V_0 + V_1 s + V_2 s^2 + ...\right] = E \tag{13}$$

and identifying like powers of s in the LHS and RHS [3] leads to

$$GV_0 = E \tag{14}$$

and

$$sCV_0 + sGV_1 = 0. \tag{15}$$

Equation (14) shows that V_0 is the DC solution obtained by solving the network while the capacitances have been removed. Rewriting Equation (15) as

$$GV_1 = -CV_0 \tag{16}$$

shows that the V_1 term of the expansion of V can be obtained by solving the same DC system as in Equation (14) but with the source term E replaced by a term depending on the circuit capacitances and the DC solution. Continuing the scheme as started with Equation (14) actually leads to a recursion for V_i, $i = 0...\infty$ [3].

Furthermore, if instead of Equation (12) we would start with a general, moment-expanded admittance matrix,

$$\left[Y_0 + Y_1 s + Y_2 s^2 + ...\right] V = E, \tag{17}$$

a similar analysis as above will show that V_m is completely determined by $Y_0...Y_m$.

3 RC-in RC-out Reduction

3.1 Introduction

In this section, we will introduce Gaussian elimination of RC networks in such a way that the realization problem is trivial or actually non-existent since all the manipulations are done using resistances and capacitances. Thus, the result is an RC circuit similar to the input, and these methods are called *RC-in RC-out* reduction methods. Such methods are highly appreciated in practice, because they fit very well in a typical chip design flow, relates to a designer's intuition and are easy for other tools to handle. Obviously, there are no problems with passivity and stability.

The results can of course only be approximate in general. Nevertheless, we will see that for an important class of RC networks it is easy to preserve the so-called Elmore delay mutually between all port nodes upon complete elimination of all internal nodes. The Elmore delay [10] actually is the 1st moment of the impulse response, and is definitely the most often used metric for analysis and synthesis of RC interconnect networks [13]. Since we took E in (12) as an impulse source, the Elmore delay is given by V_1 in (16).

Alternatively, the elimination process may be carried out incompletely, in such a way that some strategic internal nodes of the circuit are preserved. Thus, the remaining, partially reduced, RC network is considered the final result of the procedure. Of course, such a network can offer a better approximation to the transfer properties or frequency response of the original, unreduced, network at the cost of a greater number of nodes.

Below, we will first discuss reduction of all internal nodes, and subsequently we will discuss methods for incomplete elimination.

3.2 Complete Reduction

In this section we will show that for a mildly restricted class of RC networks, Gaussian elimination can be developed into a procedure that produces a reduced RC network, in the sense that it has fewer or no internal nodes, with unchanged DC and Elmore delay properties between each pair of nodes that remains after the elimination. For the restricted class of RC networks considered, we will show that we do not need to match the first moment of each entry of the $Y(s)$ matrix completely and that we can use this extra freedom to have the reduced $Y(s)$ matrix correspond exactly to a realizable reduced RC circuit. This means that all elements have positive real values. We refer to this method of complete reduction while preserving the Elmore delay properties as Elmore Preserving Reduction (EPR). This method was introduced (using a different approach and proof strategy) in [27] and [28].

We begin with so-called non-leaky RC circuits, which are RC circuits without grounded resistors, which in practice covers almost all interesting cases. We will call DC-connected components of such a circuit a *conductor*. A circuit can consist of multiple conductors, but the Elmore delay is only defined between two nodes on the same conductor. (Cross-talk is defined between multiple conductors.) We allow grounded and non-grounded (floating) capacitors, but for purposes of Elmore delay modeling we distinguish between floating capacitors between 2 conductors (*cross-coupled capacitors*) and floating capacitors on the same conductor (*self-coupled capacitors*). We now have the following lemma:

Lemma 1. *For computing the Elmore delay between any two nodes on the same conductor, self-coupled capacitors have no effect and cross-coupled capacitors are invariant from grounded capacitors.*

Proof. First consider self-coupled capacitors. Consider the contribution of a single self-coupled capacitor c_{ij} to the RHS of (16). We can write

$$-CV_0 = -\begin{bmatrix} \cdots & \cdots & \cdots & \cdots & \cdots \\ \cdots & +c_{ij} & \cdots & -c_{ij} & \cdots \\ \cdots & \cdots & \cdots & \cdots & \cdots \\ \cdots & -c_{ij} & \cdots & +c_{ij} & \cdots \\ \cdots & \cdots & \cdots & \cdots & \cdots \end{bmatrix} \begin{bmatrix} \cdots \\ v_{0,i} \\ \cdots \\ v_{0,j} \\ \cdots \end{bmatrix} = -\begin{bmatrix} \cdots \\ c_{ij}(v_{0,i} - v_{0,j}) \\ \cdots \\ c_{ij}(v_{0,j} - v_{0,i}) \\ \cdots \end{bmatrix}$$

where $v_{0,i}$ is the ith component of V_0, which is the DC solution at node i. Clearly, this contribution to the RHS vanishes if $v_{0,i} = v_{0,j}$, which above is the case for the non-leaky RC networks considered here. Hence, in such networks, self-coupled capacitors don't contribute to the Elmore delay since they don't generate source terms in the RHS of Equation (16).

Now consider cross-coupled capacitors. Without loss of generality we take node i on the switching conductor and node j on the quiet conductor. Similar arguments as above lead to

$$-CV_0 = -\begin{bmatrix} \cdots & \cdots & \cdots & \cdots & \cdots \\ \cdots & +c_{ij} & \cdots & -c_{ij} & \cdots \\ \cdots & \cdots & \cdots & \cdots & \cdots \\ \cdots & -c_{ij} & \cdots & +c_{ij} & \cdots \\ \cdots & \cdots & \cdots & \cdots & \cdots \end{bmatrix} \begin{bmatrix} \cdots \\ v_{0,i} \\ \cdots \\ v_{0,j} \\ \cdots \end{bmatrix} = -\begin{bmatrix} \cdots \\ c_{ij}v_{0,i} \\ \cdots \\ -c_{ij}v_{0,i} \\ \cdots \end{bmatrix} \tag{18}$$

since $v_{0,j}$ is 0 because node j is on the quiet conductor. Furthermore, the matrix G in Equation (16) may be partitioned as

$$G = \begin{bmatrix} G_{11} & 0 \\ 0 & G_{22} \end{bmatrix}$$

where the 0 blocks originate from the conductors being DC-isolated. Trivially, the inverse of G has the same structure. Now, upon formally solving (16) by application of G^{-1} to the RHS given by (18) it follows that for the first moments of the voltages on the switching conductor, c_{ij} is playing the same role as a grounded capacitor. □

Note that, in general, the non-switching side of a cross-coupled capacitor is not grounded but connected to a (quiet) voltage source through some path with non-zero resistance. Conveniently, the second part of the lemma makes clear that the resistance in this path has no influence on the Elmore delay of the switching conductor. Actually, this observation is consistent with the fact that in an RC-tree, only the resistances on the path between input and output node contribute to the Elmore delay, all off-path resistances can be shorted [9].

Lemma 1 implies that any MOR procedure for preserving the Elmore delay can allow general non-leaky RC circuits, but can start by just removing self-coupled capacitors and can also ignore any (positive or negative) self-coupled capacitors that are generated during the elimination procedure. Also, cross-coupled capacitors don't need any special treatment. Thus, without loss of generality, we can consider non-leaky RC-networks with only grounded capacitors to develop our Elmore-preserving elimination technique below.

For that purpose, consider a node x in a circuit as in Figure 1 where all the branches are conductances g_i but add a ground node and a capacitor with value c between node x and ground, which has index 0. This grounded capacitor has an admittance sc. This situation is sketched in Figure 2.

Upon performing $Y\Delta$ transformation of node x using such frequency-domain admittances, we get

$$y_{kj} = \frac{g_k g_j}{\sum_i g_i + sc}$$
$$y_{k0} = \frac{g_k sc}{\sum_i g_i + sc}$$

where $i, k, j \in 1 \cdots n$. Neither y_{kj} nor y_{k0} is directly realizable—these quantities don't correspond to a resistor or capacitor. However, upon expanding them about $s = 0$ and retaining only the terms until first order, we get

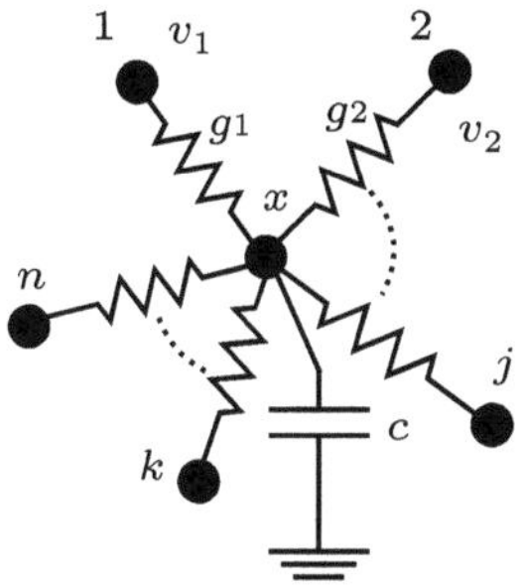

Fig. 2. Resistive node with a grounded capacitor.

$$y_{kj} = \frac{g_k g_j}{\sum_i g_i} - \frac{s c g_k g_j}{(\sum_i g_i)^2} + O(s^2) \tag{19}$$

$$y_{k0} = \frac{s c g_k}{\sum_i g_i} + O(s^2). \tag{20}$$

This result preserves the DC response and the first moment of the impulse response, V_0 and V_1 in (13). In fact, we can recognize y_{k0} as a capacitor, a fraction of c in proportion to the ratio of g_k to the sum of all conductances connected to node x. Also, we can recognize the first term in the RHS of y_{kj} as a conductor, allowing trivial realization. However, the second term in this RHS corresponds to a *negative* capacitor. Nevertheless, because of Lemma 1 this capacitor does not contribute to the Elmore delay V_1 (although in the notation of (17) it contributes to Y_1). Thus it can be dropped and it doesn't need realization to preserve the Elmore delay. As such, we actually have an *RC-in, RC-out* procedure for $Y\Delta$ transformation that preserves the Elmore delay. We call this the *EPR* method for Elmore Preserving Reduction. See Figure 3 for an illustration of all the steps for a circuit with 2 resistors and a grounded capacitor.

From a circuit point of view, the EPR method can be described as follows. For a non-leaky RC circuit the EPR method first drops all self-coupling capacitors and redistributes the ground capacitance of a node to be deleted to its neighboring nodes using a weight factor being the ratio of the conductance to the neighbor node and the sum of all conductances connected to the victim node. As such, the largest fraction of the capacitance to be distributed moves to the node that is most strongly coupled to the victim node, and so that the sum of all ground capacitances before and after redistribution is the same. The same algorithm is also used for coupling capacitances to other conductors. The 'other' end of a coupling capacitor is preserved, and this step thus actually splits and distributes coupling capacitors. Subsequently, the node (which is then free of any capacitors) is eliminated using Gaussian elimination. This elimination step is repeated for all internal nodes, until only port nodes remain. The final result is a smaller circuit where the DC resistance as well as the Elmore delay between each (directed) pair of remaining nodes is unchanged. Moreover, the total capacitance to ground as well as between each pair of conductors is also preserved. Consequently, the reduced model is accurate for DC up to a frequency roughly corresponding to the first pole of the system, since only the first dominant time constant is matched.

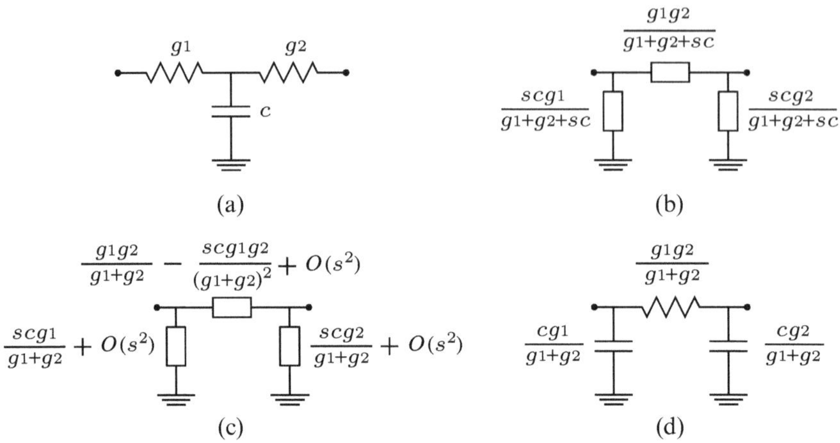

Fig. 3. Example circuit for Elmore preserving reduction (a), its frequency domain representation with node 1 eliminated (b), the same form with a series expansion of the admittance terms (c), its reduced form (c) and its realizable reduced form that preserves the Elmore delay properties (d).

The next subsection will present a way to maintain accuracy until higher frequencies by preserving some internal nodes of the network.

3.3 Partial Elimination

EPR preserves the first moment of the impulse response, which is in fact sufficient for many applications, mostly in RC interconnections in digital circuits. This is of course related to the essentially low-pass behavior of such interconnections. Rather than trying to preserve more moments while still doing a complete elimination (this will be investigated in the next section), we will now consider partial elimination procedures. The idea is that the original, unreduced, circuit is completely accurate, and that inaccuracies are introduced and grow with each node that is eliminated. By not eliminating all nodes but only a judiciously chosen subset of them, the original circuit behavior can be better preserved while hopefully the amount of reduction is not hampered too much.

Such a procedure was first proposed in [7]. It operates by recognizing that, since EPR preserves the first moment of the impulse response, any error is in the 2nd and higher moments. This error grows with increasing frequency f, since f is squared in the contribution of the 2nd order terms. Hence, by having a user-specified highest frequency of interest (f_s) and evaluating the magnitude of the 2nd moment at f_s *if the node would be eliminated*, its actual elimination can be prevented if the resulting error would exceed a threshold.

As such, the amount of reduction is controllable by f_s. At low values, all error weights are low and all nodes can be eliminated—in the limit the DC behavior is always accurate since the 2nd moment (actually, all moments from order 1 and up) are inconsequential. At growing values of f_s, more and more nodes of the circuit need to be preserved for maintaining accuracy of the reduced network at frequencies from DC to f_s.

More specifically, the error weight is calculated for all nodes before elimination of any node. Then, the error weights are ranked and the node with the smallest error weight is reduced first. Subsequently, some error weights (those of the original neighbors of the node that was deleted) have to be recalculated, and again the node with the smallest error weight is reduced, etc., until the smallest error weight exceeds a threshold in which case the reduction stops. The remaining nodes are apparently essential for an accurate model from DC to f_s.

Actually, the above Selective Node Elimination (SNE) procedure is heuristic in nature, and the precise definition of the error weight can be done in different ways. This issue was investigated in [9], the result being that almost the simplest definition is generally very effective. As such, the error weight is the maximum of all *shunt errors* for a node, the latter being defined as:

$$\delta_{kj} = \lim_{s \to 2j\pi f S} \left| \frac{s^2 y_{kj}^{(2)}}{y_{kj}^{(0)} + s y_{kj}^{(1)}} \right|$$

where $y_{kj}^{(l)}$ is the l-th moment of y_{kj} after (supposed) elimination of a node x which is a common neighbor of nodes k and j before elimination. Here, the name *shunt error* relates to the fact that in general, there can already be a resistor and/or a capacitor between node k and j before elimination of node x, and after elimination an error is made since $y_{kj}^{(2)}$ can not be represented using Rs and Cs. Furthermore, we note that a further simplification by only looking at the shunt errors to ground (i.e. index $j = 0$) is typically also effective and further reduces the calculation time for the error weight from being quadratic in the degree of a node to only linear.

By way of example, consider a series RC circuit. The physical (IC layout) implementation is shown in Figure 4(a). While the ideal behavior of this device is

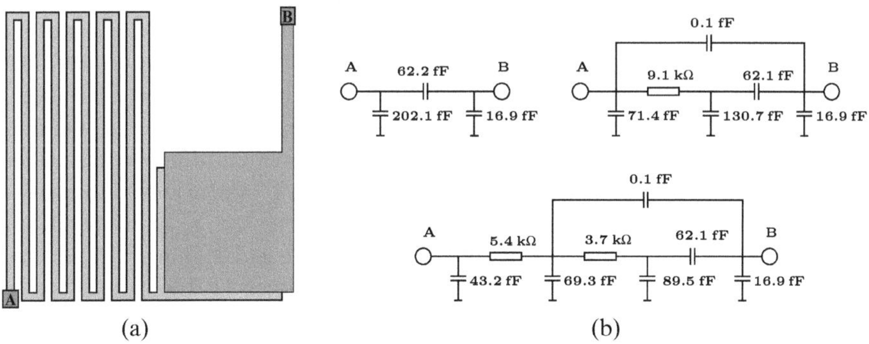

Fig. 4. (a) Layout of a poly resistor (light) on top of which a metal plate capacitor (dark). (b) Circuits extracted from layout of RC network, in order of increasing complexity: EPR model, f_s = 100 MHz, f_s = 200 MHz

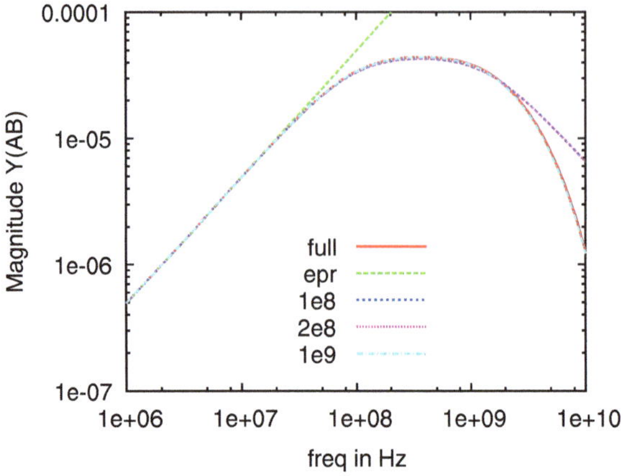

Fig. 5. Magnitude of the transadmittance of the RC structure.

showing an impedance of $r + 1/sc$, with r and c the nominal values, its physical realization shows far more complicated behavior. To analyze this behavior, a detailed model was computed using the SPACE layout to circuit extractor [25], employing the *boundary element method* for the parasitic capacitances among all segments, and a *finite element method* for the resistance. In this way, a discretized model with 109 nodes, 164 resistances and 734 capacitances was extracted.

This detailed circuit was reduced using both the EPR method, as well as using the SNE method. Figure 4(b) shows some resulting networks, in particular for the EPR method and for 2 different frequency settings of the SNE method. Also, Figure 5 shows the magnitude of its transadmittance for the full (unreduced) network, the result of the EPR procedure and for three different SNE frequencies. (Actually, note that when the f_s setting is low enough, the SNE method reduces to the EPR method.) Clearly, the SNE method preserves the accuracy of the transadmittance until frequencies at least as high as the specified f_s. The phase response was not plotted since it does not offer any additional insights over the magnitude response.

While the SNE method [7] aims at controlling the order and extend of node elimination by quantification of the error associated with the second moment of the impulse response, another technique called TICER was developed later in [21]. This method does not directly refer to any moments, rather it defines for each node a so called nodal time constant being the ratio of the total capacitance to the total admittance attached to a node. While its original derivation was different, the method can be understood by noting that the reactive term (the 1st order term) in Equation (19) is negligible compared to the resistive term when $sc << \sum_i g_i$. Generalizing for the case with multiple capacitors attached to the node and defining $\sum_i c_i$ as their sum, we have the so-called quick node elimination rule of TICER: a node can be eliminated when $s\frac{\sum_i c_i}{\sum_i g_i} << 1$, and the result is exactly the same network as implied by the SNE method and as illustrated for a simple T-network in Figure 3.

However, [21] does not mention any ordering of the elimination except for giving preference to nodes of low degree. This would be advantageous from a CPU-time perspective, but may not allow the largest reduction.

4 Elimination for Synthesis

Rather than complete or selective elimination while still requiring that the result should be maintained as an RC circuit, it is also feasible to work with higher-order expansions and have a subsequent realization step. Here, the requirements for realization are a slightly relaxed as compared to the general realization problem, since only 1-port admittances need to be synthesized with magnitude and phase behavior specified as a function of s. Historically, the first of such techniques was presented in [8], it represented the admittances as moment series in s. Later, reference [18] presented a technique that uses rational functions for the higher order admittances, which avoids the well known ill-conditioning problem of moment series. As such, it would also be better suited for reduction of RLC circuits.

4.1 Reduction Using Moment Series

First, assume a single conductor in an RC network. Thus, all nodes are connected via one or more resistances. Some nodes will be eliminated such that the result is exact up to a user-specified order where each admittance is presented as a Taylor series expression around $s = 0$. As in Figure 1 and Equation (3), Gaussian elimination again produces admittances in the reduced network as

$$y_{kj} = \frac{y_k y_j}{\sum_i y_i}. \tag{21}$$

However, now we take the admittances to be expressed as

$$y(s) = y^{(0)} + y^{(1)}s + \ldots = \sum_{l=0}^{\infty} y^{(l)} s^l$$

where we use the superscript $^{(i)}$ notation rather than the subscript notation for consistency with the other moment series expressions. Trivially, resistors and capacitors in the original network correspond to such a series with respectively only the first and second moment being non-zero. With (21) we get

$$\sum_{l=0}^{\infty} y_{kj}^{(l)} s^l = \frac{\sum_{l=0}^{\infty} \left(\sum_{p=0}^{l} y_i^{(l)} y_j^{(l-p)} \right) s^l}{\sum_{l=0}^{\infty} \left(\sum_i y_i^l \right) s^l}$$

or

$$\sum_{l=0}^{\infty}\left(\sum_{p=0}^{l} y_i^{(l)} y_j^{(l-p)}\right) s^l = \left(\sum_{l=0}^{\infty} y_{kj}^{(l)} s^l\right)\left(\sum_{l=0}^{\infty}\left(\sum_i y_i^l\right) s^l\right)$$
$$= \sum_{l=0}^{\infty}\left(\sum_{p=0}^{l}\left(y_{kj}^{(p)} \sum_i y_i^{(p-l)}\right)\right) s^l \tag{22}$$

The lth moment can be easily extracted, which is itself a sum over p running from 0 to l. This gives rise to a recursive procedure for $y_{kj}^{(l)}$ by separating-off the lth term in the RHS. In this way, Equation (22) leads to

$$\sum_{p=0}^{l} y_i^{(l)} y_j^{(l-p)} = \sum_{p=0}^{l-1}\left(y_{kj}^{(p)} \sum_i y_i^{(p-l)}\right) + y_{kj}^{(l)} \sum_i y_i^{(0)}$$

and reordering gives the following recursive expression for $y_{kj}^{(l)}$:

$$y_{kj}^{(l)} = \frac{\sum_{p=0}^{l} y_i^{(l)} y_j^{(l-p)} - \sum_{p=0}^{l-1}\left(y_{kj}^{(p)} \sum_i y_i^{(p-l)}\right)}{\sum_i y_i^{(0)}} \tag{23}$$

Because we assumed that all nodes were resistively coupled, the denominator never becomes zero. In fact, it is only required that each node to be deleted is connected to at least one resistor. If that would not be the case, the recursion can be rewritten to start at the first moment and the denominator would sum all capacitors attached to the node [8].

Hence, the recursive procedure from Equation (23) can be used to eliminate all internal nodes in a circuit, resulting in a reduced netlist where all connections are generalized admittances expressed as a Taylor expansion of $y(s)$. For practical purposes, this expression is usually truncated at an order β as follows:

$$y_{kj} = \sum_{l=0}^{\beta} y_{kj}^{(l)} s^l \tag{24}$$

The zero and first order moments correspond to a parallel connection of a resistor (of value $1/y_{kj}^{(0)}$) and a capacitor (of value $y_{kj}^{(1)}$) respectively. The higher order moments are non-physical. Therefore, it was proposed in [8] to subsequently transform such admittances into a rational function using Padé matching. Then, time domain analysis would be possible after network synthesis, or using the inverse Laplace transform.

However, moments based expressions are not well conditioned, and in fact this method is suffering from disadvantages similar to those of the AWE method [3, 17]. In particular, only a very small number of moments carry significant information. Moreover, note that truncation of such a series, in general, may render it unstable. In conclusion, this method is less interesting from a practical point of view. See the next subsection for a more viable alternative.

4.2 Reduction Using Rational Functions

In this subsection, an alternative method will be presented that is more difficult to implement, but gives better results. This method was recently published in [18]. The method processes the admittance functions in rational polynomial form:

$$y(s) = \frac{\sum_{i=0}^{\alpha} n_i s^i}{\sum_{j=0}^{\alpha} d_j s^j} \tag{25}$$

This method performs model order reduction by truncating the numerator and denominator polynomials at a certain degree, indicated by α in the equation above. This is actually a proper truncation, in the sense that upon increasing α to $\alpha' > \alpha$, a more complex model will arise, but with the first α terms identical to that of the order α model.

During the elimination, care has to be taken to cancel common factors in the numerators and denominators from (25). This is necessary in order to avoid numerical explosion of the coefficients. Such common factors between numerator and denominator arise directly upon elimination of a node where all the admittances connected to it have the same (common factor in the) denominator. These are referred to as *Type I* common factors, and are relatively easy to detect and correct.

Furthermore, the method takes into account that there are also hidden common factors between rational functions resulting from the elimination of neighboring nodes. These are more difficult to detect, because they result in common factors which appear *explicitly* in the denominator, but only *implicitly* in the numerator. The method also identifies these common factors (referred to as *Type II* common factors) and cancels them to keep the relevant data as clean as possible.

In [20], the common factor mechanism was illustrated with a simple example that starts with an RC network as in Figure 1 with $n = 3$. The demonstration proceeds by performing the elimination of node x as follows

$$\begin{aligned} y_{12} &= \frac{y_1 y_2}{y_1+y_2+y_3} = \frac{y_1 y_2}{\Sigma} \\ y_{13} &= \frac{y_1 y_3}{y_1+y_2+y_3} = \frac{y_1 y_3}{\Sigma} \\ y_{23} &= \frac{y_2 y_3}{y_1+y_2+y_3} = \frac{y_2 y_3}{\Sigma} \end{aligned} \tag{26}$$

We observe that y_{12}, y_{13} and y_{23} share a common denominator $y_1 + y_2 + y_3$, which we will indicate with Σ for more convenient notation later on. The result of this elimination step being a circuit with 3 nodes, we can, w.o.l.g., continue by eliminating node 3 as well and we proceed in 2 steps. First, we calculate the elimination formula for node 3:

$$y_{12}' = \frac{y_{13} y_{23}}{y_{13} + y_{23}}$$

Writing it out in full

$$y_{12}' = \frac{\frac{y_1 y_3}{\Sigma} \cdot \frac{y_2 y_3}{\Sigma}}{\frac{y_1 y_3}{\Sigma} + \frac{y_2 y_3}{\Sigma}}$$

we observe that Σ is a common factor in both numerator and denominator, which allows it to be canceled. In [18], this common factor is referred to as a *Type I* common factor. After cancellation, the result is

$$y'_{12} = \frac{y_1 y_2 y_3^2}{\Sigma \cdot (y_1 y_3 + y_2 y_3)}$$

This result is then placed in parallel with the already existing admittance y_{12}

$$y''_{12} = y_{12} + y'_{12} = \frac{y_1 y_2}{\Sigma} + \frac{y_1 y_2 y_3^2}{\Sigma \cdot (y_1 y_3 + y_2 y_3)} \quad (27)$$

Evaluating this expression, we arrive at

$$y''_{12} = \frac{y_1^2 y_2 y_3 + y_1 y_2^2 y_3 + y_1 y_2 y_3^2}{\Sigma \cdot (y_1 y_3 + y_2 y_3)}$$

which can then be rewritten as

$$y''_{12} = \frac{(y_1 + y_2 + y_3) \cdot (y_1 y_2 y_3)}{\Sigma \cdot (y_1 y_3 + y_2 y_3)}$$

Recall that $y_1 + y_2 + y_3 = \Sigma$, which results in

$$y''_{12} = \frac{\Sigma \cdot (y_1 y_2 y_3)}{\Sigma \cdot (y_1 y_3 + y_2 y_3)} = \frac{y_1 y_2 y_3}{y_1 y_3 + y_2 y_3} \quad (28)$$

Here we observe that Σ also appears as a factor in the numerator, but only implicitly. In [18], this common factor is referred to as a *Type II* common factor.

The fact that the Type II common factor occurs only implicitly in the numerator, makes it harder to detect. Fortunately, it is known under which conditions it appears: when 2 admittances connected to a node that is being eliminated share factors in their denominator (viz. y_{13} and y_{23} in Equation (26)) and these same factors are also present in the numerator of the existing admittance (viz. y_{12}) with which the resulting admittance will be placed in parallel (viz. Equation (27)), then these factors will implicitly appear in the numerator of the final result (viz. y''_{12} in Equation (28)).

When the elimination procedure has finished, the resulting rational function cannot immediately be processed by a regular circuit simulator. Therefore, the method synthesizes a network from the rational function through Brune synthesis [1] after stabilization [19], because truncation does in general not preserve stability. Subsequently, the resulting network can be simulated with a circuit simulator.

4.3 Example

In this subsection, taken from [20], we will perform a frequency domain comparison of the methods from [8] and [18]. Consider the RC circuit of Figure 6, representative for an integrated resistor. While a pure resistive behavior would be the design goal, it actually suffers from capacitive parasitics that influence its high-frequency characteristics.

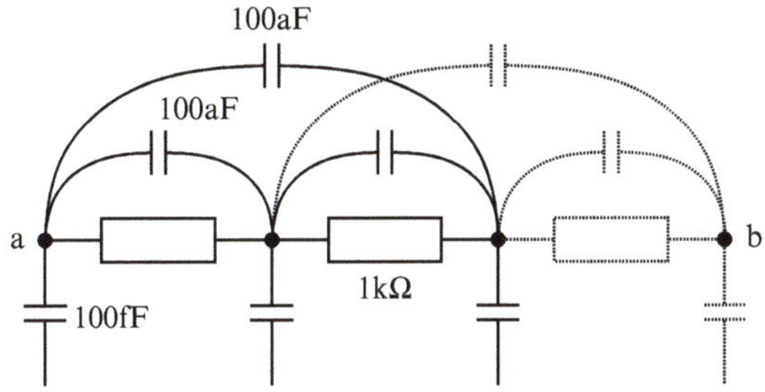

Fig. 6. RC circuit with terminals a and b. Only 2 internal nodes are drawn, but the actual circuit has 7 internal nodes; it contains 6 additional sections of the type drawn in dotted lines. All resistors are 1kΩ, ground capacitances are 100fF and coupling capacitances are 100aF.

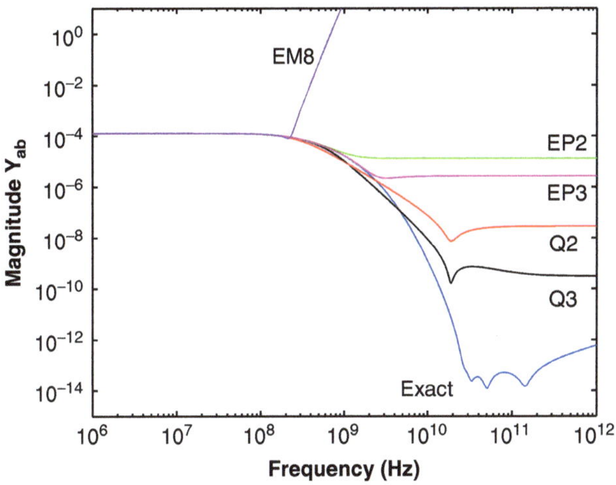

Fig. 7. Magnitude of the transmittance between nodes a and b from Figure 6. Legend: Q_i = Qin ith order, EP_i = Elias-Padé ith order, EM_8 = Elias-Moment series 8th order

Figure 7 shows the magnitude of y_{ab} as a function of the frequency, obtained by using the different methods. Here, the curve labeled Exact is obtained using a Spice frequency domain sweep of the network in Figure 6. The curves labeled Q_i are obtained using the method from Qin and Cheng [18], with order (α in (25)) i. The curve labeled EM_8 is obtained by using the 8th order moments expansion in Equation (24) (i.e. $\beta = 8$). The curves labeled EP_i are from the i-th order Padé approximations based on the EM_8 result.

A discussion of these results is as follows.

It is clear that the moment expansion EM_8 only has a limited range of accuracy. Furthermore, it is only marginally better than a 6th order moments expansion (not shown). This can actually be expected from the properties of moment representations [3]. On the other hand, the good matching properties of the Padé technique for electrical circuits, and RC circuits in particular, are obvious from the EP_2 and EP_3 curves. While they are actually derived from the EM_8 result, they match the exact result over a greater frequency range. However, Q_2 and Q_3 are still better, while they actually are of the same complexity as the EP_2 and EP_3 results, respectively.

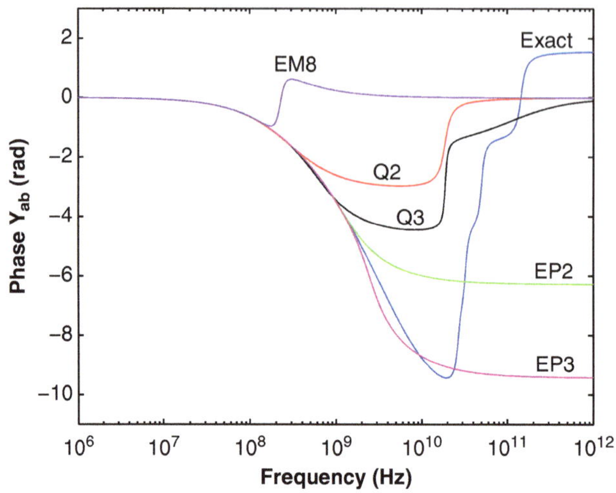

Fig. 8. Phase of the transmittance between nodes a and b from Figure 6. Legend: Q_i = Qin ith order, EP_i = Elias-Padé ith order, EM_8 = Elias-Moment series 8th order

Figure 8 shows the phase behavior. Here, the result is not so clear. The global shape of the phase plot is better matched by the Q_i curves than by the EP_i curves, while the latter actually follow the exact curve over a greater range.

4.4 Conclusion

In this section, we have compared two methods of producing higher order interconnect models using complete Gaussian elimination of all non-terminal nodes, and using rational or moment based approximations (in s) of the resulting admittances. In the frequency domain, the rational approximation approach seems to work better. Furthermore, since the moments-based approach is in fact an expansion around $s = 0$, it actually presupposes a low-pass character of the networks. However, as was actually demonstrated in [18], reduction using rational functions is also capable of reducing RLC networks rather than RC networks.

5 Computational Aspects

One of the most striking issues related to the performance of direct methods for solving systems of equations is that of elimination ordering. Consider Figure 9, where the small dots are internal nodes to be eliminated and the large dots are port nodes to be preserved. This figure illustrates 2 sequences of elimination, a partial sequence where the center node with the highest degree is eliminated first, and a full sequence where each time a node with the lowest degree is chosen to be eliminated. Clearly, the latter sequence is better than the first one, because of the so-called *fill-in* created from eliminating a high-degree node. When elimination of a high-degree node

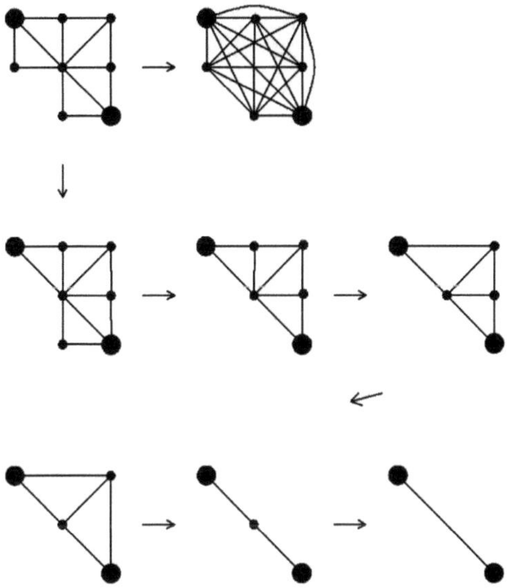

Fig. 9. Illustration of the effect of different elimination orderings.

produces fill-in, the subsequent steps also become more expensive. In general, the elimination of a node with degree d takes $O(d^2)$ time and can produce $O(d^2)$ fill-in branches.

Since the problem of computing an optimal elimination order is NP-complete [30], heuristic methods are necessary. Indeed, many heuristics have been suggested (see [6] for a summary and comparison) and one of the best heuristics for the type of problems considered (i.e. symmetric, sparse and positive definite) is the minimum degree heuristic [16,23]. With this heuristic, nodes are eliminated in order of ascending elimination degree d. Ties are broken arbitrarily.

In this respect, it can be noted that the EPR method from Section 3.2 as well as the moment-based and rational-function based methods from Sections 4.1 and 4.2 would allow optimized elimination orderings. However, the Selective Node Elimination technique from Section 3.3 (or TICER [21], for that matter) does not allow such ordering since the main ordering is on the error weight: nodes with a high error weight should be eliminated before other nodes.

While some techniques indeed allow optimizing the elimination order, it would be even more advantageous if these could be combined with the process of actually computing the initial circuit. In fact, as soon as for a node all its neighbors are known, a node can be eliminated. Doing so, a process which is called 'on-the-fly' reduction, would be optimal in the number of nodes that resides in the core memory of the computer. However, because of the fill-in associated with eliminating nodes of a high degree, the actual memory requirements as well as CPU time could be much higher than would be the case for the optimal elimination ordering. In general, the order in which the nodes become available for elimination is not the same as their optimal or near-optimal elimination ordering.

For example, the SPACE layout to circuit extractor [25], generates the nodes in an order that is determined by the scanline algorithm that processes the layout. This scanline moves from left to right over the layout, and at each scanline stop the scanline is traversed from bottom to top. During this scanline processing, all relevant calculations are performed using as much as possible only the spatially localized layout data.

With respect to optimizing the elimination order during on-the-fly generation and reduction, reference [24] shows that a relatively short, fixed-length priority queue of nodes ordered by their degree is sufficient to practically get the performance of elimination in minimum-degree order. Each time when, during the build-up of the detailed RC network, a node becomes ready for elimination (because it is an internal node and all its neighbors are known), it is only inserted into the priority queue. If the queue, which has a fixed size, would be full prior to insertion, first a node with the lowest degree is removed from the queue and subjected to Gaussian elimination. At the end, all remaining nodes in the queue are eliminated. This technique is called *Delayed Frontal Solution*.

Figure 10(a) shows the effect of the length of the priority queue on the elimination degree d, a queue length of 0 means immediate elimination as soon as a node becomes ready, increasing the queue size clearly reduces the average value of d. The net effect of this on the total memory and CPU time is illustrated in Figure 10(b). Evidently, a small queue size can already give a significant speedup and a slightly reduced amount of *total* memory, while a large queue can give more than an order of magnitude speedup at the cost of extra memory.

Additional techniques for speeding-up elimination-based MOR have been presented in [29]. These techniques basically partition a problem into a number of smaller problems to be solved independently, but another view on this is that they eliminate fill-in between nodes from different sections. The technique is illustrated in Figure 11; (a) shows a section of a layout while (b) shows the corresponding combined rectangular/triangular mesh for resistance extraction using the Finite Element

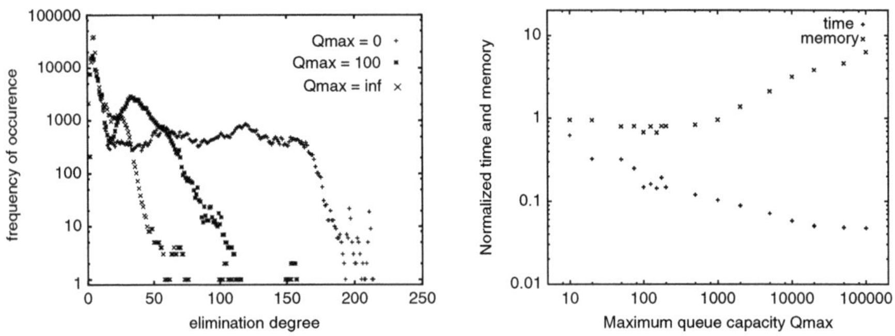

(a) (b)

Fig. 10. Elimination degree frequency for a 1753 transistor chip while extracting all resistances including well resistances for three different values of Q_{max} (a) and normalized computation time and memory as a function of Q_{max} (b) [24].

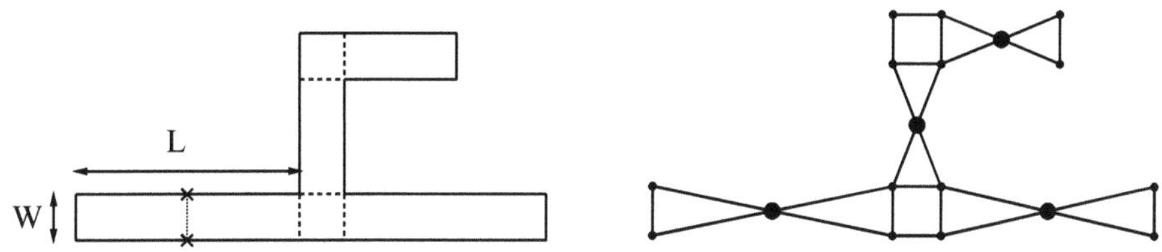

Fig. 11. Illustration of using equipotential lines to introduce articulation nodes in the resistance network obtained from a FEM discretization.

method. A linear finite element mesh for the Laplace/Poison equation is actually equivalent to a resistance network [15, 26], which can consequently be solved using Gaussian Elimination.

The nodes of the resistance network are given by the bullets, while the thick bullets are special and are called *articulation nodes*. These are not in the discretization by nature, but are obtained from contracting two nodes that are on opposite sides of this articulation node along the long side of a square conducting region. The rationale for this is that the potential along a cross-section of the layout that is perpendicular to the current flow is constant. Such a situation is depicted in Figure 11(a). If the ratio L/W is large enough, the current flow is perpendicular to the dotted line joining the crosses. While a valid and accurate resistance network can be created by putting mesh nodes at the location of the crosses with an edge (resistor) joining them, these two nodes can be contracted without losing accuracy. Applying this technique for each 'long' rectangle in Figure 11(a) would result in the articulation nodes of Figure 11(b).

The significance of such articulation nodes lies in the fact that they can be eliminated after all the other nodes. (This can be nicely combined with the ordering by degree by giving articulation nodes the lowest priority.) In that case, no fill-in can be generated that spans two articulation groups or, from another viewpoint, each articulation group can be solved independently with a very low cost of combining the partial solutions. Hence, a $O(N^p)$ problem with $p > 1$ is replaced by m problems of size $O(N/m)$ for a complexity gain of $O(m^{(p-1)})$. Also, if the maximum size of each articulation group is bounded from above, the solution becomes linear in m and hence in N (the problem size). This boundedness of the group size is actually very often achieved or nearly achieved by nature of the typical IC interconnect geometries. Additional articulation nodes can furthermore be generated if there is a large difference between conductivity of connected interconnect segments [29].

Please note that Delayed Frontal Solution and Articulation Node Partitioning ideally are combined. The result is shown in Figure 12, which clearly illustrates that the elimination degree d is typically much lower when these techniques are applied. The lower elimination degree causes a significant reduction in CPU time (more than proportional) and memory.

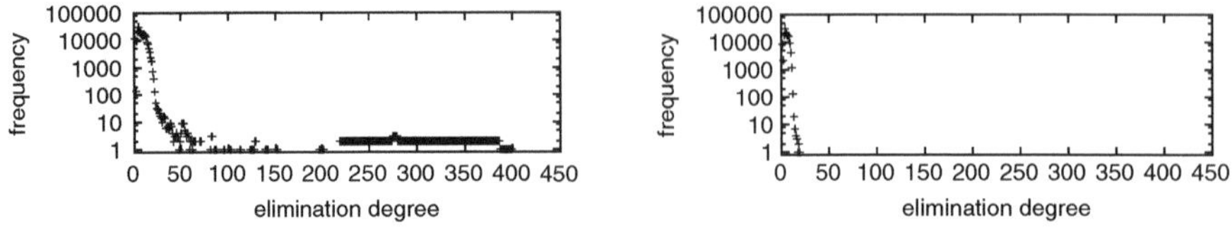

(a) (b)

Fig. 12. Elimination degree frequency for a 1467 transistor design without using articulation nodes (a) and when using equipotential lines and equipotential areas for introducing articulation nodes (b) [29].

6 Conclusion

This chapter has focused on direct techniques for Model Order Reduction of RC circuits. In particular, Gaussian Elimination, or equivalently, Y transformation, was discussed.

One of the main strong points of direct techniques relates to what is called 'on-the-fly' reduction as has been discussed in Section 5. In practice, such a technique can drastically reduce the amount of circuit data to be stored for layout to circuit extraction as compared to outputting a full, unreduced, netlist, enabling the handling of much bigger designs using the same amount of disk storage. In many cases, the amount of reduction achievable by direct techniques is sufficient for subsequent simulation, while the computation time for performing the reduction is

Additional advantages of direct (elimination-based) techniques for MOR include performance - these techniques can exhibit linear or close to linear CPU time and sub-linear memory requirements, as has been explained in Section 5. Also, in contrast to the projection-based techniques, these techniques do not have problems with networks with many ports while simultaneously the realization problem is non-existent or trivial.

In some cases, particularly when using the articulation node technique, the resulting models preserve sparsity of the input. In other cases, sparsity is lost in theory (e.g. the EPR technique produces a full graph network on the port nodes), but can be easily restored using a heuristic simplification procedure. Such a procedure was not discussed in this chapter, but it is implemented in the SPACE layout to circuit extractor [25]. Such heuristic procedures work by dropping or combining large shunt paths or small series paths.

While the techniques that were developed in this chapter do seem to be very useful on their own, please note that they can also work in combination with iterative, projection based, techniques, for example as a pre-processing step.

Acknowledgements

All the the knowledge and results as developed in this chapter naturally resulted from working together. For important contributions to this work, the author wishes in particular to gratefully acknowledge Arjan van Genderen, Peter Elias and Eelco Schrik.

References

1. V. Belevitch. *Classical Network Theory*. Holden Day, 1968.
2. Tsung-Hao Chen, Jeng-Liang Tsai, Charlie C-P Chen, and T Karnik. HiSIM: hierarchical interconnect-centric circuit simulator. In *Computer Aided Design, 2004. ICCAD-2004. IEEE/ACM International Conference on*, pages 489–496, November 2004.
3. E. Chiprout and M.S. Nakhla. *Asymptotic Waveform Evaluation and Moment Matching for Interconnect Analysis*. Kluwer Academic Publishers, 1994.
4. F. J. R. Clement. *Computer Aided Analysis of Parasitic Substrate Coupling in Mixed Digital-Analog CMOS Integrated Circuits*. PhD thesis, EPFL, Lausanne, 1995.
5. Richard W. Cottle. Manifestations of the schur complement. *Linear Algebra and its Applications*, 8(3):189–211, June 1974.
6. I.S. Duff, A.M. Erisman, and J.K. Reid. *Direct Methods for Sparse Matrices*. Oxford Science Publications, 1986.
7. P. J. H. Elias and N. P. van der Meijs. Extracting circuit models for large RC interconnections that are accurate up to a predefined signal frequency. In *Proc. 33rd Design Automation Conf.*, pages 764–769, Las Vegas, Nevada, June 1996.
8. P. J. H. Elias and N. P. van der Meijs. Including higher-order moments of RC interconnections in layout-to-circuit extraction. In *Proc. European Design and Test Conf.*, pages 362–366, Paris, France, March 1996.
9. P.J.H. Elias. Theory of selective node elimination. Technical Report Unpublished Notes, Delft University of Technology, Delft, The Netherlands, 1995.
10. W.C. Elmore. The transient response of damped linear networks with particular regard to wideband amplifiers. *J. Applied Physics*, 19:55–63, January 1948.
11. M.K. Enns and J.J. Quada. Sparsity-enhanced network reduction for fault studies. *Power Systems, IEEE Transactions on*, 6(2), May 1991.
12. G. H. Golub and C. F. Van Loan. *Matrix Computations*. The Johns Hopkins University Press, 1989.
13. R. Gupta, B. Tutuianu, and L.T. Pileggi. The Elmore delay as a bound for RC trees with generalized input signals. *IEEE Transactions on Computer-Aided Design of Integrated Circuits and Systems*, 16(1):95–104, 1997.
14. T.A. Johnson. Resistive network simplification technique. *IBM Technical Disclosure Bulletin*, 24(5):2647–2649, Oct 1981.
15. P. Kazil and P. Dewilde. A simple and fast method for obtaining resistance of VLSI interconnect. In *Proc. IEEE Int. Conf on Computer Design*, pages 342–345, October 1986.
16. H.M. Markowitz. The elimination form of the inverse and its application to linear programming. *Managment Sci.*, 3:255–269, 1957.
17. L.T. Pillage and R.A. Rohrer. Asymptotic waveform evaluation for timing analysis. *Computer-Aided Design of Integrated Circuits and Systems, IEEE Transactions on*, 9(4):352–366, 1990.
18. Zhanhai Qin and Chung-Kuan Cheng. Realizable parasitic reduction using generalized $Y\Delta$ transformation. In *Design Automation Conference, 2003. Proceedings*, pages 220–225, June 2003.
19. E.L. Peterson R.J. Duffin and C. Zener. *Geometric Programming: Theory and Application*. John Wiley and Sons, New York, 1967.
20. E. Schrik and N. P. van der Meijs. Comparing two $Y\Delta$ based methodologies for realizable model reduction. In *ProRISC IEEE 14th Annual Workshop on Circuits, Systems and Signal Processing*, November 2003.

21. B.N. Sheehan. TICER: Realizable reduction of extracted RC circuits. In *Computer-Aided Design, 1999. Digest of Technical Papers. 1999 IEEE/ACM International Conference on*, pages 200–203, 1999.
22. W.F. Tinney and J.M. Bright. Adaptive reductions for power flow equivalents. *IEEE Transactions on Power Systems*, PWRS-2:351–360, 1987.
23. W.F. Tinney and J.W. Walker. Direct solutions of sparse network equations by optimally ordered triangular factorization. *Proc. IEEE*, 55:1801–1809, 1967.
24. N. P. van der Meijs and A. J. van Genderen. Delayed frontal solution for finite-element based resistance extraction. In *Proc. 32nd Design Automation Conf.*, pages 273–278, San Francisco, California, June 1995.
25. N.P. van der Meijs, A.J. van Genderen, et al. Space layout to circuit extractor home page, March 2007. URL: http://space.tudelft.nl.
26. A. J. van Genderen. *Reduced Models for the Behavior of VLSI Circuits*. PhD thesis, Delft University of Technology, Delft, The Netherlands, October 1991.
27. A. J. van Genderen and N. P. van der Meijs. Extracting simple but accurate RC models for VLSI interconnect. In *Proc. Int. Symp. on Circuits and Systems*, pages 2351–2354, Helsinki, Finland, June 7-9 1988.
28. A. J. van Genderen and N. P. van der Meijs. Reduced RC models for IC interconnections with coupling capacitances. In *Proc. IEEE 3rd European Design Automation Conf.*, pages 132–136, Brussels, Belgium, March 1992.
29. A. J. van Genderen and N. P. van der Meijs. Using articulation nodes to improve the efficiency of finite-element based resistance extraction. In *Proc. 33rd Design Automation Conf.*, pages 758–763, Las Vegas, Nevada, June 1996.
30. M. Yannakis. Computing the minimum fill-in is np-complete. *Siam J. Alg. Disc. Meth.*, 2:77–79, 1981.

Reduced Order Models of On-Chip Passive Components and Interconnects, Workbench and Test Structures

Daniel Ioan and Gabriela Ciuprina

Politehnica University of Bucharest, Romania
lmn@lmn.pub.ro

1 Extraction of the EM-FW State Models for Passive Components

The models of passive components have to describe all relevant electromagnetic field effects at high frequency encountered inside these devices. These effects are quantified by the Maxwell equations of the electromagnetic field in full wave (FW) regime. Therefore, at the first level of approximation, the model of a passive device is defined by an electromagnetic (EM) field problem, formulated by Maxwell partial differential equations with appropriate boundary and initial conditions. This problem defines a consistent I/O system which has a unique response, described by the output signals, for any input signal applied as terminal excitations. This system with distributed parameters has an infinite dimension state space, but a finite number of inputs and outputs related to the device terminals.

The next level of approximation in the modeling process (Fig. 1) results by applying a numerical method to discretize the continuous model defined above. This step associates a simpler ODE to the original PDE model, actually a system of DAE. It is an important step ahead, reducing the infinite dimensional state-space which is specific to distributed systems to a finite one. However, the size of the state-space is still too large for the designers needs. It has an order similar to the number of DOFs associated to the cells, finite elements used to discretize the computational domain. That is why a third step is necessary (Fig. 1), aiming to reduce this order, and to generate a "compact" model, e.g. a small SPICE circuit, which preserves the behaviour of the passive component, from terminals point of view, for instance the input-output relationship.

The basic equations used to model the electromagnetic field effects in any device without movement, including the integrated passive devices are the **Maxwell equations**:

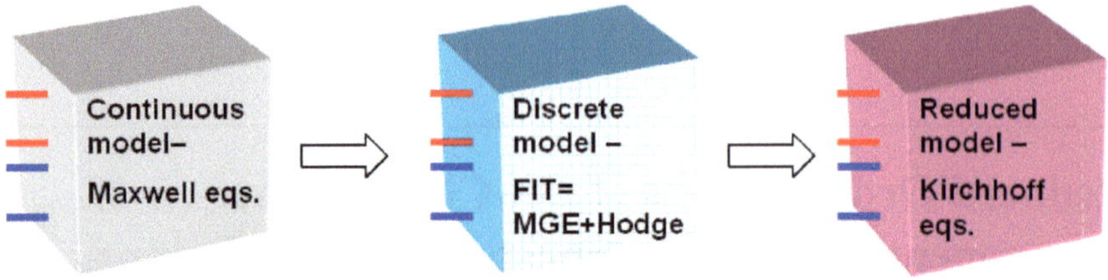

Fig. 1. Three levels of abstraction for a component model and its corresponding equations.

$$\operatorname{curl} \mathbf{H} = \mathbf{J} + \frac{\partial \mathbf{D}}{\partial t}, \tag{1}$$

$$\operatorname{curl} \mathbf{E} = -\frac{\partial \mathbf{B}}{\partial t}, \tag{2}$$

$$\operatorname{div} \mathbf{D} = \rho, \tag{3}$$

$$\operatorname{div} \mathbf{B} = 0. \tag{4}$$

These equations have to be complemented with the constitutive equations, which describe the material behaviour, from electromagnetic point of view. In the linear and isotropic materials, these constitutive equations have the simplest form:

$$\mathbf{D} = \varepsilon \mathbf{E}, \tag{5}$$

$$\mathbf{B} = \mu \mathbf{H}, \tag{6}$$

$$\mathbf{J} = \sigma \mathbf{E}. \tag{7}$$

The three material "constants" have nonnegative values, dependent on position: $\varepsilon(M)$, $\mu(M)$ and $\sigma(M)$. Usually, these functions are constant on sub-domains, separated by discontinuity interfaces between homogeneous materials. The solutions of these equations are the following "fields": $\mathbf{E}(M,t)$, $\mathbf{D}(M,t)$, $\mathbf{B}(M,t)$, $\mathbf{H}(M,t)$, $\mathbf{J}(M,t)$, $\rho(M,t)$, vector or scalar functions of position (M) and time (t). The correct formulation of the continuous field problem (1)÷(7) consists of the identification of the appropriate boundary conditions able to allow the consistent field-circuit coupling. To conclude, the extraction of reduced order models for passive components requires the solution of the correct formulated EM field problem, particularly associated to the concept of Electric Circuit Element defined below [IM99].

An **Electric Circuit Element** *(ECE) is a simply connected domain bounded by a fixed surface Σ comprising n' disjoint parts $S'_1, S'_2, \ldots S'_n$, called electric terminals on which:*

$$\mathbf{n} \cdot \operatorname{curl} \mathbf{E}(P,t) = 0 \quad \text{for } (\forall)\, P \in \Sigma, \tag{8}$$

$$\mathbf{n} \cdot \operatorname{curl} \mathbf{H}(P,t) = 0 \quad \text{for } (\forall)\, P \in \Sigma - \cup S'_k, \tag{9}$$

$$\mathbf{n} \times \mathbf{E}(P,t) = \mathbf{0} \qquad \text{for } (\forall)\, P \in \cup S'_k, \tag{10}$$

where $\mathbf{n}$ *is the normal unitary vector of Σ in the point P (Fig. 2).*

Condition (8) **interdicts the inductive couplings** through the device boundary, between the domain and its environment. Condition (9) **interdicts the conductive**

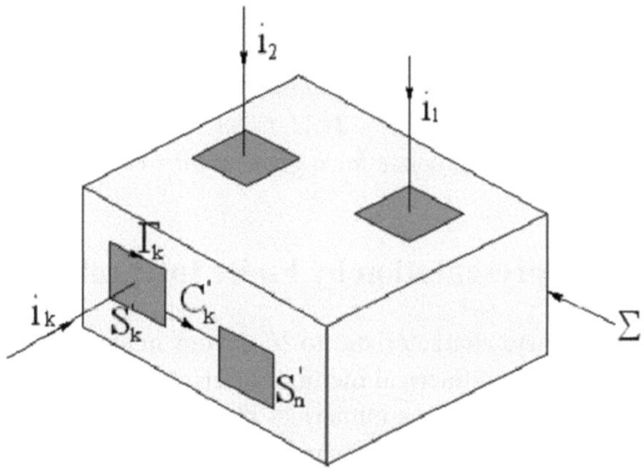

Fig. 2. ECE - Electric Circuit Element with multiple terminals.

and capacitive couplings through the device boundary, excepting for the electric terminals. Condition (10) **interdicts the variation of the electric potential over every electric terminal**, allowing the connection of the ECE terminals to the nodes of any external electric circuit. With these boundary conditions, the interaction between ECE and the environment is completely described by $2n'$ scalar variables, two for each terminal: one current and one voltage.

• For each electric terminal k, its **current** is defined as the total current (conduction and displacement) flowing through it:

$$i_k(t) = \int_{\Gamma_k} \mathbf{H} \cdot \mathbf{dr}, \tag{11}$$

where $\Gamma_k = \partial S'_k$ is a closed curve, the boundary of the terminal surface S'_k. We assume that Γ_k are oriented so that associated normal $\mathbf{n}$ of S'_k is inwards oriented. Due to (9), the sum of all terminal currents is zero and KCL is a consequence.

• For each electric terminal k, its **voltage** is defined as the integral

$$v_k(t) = \int_{C_k} \mathbf{E} \cdot \mathbf{dr} \tag{12}$$

along an arbitrary curve C_k, included in Σ which is a path between a point on S'_k and a point on a reference terminal, let's say S'_n. The condition (8) ensures the consistent definition of the terminal voltage, its independence from the shape of C_k and KVL as a consequence. The following uniqueness theorem is fundamental for the correct formulation of the EM field problem with appropriate boundary conditions for the models extraction [RTT66].

The electromagnetic power transferred by any ECE through its boundary from outside to inside of it is given by the expression

$$P = \sum_{k=1}^{n'-1} v_k i_k. \tag{13}$$

As a consequence, the EM field problem associated to an ECE with equations (1) ÷ (7), boundary conditions (8) ÷ (10), zero initial conditions, having some terminals excited by known voltages and the rest by known currents has a unique field solution: $\mathbf{E}(M,t)$, $\mathbf{D}(M,t)$, $\mathbf{B}(M,t)$, $\mathbf{H}(M,t)$, $\mathbf{J}(M,t)$, $\rho(M,t)$, *for* $(\forall)M \in D$, $t > 0$, *and therefore ECE has a unique response for a given arbitrary terminal excitations.*

2 Finite States Representation by Finite Integrals Technique

The Manhattan geometry, characteristic to IC layout makes the Finite Integration Technique (FIT) a suitable numerical method for electromagnetic field computation in that kind of structures. FIT is a numerical method able to solve field problems based on spatial discretization "without shape functions". Two staggered orthogonal (Yee type) grids are used as discretization mesh [CW01]. The centers of the primary cells are the nodes of the secondary cells and the secondary grid is not extended outside the primary grid. The degrees of freedom used by FIT are not local field components as in FEM or in FDTD, but the global variables i.e. electric and magnetic voltages $\mathbf{u}_e$, $\mathbf{u}_m$ and magnetic and electric fluxes φ, ψ assigned to the grid elements: edges and faces, respectively. They are associated to these grids elements in a coherent manner (Fig. 3). Applying the global form of electromagnetic field equations on the mesh elements (elementary faces and their borders), a system of DAE, called Maxwell Grid Equations (MGE) is obtained:

$$\operatorname{curl}\mathbf{E} = -\frac{\partial \mathbf{B}}{\partial t} \quad \Rightarrow \quad \int_{\Gamma} \mathbf{E}\cdot \mathrm{d}\mathbf{r} = -\iint_{S_\Gamma} \frac{\partial \mathbf{B}}{\partial t}\cdot \mathrm{d}\mathbf{A} \quad \Rightarrow \mathbf{C}\mathbf{u}_e = -\frac{\mathrm{d}\varphi}{\mathrm{d}t} \tag{14}$$

$$\hookrightarrow \operatorname{div}\mathbf{B} = 0 \quad \Rightarrow \quad \iint_{\Sigma} \mathbf{B}\cdot \mathrm{d}\mathbf{A} = 0 \quad \Rightarrow \quad \mathbf{D}'\varphi = \mathbf{0} \tag{15}$$

$$\operatorname{curl}\mathbf{H} = \mathbf{J} + \frac{\partial \mathbf{D}}{\partial t} \Rightarrow \int_{\Gamma} \mathbf{H}\cdot \mathrm{d}\mathbf{r} = \iint_{S_\Gamma}\left(\mathbf{J} + \frac{\partial \mathbf{D}}{\partial t}\right)\cdot \mathrm{d}\mathbf{A} \quad \Rightarrow \mathbf{C}'\mathbf{u}_m = \mathbf{i} + \frac{\mathrm{d}\psi}{\mathrm{d}t} \tag{16}$$

$$\hookrightarrow \operatorname{div}\mathbf{D} = \rho \quad \Rightarrow \quad \iint_{\Sigma} \mathbf{D}\cdot \mathrm{d}\mathbf{A} = \iiint_{\mathcal{D}_\Sigma} \rho\, \mathrm{d}v \quad \Rightarrow \mathbf{D}\psi = \mathbf{q} \tag{17}$$

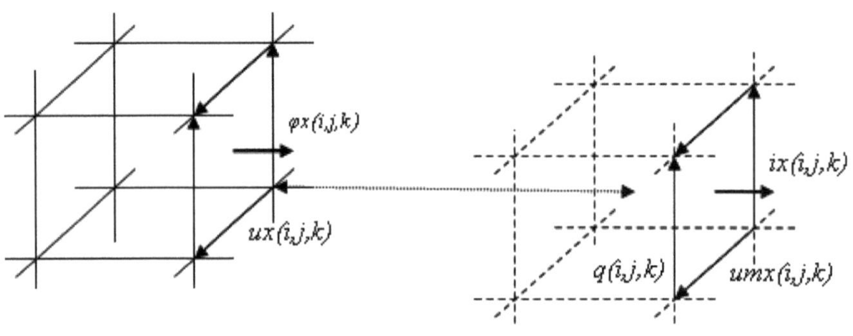

Fig. 3. Dofs for FIT numerical method in the two dual grids cells.

$$\hookrightarrow \operatorname{div}\mathbf{J} = -\frac{\partial \rho}{\partial t} \quad \Rightarrow \quad \iint_{\Sigma} \mathbf{J}\cdot \mathbf{dA} = -\iiint_{\mathcal{D}_{\Sigma}} \frac{\partial \rho}{\partial t}\,\mathrm{d}v \quad \Rightarrow \mathbf{Di} = -\frac{\mathrm{d}\mathbf{q}}{\mathrm{d}t} \quad (18)$$

FIT combines MGE with the Hodge's operators, which describe the material behavior

$$\begin{aligned} &\mathbf{B} = \mu\mathbf{H}, && \mathbf{D} = \varepsilon\mathbf{E}, && \mathbf{J} = \sigma\mathbf{E} \quad \Rightarrow \\ \Rightarrow \quad &\varphi = \mathbf{M}_{\mu}\mathbf{u}_m = \mathbf{M}_{\nu}^{-1}\mathbf{u}_m, && \psi = \mathbf{M}_{\varepsilon}\mathbf{u}_e, && \mathbf{i} = \mathbf{M}_{\sigma}\mathbf{u}_e. \end{aligned} \quad (19)$$

The main characteristics of the FIT method are:

- There is no discretization error in the MGE fundamental equations. All numerical errors are hold by the discrete Hodge operators.
- An equivalent FIT circuit (Fig. 4), having MGE+Hodge as equations may be easily build. The graphs of the two constituent mutually coupled sub-circuits are exactly the two dual discretization grids; therefore the complexity of the equivalent circuit has a linear order with respect to the number of grid-cells [ICR06a].
- MGE are:
  - **sparse**, having maxim 6 non-zero entries per row,
  - **metric-free**: $\mathbf{C}$ - the discrete-curl and $\mathbf{D}$ - the discrete-div operators have only 0, +1 and -1 as entries,
  - **mimetic**: in Maxwell equations curl and div operators are replaced by their discrete counterparts $\mathbf{C}$ and $\mathbf{D}$, and
  - **conservative**: the discrete form of the discrete charge conservation equation is a direct consequence of both Maxwell and as well as of the MGE equations.

Due to these characteristics the numerical solutions have **no spurious modes**.

The size of DOF vectors are:

- electric voltages $\mathbf{u}_e$ equal to the number of branches in the primary grid N_b;

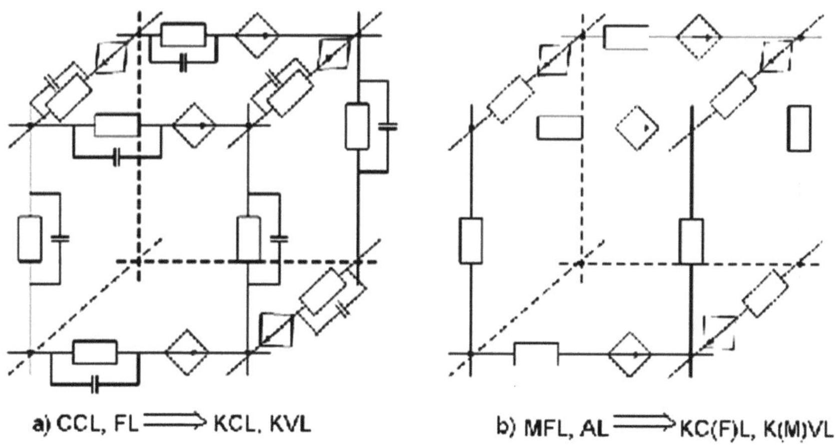

Fig. 4. Electric (left) and magnetic (right) equivalent FIT circuits.

- magnetic voltages $\mathbf{u}_m$ equal to the number of branches in the secondary grid N_b';
- magnetic fluxes φ equal to the number of faces in the primary grid N_f;
- electric fluxes ψ equal to the number of faces in the secondary grid N_f'.

In order to avoid confusions with the capacitance matrix $\mathbf{C}$ for the circuit in Fig. 4, in the rest of the paper the discrete-curl operator will be denoted by $\mathbf{B}$. Actually, it is the topological matrix of branches-loops incidence in the equivalent FIT electrical circuit, while the $\mathbf{B}'$ matrix is branches-loops incidence in the equivalent FIT magnetic circuit.

Details about how the elements of the matrices within MGE and Hodge operators can be computed and stored are presented for instance in [ICR06a]. As an example, typical expressions of Hodge operators are

$$M_{\sigma_{jk}} = \frac{1}{l_k} \sum_{j=1}^{4} \sigma_j A_j, \tag{20}$$

$$M_{\nu_{jk}} = \frac{1}{A_k} \sum_{j=1}^{2} \frac{l_j}{\mu_j}, \tag{21}$$

$$M_{\varepsilon_{jk}} = \frac{1}{l_k} \sum_{j=1}^{4} \varepsilon_j A_j, \tag{22}$$

where A is the area and l is the length, as they are represented in Fig. 5.

Due to the fact that the elements of Hodge operators $\mathbf{M}_\sigma$, $\mathbf{M}_\nu$, $\mathbf{M}_\varepsilon$ have dimensions of electric conductance, magnetic reluctance and electric capacitance, they will be denoted by $\mathbf{G}_e$, $\mathbf{R}_m = \mathbf{G}_m^{-1}$, and $\mathbf{C}_e$, respectively. With these notations, the constitutive relationships become:

$$\varphi = \mathbf{G}_m \mathbf{u}_m, \tag{23}$$

$$\psi = \mathbf{C}_e \mathbf{u}_e, \tag{24}$$

$$\mathbf{i} = \mathbf{G}_e \mathbf{u}_e. \tag{25}$$

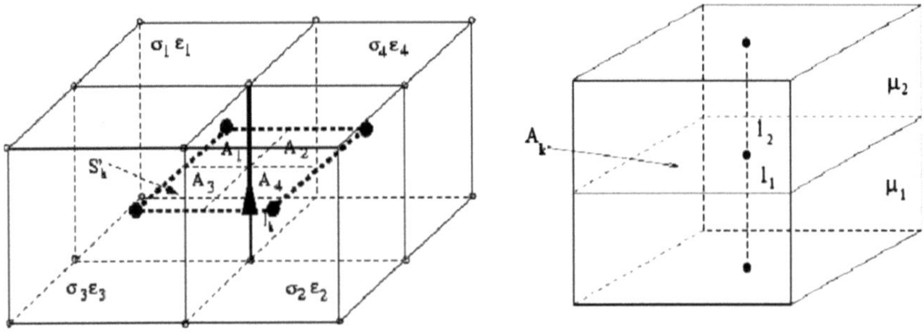

Fig. 5. Discretization of the Hodge operators.

With these new notations, the MGE relationships (14) and (16) become:

$$\frac{\mathrm{d}\boldsymbol{\varphi}}{\mathrm{d}t} + \mathbf{B}\mathbf{u}_e = \mathbf{0}, \tag{26}$$

$$\frac{\mathrm{d}\boldsymbol{\psi}}{\mathrm{d}t} + \mathbf{i} - \mathbf{B}'\mathbf{u}_m = \mathbf{0}. \tag{27}$$

By eliminating the flux vectors, similar equations to the equations of an electric circuit are obtained:

$$\mathbf{G}_m\frac{\mathrm{d}\mathbf{u}_m}{\mathrm{d}t} + \mathbf{B}\mathbf{u}_e = \mathbf{0}, \tag{28}$$

$$\mathbf{C}_e\frac{\mathrm{d}\mathbf{u}_e}{\mathrm{d}t} + \mathbf{G}_e\mathbf{u}_e - \mathbf{B}'\mathbf{u}_m = \mathbf{0}, \tag{29}$$

with the following matrix representation:

$$\mathbf{C}\frac{\mathrm{d}\mathbf{x}}{\mathrm{d}t} + \mathbf{G}\mathbf{x} = \mathbf{0},$$
where
$$\mathbf{C} = \begin{bmatrix} \mathbf{G}_m & \mathbf{0} \\ \mathbf{0} & \mathbf{C}_e \end{bmatrix}, \quad \mathbf{G} = \begin{bmatrix} \mathbf{0} & \mathbf{B} \\ -\mathbf{B}' & \mathbf{G}_e \end{bmatrix}, \quad \mathbf{x} = \begin{bmatrix} \mathbf{u}_m \\ \mathbf{u}_e \end{bmatrix}. \tag{30}$$

The system (30) is very sparse, each row of the matrix $\mathbf{C}$ containing no more than one nonzero entry. By a suitable ordering of equations, the non-zero elements may be placed on the diagonal position of matrix, all non-diagonal elements being zero. The matrix $\mathbf{G}$ contains in each row no more than five nonzero elements: four with integer values (two +1 and two -1) and a real non-negative number.

The system (30) of DAE is far to be the complete I/O state-representation of our passive device, first of all, because the number of equations (E) is not equal to the number of variables (V). As a consequence, $\mathbf{C}$ and $\mathbf{G}$ are non-square matrices. The number of equations in (30) $E = N_f + N_f'$ is equal to the total number of elementary faces in both primary and secondary grids, while the number of variables $V = N_b + N_b'$ is equal to the total number of branches in both primary and secondary grids, therefore the size of both $\mathbf{C}$ and $\mathbf{G}$ matrices is $(N_f + N_f') \times (N_b + N_b')$ with $E < V$.

From the total number of electrical branches N_b, a number of N_{bB} branches are removed, corresponding to the branches placed on the boundary. The remaining number of $(N_b - N_{bB})$ branches is equal to the total number of the elementary magnetic faces. Meanwhile, to each magnetic branch corresponds an electric face, therefore without the electric boundary faces and branches, the number of equations $E' = (N_f - N_{fB} + N_f')$ is equal to the number of state variables $V' = (N_b - N_{bB} + N_b')$. Consequently, in this case, the deficit of equations in the system (30) is equal to the number of border branches N_{bB}. According to the Euler relation $N_{bB} = (N_{nB} - 1) + (N_{fB} - 1)$ is equal to the number of the boundary nodes excepting one plus the number of the boundary faces, excepting one. This deficit of equations $V' - E' = N_{bB}$ will be covered by the discrete form of the boundary conditions. The FIT discretization of the Maxwell equations frequently described in literature does not explain the most difficult part of the algorithm, namely the set-up of the appropriate boundary conditions and their discretization, without which it is not possible to obtain a state space representation of the device, compatible with an external electric circuit, to which the device is connected.

3 State Representation of the Boundary Conditions

The first ECE boundary condition (8):

$$\mathbf{n} \cdot \operatorname{curl} \mathbf{E}(P,t) = \frac{\mathrm{d}B_n}{\mathrm{d}t} = 0 \quad \text{for} \, (\forall) \, P \in \Sigma, \tag{31}$$

is automatically satisfied when the normal component of the magnetic flux density is zero on the boundary. This happens naturally, because in our version of FIT there are no branches of the secondary (magnetic) grid crossing the boundary of the computational domain covered by the primary-electric grid. The discrete form of (31) is represented by the KVL written on the $(N_{fB} - 1)$ boundary faces excepting one, where the KVL is a consequence. Therefore (30) keeps its form with both $\mathbf{C}$ and $\mathbf{G}$ matrices of size $(N_f + N_f') \times (N_b + N_b')$, but with zero values for $(N_{fB} - 1)$ diagonal entries of $\mathbf{G}_m$.

The second ECE boundary condition (9):

$$\mathbf{n} \cdot \operatorname{curl} \mathbf{H}(P,t) = J_t = 0 \quad \text{for} \, (\forall) \, P \in S_e = \Sigma - \cup S_k', \tag{32}$$

is satisfied when the normal component of the total current density (conduction plus displacement) is zero on the device boundary S_e, excepting on the device terminals. As a consequence, no current will be injected from outside:

$$i_k = 0, \quad \text{for any node } k \text{ on } S_e. \tag{33}$$

The third ECE boundary condition (10):

$$\mathbf{n} \times \mathbf{E}(P,t) = \mathbf{0} \qquad \text{for} \, (\forall) \, P \in S_1', S_2', \ldots, S_n' \tag{34}$$

is satisfied if the electric voltages along any terminal branches is zero:

$$u_{eb} = 0, \quad \text{for any branch } b \text{ belonging to any terminal } S_k'. \tag{35}$$

All equations associated to the faces placed on terminals can be eliminated, reducing the number of rows of matrices $\mathbf{C}$ and $\mathbf{G}$. The voltages $\mathbf{u}_{eb}$ (and the corresponding columns of matrices $\mathbf{C}$ and $\mathbf{G}$) can be eliminated from the set of the state variables as well.

For terminals excited in current, the non-homogeneous boundary condition can be expressed by means of Hodge operators as:

$$i_{Tk}(t) = \sum_{m \in S'_k} i_m = \sum_{m \in S'_k} \left(C_{e_m} \frac{du_{e_m}}{dt} + G_{e_m} u_{e_m} \right), \tag{36}$$

where $i_{Tk}(t)$ is the total current injected in the terminal k. The sum is done over all electrical branches (belonging to the boundary or orthogonal to it) directly connected to the terminal k. The matrix representation of (36) is:

$$\mathbf{S}_C \frac{d\mathbf{u}_e}{dt} + \mathbf{S}_G \mathbf{u}_e = \mathbf{i}_T, \tag{37}$$

where $\mathbf{i}_T$ is the vector of the excitation currents injected in terminals. Denoting by $\mathbf{S}$, the branch-terminal connexion matrix (each row, associated to a terminal has elements of values +1 or -1 in the columns corresponding to branches connected to that terminal), we can write:

$$\mathbf{S}_C = \mathbf{S}\mathbf{C}'_e, \quad \mathbf{S}_G = \mathbf{S}\mathbf{G}'_e, \tag{38}$$

where $\mathbf{C}'_e$ and $\mathbf{G}'_e$ are square diagonal matrices of Hodge operators, with the size equal to the total number of primary branches including those on boundary (size of $\mathbf{u}_e$).

The boundary condition (33), can be written as (36) with zero left hand side. This is equivalent to the extension of the matrix $\mathbf{S}$, by adding a row for any node on the boundary surface (excepting for the terminals). In this manner, the Kirchhoff current law on all boundary nodes is imposed, excepting for the reference node, for which KCL is a consequence. 0 By combining (30) with the boundary conditions (37) the semi-state descriptor representation of the system associated to the passive integrated component is obtained:

$$\mathbf{C}\frac{d\mathbf{x}}{dt} + \mathbf{G}\mathbf{x} = \mathbf{w}, \tag{39}$$

where

$$\mathbf{C} = \begin{bmatrix} \mathbf{G}_m & \mathbf{0} \\ \mathbf{0} & \mathbf{C}_e \\ \mathbf{0} & \mathbf{S}_C \end{bmatrix}, \quad \mathbf{G} = \begin{bmatrix} \mathbf{0} & \mathbf{B} \\ -\mathbf{B}' & \mathbf{G}_e \\ \mathbf{0} & \mathbf{S}_G \end{bmatrix}, \quad \mathbf{w} = \begin{bmatrix} \mathbf{0} \\ \mathbf{0} \\ \mathbf{i}_T \end{bmatrix} = \mathbf{B}_i \mathbf{i}_T, \tag{40}$$

with $\mathbf{C}$ and $\mathbf{G}$ square matrices and $\mathbf{i}_T$ as the input vector of terminal excitation currents. If instead of KCL (36) for the boundary nodes, the nodal equations (or the equations of the tree voltages) are used, then the $\mathbf{C}$ matrix becomes a symmetric positive definite one (and the state space becomes a minimal one).

In this final form of state equations, the following relaionships are also embedded as it expected: the constitutive relations of all electric and magnetic branches, Faraday and Ampere-Maxwell laws over all mesh-grid, Kirchhoff voltage and current/flux laws in all nodes (in internal nodes as a consequence) and finally, the terminal excitation conditions.

If all floating **terminals of the device are current-excited**, then the output signals are the electrical voltages of terminals:

$$u_{Tk} = \sum_{m \in C_k} u_{e_m}, \tag{41}$$

where C_k is a set of branches which compose a path from the current terminal k to the reference terminal. The matrix form of (41) is

$$\mathbf{u}_T = \mathbf{S}_u \mathbf{u}_e, \tag{42}$$

where $\mathbf{u}_T$ is the system response and $\mathbf{S}_u$ is the branch-to-path connection matrix (+1 or -1 for boundary branches comprising the path). In conclusion, in the current-excitation case, the complete LTI system is defined by semi-state equations (39) and the output relations (42):

$$\mathbf{C}\frac{\mathrm{d}\mathbf{x}}{dt} + \mathbf{G}\mathbf{x} = \mathbf{B}_i \mathbf{i}_T, \qquad \mathbf{u}_T = \mathbf{S}_u \mathbf{u}_e, \tag{43}$$

with $\mathbf{C}$ and $\mathbf{G}$ given by (40). The transfer function of the MIMO system (43) is the impedance matrix $\mathbf{Z}$ of the passive component.

In the case of **voltage-excited components**, (37) may describe the output and (41) may describe the input relationships. However, there is still another representation of the voltage-excited device, as a standard LTI system, without time derivative in the output equations. It is generated by the definition (11) of the terminal current. The discrete form of this relationship is:

$$i_{Tk}(t) = \sum_{m \in \Gamma_k} u_{em} \tag{44}$$

where Γ_k is a set of branches defining the contour of the terminal k. The matrix form of (44) is

$$\mathbf{i}_T = \mathbf{S}_i \mathbf{u}_e, \tag{45}$$

where $\mathbf{i}_T$ is the vector of response currents and $\mathbf{S}_i$ is the branch-to-contours connection matrix (+1 or -1 for boundary branches comprising the contour). In conclusion, when all device terminals are voltage-controlled, the complete LTI system can be defined alternatively by the semi-state equations (39) with $\mathbf{w} = \mathbf{B}_u \mathbf{u}_T$ and the output relations (45):

$$\mathbf{C}\frac{\mathrm{d}\mathbf{x}}{dt} + \mathbf{G}\mathbf{x} = \mathbf{B}_u \mathbf{u}_T, \qquad \mathbf{i}_T = \mathbf{S}_i \mathbf{u}_e, \tag{46}$$

with $\mathbf{C}$ and $\mathbf{G}$ given by (40), $\mathbf{u}_T$ the vector of excitation voltages and $\mathbf{i}_T$ the response vector of terminal currents. The transfer function of the MIMO system (46) is the admittance matrix $\mathbf{Y}$ of the device.

If the **component is hybrid excited**, the associated LTI system is defined by:

$$\mathbf{C}\frac{\mathrm{d}\mathbf{x}}{\mathrm{d}t} + \mathbf{G}\mathbf{x} = \mathbf{B}'\mathbf{w}, \qquad \mathbf{y} = \mathbf{L}\mathbf{x}, \tag{47}$$

with

$$\mathbf{x} = \begin{bmatrix} \mathbf{u}_m \\ \mathbf{u}_e \end{bmatrix}, \mathbf{w} = \begin{bmatrix} \mathbf{u}_T{}' \\ \mathbf{i}_T{}'' \end{bmatrix}, \mathbf{B}' = \begin{bmatrix} \mathbf{B}'_u & \mathbf{0} \\ \mathbf{0} & \mathbf{B}'_i \end{bmatrix}, \mathbf{y} = \begin{bmatrix} \mathbf{i}'_T \\ \mathbf{u}_T{}'' \end{bmatrix}, \mathbf{L} = \begin{bmatrix} \mathbf{S}_i & \mathbf{0} \\ \mathbf{0} & \mathbf{S}_v \end{bmatrix}, \quad (48)$$

where the matrices $\mathbf{C}$, $\mathbf{G}$, $\mathbf{B}'$, $\mathbf{L}$, input vector $\mathbf{w}$ and output vector $\mathbf{y}$ are defined accordingly, by combining (43) and (46). Regardless the excitation mode, the state variables are the voltages along the all electric and magnetic grid branches, excepting those on terminals. The transfer function of the MIMO system (47) is the hybrid matrix $\mathbf{H}$ of the passive component:

$$\begin{aligned} & \mathbf{y} = \mathbf{H}\mathbf{w} = \mathbf{L}\mathbf{x} = \mathbf{L}\left(s\mathbf{C}+\mathbf{G}\right)^{-1}\mathbf{B}'\mathbf{w} \quad \Rightarrow \\ \Rightarrow \quad & \mathbf{H} = \mathbf{L}\left(s\mathbf{C}+\mathbf{G}\right)^{-1}\mathbf{B}'. \end{aligned} \quad (49)$$

If the first m out of total n terminals are controlled in voltage and the rest $n-m$ are controlled in current, the hybrid matrix has the following block structure:

$$\mathbf{H} = \begin{bmatrix} \mathbf{H}_{11} & \mathbf{H}_{12} \\ \mathbf{H}_{21} & \mathbf{H}_{22} \end{bmatrix}, \mathbf{y} = \mathbf{H}\mathbf{w} = \begin{bmatrix} \mathbf{i}_T{}' \\ \mathbf{u}_T{}'' \end{bmatrix} = \begin{bmatrix} \mathbf{H}_{11} & \mathbf{H}_{12} \\ \mathbf{H}_{21} & \mathbf{H}_{22} \end{bmatrix} \begin{bmatrix} \mathbf{u}_T{}' \\ \mathbf{i}_T{}'' \end{bmatrix} \quad (50)$$

where $\mathbf{H}_{11} = \mathbf{Y}$ is the square admittance matrix of size m, $\mathbf{H}_{22} = \mathbf{Z}$ the square impedance matrix of size $(n-m-1)$, while $\mathbf{H}_{12}$ is the voltage transfer coefficient and $\mathbf{H}_{21}$ is the current transfer coefficient. In the degenerate case, when $m = 0$, the matrix $\mathbf{H}$ becomes the impedance matrix, and when $m = n-1$, then $\mathbf{H}$ becomes the admittance matrix of the component. The number of inputs is equal to the number of outputs. Each device terminal, excepting the reference terminal has either its voltage or its current as an input scalar signal, and its current and respectively its voltage as an output.

4 ROM WorkBench

The next step after the extraction of the state space model for passive components is the order reduction. In order to decide which ROM technique is the most appropriate for this type of model, a new tool called *ROM Workbench* was conceived in the frame of the FP6/IST/CODESTAR project [webb]. Its aim is to allow the user to reduce models by means of as many ROM techniques as possible, and to compare the results. The ROM Workbench consists of a set of benchmark problems, a set of model order reduction methods and criteria for results evaluation and comparison. Fig. 6 shows the main blocks of the ROM Workbench. The models that can be read in are of various types: linear time invariant systems given by state (or semi-state) space matrices, frequency responses described by the variation of impedance, admittance or scattering parameters with respect to the frequency, circuits given as net-lists or transfer functions given by poles and residues. Once read into the database of models, different actions can be carried out. The most important is the reduction but other actions are also useful, such as conversions between different

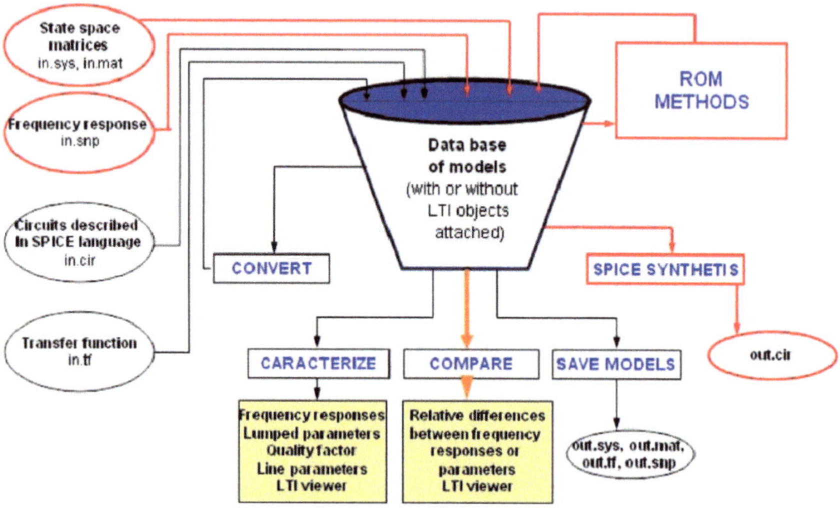

Fig. 6. Main blocks of the ROM Workbench.

types of representations, characterization/visualization of a system (plots or computation of lumped or line parameters, etc), or synthesis of the equivalent circuit. A very important facility is that the systems can be compared, the comparison between the original model and reduced models obtained from different reduction techniques allowing one to decide which method performs best for a given problem.

The reduction can be carried out by means of various methods. These methods include: explicit moment matching, Krylov subspace techniques [CPO02], Laguerre techniques [KZ99], combined techniques (such as two step Lanczos, or Krylov based followed by a truncated balanced reduction), and truncated balanced realization procedures [webc]. The vector fitting method proposed in [GS99], which finds the transfer function matching a given frequency characteristic is included in the ROM Workbench as an order reduction method.

The workbench is able to compare responses obtained for different systems. The comparison can be carried out either on the time responses (step, impulse, etc.) or on the frequency responses (Bode, Nyquist, Smith, etc). Lumped parameters, quality factors or line parameters can also be compared.

The available measurements data for the tested devices are usually the scattering (**S**) parameters. That is why, in the results that follows the criteria used for comparison is an error estimator based on the Frobenius norm $\|.\|_F$, computed as $\mathrm{rms}\|\mathbf{S}_{\mathrm{ref}} - \mathbf{S}_{\mathrm{an}}\|_F / \max_f \|\mathbf{S}_{\mathrm{ref}}\|_F$, where $\mathbf{S}_{\mathrm{ref}}$ are the scattering parameters for the reference system (for instance the measured data), and $\mathbf{S}_{an}$ are the scattering parameters of the analyzed system, for instance the reduced order model.

The first implementation of the ROM workbench was under Matlab, with GUI. This implementation proved to be very useful, as a prototyping tool in the research as well as for the educational purposes (see for instance the MOR course `http://www.win.tue.nl/casa/meetings/special/mor/`). Other implementations do not use any graphical interface at all, being appropriate for linking the reduction step to the EM simulator, especially in the case of large scale problems that need high computational resources. For instance, in the CODESTAR project various tools of the ROM Workbench became stand alone programs using the MATLAB Compiler Toolbox. Future versions of the ROM Workbench will include features related to parameterization, its further development being supported in the frame of the CHAMELEON-RF project [weba].

5 All Levels Reduced Order Modelling

The ROM WorkBench procedures were applied to a series of test structures, relevant to the Codestar project, which will be described in the next section. The reduced order method which behaved best in these study cases was the one based on Vector Fitting [GS99]. Another important conclusion of the numerical experiments was that *a-posteriori* order reduction is not enough effective, if it is not accompanied by *a-priori* and *on-the-fly* reduced order techniques. The idea to apply any reduction as soon as possible in the model extraction process lead to a strategy called **ALROM - All Levels Reduced Order Modeling** [ICRS06], which basically uses the following steps to model the passive components and interconnects:

A) Grid calibration. The minimal orthogonal grid necessary to define the material sub-domains is successively refined, until the equivalent capacitances of the passive component are accurate enough. The **dual Finite Integration Technique (dFIT)** used to solve the electrostatic field equations and to extract the capacitances provides lower and upper bounds of the exact solution [IRC04]. These bounds are used to control the accuracy of the numerical solution by means of a multi-grid approach.

B) Virtual boundary calibration. The computational domain is successively extended, until the inductance extracted by averaging the dual Neumann and Dirichlet boundary conditions is accurate enough [ICR06b]. Actually, an Equivalent Layer on the boundary of the computational domain is used to model the Open Boundary of the magneto-static field. Choosing for the material constant of ELOB (Equivalent Layer of Open Boundary) the magnetic relative permeability $\mu_r \ll 1$, the scalar magnetic potential satisfies the Neumann boundary condition, while for $\mu_r = M \gg 1$ the potential satisfies the Dirichlet boundary condition. These two dual boundary conditions are used to control the accuracy of the numerical solution because they provide lower and upper bounds of the inductances extracted from the exact field solution in the un-bounded domain.

C) Frequency analysis. By using the grid resulted in previous steps, after refining and extension process, the frequency dependent matrix of the component is computed in a minimal set of frequency samples, solving the linear complex system (49), a FIT consequence of Maxwell equations. To build numerical scissors for the exact solution, a practical approach we propose is to use the dual (complementary) solutions, solving the Maxwell Grid Equations two times, and computing the admittance matrix using the two dual-staggered grids and two type of boundary conditions, for a sequence of frequency samples ω:

- $\mathbf{Y}_p(\omega)$ is computed by FIT on the primary grid with ELOB parameters $\varepsilon_r = M \gg 1$, $\mu_r = 1$;
- $\mathbf{Y}_s(\omega)$ is computed by FIT on the secondary grid with ELOB parameters $\varepsilon_r = 1$, $\mu_r = M \gg 1$.

By averaging the two admittance, a numerical solution $\mathbf{Y}_a(\omega) = (\mathbf{Y}_p(\omega) + \mathbf{Y}_s(\omega))/2$ is generated, which provides a better accuracy than any of the two direct extracted admittances $\mathbf{Y}_p(\omega)$ or $\mathbf{Y}_s(\omega)$, at least at low frequencies. In the case of interconnects, the p.u.l. frequency dependent line parameters $\mathbf{Z}_l(\omega) = \mathbf{R}_l(\omega) + j\mathbf{L}_l(\omega)$ and $\mathbf{Y}_l(\omega) = \mathbf{G}_l(\omega) + j\mathbf{C}_l(\omega)$ are extracted using a similar technique, but solving the 2D EMQS equation of the TM field [ICRS06].

D) Length extension. This is an optional step, applied in the case of the TL model for interconnects. The frequency characteristic $\mathbf{Y}(\omega)$ of the real length line is computed by transmission line equations in an extended set of adaptive frequency samples, using an appropriate interpolation of p.u.l. parameters.

E) Optimal order of the compact model. In this algorithm step, the frequency characteristic $\mathbf{Y}(\omega)$ of the analyzed component is approximated by rational functions using the Vector Fitting procedure [GS99] and then a SPICE equivalent circuit is synthesized by the Differential Equation Macromodel [PR04]. The couple of these procedures is iterated, successively increasing the order of the extracted model. Compact models of increasing order and their equivalent circuits are extracted and simulated in the frequency domain with SPICE, until the result is close to $\mathbf{Y}(\omega)$, on the frequency range of interest. In this way, the compact model and its SPICE equivalent circuit for the given components having an optimal order are generated.

F) Validation. Based on the results of the SPICE simulation in frequency domain, the scattering parameters $\mathbf{S}(\omega)$ are computed and compared with the measurements, for a series of test structures, of practical interest - the Codestar benchmarks, presented below.

The control of the solution accuracy plays a crucial role in *a-priori* order reduction. Use of an optimal grid minimizes the number of required DOFs. By extending or refining more the optimal grid, the order of the macro-model generated by discretization is increased, whereas by reducing the grid size the solution accuracy becomes too low to be acceptable.

6 Test Structures

6.1 Meander Resistor (RPOLY2_ME Benchmark)

A semi-state space model for the meander on-chip resistor depicted in Fig. 7 has been computed using the finite integration technique described above. For a grid having 368200 nodes, a macro-model having 19510 degrees of freedom has been obtained using a-priori order reduction techniques.

After the computation of its frequency response on the range 1-20 GHz, a reduced order model of order four was obtained using the vector fitting algorithm. This reduced model was then synthesized and the equivalent circuit thus obtained has been

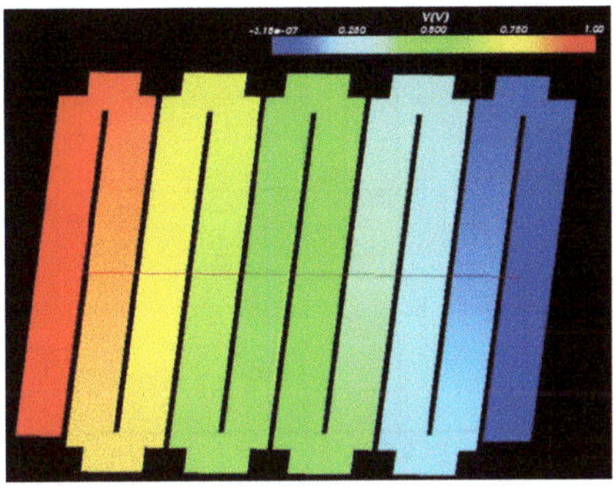

Fig. 7. Meander resistor benchmark.

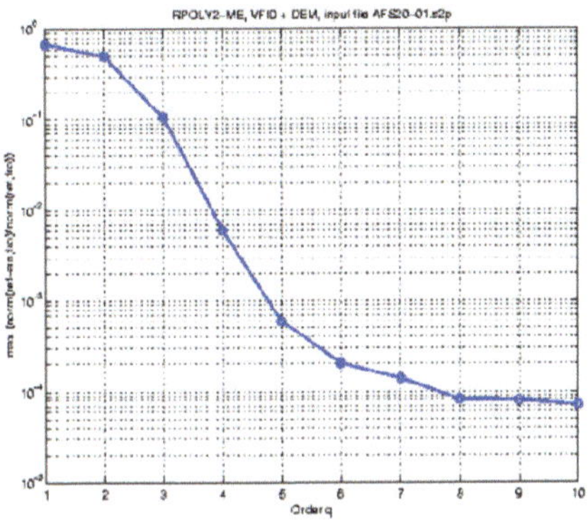

Fig. 8. Relative error versus the order of the reduced system.

simulated with SPICE. The relative error of **S** parameters between the simulation of the initial model and the reduced one is very low, about 0.16 %. The reduction time was also very low, less than 1 second. Fig. 8 shows the relative error between the macromodel and the reduced model for increasing order of the reduced system. Even for very low orders, vector fitting was able to find accurate reduced models.

Fig. 9 shows the comparison between the measurements (real part of parameter S_{11}) and the simulation of the reduced order model. Fig. 10 shows the same comparison for the extracted resistance.

6.2 Capacitor (CMIM Benchmark))

Another benchmark tested was a metal-insulator-metal capacitor (Fig. 11). A grid having 833280 nodes have been used, conducing to a macro-model of order 29925. The macro-model has been reduced to a model of order four, using the vector-fitting algorithm. The relative error between the initial model and the reduced one is 0.2 %, obtained in less than 1 second. The relative error between the measurements and the reduced model is 3.75 %.

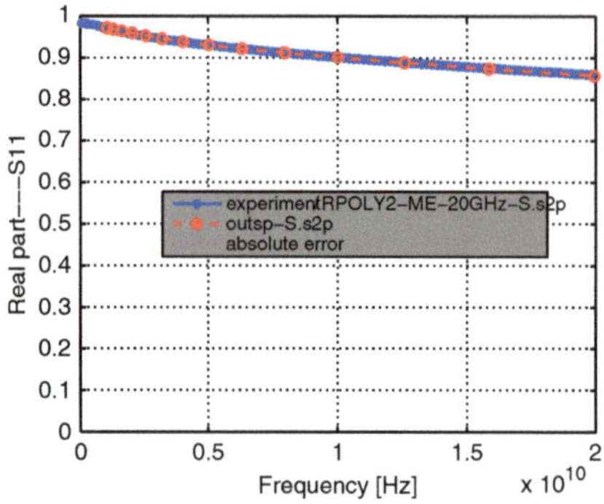

Fig. 9. Parameter S_{11} - measurement and simulation of reduced model.

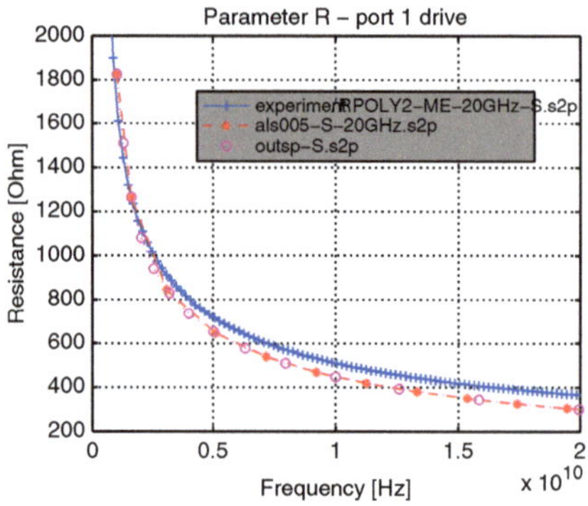

Fig. 10. Rezistence extracted from measurements, simulation of initial and reduced models.

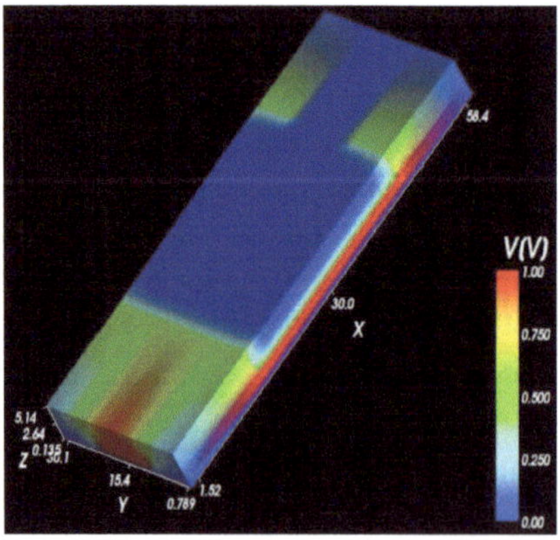

Fig. 11. Capacitor benchmark.

Figs. 13 and 14 show the comparison between measurements (S_{12} real and imaginary parts) and simulation of reduced model. Fig. 12 shows the variation of the extracted capacitance with respect to the frequency.

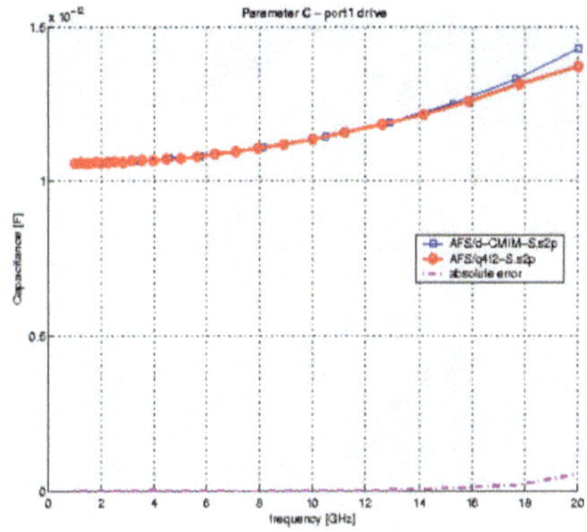

Fig. 12. Capacitance extracted from the simulation of initial model and of reduced model.

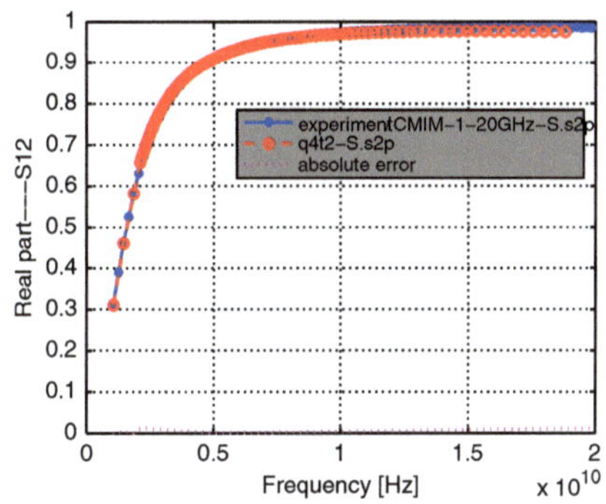

Fig. 13. Parameter S_{12} - real part, experiment and reduced model.

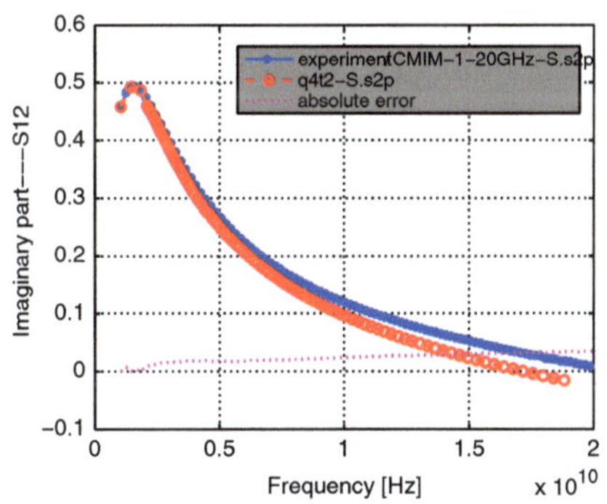

Fig. 14. Parameter S_{12} - imaginary part, experiment and reduced model.

6.3 SP-SMALL Benchmark

In the case of an inductor benchmark (Fig. 15), a grid having 596068 nodes generates a macro-model of order 9614. This has been reduced also to a model of order four,

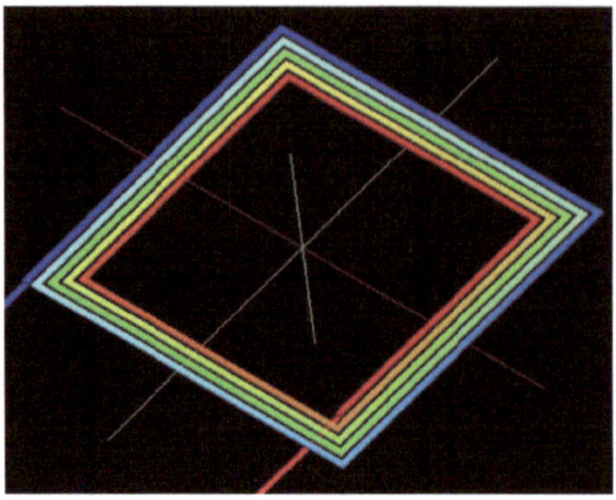

Fig. 15. Inductor benchmark.

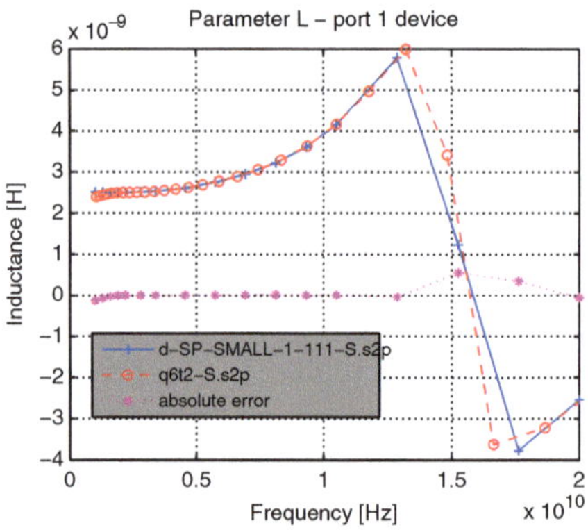

Fig. 16. Extracted inductance: simulation of macromodel and reduced model.

using the vector-fitting algorithm. The relative error between the initial model and the reduced one is 0.5 %, obtained in less than 1 second. Figs. 17 and 18 show the comparison between the macro-model frequency response (S_{11} real and imaginary parts) and simulation of reduced model. Fig. 16 shows the variation of the extracted inductance with respect to the frequency.

6.4 Coplanar Line

A coplanar line discretized with a grid having 2866441 nodes generates a macromodel of order 19972. The vector-fitting procedure was also the most successful reduction procedure for this case, generating a reduced order model with ten degrees of freedom, in less than 1 second. The error between the initial macro-model and the reduced model is 1.3 %, for a frequency range 1-30 GHz. This corresponds to a relative error between measurements and reduced order model of 5.5 %. Fig. 19 shows how the order increasing improves frequency response of the reduced order model.

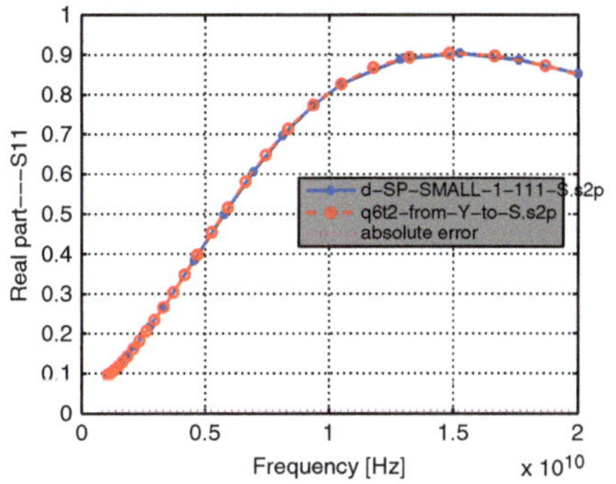

Fig. 17. Parameter S_{11} - real part.

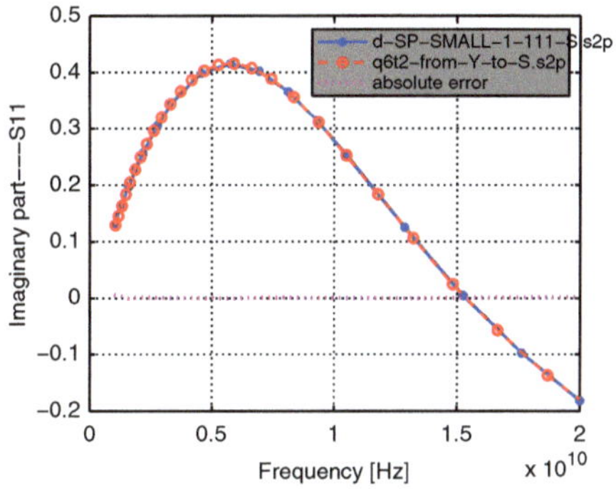

Fig. 18. Parameter S_{11} - imaginary part.

Files corresponding to these four benchmarks (macromodels, reduced order models and the measured frequency characteristics) may be found at `www.lmn.pub.ro/BookMOR`.

7 Conclusions

The proposed strategy proved to be an efficient methodology for modeling and simulation of on-chip passive components.

An important step in the modeling process is represented by the correct formulation of the EM field problem in mathematical terms. The proper boundary conditions and the solution uniqueness theorem allow the consistent definition of a dynamical system with distributed parameters. The next step is to reduce the system space-state at a finite dimension, applying a numerical method. FIT was the numerical method suitable for the class of integrated passive components. In order to extract the models for these components as state-space representations the discrete form of the boundary conditions plays a crucial role. The electric and magnetic voltages along the edges of the dual FIT grids are the state variables of the model.

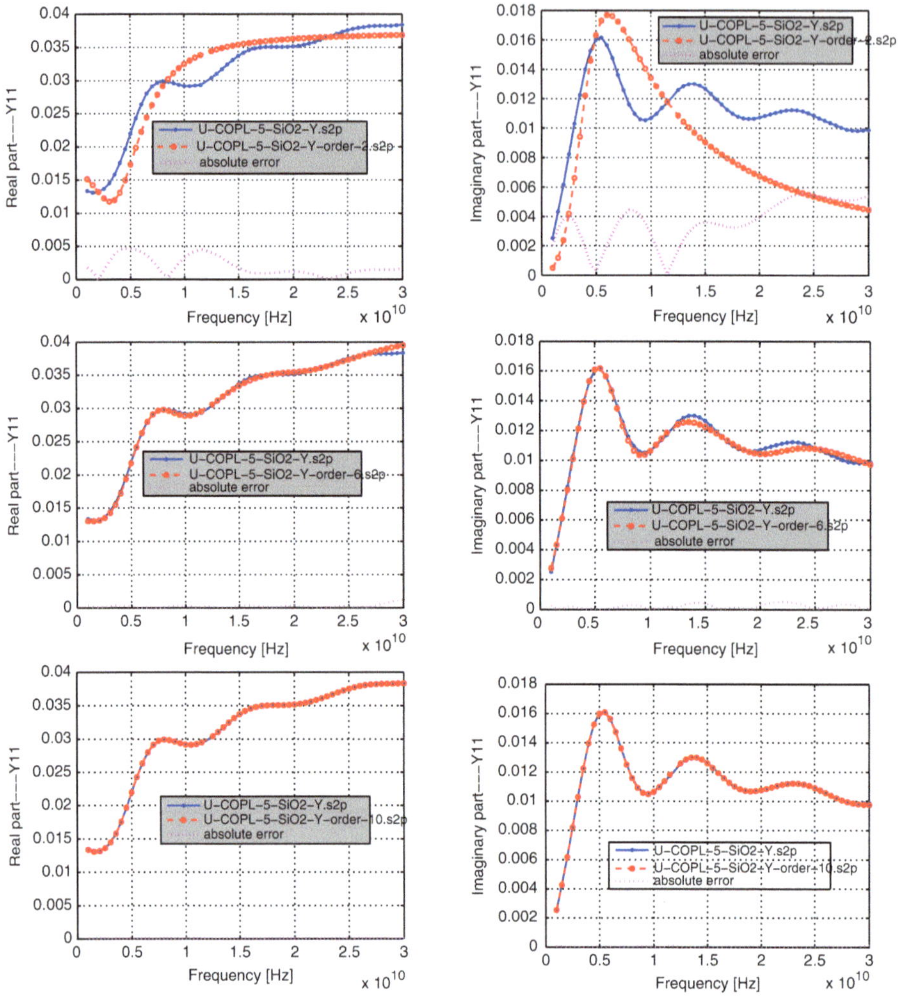

Fig. 19. Comparison between measurements (admittance) and reduced order model: left figures: real parts for measurements and reduced models of order (from top to bottom) 2, 6, 10; right figures - similar for imaginary parts.

When applying reduction procedures for a given model, an environment such as the ROM Workbench is very useful due to its flexibility. It allowed us to conclude that the vector fitting procedure is the best method to reduce electromagnetic full-wave models of passive components considered as Codestar benchmarks. For this class of problems, the reduction conduces to very low orders (less than 10) with an extremely low computational effort (less than 1 sec), the relative error between the simulation result and its reduced order model being less than 1.

Acknowledgment

The authors would like to acknowledge the support of Austriamicrosystems (Ehrenfried Seebacher) for providing, within the common European Codestar research project [webb], the experimental data used in this paper.

References

[CPO02] Mustafa Celik, Lawrence Pileggi, and Altan Odabasioglu. *IC Interconnect Analysis*. Kluwer Academic Publishers, 2002.

[CW01] M. Clemens and T. Weiland. Discrete electromagnetism with the finite integration technique. *Progress in Electromagnetics Research (PIER)*, 32:65–87, 2001. http://ceta.mit.edu/PIER/pier.php?volume=32.

[GS99] B. Gustavsen and A. Semlyen. Rational approximation of frequency domain responses by vector fitting. *IEEE Trans. Power Delivery*, 14(3):1052–1061, 1999.

[ICR06a] D. Ioan, G. Ciuprina, and M. Rădulescu. Algebraic sparsefied partial equivalent electric circuit - aspeec. *Scientific Computing in Electrical Engineering (A.M. Anile, G. Al and G. Mascali Eds*, 9:45–50, 2006.

[ICR06b] Daniel Ioan, Gabriela Ciuprina, and Marius Rădulescu. Absorbing boundary conditions for compact modeling of on-chip passive structures. *COMPEL: The International Journal for Computation and Mathematics in Electrical and Electronic Engineering*, 25(3):652–659, 2006.

[ICRS06] D. Ioan, G. Ciuprina, M. Rădulescu, and E. Seebacher. Compact modeling and fast simulation of on-chip interconnect lines. *IEEE Transactions of Magnetics*, 42(4):547–550, 2006.

[IM99] D. Ioan and I. Munteanu. Missing link rediscovered: The electromagnetic circuit element concept. *JSAEM Studies in Applied Electromagnetics and Mechanics*, 8: 302–320, 1999.

[IRC04] D. Ioan, M. Rădulescu, and G. Ciuprina. Fast extraction of static electric parameters with accuracy control. *Scientific Computing in Electrical Engineering*, 4:248–256, 2004.

[KZ99] L. Knockaert and D. De Zutter. Laguerre-svd reduced order modeling. *Electrical Performance of Electronic Packaging*, pages 249–252, 1999.

[PR04] T. Palenius and J. Roos. Comparison of reduced-order interconnect macromodels for time-domain simulation. *IEEE Trans. Microwave Theory and Techniques*, 52(9):2240–2250, 2004.

[RTT66] R. Răduleţ, A. Timotin, and A. Tugulea. Introduction des parametres transitoires dans l'tude des circuits electrique lineaires ayant des elements non filiformes et avec pertes suplimentaires. *Rev. Roum. Sci Techn. - Electrotech. et Energ.*, 11(4): 565–639, 1966.

[weba] CHAMELEON-RF website. www.chameleon-rf.org.

[webb] CODESTAR website. www.imec.be/codestar.

[webc] SLICOT website. The control and systems library slicot, www.slicot.com.

Index

ad-exponential, 228
adjoint map, 215
algebraic Lyapunov equation, 221, 233, 234
Arnoldi, 41, 374–377, 386, 387
Arnoldi process, 21
ASM, 162
asymptotic waveform evaluation, 18
asymptotically stable, 135
augmented Lagrangian, 201
AWE, 18

balanced realization, 221
balanced system, 138
balanced truncation, 23, 137
basis matrix, 62
block Arnoldi process, 63
block Krylov sequence, 62
block Krylov subspace, 61
Boundary Conditions, 379
Boundary Value Problem, 379
branch constitutive relations, 52

canonical Gramian, 219
Chebyshev, 389–392
circuit analysis, 195
compact modelling, 7
completion, 247
complex frequency hopping, 19
condition number, 34
Congruence Transform, 361, 372, 377
conjugate gradients, 197
constraint preconditioner, 202
constraned optimization, 195
control variable, 158
controllability Gramian, 23, 138
convolution, 368

DCPB*, 164
deflated block Krylov sequence, 62
deflation, 62
dominant pole, 178
DPA, 180

equivalent first-order system, 57
error
  minimum norm, 40
evaluator functional, 230
event, 215

first-order formulation, 57
first-order time-invariant linear dynamical system, 57
first-order transfer function, 60
Frobenius conditions, 239

generalized Lyapunov equation, 138
gradient, 227
Gram–Schmidt
  modified, 42

Hankel matrix, 238
Hankel singular values, 138
Hermitian matrix, 35
Hessenberg matrix, 35
high-speed, 361
Hilbert space, 378, 380, 392, 399
Hilbert subspace, 383, 387
Householder reflection, 43
HSS preconditioning, 205

IFISS, 206
impulse response function, 14
incidence matrix of a graph, 51
incompressible flow, 196
Initial Value Problem, 379
input-output system, 10
integro-DAEs, 54
integro-differential-algebraic equations, 54
interconnect, 361–363, 366, 367
interior point methods, 196

Jacobian problem, 239
Jordan block, 36
Jordan form, 36

Kirchhoff's current law, 52
Kirchhoff's voltage law, 52
Krylov space, 18
Krylov subspace methods, 195

Lagrange multipliers, 196
Lagrangian, 196
Laguerre polynomial, 22
Lanczos, 42
Lanczos method, 19, 44, 197
Laplace transform, 22
Laplace-domain, 55
Leverrier, 237
Lie-derivative, 227
Lie-exponential, 230, 232
Lie-product, 227
linear dynamical system, 54
Linear system
  sensitivity, 35
look-ahead Lanczos, 20
LTI, 221
LTI system, 11
LTV system, 11
lumped-element circuit model, 51
Lyapunov equation, 24

M-balanced, 224
Markov parameters, 140
Matrix norm, 34
matrix pencil, 15, 57
matrix PVL, 20
Matrix-Padeé Via Lanczos, 56
Mayer-Lie, 239
MINRES, 197
mixed finite elements, 195
modal approximation, 177
modal equivalent, 178
modal truncation, 23
model order reduction, 55, 201
modeling of electronic circuits, 51
modified Gram-Schmidt, 22
moment, 18
moment matching approximation, 139
moment of transfer function, 13
moments, 139
MPVL, 56
multigrid, 206

Normal matrix, 37

observability Gramian, 23, 138, 214
observability map, 216
observability matrix, 223, 232, 239
operational MOR, 7
Optimal Hankel Norm Reduction, 25
order reduction via projection, 61
orthogonal basis, 377, 394, 395

Padé approximation, 139
Padé-type approximation property, 63, 66
Padé-via-Lanczos, 19
passive, 56
passivity, 11, 16, 17, 56
Petrov-Galerkin, 40
POD, 220
pole, 14, 178
positive definite, 35
positive definite matrix, 16
positive real, 56
positive real matrix, 16
positive realness, 56
preconditioning, 195
PRIMA, 21
PRIMA reduced-order model, 62
projection, 11
projection method, 11
projection of dynamics, 220
projection theorem, 63
Proper Orthogonal Decomposition, 27
proper orthogonal decomposition, 247
proper transfer function, 135
pseudo-balancing, 244
PVL, 19

QR method, 22
quasi-minimum residuals, 197
quasi-triangular matrix, 36

RCL circuit equations, 52
RCL structure, 56
reachability Gramian, 214
reachability map, 215
reachability matrix, 222, 229, 239
reachable aims
  coarse model, 159
reduced-order model, 55
regular matrix pencil, 58
regular pencil, 135
residual
  minimum norm, 40
residue, 14, 178
right coprime factors, 143
Ritz value, 37
Ritz vector, 37
Ritz-Galerkin, 40

saddle point problem, 195
SADPA, 181
Schilders' factorization, 203
Schur complement, 201
Schur decomposition, 35
  real, 36
Selective Node Elimination, 26
selective node elimination, 26
shift-and-invert, 21
SIB, 219
sign-symmetry, 221, 224
singular value decomposition, 22
sliding interval balancing, 219
SPRIM, 63, 65
SPRIM for general RCL circuits, 65
stability, 11, 15
stable pencil, 135
stable transfer function, 135
state variable, 10
state-space dimension, 54, 57
state-space vector, 54
Stokes problem, 207
structure preservation, 56
SVD, 22
SyMPVL, 56
SymPVL, 20
SyPVL, 56

TBR, 23
Telegraphers equations, 368, 385, 394, 399
transfer function, 13, 55
transition matrix, 214
transmission line, 362, 363, 368, 369, 371, 374, 397–399
tridiagonal matrix, 35
Truncated Balanced Realization, 23
two-step Lanczos, 20

uncertainty equivalence, 247
uncorrelated realization, 241
Uzawa method, 205

vector fitting, 30
Very Large Scale Integration, 361

zero, 188

MIX
Papier aus verantwortungsvollen Quellen
Paper from responsible sources
FSC® C105338

If you have any concerns about our products,
you can contact us on
ProductSafety@springernature.com

In case Publisher is established outside the EU,
the EU authorized representative is:
Springer Nature Customer Service Center GmbH
Europaplatz 3, 69115 Heidelberg, Germany

Printed by Libri Plureos GmbH
in Hamburg, Germany